Membranes and Sensory Transduction

Membranes and Sensory Transduction

Edited by

Giuliano Colombetti

and

Francesco Lenci

Institute of Biophysics
National Research Council
Pisa, Italy

Plenum Press • New York and London

Library of Congress Cataloging in Publication Data

Main entry under title:

Membranes and sensory transduction.

Includes bibliographical references and index.
Contents: Membranes/Alessandra Gliozzi and Ranieri Rolandi—Biochemistry of chemosensory behavior in prokaryotes and unicellular eukaryotes/Barry L. Taylor and Sharon M. Panasenko—Mechanosensory transduction in protozoa/Yutaka Naitoh—[etc.]
1. Membranes (Biology) 2. Senses and sensation. I. Colombetti, Giuliano. II. Lenci, Francesco. [DNLM: 1. Cell membrane—Physiology. 2. Sense organs—Ultrastructure. 3. Sense organs—Physiology. 4. Cell membrane—Ultrastructure. WL 700 M533]
QH601.M4816 1984 574.1′8 84-3373
ISBN 0-306-41439-2

A Division of Plenum Publishing Corporation
233 Spring Street, New York, N.Y. 10013

Printed in the United States of America

To my wife Sonia and my daughter Sara
G. C.

In memory of my father
F. L.

Contributors

Barry Bean Department of Biology, Lehigh University, Bethlehem, Pennsylvania 18015, USA

Giuliano Colombetti Istituto di Biofisica, Consiglio Nazionale delle Ricerche, 56100 Pisa, Italy

Donna R. Fontana MSU-DOE Plant Research Laboratory, Michigan State University, East Lansing, Michigan 48824; *present address:* Department of Physiological Chemistry, Johns Hopkins University, Baltimore, Maryland 21205, USA

Alessandra Gliozzi Istituto di Scienze Fisiche dell'Università di Genova, 16146 Genoa, Italy

Donat-P. Häder Fachbereich Biologie Marburg Universität, Lahnberge, D-3350 Marburg/Lahn, Federal Republic of Germany

Wolfgang Haupt Institut für Botanik und Pharmazeutische Biologie, University of Erlangen-Nürnberg, D-8520 Erlangen, Federal Republic of Germany

Francesco Lenci Istituto di Biofisica, Consiglio Nazionale delle Ricerche, 56100 Pisa, Italy

Yutaka Naitoh Institute of Biological Sciences, University of Tsukuba, Ibaraki 305, Japan

Sharon M. Panasenko Chemistry Department, Pomona College, Claremont, California 91711, USA

Kenneth L. Poff MSU-DOE Plant Research Laboratory, Michigan State University, East Lansing, Michigan 48824, USA

Ulrich Pohl Max-Planck Institut für Molekulare Genetik, D-1000 Berlin 33, Federal Republic of Germany

Ranieri Rolandi Istituto di Scienze Fisiche dell'Università di Genova, 16146 Genoa, Italy

Vincenzo E. A. Russo Max-Planck Institut für Molekulare Genetik, D-1000 Berlin 33, Federal Republic of Germany

Barry L. Taylor Department of Biochemistry, School of Medicine, Loma Linda University, Loma Linda, California 92350, USA

Gottfried Wagner Botanisches Institut I, D-6300 Giessen, Federal Republic of Germany

Bruce D. Whitaker Department of Biochemistry, University of Wisconsin, Madison, Wisconsin 53706, USA

Preface

The main purpose of this book is to unify approaches and ideas in the field of aneural sensory transduction. This field has recently come to the attention of several research groups in various disciplines, and their number seems to be growing. Unfortunately, because of the diverse scientific backgrounds of the researchers in the field, the apparent heterogeneity of experimental techniques (i.e., behavioral response analysis, sophisticated biochemical and genetic manipulations, conventional and pulsed laser spectroscopy) and theoretical approaches may be discouraging, for both the experienced worker and the newcomer. Actually, this heterogeneity is more apparent than real, and unifying concepts, approaches, and ideas already exist, particularly with respect to all the questions concerning the role of membranes and their properties (such as ion permeability, electric potentials, and active transport) in the various steps of sensory perception and transduction processes.

It is currently accepted that most, if not all, the fundamental facts in molecular sensory physiology of aneural organisms, be they chemosensory, photosensory, or geosensory, can ultimately be understood in terms of a few basic ideas. Each chapter of this book emphasizes and clarifies the role of membrane properties and phenomena in the particular sensory response examined. Of course, in some cases, this task has been rather complex because of the limited amount of experimental data clearly supporting a membrane-based model of sensory transduction.

This book opens with a general introduction to dynamic structure and transport properties of biological membranes (Chapter 1); a section of the chapter is dedicated to model photosensitive membranes, with the aim of discussing some of the features of artificial systems that can mimic those of major photoresponses. Chapter 2 deals with chemosensory processes in unicellular eukaryotes and prokaryotes, with particular attention to the biochemical aspects of these phenomena, whereas Chapter 3 presents mechanoresponses in protozoa, one of the sensory systems in which the role of membranes has been clearly ascertained and thoroughly investigated. The perception and sensory transduction of thermal stimuli are reviewed and discussed in Chapter 4, which first examines the cellular structures and processes affected by temperature, and then presents the most significant biological examples. Chapter 5 offers a critical analysis of the most recent observations and hypotheses on the mechanisms of geotaxis in free-swimming cells, along with a detailed discussion of the relative contributions of passive hydrodynamic and physiological mechanisms of orientation with respect to gravity. The last three chapters are devoted to photosensory processes. Chapter 6 focuses on the possible role of membrane properties and functions in the basic processes of photobehavior of freely motile microorganisms. U. Pohl and V. E. A. Russo (Chapter 7) have dedicated their contribution to the late Nobel laureate Max Delbrück, whom we all remember. In addition to critically discussing what is known about photosensory transduction in phototropism, they give a comprehensive presentation of the basic phototropic phenomena. Finally, Chapter 8 illustrates light-oriented chloroplast movement, as an excellent subject to study the involvement of membranes in perception and transduction of, and in response to, photic stimuli.

Some of the discussions include a measure of speculation, but should help in understanding and interpreting experimental results as well as in planning new experiments. We do hope that this book will also provide a few hints for further research in this field.

Of course, we are fully responsible for the scientific coordination of the book, whereas the authors, to whom we are sincerely grateful and whom we have asked for contributions and original opinions, are responsible for their specific statements and viewpoints.

Giuliano Colombetti
Francesco Lenci

Contents

Chapter 1

Membranes: Structure and Function

Alessandra Gliozzi and Ranieri Rolandi

Chapter 3

Mechanosensory Transduction in Protozoa

Yutaka Naitoh

Chapter 4

Temperature Sensing in Microorganisms

Kenneth L. Poff, Donna R. Fontana, and Bruce D. Whitaker

Chapter 5

Microbial Geotaxis

Barry Bean

Chapter 6

Photosensory Responses in Freely Motile Microorganisms

Francesco Lenci, Donat-P. Häder, and Giuliano Colombetti

Chapter 7

Phototropism

Ulrich Pohl and Vincenzo E. A. Russo

Chapter 8

Chloroplast Movement

Wolfgang Haupt and Gottfried Wagner

Membranes and Sensory Transduction

Chapter 1

Membranes

Structure and Function

Alessandra Gliozzi and Ranieri Rolandi

1. THE DYNAMIC STRUCTURE OF CELL MEMBRANES

1.1. Introduction

The membrane surrounding the living cell is not only a passive barrier separating the interior of the cell from its environment, it must also make possible the selective interaction of these two regions, between which substances and information are continually exchanged. To better understand the dynamics of these processes, it seems useful to give a description of chemical composition, architecture, and topography of the plasma membrane.

1.2. Chemical Components

The major membrane components are proteins and lipids. Lipids are amphipathic molecules with a hydrophilic or polar head and a hydrophobic or apolar tail. Nearly all body lipids are built around the three-carbon glyceryl framework, as shown in Fig. 1. In phospholipids, the most common class of membrane lipids, two fatty acid chains are attached to two carbons, while a phosphate, usually with an added group, is attached to the third one. Usually

Alessandra Gliozzi and Ranieri Rolandi • Istituto di Scienze Fisiche dell'Università di Genova, 16146 Genoa, Italy.

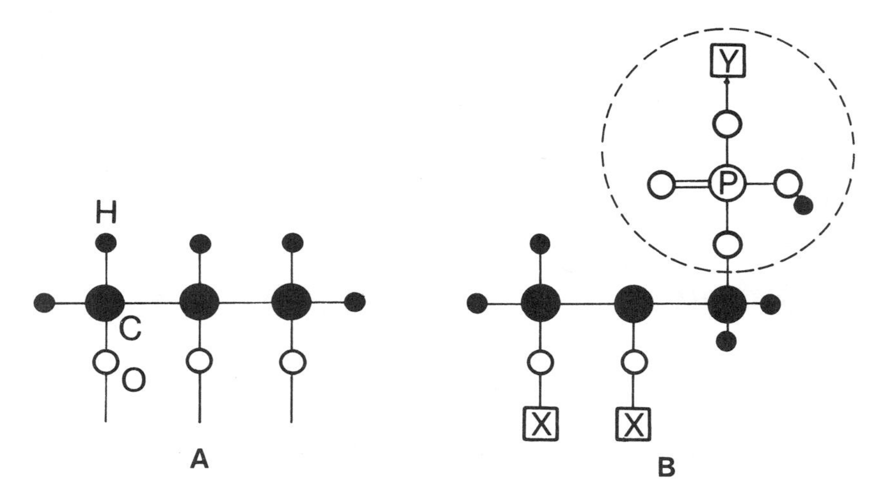

Figure 1. (A) Glyceryl framework. (B) Structure of a phospholipid. The hydrophobic part of the molecule is made up of two fatty acids, X, which are generally different, while the polar part of the molecule, indicated by the dashed circle, consists of a phosphate group, usually with an added group Y.

a phospholipid from a biological membrane contains a saturated and an unsaturated fatty acid. Thus the membrane lipids differ in the nature of either the fatty acid, or the polar group, or both.

Lipid molecules are present in a much larger amount than are proteins; however, taking into account the larger mass of the latter, most membranes contain as much by weight of lipid as of protein. For a particular type of membrane there are only few kinds of lipids, whereas there is a great variety of proteins, some of which are no more than a few molecules per membrane. Very few proteins have been fully characterized in terms of their chemical composition or molecular structure, one important reason being the difficulty in extracting intact proteins from their associated lipids. Lipids are responsible for the electrical isolation of the cell, while specialized proteins provide the pathways through which permeation occurs.

Another important component of the plasma membrane is cholesterol, a neutral lipid, the main function of which seems to be that of keeping lipids in a fluid condition when the temperature is low and of defluidizing the hydrophobic chains at high temperature. Cholesterol would thus appear to act as a fluidity buffer.

Virtually all plasma membranes contain another class of substances, the sugars. Nearly all cells are more or less sugar coated. Membrane sugars almost never occur in the form of simple sugars, such as glucose, but rather as oligosaccharides or as longer polymers, i.e., polysaccharides, anchored to lipids or to proteins to form glycolipids or glycoproteins, respectively. These substances play important roles in cell biology, being implicated in such processes as cell recognition and many immunological reactions.

1.3. Membrane Structural Framework

There is no doubt that lipid and protein organization is basic to membrane function. The first successful experimental approach to structure studies was performed by two Dutch investigators, Gorter and Grendel (1925). These workers extracted the lipids from the plasma membrane of a known quantity of red blood cells; this amount of lipids, spread in a monolayer, was shown to occupy a double surface with respect to the red blood cells membrane, hence the conclusion that lipids in the plasma membrane are arranged in a bilayer structure.

The same lipid arrangement was proposed independently, about 10 years later, by Danielli and Davson (1935), on the basis of physicochemical considerations on the amphipathic nature of the lipid molecule. In fact, the simplest stable structure is one in which two monolayers are apposed by hydrophobic interaction of the apolar chains, exposing the two hydrophilic sides to the external polar medium.

Later studies making use of more sophisticated techniques, such as high-resolution electron microscopy, x-ray diffraction, and electrical and optical measurements, have shown that such forecasting is quite general. In fact, in spite of the great variety of its functions, the basic structure of the plasma membrane was found always to be the same, i.e., a lipid leaflet, about 40 Å thick, formed by the hydrophobic apposition of two monolayers, as sketched in Fig. 2. At physiological temperatures the lipids are in a fluid liquid-crystal state, while at lower temperatures they undergo a phase transition to a gel configuration. The lipid leaflet is thus the structural framework of the membrane, permitting the anchorage of the proteins. Such basic structure is not only peculiar to the plasma membrane of animal and vegetable cells, but to the inner membrane of virtually all the organelles occurring in the cytoplasm, such as mitochondria, endoplasmic reticulum, lysosomes, and Golgi apparatus. The single known exception to this general rule seems to be the plasma membrane of the thermophilic *Archaebacteria.* The plasma membrane of this group of bacteria, which, like all the other *Archaebacteria,* is thought to represent one of the first forms of life on earth, in the pre-Cambrian era, is characterized by unusual isoprenoid bipolar lipids (Bu'Lock *et al.,* 1982; De Rosa *et al.,* 1980). Recent work on thin films formed from lipids of the archaebacterium *Caldariella acidophila* indicates that they are arranged in a single monolayer structure (Gliozzi *et al.,* 1982*a,b,* 1983).

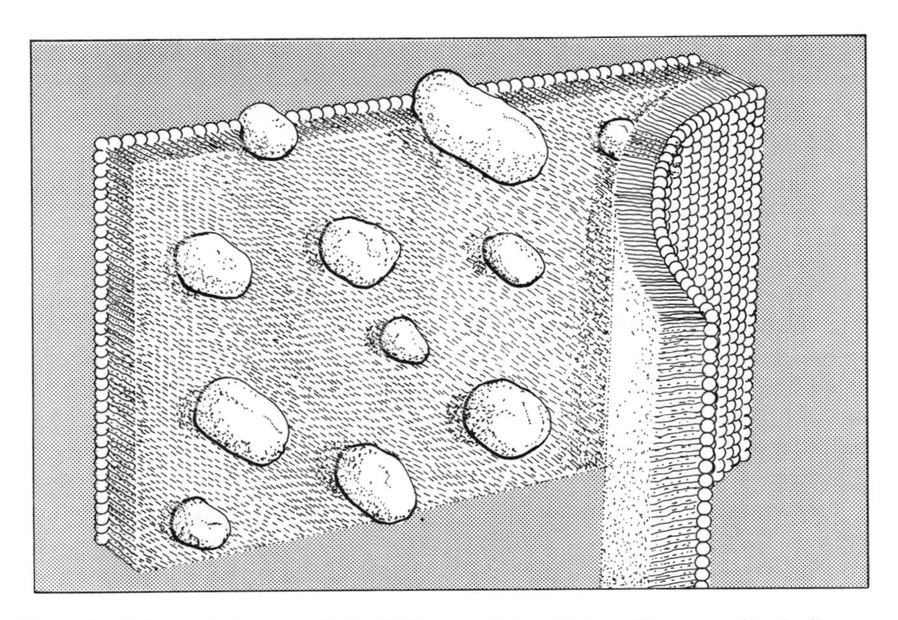

Figure 2. Schematic diagram of a lipid bilayer split in mid-plane. The protruding bodies represent the proteins.

A peculiar feature of all cells is asymmetry of the two monolayers, both in lipids as well as in protein composition. Moreover, sugars are always linked to the outer side of the cell membrane, as illustrated in Fig. 5. Such chemical asymmetry reflects the requirement that the inner and the outer part of a cell membrane must serve different physiological functions.

1.4. Danielli–Davson and Fluid Mosaic Models

A problem that arises immediately is how the proteins interact with the bilayer structure described above, and what the structure of the plasma membrane to serve the various physiological functions is. Many models have been proposed to illustrate the architecture of the plasma membrane. Of them, the most important historically is the pioneering Danielli–Davson model (1935), the so-called unit-membrane model. In a review on the development of this membrane model, Danielli (1975) emphasizes that "the basic features of this model, have survived all efforts to demolish them" giving us an idea of the heated discussions underlying membrane models. According to the final version of this model, the membrane is composed of six layers. At the center of the model is the lipid bilayer, covered on both sides by two layers of unfolded proteins, with an additional layer of loosely attached, adsorbed proteins. The ionic and the molecular transport was supposed to occur through "pores" lined with proteins and extending through the lipid bilayer. Figure 3 is an illustration of this model. Support for this model was thought to be given by electron micrographs of the cell membrane in which the two external lines were identified as the protein layer, while the inner one was thought to be the lipid bilayer. We know that this picture is not correct and that such image is peculiar to the fixation methods.

The Danielli–Davson model was certainly based on a brilliant intuition; however, at that time it could not be based on a precise knowledge of the structural and the physicochemical properties of the membrane components and of the interactions between these molecules. Such knowledge became possible for the development of various techniques, such as x-ray diffraction, high-resolution electron microscopy, Raman spectroscopy, and nuclear magnetic resonance.

One of the first models that introduced a fundamentally different concept, that of long-range order in biomembranes, was introduced by Changeux *et al.* (1967). According to this model, biomembranes comprise two-dimensional crystalline lattices, consisting of lipoprotein subunit, protomers, in regular repetitive arrays. In subsequent years much emphasis was given to the finding

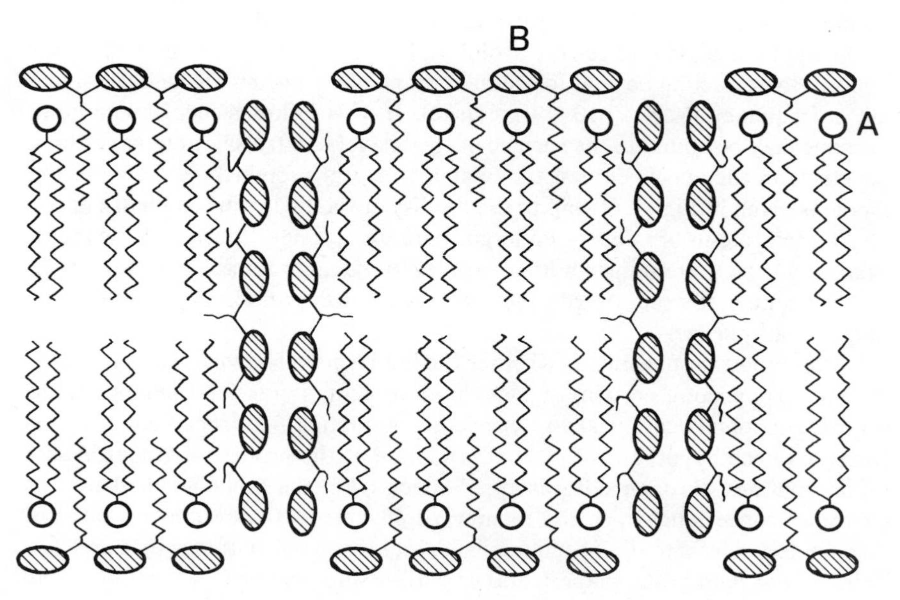

Figure 3. Representation of the Danielli–Davson model of the plasma membrane. (A) Lipids; (B) proteins.

that lipids in the membrane are in a liquid crystal configuration. In particular, experiments on immunological reactions (Frye and Edidin, 1970) showed that by fusing mouse and human cells *in vitro* and following the antigen redistribution on the cell surface (by means of fluorescent antibodies) they were intermixing after about 40 min when cells were at physiological temperature, while the rate of intermixing was much slower by cooling to 4°C. Also significant have been the experiments on freeze-fracture electron microscopy, which showed the presence of intramembrane particles representing regions in which protein or protein complexes penetrate into or through the membrane (Fig. 4). Typical surface concentrations of particles are 3×10^8 particles/cm^2.

Taking into account these results and taking advantage of the deeper knowledge of protein structure, Singer and Nicholson (1972) proposed the fluid mosaic model. According to this model, the membrane comprises a lipid bilayer, fluid at physiological temperatures, the molecules of which are free to undergo translational diffusion. In fact, data on artificial lipid bilayers have shown that lipids in a fluid condition are able to exchange their position with that of neighboring ones. By the spin-label technique, it has been shown that the frequency of this lateral displacement is of the order of 10^7 sec^{-1} for biological membranes and synthetic lipids at a temperature above their thermal phase transition (Träuble and Sackmann, 1972, 1973). This means that a lipid molecule in a fluid lipid matrix remains at a given site in the lipid lattice for no longer than $\sim 10^{-7}$ sec. The corresponding coefficient for lateral diffusion of the lipid molecule is of the order of 3×10^{-8} cm^2 sec^{-1}. The flip–flop movement in the direction normal to the membrane plane occurs instead at a much lower frequency ($<2 \times 10^{-5}$ times/sec). In this bidimensional continuum, proteins may be bound to the surface of the bilayers (extrinsic proteins) or may penetrate in the apolar core of the bilayer (integral proteins). A sketch of the model is given in Fig. 5. Owing to the fluidity of the hydrocarbon chains of the lipids, the proteins are able to undergo free translational diffusion in the membrane, and the protein distribution parallel to the membrane plane is random. The fundamental structure can thus be visualized as proteins floating, like icebergs, in a lipid sea.

An important question is whether the lipid is merely an essentially passive structure supporting proteins or some lipid–protein interaction fundamental to protein function occurs instead. A specific interaction with specific lipids contrasts with the hypothesis of a random motion of the proteins, without loss of their functions, through different lipid species. On the other hand, there are examples concerning especially membrane enzymes, of interactions with specific lipids essential to their functioning. A tentative explanation is that proteins move in the membrane plane together with a surrounding lipid annulus that contains the lipids specific to the protein interactions. However, the existence of a lipid annulus is controversial; recent work tends to reject such hypothesis in most cases.

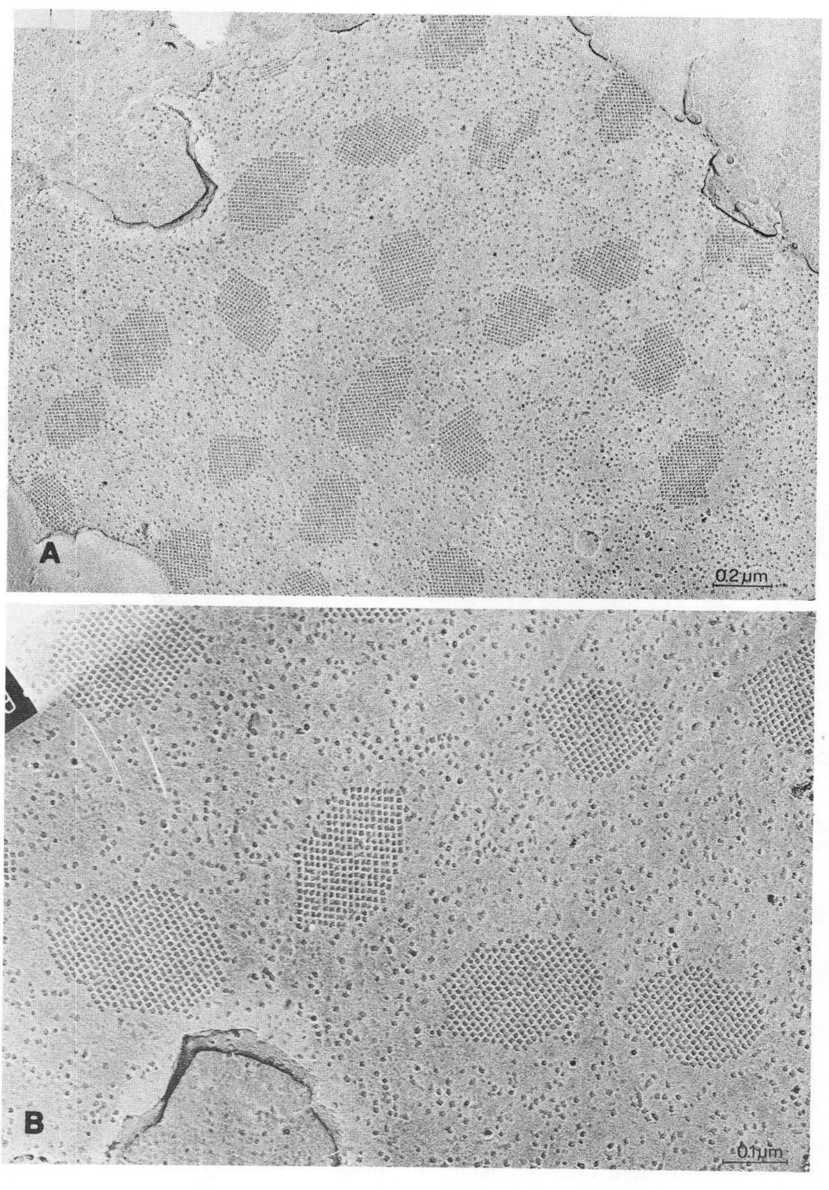

Figure 4. Freeze-fractured *Chlamidobotrys stellata* plasma membrane at two different magnifications, (A) ×110,000; (B) ×200,000. The cells were frozen using ultrarapid cryofixation without any pretreatment, then fractured and rotary shadowed with W-Ta in Balzer's BAF 301 freeze-etching unit. Note the presence of ordered arrays of intramembrane particles separated by domains of dispersed particles on a very smooth background. (Courtesy of T. Gulik, Centre de Genétique Moleculaire, Gif-sur-Yvette, France.)

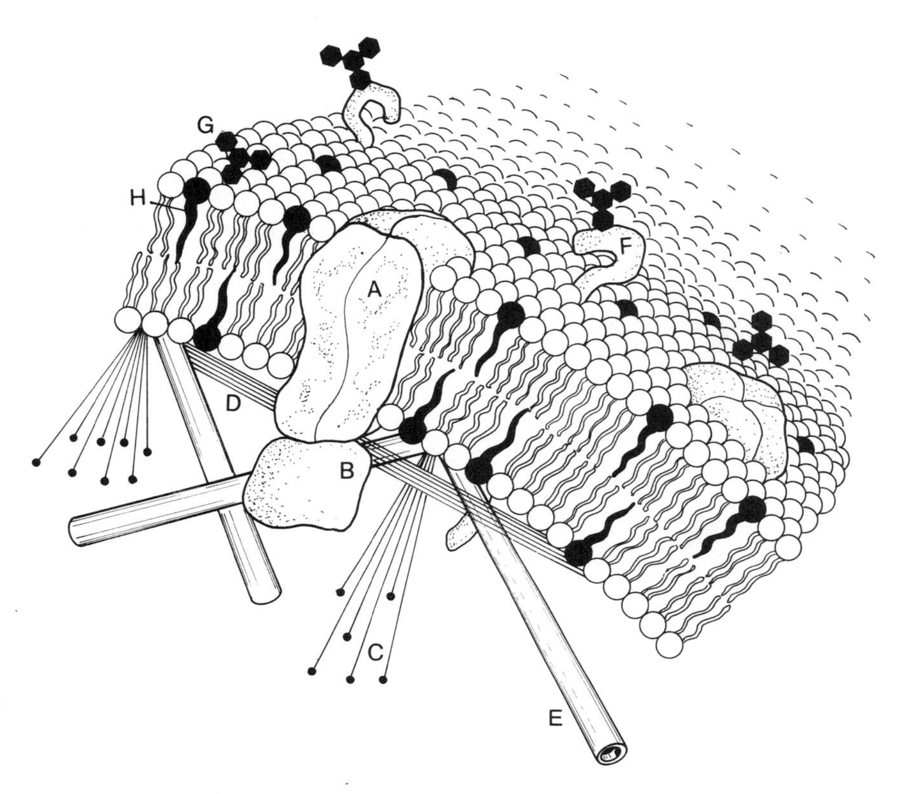

Figure 5. Representation of the fluid mosaic model of the plasma membrane. The membrane-associated microtubule and microfilament system involved in transmembrane control is also shown. (A) Integral protein; (B) peripheral protein; (C) myosin molecules; (D) actin filaments; (E) microtubules; (F) glycoprotein; (G) glycolipid; (H) cholesterol. (From Singer and Nicholson, 1972; Nicholson, 1976.)

1.5. Laboratory Models

The most peculiar feature of the plasma membrane is the fact that, in spite of the great variety of its functions, the basic structure is always the same. This fact has stimulated many investigators to study lipid–protein interactions in a variety of model systems. Not only have the models provided valuable information on the properties of biomembranes, but have often suggested new ideas on experiments to be performed in biological membranes.

1.5.1. Lipid Monolayers

The simplest model of cell membrane is a lipid monolayer. This is generally produced by spreading the phospholipid in a trough containing water solution. The trough is equipped with a moving barrier designed to sweep the surface and compress the film. Because molecules of these substances are amphipathic, they will tend to form monolayers, with hydrophilic ends in contact with water and hydrophobic ends with air.

Monolayers have been used to study a variety of membrane phenomena. The first biological application of the monolayer technique as a membrane model is the previously mentioned Görtel and Grendel experiment. The parameters that are usually measured with a monolayer technique are the surface tension and, from the pressure-area isotherms, the packing characteristics expressed as area per molecule. The critical role of the surface tension in controlling incorporation of proteins, when reconstitution experiments are approached, has been stressed (Schindler, 1980; Schindler and Quast, 1980). The electrostatic potential at the monolayer–solution interface is also a measurable parameter; it may be a very important factor in determining the ionic transport in lipid bilayers, as discussed in Section 2.6.

1.5.2. Planar Lipid Bilayers

Artificial lipid bilayers were obtained by Mueller *et al.* (1962). According to their technique, a small amount of lipid is smeared across a hole a few millimeters in diameter, in a Teflon partition separating two aqueous solutions (Fig. 6A). The membrane formation can be followed, through a microscope, illuminating the front of the membrane with a white light. The membrane is initially thousands of Ångstroms in thickness, as deduced by the appearance of colored interference fringes. The formation of black spots, corresponding to negative interference regions, reveals the spontaneous thinning of the membrane. In its final state, the membrane appears optically black, bounded by a clear annulus corresponding to the Plateau–Gibbs border. Thickness measurements indicate that the film is a bimolecular leaflet, consisting of a nonpolar region, containing the lipid hydrocarbon chains and some solvent, sandwiched between the polar head groups of the lipids.

Physicochemical properties of black lipid membranes (BLM), compared with those of biological membranes, are summarized in Table I. Table I shows that while many properties, such as capacitance, refraction index, water permeability, and dielectric breakdown, are of the same order of magnitude as that of biological systems, a dramatic discrepancy, of many orders of magnitude, is displayed by membrane conductance. This fact is not surprising, since a pure lipid leaflet behaves like an electrical insulator. In fact, as previously discussed, the transport properties of a membrane seem to be related essentially to the protein components. Chemical substances able to mimic the biological behavior were thus inserted in artificial bilayers; the induced transport properties were compared with those of biological membranes. This argument is dealt with in Sections 3.11 and 3.12 in an analysis of the molecular basis of

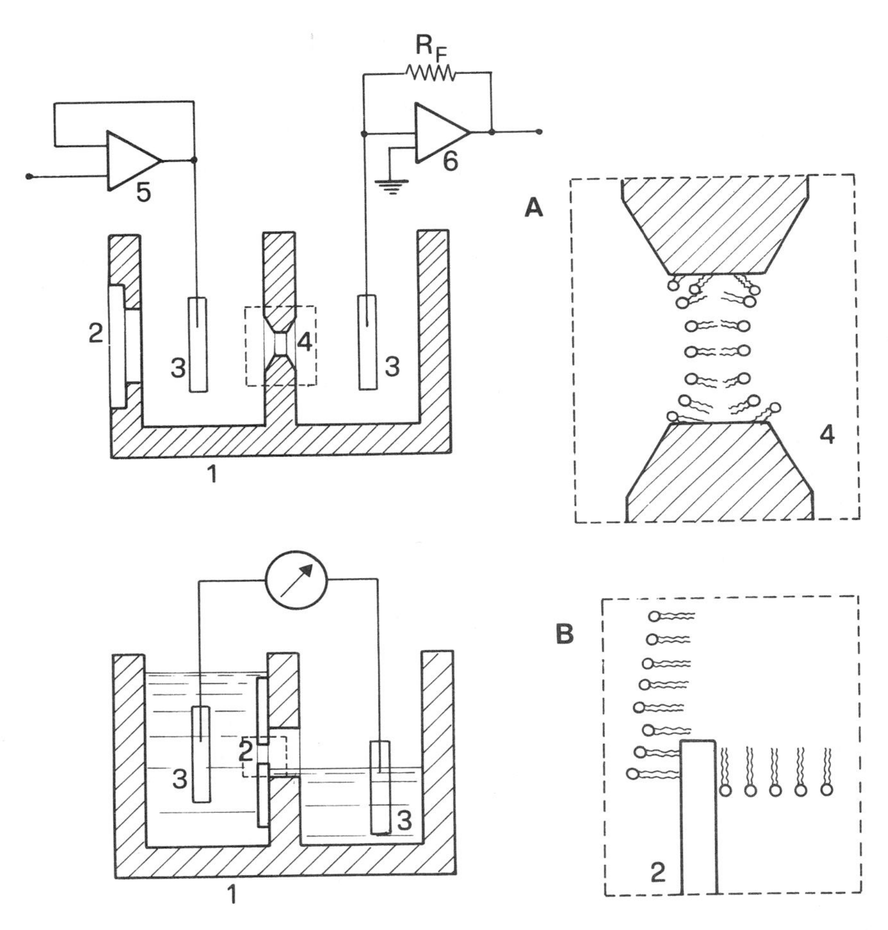

Figure 6. (A) Schematic representation of the experimental setup for the formation of black membranes with the classical Mueller *et al.* (1962) technique: (1) Teflon cell; (2) glass window; (3) electrodes; (4) hole for the membrane formation (the magnification to the right is obviously not drawn to scale and does not represent the real situation for the Plateau–Gibbs border); (5) voltage-clamp system; (6) current measurement system. (B) Schematic representation of the experimental setup for the formation of black films with the Montal and Mueller (1972) technique: (1) Teflon cell; (2) teflon partition; (3) electrodes.

Table I
Comparison Between Natural and Artificial Membrane Properties

Property	Natural membranes	BLM[a]
Hydrocarbon thickness (Å)	20–40	25–70
Capacitance ($\mu F/cm^2$)	0.8–1.5	0.3–0.9
Specific conductance (S)	10^{-2}–10^{-5}	10^{-3}–10^{-9}
Dielectric breakdown (mV)	100–400	100–500
Refraction index	1.6	1.3–1.6
Water permeability ($\mu m/sec$)	0.25–400	8–50

[a]BLM, Black lipid membrane.

excitability. Physicochemical studies on BLM structure have shown that solvent dissolved in the membrane determines to a noticeable extent the characteristics of the bilayer, with formation of microlenses, lipid-solvent stoichiometry, and solvent phase transition being rather important phenomena. A detailed review on studies of the lipid bilayer as a model system for biological membranes has been published elsewhere (Gliozzi, 1980*a*). From these studies it appears of crucial importance to build up solvent-free bilayers. To this purpose, White (1978) was able to form BLM containing negligible amounts of alkyl solvent. The membranes were formed from glyceryl mono-oleate dispersed in squalene.

An alternative method of formation of lipid bilayers, first described by Montal and Mueller (1972) and recently modified by Schindler (1980), is shown schematically in Fig. 6B. According to the Montal and Mueller method, a small amount of a hexane–lipid solution is spread over the free aqueous surface in the two half-cells. By evaporation of the hexane, two monolayers are formed. According to Schindler's method, spontaneous monolayer formation occurs from vesicles having a radius of >50 nm. Vesicles are filled into the two compartments of the membrane cell, just below the aperture of the frame. In both cases a lipid bilayer is formed across the hole, by increasing the level of the solution. Formation of a planar membrane occurs by hydrophobic apposition of the two monolayers.

An important structural parameter of these membranes is the thickness of the hydrophobic core, which may be evaluated by the electrical capacity of the membrane. With bilayers made from monolayers of a number of lipids, values of specific capacity of 0.6–0.9 $\mu F/cm^2$, which is close to the capacity of many biological membranes, were observed (Benz *et al.*, 1975). The corresponding dielectric thickness of these films strongly suggests that they are almost free of solvent (Benz *et al.*, 1975; White, 1978).

1.5.3. Liposomes

A further model of biological membranes is the liposome, prepared from a wide spectrum of synthetic as well as natural lipids. The lipids dispersed in a volatile solvent (e.g., chloroform) are evaporated to dryness in a rotating test tube. Adding a water solution in a proportion over 50%, they form liposomes—onionlike structures consisting of stacked lamellae of lipids in a bilayer configuration, with water filling the spaces between the lamellae. The main advantage of liposomes over black membranes is that they are present in a higher quantity and have a well-defined composition (no solvent is present); however, electrical properties cannot be measured. Ultrasonic irradiation converts these structures to simpler vesicles, in which a single bilayer surrounds an aqueous compartment of 200–500 Å in diameter. The number of lipid molecules per vesicle is evaluated to be 1.9×10^3–2.7×10^3. Of these, 65% are present in the external layer, owing to the very high curvature of the vesicle. The liposome model has helped elucidate a number of problems in membrane physiology. One of the most interesting of these problems is the study of protein–lipid interactions in reconstituted systems.

2. KINETIC AND THERMODYNAMIC APPROACH TO MEMBRANE TRANSPORT PROPERTIES

2.1. Equilibrium State

The equilibrium state can be analyzed from both a thermodynamic and statistical mechanical viewpoint. According to the thermodynamic approach, at equilibrium the electrochemical potential of any species i must be equal in any phase to which the species has access. Thus, for two phases 1 and 2

$$\tilde{\mu}_i^{(1)} = \tilde{\mu}_i^{(2)} \tag{1}$$

The electrochemical potential $\tilde{\mu}_i$, for an ideal diluted solution is

$$\tilde{\mu}_i = \mu_i(\mathrm{o}) + RT \ln c_i + p\overline{V}_i + z_i F\psi \tag{2}$$

where $\mu_i(\mathrm{o})$ is the standard electrochemical potential, c_i is the concentration of the ith species, $\overline{V}$ is its partial molar volume, p is the hydrostatic pressure, z is the valence, ψ is the electrostatic potential, and R, F, and T are the gas constant, the Faraday constant, and absolute temperature, respectively. The potential energy of species i could change for other work terms, e.g., gravitational energy, magnetic energy. In such a case, the relevant work terms must be added to Eq. (2).

The statistical mechanical approach implies that at thermal equilibrium the particles satisfy the Boltzmann distribution. This means that, considering only a one-dimensional situation, the concentration of the ith species at a distance (x) from the plane of the membrane is

$$c_i(x) = c_\infty \exp(-w_i(x)/RT) \quad (3)$$

where $w_i(x)$ is the potential energy per mole of the ith species and c_∞ is its concentration at a point defined as zero potential energy. If w_i is purely electrostatic energy, then Eq. (3) becomes

$$c_i(x) = c_\infty \exp(-z_i F\psi(x)/RT) \quad (4)$$

Condition (1) on the electrochemical potential can also be derived by statistical mechanical considerations showing that the statistical mechanical approach of a system at equilibrium is equivalent to the thermodynamic approach. In fact, considering the logarithm of the Boltzmann distribution for two phases 1 and 2 and rearranging the terms, one gets

$$\ln c_i^{(1)}(x) + \frac{w^{(1)}(x)}{RT} = \ln c_i^{(2)}(x) + \frac{w^{(2)}(x)}{RT} \quad (5)$$

If $w(x)$ is only composed of internal energy and electrostatic energy terms, Eq. (5) becomes

$$RT \ln c_i^{(1)}(x) + P^{(1)}\overline{V}_i + z_i F\psi^{(1)}(x) = RT \ln c_i^{(2)}(x) + P^{(2)}\overline{V}_i + z_i F\psi^{(2)}(x) \quad (6)$$

which is coincident with Eqs. (1) and (2).

2.2. Nonequilibrium States

States of equilibrium can be treated in a simple and comprehensive manner, whereas it is necessary to employ more complex methods in order to obtain satisfactory solutions to problems associated with nonequilibrium processes. Even though the thermodynamic and kinetic (or statistical mechanical) approaches are equivalent at equilibrium, when ion transport is considered, the two approaches (although *a priori* equivalent) do not necessarily provide the same kind of information. In fact, the thermodynamic description, although more general, does not give the molecular mechanism of ion transport, whereas the kinetic approach, being based on a particular model, loses its general validity. The theory of membrane transport has thus been developed through both thermodynamic and kinetic treatment, leading to the Nernst–Planck flux equations and the Eyring reaction-rate theory. A brief outline of both approaches is presented in the following sections.

2.3. The Nernst–Planck Flux Equations

According to nonequilibrium thermodynamics, the phenomenological force acting on a single ion i in a continuous system is the negative gradient of its electrochemical potential $-$ grad $\tilde{\mu}_i$. Considering a unidirectional flux in the x direction, under isothermal and isobaric conditions, we get from Eq. (2)

$$\frac{d\tilde{\mu}_i}{dx} = RT \frac{d}{dx} \ln c_i + z_i F \frac{d\psi}{dx} \tag{7}$$

If coupling between fluxes can be neglected, then the average diffusion velocity vvv_i of a constituent can be assumed to be proportional to its driving force. For a unidirectional flux, we get

$$v_i = -u_i \frac{d\tilde{\mu}_i}{dx} \tag{8}$$

where u_i is the molar mobility of species i, which is related to the diffusion coefficient by the simple relationship

$$D_i = RTu_i \tag{9}$$

Recalling the definition of flux J_i,

$$J_i = c_i v_i \tag{10}$$

and combining it with Eqs. (7) and (8), we get the Nernst–Planck flux equations:

$$J_i = -u_i c_i \left(\frac{RT}{c_i} \frac{dc_i}{dx} + z_i F \frac{d\psi}{dx} \right) \tag{11}$$

Equation (11) is the basis of the thermodynamic description of ion transport. Some simple cases are considered in Section 2.4.

For a nonelectrolyte ($z_i = 0$), Eq. (11) becomes the well-known Fick's law of diffusion:

$$J_i = -D_i \frac{dc_i}{dx} \tag{12}$$

If the diffusing species is ionic and is in equilibrium, so that $J_i = 0$, we get

$$-u_i c_i \left(RT \frac{d}{dx} \ln c_i + z_i F \frac{d\psi}{dx} \right) = 0 \tag{13}$$

Integrating this equation from ∞ to x, and recalling that $\psi(\infty) = 0$, one obtains Eq. (4), i.e., the Boltzmann distribution of ionic species. Moreover, integrating Eq. (13) across a membrane of thickness δ and indicating as $\psi(0)$ and $\psi(\delta)$ the potential in the membrane phase at the left and right boundaries, respectively, we get the Nernst equilibrium potential:

$$\psi(\delta) - \psi(0) = -\frac{RT}{z_i F} \ln \frac{c_i(\delta)}{c_i(0)} \tag{14}$$

Note that $c(0)$ and $c(\delta)$ are the boundary concentrations inside the membrane. These same variables may be related to the concentrations of the bathing media, as in Section 2.4.2, Eq. (17). For a more detailed analysis of the physical implications of the Nernst–Planck equation in various physical situations, the reader is referred to the review paper by Finkelstein and Mauro (1974).

2.4. Specific Cases

2.4.1. Goldman–Hodgkin–Katz Flux Equation

The Nernst–Planck equation has been used to describe the ionic fluxes across biological membranes in steady-state conditions. The following simplifying assumptions have been employed (Goldman, 1943; Hodgkin and Katz, 1949):

1. The regions through which diffusion occurs are macroscopically homogeneous, and therefore the mobility can be considered uniform throughout the thickness.
2. Coupling between flows of different ions or between ions and solvent can be neglected; behavior often referred to as independence principle.
3. The electric potential is supposed to fall linearly within the membrane. This assumption implies that the volume charge density at any point inside the membrane $\rho(x)$ is so small that it can be considered approx-

imately zero; a direct consequence of the Poisson equation, treated in Eq. (21).

By use of assumption no. 3, one can substitute in Eq. (11) $d\psi/dx$ with $\Delta\psi/\delta$, where

$$\Delta\psi = \psi(\delta) - \psi(0)$$

Equation (11) thus becomes a linear first-order differential equation, which can be easily solved, imposing a fixed value of concentration at the two boundaries. Rearranging the terms, an explicit expression of the flux can be found in the classical Goldman–Hodgkin–Katz flux equation:

$$J_i = \frac{RTu_i}{\delta} U \frac{c_i(0) - c_i(\delta) \exp U}{\exp U - 1} \tag{15}$$

where the reduced potential

$$U = \frac{z_i F \Delta\psi}{RT}$$

has been introduced.

This equation is widely employed in physiological work, in which δ indicates the inside and 0 the outside of the cell. Note that Eq. (15) is often given with an opposite sign, because outward flux, in the direction $\delta \rightarrow 0$, is defined as positive.

Flux equation (15) shows the relationship between fluxes and potential, which corresponds to a passive configuration of fluxes, i.e., to fluxes driven only by the gradient of their electrochemical potentials. If active fluxes are present, the total flux of an ion is given by the algebraic sum of active and passive fluxes. In the case in which permeation of these ions occurs through the same pathway, the ionic pump may alter the passive configuration of concentration, making Eq. (15) no longer valid. However, it may be a good approximation if active fluxes are one order of magnitude lower than the passive ones.

If $J_i = 0$, Eq. (15) reduces to Eq. (14), the Nernst potential, which must occur at equilibrium.

2.4.2. Membrane Potential

At this point, we want to derive the potential difference established across a membrane permeable to many species, in steady-state conditions, when no external potential is applied. We shall consider the case in which only monovalent species do permeate through the membrane. This is a particularly interesting case for describing many physiological situations, in which Na^+, K^+, and Cl^- are the most important permeant ions. The equation that is widely employed in physiological works was originally attributable to Goldman (1943) and successively rederived by Hodgkin and Katz (1949). It was obtained from Eq. (15), by setting $z_i = \pm 1$.

The condition that the membrane is in a free diffusion state, i.e., that there is no current source, yields

$$I = F(\Sigma_i z_i J_i^{\mathrm{p}} + \Sigma_i z_i J_i^{\mathrm{a}}) = 0 \tag{16}$$

where I is the current density and the indices p and a indicate passive and active fluxes, respectively.* If the fluxes resulting from active transport do not carry a net current, i.e.,

$$\Sigma z_i J_i^{\mathrm{a}} = 0$$

(the so-called nonelectrogenic hypothesis), the J_is are only the passive fluxes given by Eq. (15).† The resulting resting potential is represented in the Goldman equation:

$$\Delta\psi = \frac{RT}{F} \ln \frac{\Sigma_{\mathrm{a}} u_{\mathrm{k}} c_{\mathrm{k}}(0) + \Sigma_{\mathrm{a}} u_{\mathrm{a}} c_{\mathrm{a}}(\delta)}{\Sigma_{\mathrm{k}} u_{\mathrm{k}} c_{\mathrm{k}}(\delta) + \Sigma_{\mathrm{a}} u_{\mathrm{a}} c_{\mathrm{a}}(0)} \tag{17}$$

When $\Delta\psi$ refers to a cell membrane it indicates the inside (δ) minus the outside (0) potential. It is also called, in physiological literature, the reversal potential. Subscripts k and a indicate cations and anions, respectively. The Goldman equation has thus been derived under the same assumptions as for the flux equation, i.e., uniform mobility, no coupling between fluxes, and linear electrical field. To these, the condition on active fluxes must be added.

Most of the assumptions mentioned in Section 2.4.1 have hard application to living biological systems. For example, the Na^+ and K^+ channels in nerve membrane may be complex permanent pores formed by protein molecules, where many inhomogeneities may exist in the electrical field, caused by

*In Eq. (16), the letter "a" represents an active (versus passive) flux. This is not to be confused with the letter "a" described in Eq. (17), which refers to an anion (versus cation). Later, in Eqs. (31) and (32), another letter a is introduced, where it is a variable standing for transmission factor.

†The use of Eq. (15) when active fluxes are present and are not spacially separated from the passive ones may be incorrect, as pointed out in Section 2.4.1.

charges and dipoles of the protein. Moreover, as previously anticipated, even in the case in which the active fluxes do not carry a net current, if permeation pathways of passive and active fluxes are the same, the ionic pump may alter the membrane potential by displacing the concentration profiles from their passive configuration. Schwartz (1971) showed that a nonelectrogenic pump will not directly alter the membrane potential only if active and passive paths are separate and, in effect, parallel. Moreover, Schwartz showed that the Goldman equation (17) was still valid under much less restrictive assumptions than those originally employed. Schwartz provided a more general expression of the membrane potential that took into explicit account the active fluxes, without any assumption about either the nature of the electrical field or the homogeneity of the membrane.

A further problem arises when concentrations are considered. In fact, Eq. (15) refers to concentrations at the two membrane–solution interfaces in the membrane phase. These must be related to the only known physical quantities, which are concentrations in the bulk solutions. Hodgkin and Katz (1949) introduced a single, simple, linear partition coefficient β_i, relating the membrane concentrations across the two membrane–solution interfaces:

$$\beta_i = \frac{c_i(\text{o})}{c_i'(\text{o})} = \frac{c_i(\delta)}{c_i'(\delta)} \tag{18}$$

where $c_i'(\delta)$ and $c_i'(\text{o})$ are the bulk concentrations inside and outside the cell membrane, respectively.

Thus, permeability coefficients P_i defined as

$$P_i = \beta_i RT u_i/\delta \tag{19}$$

appear in their equation in place of the u_i values of Eq. (17). According to Eq. (19), the selective permeability for a particular ion can be increased by increasing its diffusion coefficient or its partition coefficient or by decreasing the length of its diffusion path.

In the following discussion we report results that make use of Eqs. (15) and (17) in conjunction with Eqs. (18) and (19). Therefore, permeabilities, or in simpler cases permeability ratios, appear instead of mobilities.

The use of Eq. (18) implies that the electrostatic effects at the membrane–solution interfaces may be neglected. Indeed, as discussed subsequently, ion charges in the membrane surface contribute to electrostatic potential and then to nonuniform ionic concentration profile in the aqueous medium surrounding the membrane, as shown in Fig. 7 (*top*). For a charged membrane, the bulk concentration approaches that at the membrane–solution interface only on increasing the ionic strength, as illustrated in Fig. 7 (*bottom*). Therefore, the

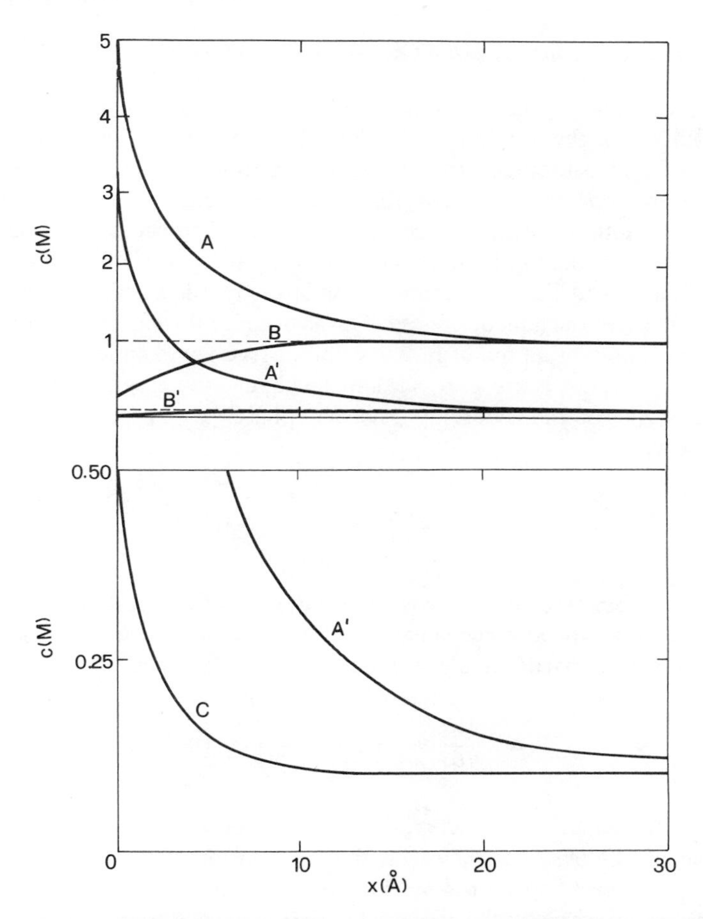

Figure 7. Molar concentration [c(M)] profiles at the membrane–solution interface calculated as a function of distance from a negatively charged membrane (σ = 1 electronic charge/150 $Å^2$). (Top) Cation (A,A′) and anion (B,B′) concentrations for a 1 *M* and a 0.1 *M* symmetrical salt solution, respectively. Equations (3), (25), and (29) have been used. (See text.) (Bottom) Curve A′ is replotted in an expanded scale, to show the effect of increasing the ionic strength. Curve C is calculated assuming the same conditions as for A′, but at higher values of ionic strength (1 *M*), obtained by adding a different symmetrical electrolyte of the same valence z to the solution.

ionic concentration at the membrane boundary, rather than that in bulk solution, should be placed in Eq. (17), in considering low-concentration solutions. Moreover, differing surface potentials, which might arise owing to the asymmetry of the plasma membrane, might also give the effect of asymmetrical partition coefficients.

2.5. Partition Coefficient at an Oil–Water Interface

We have already faced the problem of relating the concentrations inside the membrane to those in bulk solution, which are the only physical quantities accessible for measurement. In the classical treatment of the Goldman–Hodgkin–Katz equation we have seen that a simple partition coefficient, relating external and internal concentrations, was supposed to occur. This coefficient refers to the partitioning in ionic pores. In fact, the partitioning of small ions between water and liquid hydrocarbon is strongly one-sided, with almost all ions remaining in the aqueous phase. This is because the electrical contribution to the free energy of an ion of radius r and charge ze, in an ideal dielectric of dielectric constant ϵ, is $z^2e^2/2r\epsilon$, so that the change in energy per ion on going from water $\epsilon_w = 80$ to a pure hydrocarbon phase $\epsilon_m \simeq 2$ is, according to the Born theory,

$$\Delta G = -\frac{z^2e^2}{2r}\left(\frac{1}{\epsilon_w} - \frac{1}{\epsilon_m}\right) \tag{20}$$

where G is Gibbs free energy, w is water, and m is hydrocarbon.

Thus, for a typical monovalent ion of radius $r = 1$ Å, we find that $\Delta G \simeq$ 80 kcal/mole; the partition is consequently very small, since

$$\beta = \frac{c_i(0)}{c_i'(0)} = \exp\left(-\frac{\Delta G}{RT}\right) \simeq 10^{-60} \tag{21}$$

where β is the partition coefficient defined by Eq. (18).

It should be observed, however, that the ion might be hydrated, thereby substantially increasing its radius. Moreover, the hydrocarbon interior of the phospholipid bilayer is much more polar than that of a pure alkane, owing to the water penetration and the thinness of hydrocarbon region. Indirect measurements of ϵ_m inside hydrated phospholipid bilayers indicate a value well above 20 down to the C-5 position and about 5.5 at the C-12 position (Griffith *et al.*, 1974). In addition, the Born theory is derived from continuum electrostatics, assuming ideally polarizable dielectrics. However, in an electrolytic solution, the electrical field near the ions or near a charged surface is intense enough at 10^6–10^7 V·cm^{-1} to line up the adjacent water molecules almost completely. As alignment becomes complete, further polarization becomes impossible, and the dielectric constant is effectively reduced. These facts do substantially lower the electrostatic free-energy change, thereby increasing the partition coefficient given in Eq. (21).

Even with these corrections, the energy barrier preventing unhydrated alkali ions from passing through the hydrocarbon region is very high. In biological as well as artificial membranes, permeability phenomena are always allowed by some special mechanism—pore or carrier—in order to avoid completely unsolvated conditions.

2.6. Electrostatic Potentials at a Membrane–Solution Interface

We have mentioned the importance of electrical fields at membrane–solution interfaces. We would now like to give a brief discussion of the related electrical phenomena at equilibrium. Two parameters of the electrostatic potential energy barrier associated with membranes are of crucial importance in biological phenomena: (1) the potential gradient inside the membrane, which affects the activity of intrinsic membrane proteins, and (2) the potential difference between bulk electrolyte solution and membrane surface, which influences ionic surface concentrations and the interaction with soluble molecules.

About 10–20% of the lipids in the membrane of many cells bear a net negative charge. As a phospholipid in a bilayer occupies an area of about 60 $Å^2$, the average charge density of a membrane composed of 20% negative lipids is 1 electronic charge/300 $Å^2$. This results in a surface potential, which is generally evaluated with the Gouy–Chapman theory on the diffuse double layer.

The Gouy–Chapman theory is a quantitative treatment between the potential and the surface charges. The Poisson equation for a planar surface is given by

$$\frac{d^2[\epsilon_0\epsilon_w\psi(x)]}{dx^2} = -\rho(x) \tag{22}$$

where $\psi(x)$ and $\rho(x)$ are, respectively, the potential and the volume charge density at a distance x from the membrane, ϵ_0 is the vacuum permittivity, and the other symbols are as previously defined. The charge density is given by

$$\rho(x) = ze\mathcal{N}[c^+(x) - c^-(x)] \tag{23}$$

where c^+ and c^- are the concentrations of cations and anions of an electrolyte which, for simplicity's sake, is supposedly symmetrical and $\mathcal{N}$ is Avogadro's number. At equilibrium, substituting the Boltzmann distribution of ions given by Eq. (4) into Eq. (22) gives the Poisson–Boltzmann relation

$$\frac{d^2(\epsilon_0\epsilon_w\psi)}{dx^2} = 2ze\mathcal{N}c_\infty \sin h[ze\psi(x)/\kappa T] \tag{24}$$

where κ is Boltzmann's constant. The Poisson–Boltzmann equation may be integrated under the assumption that ϵ_w does not vary in proximity of the membrane, near $x = 0$.

Imposing the appropriate boundary conditions:

$$\psi(x) = \psi_0 \quad \text{at} \quad x = 0 \quad \text{and} \quad \frac{d\psi}{dx} = \psi(x) = 0 \quad \text{at} \quad x = \infty$$

the resulting potential $\psi(x)$ is:

$$\psi(x) = \frac{2\kappa T}{ze} \ln \frac{1 + \alpha \exp(-x/L)}{1 - \alpha \exp(-x/L)} \tag{25}$$

where

$$\alpha = \frac{\exp(ze\psi_0/2\kappa T) - 1}{\exp(ze\psi_0/2\kappa T) + 1} \tag{26}$$

and

$$L = \left[\frac{2e^2 \mathcal{N} c(\infty) z^2}{\epsilon_w \epsilon_0 \kappa T}\right]^{-1/2} \tag{27}$$

The parameter L is defined as the Debye length, which determines the extent of space charge region. For small potentials, $\psi(x)$ falls to $1/e$ its value at the membrane interface in a distance L.

The charge density on the surface of the membrane σ can be related to the surface potential ψ_0 by the electroneutrality condition

$$\sigma = -\int_0^\infty \rho(x)\, dx \tag{28}$$

which, combined with previous equations, gives the Gouy–Chapman equation:

$$\sigma = [8\mathcal{N} c(\infty) \epsilon_0 \epsilon_w \kappa T]^{1/2} \sin h(ze\psi_0/2\kappa T) \tag{29}$$

Figure 8 illustrates the dependence of the potential profile on the membrane charge density and on the concentration of the ionic salt solution. Considering, for instance, the charge density evaluated for a phospholipid bilayer as 1 electronic charge/300 $Å^2$ at 0.1 M concentration, a surface potential of -60 mV is expected according to the Gouy–Chapman theory.

It is worthwhile to observe that Eq. (29) has been derived under the following assumptions:

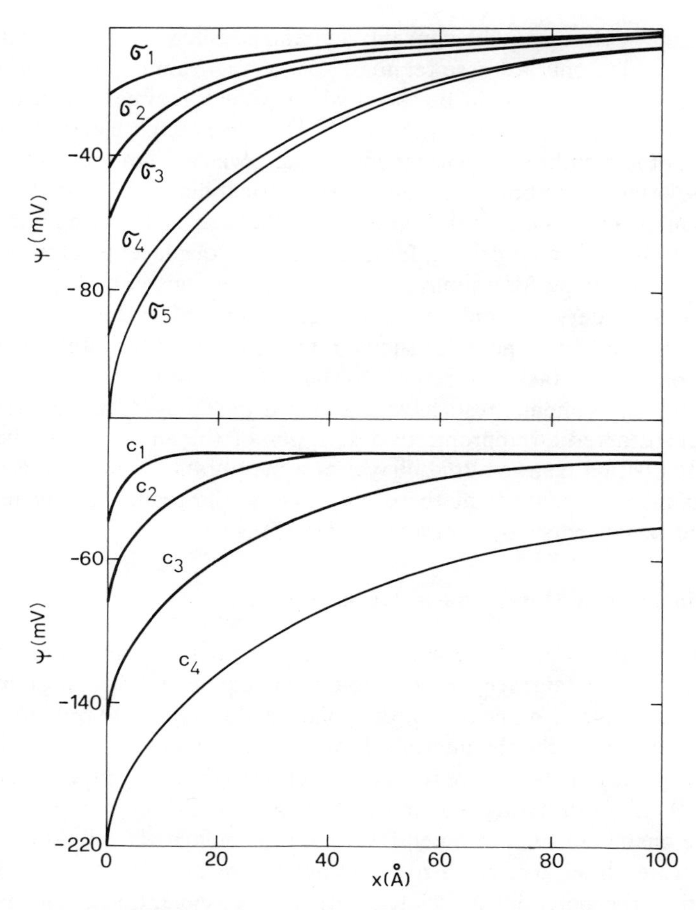

Figure 8. Potential profiles at the membrane–solution interface calculated as a function of distance from a negatively charged membrane. (Top) The charge density has been varied according to the following values (in electronic charges/Å^2): $\sigma_1 = 1/900$; $\sigma_2 = 1/600$; $\sigma_3 = 1/300$; $\sigma_4 = 1/150$; $\sigma_5 = 1/90$; $c = 10^{-1}$ *M*. (Bottom) The ionic concentration has been varied according to the following values: $c_1 = 1$ *M*; $c_2 = 10^{-1}$ *M*; $c_3 = 10^{-2}$ *M*; $c_4 = 10^{-3}$ *M*; $\sigma = 1$ electronic charge/150 Å^2. The electrostatic potential (ψ) is in millivolts.

1. The ions are point charges.
2. The dielectric constants do not vary in proximity to the membrane.
3. Image charge effects can be ignored.
4. The surface charge is uniform over the membrane.

Since both the Poisson–Boltzmann equations as well as the Gouy–Chapman theory describe electrostatic equilibrium, it is not correct to use them in conjunction with nonequilibrium flux equations. Such a problem might limit the applicability only to those regimes, not too far from equilibrium, in which small fluxes occur. It might be asked whether such regimes are relevant to a biological membrane. This is the case because very high electrical fields are present in the membrane environment. In fact, even a 10-mV potential applied to a 50-Å-thick membrane produces an electrical field of 2×10^4 V/cm. To oppose such a high electrical driving force, there must be, at equilibrium, an equally high diffusional driving force. Even small displacements from equilibrium of such driving forces may result in a net flux, given by Eq. (11), which is within the order of magnitude of biological fluxes. The use of Eqs. (25) and (29) is also justified in those diffusional regimes in which the interfacial adsorption of ions is very fast with respect to the diffusion time.

Many experimental tests have shown that the Gouy–Chapman theory can still be considered an appropriate description of the electrostatic phenomena associated with a charged lipid bilayer at a membrane–solution interface. For a comprehensive analysis of these phenomena, the reader is referred to the excellent review paper by McLaughlin (1977).

2.7. Dipoles at a Membrane–Solution Interface

Besides the effect of the net charges of the lipids or other charged species at the membrane surface, the orientation of dipoles in the water molecules adjacent to the membrane and of the polar head group could produce a potential difference between the interior of the membrane and the aqueous phase. A contribution could also derive from the ester linkages to the glyceryl backbone. One method of estimating the dipole potential associated with lipids is to measure the change in surface potential when a monolayer of a lipid is spread at an air–water interface. Such measurements indicate (Hladky and Haydon, 1973) that the potential associated with dipole orientation could be several hundred millivolts positive with respect to the aqueous solution, as diagrammed in Fig. 9A.

More recently, asymmetrical boundary potentials have been produced in a lipid bilayer membrane by various methods: (1) increasing the ionic strength only on one side of a charged membrane, (2) forming asymmetrically charged membranes from two different monolayers by use of the Montal and Mueller method, or (3) forming membranes with different dipoles at the two interfaces. In all cases, a potential gradient across the hydrophobic portion of the lipid is determined, as diagrammed in Fig. 9B. Such an electrical field can be tested by measuring the changes in conductance in the presence of monazomycin, a voltage-dependent pore-forming antibiotic (Mueller and Finkelstein, 1972). (The voltage dependency of an ionophore is discussed in Section 3.11.) A different technique might be that of measuring the voltage-dependent membrane

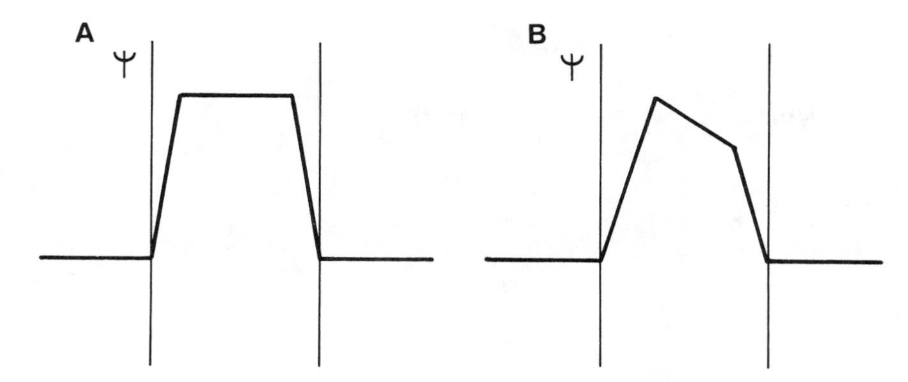

Figure 9. Sketch of the electrostatic potential profile attributable to dipoles at the membrane–solution interfaces. (A) Symmetrical situation. (B) Asymmetrical situation determining a potential gradient across a lipid core. ψ, Electrostatic potential.

capacitance (Schoch *et al.*, 1979) or the current–voltage characteristics induced by a carrier molecule (Hall and Latorre, 1976), both measurements being sensitive to the different electrostatic energy barriers.

2.8. The Eyring Theory

The Goldman–Hodgkin–Katz approach has the simplicity of a one-dimensional integration of the Nernst–Planck equations, but the disadvantage of not dealing easily with structural details within the membrane. Another approach, the Eyring rate theory approach (Eyring, 1935; Parling and Eyring, 1954), describes the membrane as a series of activation energy barriers (Fig. 10) and derives voltage-dependent rate constants for the passage of an ion across each barrier.

Considering the ionic transport that takes place across a specific channel, which might be visualized as an integral membrane protein, every dipole, charged group, or constriction in the pore can be represented by a dip or rise in the energy profile. For the sake of simplicity, we shall treat only the simple case in which a single ion is flowing through a channel. The resulting equation is shown to coincide, under appropriate assumptions, with that obtained by integrating the Nernst–Planck equations.

Assuming that the distances between minima λ are equal, the flux of an ion i over successive barriers j has the form

$$J_{ij} = \lambda(c_{j-1}k_{j-1} - c_j k_j') \tag{30}$$

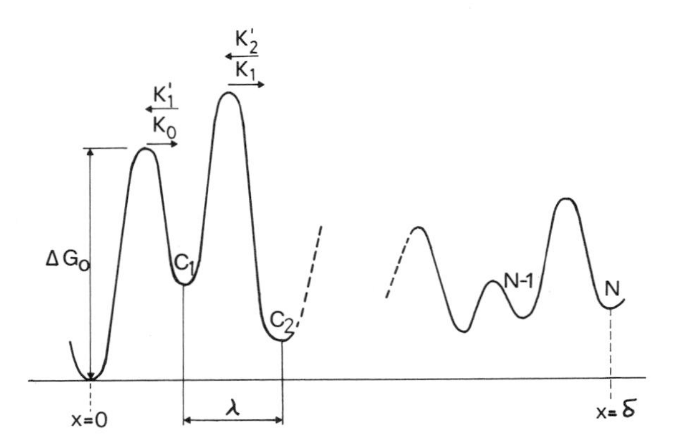

Figure 10. Schematic representation of the membrane considered as a series of activation energy barriers, according to the Eyring rate theory of transport.

where k_{j-1} and k'_j are the forward and backward rate constants, indicated in Fig. 10. These rate constants are related to the height of the jth barrier by the relationship

$$k_j = a_j \frac{\kappa T}{h} \exp\left(\frac{\Delta G_j}{RT}\right) \tag{31}$$

where a_j is a transmission factor, h is Planck's constant, and ΔG is the height of the jth barrier.

Considering that in steady state the fluxes over successive barriers are equal, $J_{ij} = J_i$, where J_i is the total flux. Summing up Eq. (30) for any barrier, the internal concentrations can be eliminated; the flux assumes the form

$$J_i = \frac{k_0\lambda\left(c_0 - c_n \dfrac{k'_1 \ldots k'_n}{k_0 \ldots k_{n-1}}\right)}{1 + \dfrac{k'_1}{k_1} + \dfrac{k'_1 k'_2}{k_1 k_2} + \cdots + \dfrac{k'_1 k'_2 \ldots k'_{n-1}}{k_1 k_2 \ldots k_{n-1}}} \tag{32}$$

It has been shown (Eyring *et al.*, 1949; Woodbury, 1971) that Eq. (32) reduces to Eq. (15), the Goldman–Hodgkin–Katz flux equation, considering the membrane composed of an infinite number of equal barriers in a constant field. For such a homogeneous membrane, Eq. (32) reduces to

$$J_i = \frac{k_i\lambda^2}{\delta} U \frac{c_0 - c_n \exp U}{\exp U - 1} \tag{33}$$

where k_i is the rate coefficient for jumping internal barriers. Equation (33) coincides with Eq. (15), if the internal mobility is identified with $u_i = k_i\lambda^2/RT$ or, in terms of membrane permeability, the external concentration is introduced in Eq. (33) by application of Eq. (19). The permeability P_i thus becomes

$$P_i = \beta_i k_i \frac{\lambda^2}{\delta} \tag{34}$$

For a detailed derivation of Eq. (33), the reader is referred to the original work of Eyring *et al.* (1949) and to the more recent work by Woodbury (1971), in which solutions for less regular profiles have been derived. The original paper by Eyring *et al.* (1949) also shows that fluxes can be expressed independently of the depth of internal wells. This finding is very important in order to make a conceptual distinction between selective transport and selective binding in ionic channels. We make use of it in discussing ionic selectivity in excitable membranes. Application of the rate theory to ionic channels in cell membranes is discussed in Section 3.

3. TRANSPORT IN EXCITABLE MEMBRANES

3.1. Introduction

The propagation of a nerve impulse in a nerve fiber is accompanied by a series of permeability changes along the surface of the membrane. The permeability changes alter the flow of ions, producing local electric currents that induce excitation in the contiguous patch of membrane. In the resting state the cell membrane potential is negative with respect to the outside. In the following discussion, the external potential $\psi(0)$ is taken as being equal to zero. A typical value for the squid axon membrane is $\psi(\delta) - \psi(o) \equiv \psi = -75$ mV. Since the major ionic components are sodium and potassium, one may compare this value with the Nernst equilibrium potential ψ of Eq. (14), which is $\psi_K = -60$ mV for K^+ ions and $\psi_{Na} = 52$ mV for Na^+ ions. These values in conjunction with the Goldman equation indicate that the resting potential is mainly determined by K^+ ions, showing that at rest the membrane has a K-selective permeability. A resting potential of -75 mV corresponds to a selectivity ratio $P_K/P_{Na} = 20$. During the rising phase of an action potential, the permeability is greatly enhanced, and the membrane becomes Na selective. The Na selectivity later decreases, and a high K permeability builds up. When the spike is over, the membrane returns to its low permeability state.

The entire process is accomplished within a time lapse on the order of few milliseconds. A qualitative picture of the time course of transmembrane potential ψ, sodium, and potassium permeability during a single nerve impulse is

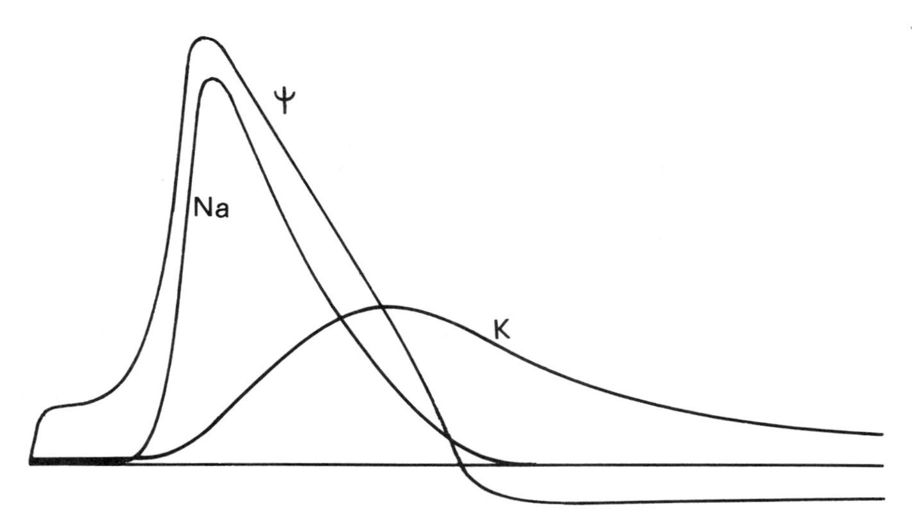

Figure 11. Qualitative representation of the time course of the transmembrane potential ψ, sodium permeability (Na), and potassium permeability (K) during an action potential. (After Schauf and Bullock, 1979.)

given in Fig. 11. The events underlying the action potential have come to be interpreted in terms of changes of membrane permeability since the beginning of this century. In fact, in his pioneering work, Bernstein (1902) proposed that excitation involves a breakdown in selective permeability to potassium. While incorrect in details, Bernstein's concept has been so fruitful that it is now possible to explain in terms of molecular models those pathways by which ions are constrained to move across cell membranes.

The questions to be answered in order to understand the mechanism of excitation can be divided into two steps. The first step concerns the phenomenological description of the single ionic currents and the kinetics of activation and inactivation of the permeability pathways. The second step is aimed at understanding, at a molecular level, the physical basis of ionic selectivity and of the gating of the ionic pathways through the membrane. The first step of the question was answered in 1952, when Hodgkin and Huxley were able to show that ionic currents can be dissected into two primary Na and K components. These workers incorporated their observation in a precise phenomenological treatment (Hodgkin and Huxley, 1952*a*,*b*). A more complete analysis and discussion of this topic may be found elsewhere (e.g., Katz, 1966; Hille, 1970). For what concerns the second step of the question, a complete answer is still unknown. In the following discussion we focus our attention on more recent work, carried out both in biological as well as in artificial membrane systems, aimed at understanding the molecular basis of selectivity and gating processes.

3.2. Description of the Ionic Currents

The specific sequence of permeability changes that gives rise to electrical excitability in squid giant axon has been derived by Hodgkin and Huxley from voltage-clamp measurements of membrane current. With the voltage-clamp technique, the membrane potential is forced to follow prescribed changes, and the current necessary to maintain such a potential is measured. Without the voltage clamp, the membrane current is partly a displacement current that charges the membrane capacity and partly an ionic current. In the voltage-clamp method, the membrane capacity is charged at a constant value and the measured current is, after an initial fast transient, an ionic current.

When the membrane is clamped at a constant voltage V_c, because ions tend to move passively down their electrochemical gradient, the values $V_c - \psi_{Na}$ and $V_c - \psi_K$ will be the Na^+ and K^+ driving forces, respectively. These will permit the prediction of whether the net movement of an ion, at a given potential, is inward or outward. An ion will thus contribute no membrane current when the membrane is clamped at equilibrium potential for that ion. Moreover, since permeability changes are almost insensitive to concentrations of the major ionic components (K^+, Na^+, Cl^-), one ionic component can be stopped, and ionic currents carried out by a single ionic species can thus be deduced. This dissection of ionic currents can be performed by using a chemical agent that blocks the ionic pathways selectively. Two chemical compounds have been of crucial importance for this purpose: a pharmacological agent, tetraethylammonium ion, (TEA), which eliminates almost completely the potassium current, and a toxin, tetrodotoxin (TTX), which eliminates the sodium current. More recently, two different peptidic fractions of scorpion venom were shown to block selectively and almost reversibly the Na^+ and K^+ currents (Carbone *et al.*, 1982).

3.3. The Hodgkin–Huxley Model

The technique used to block the ionic pathways selectively was applied by Hodgkin and Huxley (1952*a*,*b*) to a voltage-clamped squid axon, depolarized by a voltage step from the normal resting potential to a value V_c. These workers showed that the current can so be dissected into four components: a capacitive current I_c, two major ionic currents carried out by sodium and potassium ions I_{Na} and I_K, and a minor component, the leakage current I_L.

According to their theory, the aim of which was to fit in the simplest form the experimental data, the relationship between current and voltage can be written in the form

$$I = C_m \frac{dV_c}{dt} + g_{Na}(V_c - \psi_{Na}) + g_K(V_c - \psi_K) + I_L \tag{34}$$

where C_m is the membrane capacity per unit area, g_{Na} and g_K are the sodium and potassium conductances, and I_L is the leakage current, while the other symbols are as previously defined. The leakage current is small and time invariant, while I_{Na} and I_K are time dependent. The time dependence of the currents, when a voltage step is applied, is shown in Fig. 12.

The sodium current is a transient flow of predominantly Na^+ ions, moving through the so-called Na channels. Each Na channel is controlled by two gating factors: Na activation (or m^3) and Na inactivation (or h), according to the relationship

$$g_{Na} = m^3 h \bar{g}_{Na} \tag{35}$$

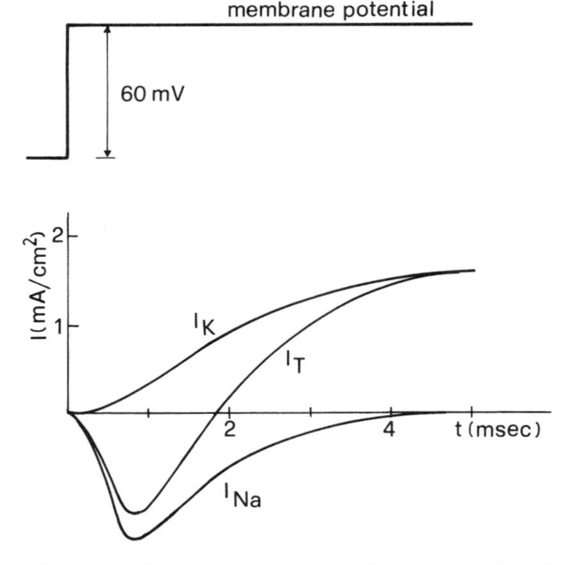

Figure 12. The ionic current, I_T, through an axon membrane clamped at 60 mV in seawater, is dissected into the two main components: the potassium component I_K obtained by substituting nine-tenths of the external sodium with choline, thus equating the activity of sodium on both membrane sides, and the sodium component I_{Na}, obtained from the difference between the two. Positive currents are those directed outside the axon. (From Hodgkin and Huxley, 1952*a*.)

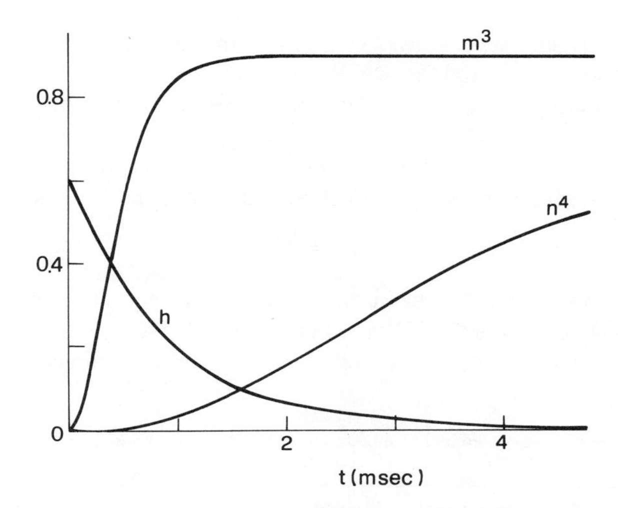

Figure 13. The time dependency of Na activation (m^3), Na inactivation (h), and K activation (n^4) after a 60-mV depolarizing step. (Calculated according to the Hodgkin–Huxley theory (1952*b*.)

where $\bar{g}_{Na}$ is a constant and m and h are continuous functions of potential and time. They are dimensionless parameters that vary between zero and 1 with first-order kinetics and rate constants that depend on the potential. The time dependence of m^3 and h after a depolarizing step is illustrated in Fig. 13.

The other major component of the ionic current is I_K, which is a maintained outward flow of K^+ ions. K channel is controlled by only one gate with an activation factor n^4, according to the relationship

$$g_K = n^4 \bar{g}_K \tag{36}$$

where, as for the parameters m and h, n is also a continuous dimensionless function, depending on potential and time, that varies between 0 and 1 and obeys the same first-order kinetic behavior. Also shown in Fig. 13 is the time dependence of n^4.

3.4. Ionic Channels

We have seen that the propagation of a nerve impulse down a nerve fiber is accompanied by a series of ionic permeability changes. In the Hodgkin–Huxley (H–H) model, ionic permeability is expressed in units of electrical conductance. The prevailing idea is that sodium and potassium permeability pathways are independent structures. They are assumed to be aqueous pores of atomic dimensions formed by proteins. The changes in ionic conductance are accomplished by the opening and closing of the pores.

The maximum sodium conductance in a squid axon when all the sodium channels are open is $g_{Na}(\max) = 120$ msec/cm^2 and the maximum potassium conductance is $g_K(\max) = 60$ msec/cm^2. These values are related to the single-channel conductance γ_j and to the pore density c_j by the relationship

$$g_j = c_j\gamma_j \tag{37}$$

Various methods have been employed to give an estimate of these two parameters. These include TTX binding, rate of TTX action, gating current, current and voltage fluctuations, and single-channel recordings. Table II sum-

Table II
Comparison of Single-Channel Conductance in Lipid Bilayers and in Natural Excitable Membranes

Type of channel	Conductance (pS)	Temperature (°C)	Reference	Observations
Molecules in BLM[a]				
EIM[b]	80, 400		Bean *et al.* (1969) Ehrenstein *et al.* (1970)	0.1 *M* KCl
Alamethicin	220, 930, 1900 2900, 3900		Gordon and Haydon (1972)	0.1 *M* KCl
Hemocyanin	200		Alvarez *et al.* (1975)	
Gramicidin	17		Hladky and Haydon (1972)	0.5 *M* NaCl
K channels				
Squid axon	12	6	Conti *et al.* (1975)	Noise spectrum
	18	5	Conti and Neher (1980)	Single-channel recording
	11	8	Gilly and Armstrong (1980)	Gating currents
Frog node	4	15	Begenisich and Stevens (1975)	Noise variance
Na channels				
Squid axon	5.8	6	Conti (1984)	Noise variance
	4.4	5	Keynes *et al.* (1980)	Gating currents
Myxicola axon	1.5	5	Bullock and Schauf (1978)	Gating currents
Frog node	8.8	13	Conti *et al.* (1980)	Noise variance
	2.9	12	Nonner *et al.* (1975)	Gating currents
Cultured rat muscle	18	19–22	Sigworth and Neher (1980)	Single channel

[a]BLM, Black lipid membrane.
[b]EIM, Excitability-inducing material.

marizes the results obtained with various techniques in different biological preparations. A more detailed description of some of these techniques is given in Section 5.

3.5. Selective Binding and Selective Permeability

For many years the selective permeability process was thought to be strictly related to the selective binding of an ion. The Eyring rate theory of transport shows, however, that the transport process is independent of the depth of the internal energy barrier, which might correspond to a binding site. A lucid analysis of the selectivity mechanism can be found in the work of Bezanilla and Armstrong (1972). In the appendix of their paper, they focus attention on the difference between selective permeation and selective binding. We shall discuss the two extreme cases according to Bezanilla and Armstrong.

Suppose that the ions K^+ and Na^+ are competing for the same group of sites S, according to the reaction

$$K^+ + S \underset{k_D}{\overset{k_R}{\rightleftarrows}} KS^+ \tag{38}$$

$$Na^+ + S \underset{k'_D}{\overset{k'_R}{\rightleftarrows}} NaS^+ \tag{39}$$

where k_R, k_D, k'_R, and k'_D are the association and dissociation rates of potassium and sodium, respectively. The ratio of KS- to NaS-occupied sites is

$$\frac{c_{KS}}{c_{NaS}} = \frac{c_K K_R K'_D}{c_{Na} K'_R K_D} \tag{40}$$

The site is K^+ selective when $k_R k'_D$ is larger than $k'_R k_D$.

Let us consider a case in which the energy barriers faced by Na^+ and K^+ ions are equal, but where K^+ is bound selectively. In such a case, supposing that $c_K = c_{Na}$ on both sides of the membrane and that $k_R = k'_R$, one gets

$$\frac{c_{KS}}{c_{NaS}} = \frac{k'_D}{k_D} \tag{41}$$

The fluxes through such a membrane from left to right will be

$$J_k = c_{KS} k_D \theta \tag{42}$$

and

$$J_{Na} = c_{NaS} k'_D \theta \tag{43}$$

where θ is the fraction of the time it takes an ion in the site to go the right side, after dissociating with the site. If binding to the site is the only factor that shows selectivity, then θ is the same for Na^+ and for K^+. Combining Eqs. (41)–(43), one gets $J_{Na} = J_K$. This result indicates that selective binding does not necessarily imply selective permeability.

In a similar vein, with simple kinetic equations, Bezanilla and Armstrong show that in the case in which equal dissociation constants are considered, the relative permeability is determined by the ratio of the association constants:

$$\frac{J_K}{J_{Na}} = \frac{k_R}{k'_R} \tag{44}$$

Thus, while for the case of carrier-mediated transport, selective binding coincides with selective transport, in the case of permeation through pores, one needs to consider selective exclusion, rather than selective binding, in order to understand the mechanism of ionic transport.

Comprehensive reviews of both sodium and potassium channel selectivity properties can be found in the literature (Hille, 1975*a*,*b*; Armstrong, 1975; Schauf and Bullock, 1979).

3.6. The Sodium Channel

The most explicit picture of an open Na channel has been drawn by the permeability studies of Hille (1973, 1975*a*,*b*). Hille determined the relative permeability of the Na channels to a number of both organic and inorganic cations in a single node of Ranvier under voltage clamp. The potassium channels were blocked with TEA. The membrane can thus be considered permeable only to ions of one sign and valence, with all permeant ions flowing in a common pathway. The zero current potential ψ becomes

$$\psi = \frac{RT}{zF} \ln \frac{\Sigma_i P_i c'_i(0)}{\Sigma_i P_i c'_i(\delta)} \tag{45}$$

where $c'_i(0)$ and $c'_i(\delta)$ represent the outside and inside concentrations, respectively. Hille's work was performed on myelinated fibers. In this preparation, neither the internal ionic concentration nor the absolute value of membrane potentials is accurately known. This difficulty was overcome by calculating permeability ratios in two different experimental situations obtained by changing only one permeant ion in the external solution. The internal concentrations are supposed to be unaltered by such a change. Therefore, Eq. (45) simplifies to

$$\psi_1 - \psi_2 = \frac{RT}{zF} \ln \frac{P_2 c_2'(0)}{P_1 c_1'(0)} \tag{46}$$

where indices 1 and 2 refer to the two permeant ions tested separately on the same nerve fiber.

On the basis of these experiments, Hille found that the sodium channels are not very selective. The permeability ratio between K^+ and Na^+, for instance, was found to be $P_K/P_{Na} = 1/12$. Moreover, a wide variety of organic cations can pass through the Na pore, but all the permeant ones can fit through a $(3 \times 5)Å^2$ aperture. Many of the permeant organic cations are as large as a partly hydrated Na^+ ion, i.e., $Na^+ \cdot H_2O$ or $Na^+ \cdot 2H_2O$.

The relative permeability of the hydroxylammonium ion $H_3N^+{-}OH$ and methyl ammonium ion $H_3N^+{-}CH_3$ is of great interest. In fact, in spite of the fact that both compounds have equal atomic radii $(4.5 \times 3.8 \times 3.8)$ Å^3, hydroxyl ammonium can pass through a sodium pore, while methyl ammonium cannot. The explanation given by Hille is that the walls of the pore are lined with oxygen atoms with which the hydroxyl hydrogen can form a hydrogen bond. The hydroxyl ammonium compound can thus pass through a pore only 3 Å wide. By contrast, the hydrogens of the methyl group cannot form hydrogen bonds, so this group cannot pass through the pore.

In conclusion, the picture given by Hille of an open sodium channel is that of a rather large (>10 Å in diameter) pore, which is narrow only at one point, indicated by Hille as the selectivity filter. Hille (1975*a*,*b*) shows that the filter can be treated in terms of the Eyring rate reaction theory, as an energy barrier, immediately preceded by a binding site. Moreover, Hille assumes that the overall channel can be treated as a sequence of four activation energy barriers, which contain no more than one ion at a time (one ion pore). Proposing such a model of an hourglass-shaped structure, Hille suggests that all the potential drop and the energy barriers might be confined in a length of 5 Å in the channel, so that fluxes are much higher than in a long, narrow pore.

3.7. The Potassium Channel

Studies of potassium channel selectivity are complicated by the observation that other ions interfere with the flow of K^+ ions through the channel. Hille and Schwartz (1978) proposed that the permeating ion must pass a sequence of energy barriers where, at variance with the model for the sodium pore, more than one ion may be in the channel at a time. Moreover, the ions are not permitted to pass by each other as they move through the channel. These assumptions lead to the phenomenon commonly referred to as single-file diffusion, or long pore effect, which has been reported in measurements of the passive movement of ions in potassium channels of nerve, muscle, and other cell membranes. These ideas are consistent with a pore model consisting of a relatively large inner mouth and a narrow part in the outer side.

Estimates of the diameter of the mouth and of the narrow part of the pore have been made on the basis of permeability studies (Bezanilla and Armstrong, 1972; Hille, 1973, 1975*b*). The mouth must be at least 8 Å in diameter, while the narrow part is permeable to ions of crystal diameters from 2.66 Å (K^+) to 2.96 Å (Rb^+, NH_4^+) and impermeable to smaller (Na^+, 1.9 Å, Li^+, 1.36 Å) as well as to larger ions (Cs^+, 3.1 Å).

To explain the exclusion of both small and large ions, it has been proposed (Bezanilla and Armstrong, 1972) that the narrow part of the pore has a range of possible diameters that extends from 2.6 to 3 Å. Large ions are thus excluded for steric reasons, while smaller ions are forced to become partly dehydrated. Selective transport will occur only if the potential energy of the dehydrated or partially dehydrated ion is lowered by interactions with charges and dipoles of the channel. Ions smaller than 2.6 Å are excluded because they do not interact favorably with the oxygens of the pore walls, as diagrammed in Fig. 14. This selectivity mechanism is very similar to that found in an ion carrier, like valinomycin, in which the steric dimensions of the internal cavity allow favorable ion–dipole interactions with the K^+ ions, but not with the smaller Na^+ ions.

3.8. Gating Currents

In all the models presented, the ionic channels are believed to be a large, complex lipoprotein in which parts of the gating structures are charged and able to move under the action of an electric field within the membrane. This movement of charges can be detected as a nonlinear component of the total membrane displacement currents. Such component is small in amplitude and,

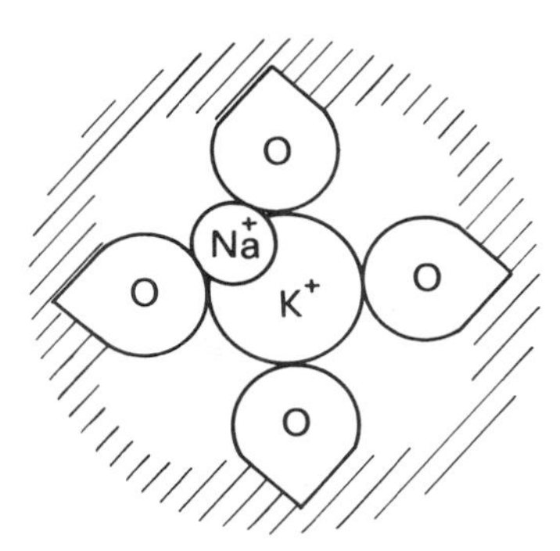

Figure 14. Schematic representation of K^+ and Na^+ in an hypothetical K^+ pore.

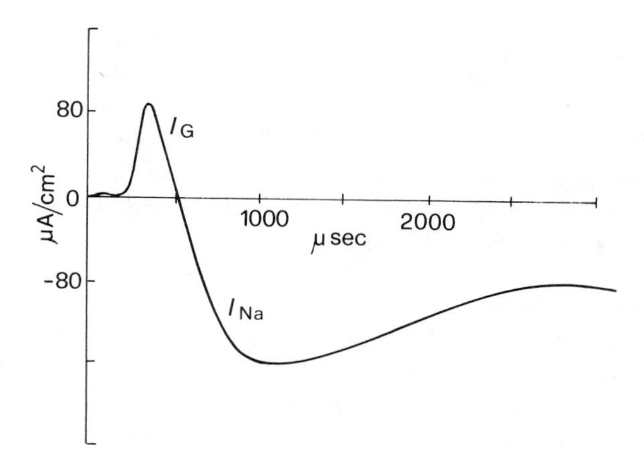

Figure 15. Gating current I_g and sodium current I_{Na} from an axon internally perfused with 290 m*M* CsF + sucrose. The current is the sum of five positive and five negative pulses of 90 mV, from an holding potential of −70 mV at 2°C. (After Armstrong, 1974.)

in order to measure it, ionic currents and the portion of capacitative current i_c, which increases linearly with dV/dt:

$$i_C = C_m \frac{dV}{dt}$$

where C_m is the membrane capacity, must be eliminated. Ionic currents can be removed by replacing Na^+ and K^+ ions with impermeant ones. In many experiments, TTX was also added to eliminate any residual Na current. To eliminate i_C, the current from a positive step is algebraically summed to that from a negative step of equal amplitude. Since gates open for positive voltage steps and not for negative ones, the asymmetry in the signal may be related to the gating process. Such gating currents, shown in Fig. 15, were first detected in squid axon by Armstrong and Bezanilla (1973), using an averaging procedure on positive and negative steps. There is much evidence that these asymmetry currents, which are not ionic, are associated with gating of the Na channel (Armstrong and Bezanilla, 1974; Keynes and Rojas, 1976; Bullock and Schauf, 1978; Keynes *et al.*, 1982).

The simplest physical model of the sodium system, suggested by the Hodgkin–Huxley equations and discussed in Section 3.2, is one in which activation of the sodium conductance depends on the three independent *m* particles; inactivation is brought about by a simple blocking *h* particle. As far as activation is concerned, however, there is no obvious way in which the voltage dependency of the asymmetry current can be equated with the sodium conductance curve. In fact, the total transfer of charge in the fast component is insufficiently steep at its midpoint and is displaced laterally along the voltage axis, as Fig. 16 makes clear (Keynes *et al.,* 1980). Regarding inactivation, no charge movement corresponding to such a process was detected, as one would expect if activation and inactivation of Na channels were to involve uncoupled separate and field-sensitive gates, as indicated in the Hodgkin–Huxley theory. The process of Na inactivation therefore seems to involve a change in channel structure that does not involve charge movement.

The process by which the gating structures relax to their initial state occurs in a longer time ($\simeq$20 msec); the associated currents are thus too small and random to be detected by these techniques.

To summarize, we may say that it is now well established in the giant axon of squid, in *Myxicola,* at the node of Ranvier in frog nerve, and in various other excitable tissues, an asymmetrical component of the membrane displacement current can be recorded, which seems to be related to conformational changes of the sodium gating system. However, when examined with sufficient detail, it becomes evident that it does not correspond in any simple way to the time course and to the voltage dependency displayed by the sodium channel. Further work is required to settle the matter.

For what concerns the K channel, no unequivocally successful recordings of K^+ gating currents have been performed to date. In fact, the kinetics are expected to be much slower and the signal-to-noise ratio poses serious ques-

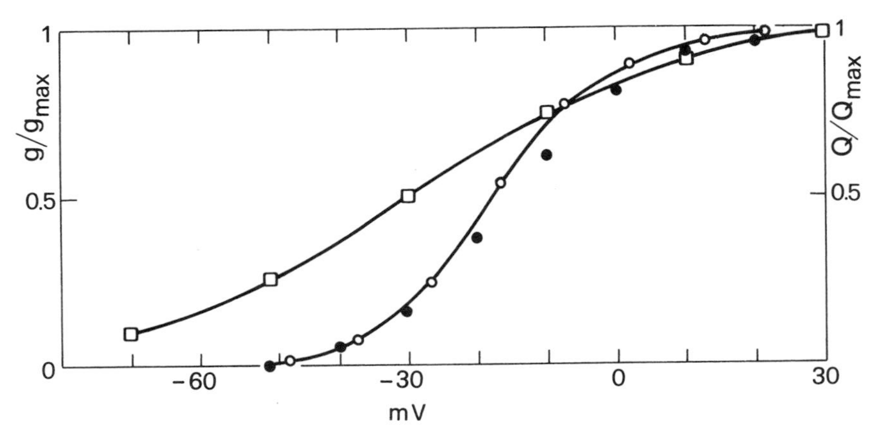

Figure 16. The conductance–voltage characteristics of sodium ions (○, ●) are compared with the fast component of the asymmetry current (□). The graph shows the normalized values, g/g_{max} and Q/Q_{max}, for conductance and transferred charges, respectively. (From Keynes *et al.,* 1980.)

tions. For more extensive treatment of this fascinating field the interested reader is referred to recent reviews by Ulbricht (1977) and Almers (1978).

3.9. Noise Analysis

Discreteness of the charge carried by the ions and randomness of the ionic movements are the primary source of fluctuations in measurements of bioelectric potentials and currents. Information can be obtained from the frequency analysis of these fluctuations, as is done for inorganic systems.

If the system is at equilibrium, part of the noise caused by thermal agitation of ions is revealed by Johnson noise, for which the power-density spectrum of the potential $S_V(\omega)$ is supplied by the Nyquist formula:

$$S_V(\omega) = 4\kappa T \mathrm{Re} Z \tag{47}$$

where ReZ is the real part of the electrical impedance Z, in which the potential fluctuations occur, while the other symbols are as previously defined. Fatt and Katz (1952) attempted to account for spontaneous activity in synaptic nerve–muscle junctions using Nyquist's theoretical treatment of Johnson noise. However, biological membranes usually cannot be considered at equilibrium, so forms of noise connected to nonequilibrium situations must be expected.

In their work on the electrical voltage noise of frog nodes of Ranvier, Verveen and Derksen (1965) created a new method of studying electrophysiology. They found that the magnitude of the fluctuations exceeds that expected from the Nyquist formula; moreover, their spectral density is not independent of frequency, as foreseen by Nyquist's theory for a membrane schematized as an *RC* parallel circuit. Such a result was confirmed by Poussart (1971), Fishman (1972), and Fishman *et al.* (1975). The excess noise is two to three orders of magnitude greater than thermal agitation noise, and the power-density spectra have a strong component of the form f^{-n}, where f is the frequency and n is ~ 1. Even if this $1/f$ noise was historically the first measured in biological membranes, its origin is still unknown. It is a quite general effect that has been measured in many electrical systems: solid-state devices, carbon resistors, glass microelectrodes, ionic solution, thick synthetic membranes, and BLM. For this generality we can suppose that $1/f$ noise is not linked to the specific mechanism that occurs in excitable biological membranes. It is not the aim of this chapter to analyze the $1/f$ noise problem, so we refer to the literature already quoted and to the reviews of Verveen and De Felice (1974), Conti and Wanke (1975), Neumcke (1978), and Conti (1983). This last and more recent review reduces the weight of $1/f$ noise in comparison with other noise forms, affirming that the predominance of the role of $1/f$ noise in biological membranes has decreased with the increasing sophistication of the noise analysis techniques.

The electrical excess noise found in excitable membranes is usually divided into transport noise and channel noise. In addition to $1/f$ noise, the shot noise is present in transport noise. This excess noise is caused by the quantal nature of the electric current (MacDonald, 1962).

In the simpler form it is assumed to result from summation of a large number of statistically independent pulses (shots) each carrying the same amount of charge. The power-density spectrum of current of the shot noise is

$$S_{\mathrm{I}}(\omega) = 2qI \tag{48}$$

where q is the charge of the particle and I is the mean current flowing. As discussed in Section 3.8, ion transport through a membrane occurs by discrete steps corresponding to the jumps of individual ions over activation barriers. Every jump of an ion is a current pulse. The jump over an energy barrier is not *a priori* a statistically independent event, e.g., it can be conditioned by the occupation of the successive valley, so the simpler model of the shot noise does not hold and a more general model is necessary (Laüger, 1975; Frehland, 1980). Power-density spectra of the shot noise expected in a squid axon membrane according to the H–H model are given by Conti (1984). The channel noise is caused by the opening and closing of individual ionic channels; its analysis could provide a way to test the validity of microscopic models of the excitable membranes. This type of interpretation was proposed by Katz and Miledi (1970, 1972) for electrical fluctuations recorded in postsynaptic membranes produced by acetylcholine (Ach) and by Siebenga and Verveen (1972) for conductance fluctuation of the axon membrane.

Stevens (1972) proposed the interpretation of channel noise power spectrum in terms of the H–H equation for the potassium conductance system using the fluctuation–dissipation theorem (Kubo, 1957), according to which the correlation function of spontaneous fluctuations of an observable and its relaxation toward the equilibrium after an initial displacement have the same time dependence. For such a theorem, noise measurements do not seem to supply new information, in comparison with the classical relaxation measurements. However, using a statistical approach, specific models can be made that assume the number of channel states, their conductance levels, and transition rates between different states. The first and perhaps the most accepted models are those directly derived from statistical interpretation of the H–H equations by Stevens (1972) and independently by Hill and Chen (1972).

Other models have been proposed, also based on hypothesis different from those of the H–H model.

Many model implications have been verified on artificial membranes (Neher and Zingsheim, 1974; Moore and Neher, 1976; Kolb and Laüger, 1978). But, if noise measurements supply a large amount of experimental results, their difficult interpretation prevents selection of a preferential model. However, there are still some data whose values are independent of the model. These data are single-channel conductances γ_i and channel densities. In fact, only assuming that i is the value of the current crossing the open channel and that no current flows in the closed channel, one obtains

$$s^2 = iI - \frac{I^2}{N} \tag{49}$$

where s^2 is the variance of the current that fluctuates around the mean value I, and N is the total number of channels. In voltage-clamp conditions, $I = (V_c - \psi)g$ and $i = (V_c - \psi)\gamma$. The value of s^2 is obtained either from the autocorrelation function A of the current or from the power spectrum in accordance with

$$s^2 = A(0) - A(\infty) \tag{50}$$

$$s^2 = \frac{1}{2\pi} \int_0^\infty S(\omega)\, d\omega \tag{51}$$

The values of single-channel conductances obtained with this method are presented in Table II, in which values from other methods are presented as well.

As an example, in order to see what kind of information noise measurements can supply on a specific model, we present two power spectra from Conti (1984). Figure 17A shows the spectrum of K^+ current fluctuations in a squid giant axon under voltage clamp. Na^+ current was removed by the addition of TTX. The dashed line is the theoretical curve obtained by summing up thermal, shot, and channel noise (according to the Nyquist, Schottky, and Hill and Chen models, respectively). The parameters involved in K channel noise are the activation n, the related relaxation time τ_n (appearing in the H–H theory as already described in Section 3.3), the single-channel conductance γ_K, and the number of channels in the clamped area N. The fit is obtained considering τ_n and γ_K, as free parameters and giving the others the values obtained from relaxation experiments. The value found for the single-channel conductance is $\gamma_K = 6.3$ pS (which is about one-half the mean value given in Table II), while $\tau_n = 18$ msec (which is about three times larger than that obtained by relaxation measurements). The number of channels can also be evaluated and gives a value of channel density of 40 channels/μm^2. Such a very small value of channel density has also been found using single-channel recording techniques (Conti and Neher, 1980) or gating current measurements (Gilly and Armstrong, 1980).

A similar procedure can be followed to get information from the spectrum

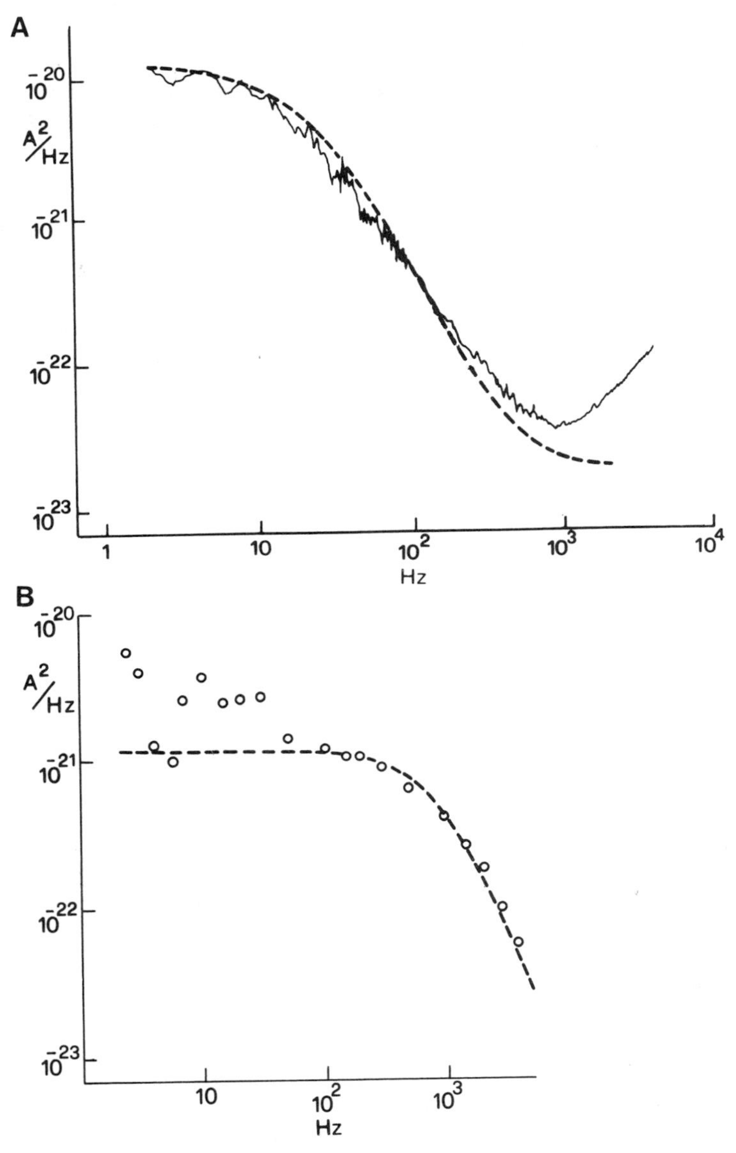

Figure 17. Power spectra of current fluctuations in a voltage-clamped squid giant axon, at − 53 mV. (A) Power spectrum of K^+ current fluctuations. Giant axon was immersed in seawater, to which 300 nanomoles of tetrodotoxin (TTX) was added. The mean potassium current was calculated as $I_K = I_{TTX} - I_2$, where I_2 is the leak current of $-16\ \mu A$, and $I_{TTX} = 3.84\ \mu A$. The theoretical curve was calculated using $\tau_n = 18$ nsec, $i_K = 0.09$ pA. (B) The power spectrum of Na current fluctuations was obtained as described in the text. The theoretical curve was calculated using parameters $\tau_m = 0.5$ msec and $i_{Na} = 0.72$ pA. (From Conti, 1984.)

of Na^+ current fluctuations shown in Fig. 17B. In this case, the free parameters are γ_{Na} and the relaxation time of the activation process, τ_m. The values for the fit are $\gamma_{Na} = 6.3$ pS in good agreement with the value given in Table II, and $\tau_m = 0.5$ msec, (as compared with 0.37 msec obtained from relaxation experiments). The channel density may be evaluated as being of the order of 100 channels/μm^2. A mean value of 300 channels/μm^2 has been found in various preparations. Figure 17B shows that the theoretical curve does not fit the experimental points at frequencies below 100 Hz. A possible explanation for such behavior has been proposed by Conti (1984).

These two examples show that the agreement between experimental data and theoretical expectations depends not only on possible theoretical models, but also on the elaboration of the experimental data. The casuistry of the interpretation of noise measurements is extended and complicated, for which the previously cited reviews provide an explanation.

3.10. Single-Channel Recordings

Macroscopic behavior of systems can be explained by various microscopic models. For example, in noise measurements, the same power spectra are obtained from channels with two conductance states that rapidly pass from one state to the other, dwelling for a measurable time in the arrival state, and from channels that rapidly open and close in an exponentially time-dependent way. Katz and Miledi (1972) used the last model to interpret noise induced by ACh in the postsynaptic membrane. In order to select among microscopic models, direct measurements of the conductance states of a single channel are needed. The first single-channel measurements were performed on artificial systems.

Bean *et al.* (1969) found the voltage-dependent conductance exhibited by BLMs treated with a particular substance, excitability-inducing material (EIM) (cf. Section 3.11) to consist of discrete steps that arise from the formation of individual ionic channels. Because in BLMs area and the number of inducing-channel molecules can be controlled, there are not many difficulties with having few channels per membrane, the transitions of which are revealed by discrete steps in current recordings. Moreover, the low background conductance of BLM, up to values of the order of 10 pS/mm^{-2}, permits recording of variations of few pS without particular electronic devices. Therefore, single-channel measurements are suitable for studying the kinetic behavior of molecules that induce channels in lipid membranes, as discussed in Section 3.11. In natural membranes the number of channels can be controlled with special techniques. Neher and Sakmann (1976), using a method consisting of the electrical isolation of a small patch of membrane by means of a glass micropipette pressed against the cell surface, succeeded in recording the ionic current of the single channel induced by ACh in the extrasynaptic membrane of denervated frog muscle fiber. The ACh-activated channel has two conductance states: closed and open. The open state has a conductance of 22 pS. The current–voltage relationship for the single channel is ohmic and channels are independent of each other; therefore, it is the voltage dependence of the dwelling time in a state that causes the nonohmic behavior of the mean current. These results fit the hypothesis of the model by which Anderson and Stevens (1973) interpreted current noise arising from voltage-clamped end plates.

Using this patch-clamp technique, single-channel recordings in a number of biological preparations have been obtained. In squid axon, perfused with low ionic strength solutions to increase the resistance of the seal between pipette and membrane, Conti and Neher (1980) recorded elementary currents of potassium channels and obtained a single-channel conductance of 18 pS. This value is larger than that obtained with other methods and reported in Table II; however, such discrepancy might be explained on the basis of the different ionic concentrations employed in this experiment. Single-channel currents recorded by Conti and Neher show a single open state for K channel, but bursts of short pulses appear. These investigators suggest the presence of a closed state in fast equilibrium with the open state, which may be described by the sequential reactions

$$C_2 \underset{k_{-2}}{\overset{k_2}{\rightleftarrows}} C_1 \underset{k_{-1}}{\overset{k_1}{\rightleftarrows}} O$$

where C_2 and C_1 are closed states and O is the open state. This reaction sequence predicts bursts of activity if k_2 is much smaller than the other rate constants. This picture, forecasting second-order kinetics, is not in agreement with the Hodgkin–Huxley model. In rat muscle cells (mioballs), Sigworth and Neher (1980) observed elementary currents of Na channels, using an improved patch-clamp technique in which the seal resistance was increased to 10^{10} Ω. According to these experiments, the sodium channel shows two conductance states. In the open state it has a conductance of 18 pS. By averaging many Na^+ single-channel records, these workers reproduced the time course of the macroscopic Na^+ current and observed a relationship between the mean open channel lifetime and the time constant of Na^+ currents, which is not in agreement with H–H model predictions.

3.11. Electrically Gated Channels in Planar Lipid Bilayers

We have seen that the basic event underlying excitation phenomena is the gating of a channel the conductance of which is dependent on the applied potential. Isolation of ionic channels from the nerve membrane and their incorporation in an artificial planar bilayer appears to be one of the most promising approaches in order to understand the physical basis of the process. However, since this procedure requires further developments of the reconstitution technique, attempts have been made to reproduce the phenomenological behavior of nerve membrane by incorporating channel-inducing molecules into planar bilayers.

The first isolated substance able to induce voltage-dependent conductance and then to mimic the excitation process in lipid bilayers is a polypeptidic compound isolated from cultures of *Enterobacter cloacae,* referred to as excitability-inducing material (EIM) (Mueller and Rudin, 1963). The EIM-doped bilayer possesses a negative differential conductance similar to that exhibited by the potassium pathway of nerve axons in high external potassium solutions (Latorre *et al.*, 1972). Discrete steps of conductance induced by small amounts of EIM were revealed by Bean *et al.* (1969).

Every step can be attributed to the opening or closing of a channel and has an amplitude of the order of 500 pS in 0.1 *M* KCl. When a membrane with only one channel is obtained, the membrane conductance fluctuates between two values, the maximum value is the open-channel conductance and the minimum is the closed-channel conductance. In oxidized cholesterol membranes, the closed-channel conductance is about one-fourth that of the open channel (Ehrenstein *et al.*, 1970). Both states of conductance are ohmic. The voltage dependency of conductance is caused by the voltage dependency of the time the channel dwells in a conductance state. Assuming the opening–closing process to be ergodic, the fraction of time the channel spends in the open state is equal to the ratio between the number of open channels n_O and the total number of channels N, and is given by (Latorre and Alvarez, 1981):

$$f(V) = \frac{n_O}{N} = \frac{1}{1 + \exp(\Delta W/RT)} \tag{52}$$

where ΔW is the difference of potential energy of the two states, which is related to the potential applied to the membrane by

$$\Delta W = \overline{Q}(V - V_O) \tag{53}$$

In Eq. (53), $\overline{Q}$ is the product of the charge translocated and the fraction of the membrane thickness traveled by the charge when the channel changes from one state to the other, and V_O is the potential at which the number of open channels is equal to the number of closed channels. Because the channels are in parallel, the macroscopic conductance is given by the sum of conductances of the open and closed channels. If γ_O is the conductance of the single open channel and γ_C is the conductance of the single closed channel, the macroscopic conductance is given by

$$g(V) = n_O\gamma_O + (N - n_O)\gamma_C \tag{54}$$

Using Eqs. (52) and (54), one gets

$$g(V) = N\left[\frac{\gamma_O - \gamma_C}{1 + \exp \overline{Q}(V - V_O)/RT} + \gamma_C\right] \tag{55}$$

The normalized function $\Delta g/g = [g(V) - N\gamma_C]/[N(\gamma_O - \gamma_C)]$ is equal to the righthand side of Eq. (52) and, as shown in Fig. 18, it fits the experimental data well. The current–potential relationship is $I = g(V) \cdot V$ which, as can easily be shown, is a function with a negative slope region, as in natural excitable membranes.

Another substance that shows a current–voltage characteristics with a negative slope region in BLMs is alamethicin (Mueller and Rudin, 1968). Alamethicin is a polypeptide produced by the fungus *Trichoderma viride*. It has antibiotic properties and is composed by 19 amino acids with a high content of hydrophobic residues. Details on its structure can be found in the extensive review of Latorre and Alvarez (1981).

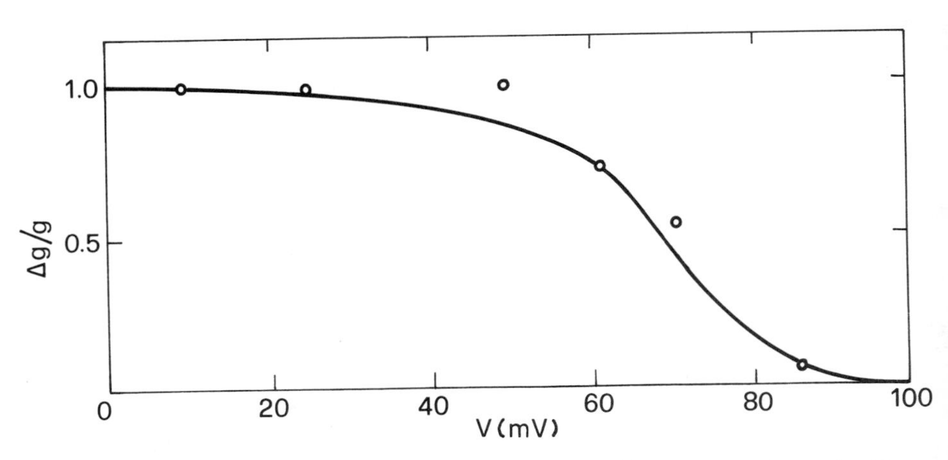

Figure 18. Conductance–voltage dependency of a black lipid membrane with excitability-inducing material added only on one side in 0.1 *M* KCl solution. The ratio of the number of open channels to the total number of channels, n_O/N, calculated from Eq. (52) (———) has the same voltage dependency as the normalized conductance (O). (From Latorre *et al.*, 1972.)

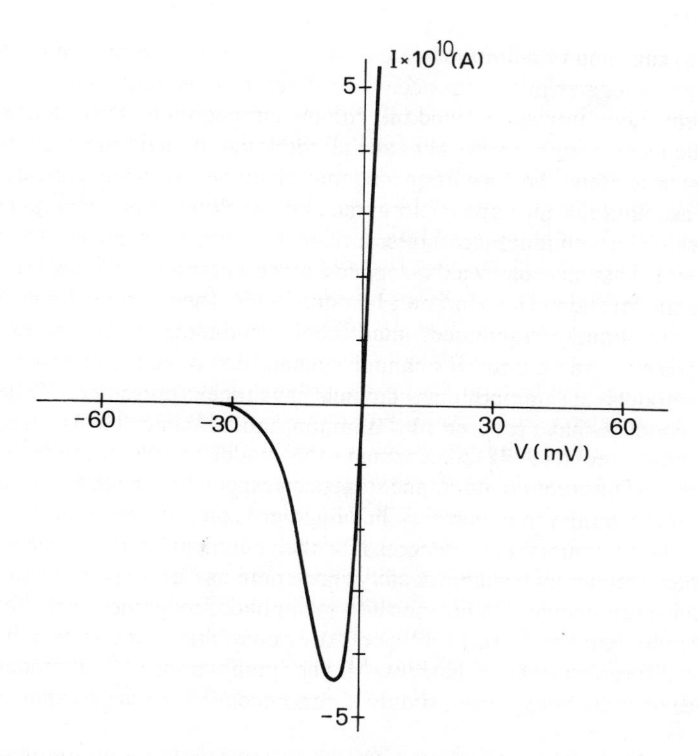

Figure 19. Voltage–current curve of a black lipid membrane with 1 *M* NaCl on both sides and 10^{-5} g/ml alamethicin added only on one side. (From Eisenberg *et al.*, 1973.)

The voltage-dependent conductance (Fig. 19) induced by alamethicin may be described by

$$g = Ac_i^{\alpha}c_s^{\eta} \exp\left(\theta \frac{eV}{RT}\right) \tag{56}$$

where A is a proportionality constant, c_i is ionic concentration, c_s is alamethicin concentration in aqueous phases, e is the electronic charge, and α, η, and θ are parameters depending on membrane composition and temperature, and the other symbols are as previously defined.

The current can be written as

$$I = VAc_i^{\alpha}c_s^{\eta} \exp\left(\frac{\theta eV}{\kappa T}\right) \tag{57}$$

a function that shows a minimum at $V = -\kappa T/\theta e$; therefore, the I–V relationship presents a region with a negative differential conductance.

Current fluctuations, obtained in voltage-clamped membranes at low alamethicin concentrations, prove that the alamethicin-induced macroscopic conductance arises from the formation of ionic channels (Gordon and Haydon, 1972). These fluctuations appear in bursts, and each burst has several conductance levels. The conductance states are mostly ohmic. Because the single-channel state distributions weakly depend on the potential, as does the mean current-burst duration (Gordon and Haydon, 1976), the strong voltage dependency of the alamethicin-induced macroscopic conductance can be explained by an increase in the number of channels when the voltage is increased.

Few models of alamethicin channel have been presented. The most accepted model is that proposed by Baumann and Mueller (1974), and independently by Boheim (1974). According to this model, the alamethicin channel is an oligomer. Different conductance states correspond to the different number of monomers forming the oligomer. The oligomer is not stable but can increase for the uptake of monomers or decrease for their removal. In the absence of an applied field, monomers, schematically represented as barrels, lie at the membrane–solution interface. When the field is applied, monomers are tilted into the membrane, and for lateral diffusion they come into contact forming bundles of a different number of barrels. All the implications of this model have not yet been made clear, even though it can account for many experimental results.

EIM and alamethicin are the substances that better mimic the behavior of nerve membranes, but many other substances form voltage-dependent channels in BLM (Ehrenstein and Lecar, 1977; Gliozzi, 1980*b*). An example is hemocyanin, a protein that transports oxygen in the blood of many invertebrates (Alvarez *et al.*, 1975; Menestrina and Antolini, 1981). Many of these substances are not related to models of nerve membranes, but have other physiological bonds. The interested reader is referred to the extensive review of Latorre and Alvarez (1981).

3.12. Ionic Carriers in Planar Lipid Bilayers

Electrical measurements on artificial bilayers have shown that macrocyclic compounds, such as valinomycin, macrotetralides, nigericin-type antibiotics, and macrocyclic polyethers, may act as mediators of selective cation transport. Therefore, their action on biological membranes, exhibiting remarkable selective permeability, is that of drastically changing the concentration difference between the two sides, which is fundamental to the biological functions. Natural membrane systems, such as nerve, mitochondria, muscle, and bacterial cell, maintain a concentration difference of Na^+ and K^+ ions by a selective cation pump, coupled to an energy source. A cation-specific carrier will induce backdiffusion of the ion, thereby reducing or dissipating its electrochemical gradient. Not only a change of concentration, but a change of membrane potential as well, will consequently be established [cf. Eq. (17)].

The study of these compounds therefore assumes a double aspect: They may help, as model systems, in understanding the salient features of selectivity, or they may be used on biological membranes in analyzing the nature of ionic flows.

The mechanism of action of valinomycin, chosen as a prototype of a carrier system, is described. Furthermore, since uncouplers of oxidative phosphorylation are extensively referred to in the following discussions, some general features on the action of a class of these molecules, which has been shown to act as proton carriers, are included.

3.12.1. Action of Cation-Specific Antibiotics

Valinomycin is a cyclic compound, pharmacologically classified as an antibiotic. It is built by an alternating sequence of α-amino acids and α-hydroxyl acids. In nonpolar media, it assumes a rather compact conformation, resembling a bracelet, the dimensions of which are 15 Å in diameter and 12 Å in height. The nonpolar chains are directed toward the exterior of the molecule, making it strongly hydrophobic, while the six oxygen carbonyls from the six ester bonds are directed toward the internal cavity. An unhydrated potassium ion fits exactly into the cavity and may establish ion–dipole interactions similar to those with the hydration shell in aqueous medium. The whole complex is then stabilized by six hydrogen bonds. In the presence of alkali metals, a complex formation takes place that follows the following selectivity sequence:

$$k_{\mathrm{Rb}} > k_{\mathrm{K}} > k_{\mathrm{Cs}} > k_{\mathrm{Na}} > k_{\mathrm{Li}}$$

where the *k*s represent the apparent stability constant of complex formation. Differences of stability constants of $\sim 10^4$ have been found between potassium and sodium (e.g., Grell *et al.,* 1975). This selectivity sequence parallels the increase in conductance induced by inserting small amounts of valinomycin, on the order of nanomoles, on both aqueous solutions of an artificial lipid bilayer. Therefore, in this case cation selective binding to the macrocyclic compound determines the selective permeability through the membrane.

The theory of carrier-induced transport is now well established (e.g., Ciani *et al.,* 1973; Laüger, 1980). The simplest treatment implies that the carrier is mainly confined inside the membrane. The transport occurs in four different steps, as illustrated in Fig. 20a:

1. Association of ion M^+ in solution(s) and carrier S in the membrane (M) at the interface, according to the heterogeneous reaction

$$S(m) + M^+(s) \underset{k_D}{\overset{k_R}{\rightleftarrows}} MS^+(m)$$

where k_R and k_D are the rate constants of association and dissociation, respectively
2. Translocation of the charged complex across the membrane, a process characterized by the rate constant k_{MS}, which may be described, according to the Eyring theory, as a jump over an activation energy barrier, driven by the external electrical field
3. Dissociation at the other interface
4. Backdiffusion of the unloaded carrier S under its own concentration gradient, a process characterized by the rate constant k_S

The kinetics of the carrier may be analyzed by studying the electrical conductance of planar bilayers. In particular, electrical relaxation techniques, such

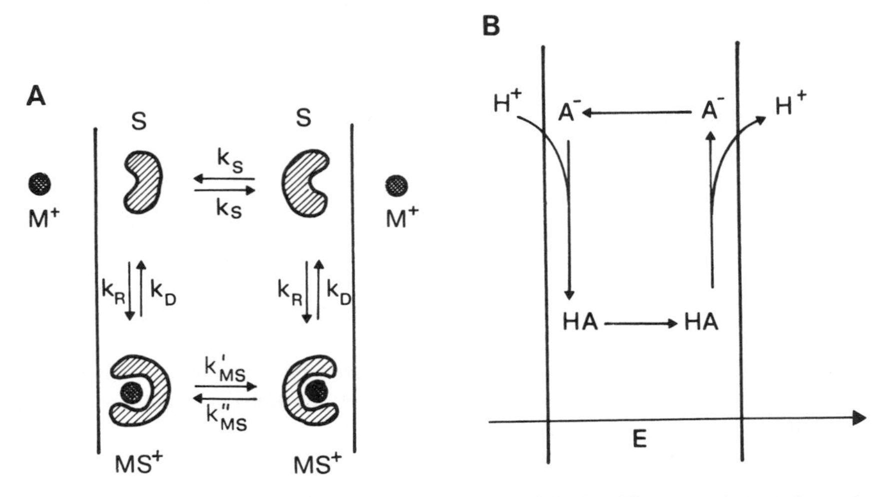

Figure 20. (A) Schematic drawing of carrier-mediated diffusion. The rate constants determining interfacial reaction and diffusion are indicated. (B) Proposed model for the action of uncouplers of oxidative phosphorylation acting as proton carriers (see text for details).

Table III
Kinetic Parameters and Partition Coefficient of Valinomycin-Mediated Cation Transport in Different Lipid Membranes[a]

Lipid	$k_R/10^4\ M^{-1}\cdot\text{sec}^{-1}$	$k_D/10^4\ \text{sec}^{-1}$	$k_{MS}/10^4\ \text{sec}^{-1}$	$k_S/10^4\ \text{sec}^{-1}$	$\gamma_S/10^3$
1,2-Dipalmitoyl-3-phosphatidylcholine*	8.2 ± 2.5	45 ± 4	9.1 ± 2	2.2 ± 0.4	28
Glycerol monopalmitoleate*	43 ± 15	13 ± 3	74 ± 7	8.5 ± 0.9	1.5
Glycerolmonooleate*	37 ± 12	24 ± 6	27 ± 3	3.5 ± 0.4	5.6
1-Oleylglycerol*	8.3 ± 1.8	15 ± 2	9.8 ± 2.6	4.1 ± 0.5	4.3
Soybean lecithin†	7	4	2	1	10

[a]The rate constants k_R, k_D, k_S, and k_{MS} and the partition coefficients γ_S for valinomycin-mediated Rb^+ transport (*) [Benz *et al.*, 1977], and for K^+ transport (†) [Gambale *et al.*, 1973], across membranes made from different lipids dissolved in *n*-decane.

as the voltage jump (Stark *et al.*, 1971) or the charge pulse method (Benz and Laüger, 1976), may be used to evaluate the individual rate constants. Table III summarizes the results obtained on different lipid membranes in transporting the indicated cation.

This kinetic analysis permits dissection of the transport process into the single steps. For instance, in the soybean lecithin/*n*-decane membrane at 1 *M* KCl all the rate constants, shown in Table III, are of the same order of magnitude. Therefore, none of the four steps is rate limiting.

The reciprocals of k_S and k_{MS} give the mean time of translocation through the membrane, and the reciprocal of k_D gives the average time of complex formation; therefore, from a knowledge of these parameters, we may evaluate the number of ions transported per unit time n/t by a single valinomycin molecule:

$$n/t = \left[\frac{1}{k_S} + \frac{1}{k_{MS}} + \frac{2}{k_D}\right]^{-1} = 10^4 \text{ ions/sec} \tag{58}$$

This value is rather low compared with the ionic transport in the axon membrane, which may be evaluated as $\sim 10^7$ ions/sec.

Two general features, peculiar to most carrier systems, are thus emerging: (1) a carrier-mediated electrodiffusion may selectively translocate ionic species, and (2) a carrier is a rather slow transport system, as compared with a channel mechanism. It therefore looks unlikely, as far as a generalization of this result is feasible, that such a mechanism is involved in the propagation of a fast signal.

3.12.2. Action of Uncouplers of Oxidative Phosphorylation

With respect to the artificial membranes, there are two different classes of weak acid uncouplers, HA. For one class the anion A^- is the charged permeant species, whereas for the other class it is an HA_2^- complex, formed between the anion A^- and the undissociated acid HA. For both classes, the current in the aqueous phases appears to be carried by buffered H^+ ions (Le Blanc, 1971; Foster and McLaughlin, 1974; Neumcke and Bamberg, 1975). As a consequence of these two different mechanisms, the conductivity at all pH values is proportional to the concentration for the A^- class, whereas it is proportional to the square of concentration for the HA_2^- class. The first class of molecules is exemplified by carbonyl cyanide-*m*-chlorophenylhydrazone (CCCP) and by carbonyl cyanide-*p*-trifluoromethoxyphenyl hydrazone (FCCP), while the second class includes the substituted benzimidazoles.

A complete analysis includes complex phenomena such as adsorption of charged molecules at the interfaces and the effect of stagnant layers on the diffusion (Hladky, 1972; McLaughlin, 1977; Ciani *et al.*, 1975). We shall limit ourselves to illustrating a model (Liberman *et al.*, 1971) according to which the A^- class of uncouplers may act as a proton carrier. The proposed transport system is represented schematically in Fig. 20B. The electric field drives an anion A^- to the opposite interface, where it combines with a proton H^+ from the solution. The undissociated salt HA will then diffuse across the membrane under its own concentration gradient and, upon dissociation, will release a proton to the external solution. As a result of this cyclic process, a proton is carried from one interface to the other. Since these molecules are known to act as uncouplers of oxidative phosphorylation in mitochondria, chloroplasts and bacteria, the proposed model is in consonance with the Mitchell chemiosmotic hypothesis. In fact, an enhancement of H^+ passive flow would dissipate the electrochemical energy accumulated by the membrane and necessary to drive adenosine triphosphate (ATP) synthesis.

4. MODEL PHOTOSENSITIVE MEMBRANES

4.1. Introduction

The photoinduced phenomena may be divided into two general classes: those in which light is used to store energy to drive biosynthetic reactions and responses in which light is revealed as a signal to provide information from the environment for perception or for regulation and control. In both cases there is a great difference in the energy involved; in fact, both in the photosynthetic process and in the light-driven proton gradient (occurring in the purple membrane of *Halobacterium halobium*) large numbers of quanta are absorbed. By contrast, in the case of neuronal phototransduction, few quanta of light are needed to trigger an action potential.

Basic mechanisms underlying the action of light on biological systems are located in the membranous portions of the cell. Some of the photoreceptor molecules involved in these processes have been identified, isolated, and characterized. An important approach for the understanding of the molecular mechanism of photoresponses is to insert these molecules in lipid bilayers and analyze conductance changes or photopotentials induced by flashes of light of various wavelengths. Much of our information about them comes from action spectra measurements, in which the relative effectiveness of different wavelengths of light in producing the response is determined.

This section surveys photosensitive model membranes intended to mimic, in an artificial system, some features of the major photoresponses, including the photosynthetic process, the light-driven ionic gradient, and phototransduction in the visual system. This treatment is necessarily brief, and a number of important details are omitted. Recent reviews that may be consulted include those by Berns (1976), Hong (1976), Montal (1979), Stoeckenius *et al.* (1979), Montal *et al.* (1981), and Pepe and Gliozzi (1983).

4.2. Chlorophyll-Containing Lipid Bilayers

The conversion of light energy and associated electron-transport reactions of photosynthesis occur in the membranous sacs called thylakoids, which are contained in the chloroplasts. The thylakoids consist of a system of lamellar membranes, arranged in stacks, in which the chlorophyll is embedded. They are approximately half-lipid and half-protein in chemical composition (e.g., Hall and Rao, 1977).

Besides chlorophylls, the two major classes of pigments are carotenoids and phycobilins. The chlorophyll molecule contains a polar porphyrin head and an hydrophobic phytol tail. It can move freely in the lipid matrix or be associated with particular proteins. These pigment protein complexes are called reaction centers. The photons captured by chlorophyll and other pigments supply the energy to split two molecules of water into a molecule of oxygen: four protons and four electrons. An electron-transport chain carries electrons from the reaction center to the outside surface of the thylakoid sac; these in turn, with the aid of a proton, reduce $NADP^+$ to NADPH. Protons will thus accumulate on the inside surface of the membrane. The resulting outflow of protons drives the synthesis of ATP from ADP and inorganic phosphate by an ATPase system. NADPH and ATP store the energy that will be used in the subsequent dark reaction converting carbon dioxide and hydrogen into glucose and water.

The first attempts to build artificial photoresponsive membranes as a model of photosynthesis date back to the early studies on BLM systems (Tien, 1968). The work was initially approached by forming lipid bilayers in an alkane solution of a crude lipid extract of chloroplasts, separating two identical aqueous solutions. Under continuous illumination and open circuit conditions, a photopotential of the order of few millivolts was recorded. The polarity was dependent on the direction of illumination, the illuminated side being negative. Later experiments showed that larger photovoltages, of the order of a hundred millivolts, could be elicited in asymmetrical conditions, by introducing in the aqueous solution a different concentration of electron acceptors and donors, such as ceric–cerous or ferric–ferrous ions. The polarity of photovoltages resulted independent of the direction of light, being only dependent on the redox composition of the aqueous phases. The acceptor-rich side was negative with respect to the donor-rich side.

Photovoltages and photocurrents were recorded either under steady-state continuous illumination or under pulsed light sources. While in pioneering work the action spectra were rather poorly correlated to the absorption spectra of the pigment, the refinement of the experimental approach has led to action spectra that perfectly match the absorption spectra (e.g., Chen and Berns, 1976; Schönfeld *et al.,* 1979).

A crucial point is to understand the physical mechanisms underlying these responses. The process occurring at the interfaces can be easily interpreted. Under the influence of light, a pigment molecule accepts an electron from a donor molecule, which is adsorbed at one interface. The reverse occurs at the other interface; i.e., a pigment molecule may donate an electron to an acceptor. How are these photoreactions coupled to give rise to an electrical signal? Two different mechanisms have been proposed. Since the two interfaces are separated by no more than 100 Å, a quantum-mechanical electron tunneling has been proposed by some workers (Berns, 1976; Tien, 1979). Alternatively, the two interfacial photoreactions may be coupled by transmembrane diffusion of pigment molecules (Hong, 1976). The latter model enables one to distinguish between continuous light responses and pulsed light responses. After interfacial photoreaction, a charged pigment can be discharged at the same interface at which it was formed, or it may diffuse across the BLM and discharge at the opposite interface. Only the latter event can be observed under continuous illumination. By contrast, if the exciting light has sufficiently high rising and falling times (<1 μsec), the interfacial photoreaction will dominate the transient photoresponses (which are of the order of 10 μsec). Both kinds of responses are shown in Fig. 21. The latter interpretation is consistent with the view (Trissl and Laüger, 1970) that a photogeneration of an electrical double layer might occur in a chlorophyll–BLM system under pulsed light stimulation. The mechanism is similar to the so-called dipole model proposed by Ullrich and Kuhn (1972) and invoked to explain the biphasic responses of various experimental models (cyanine dye–BLM, retinal–BLM, flavin–BLM).

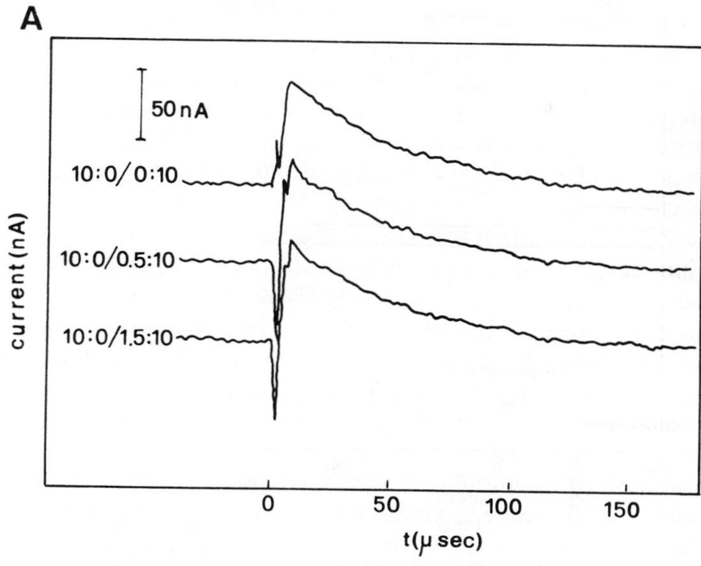

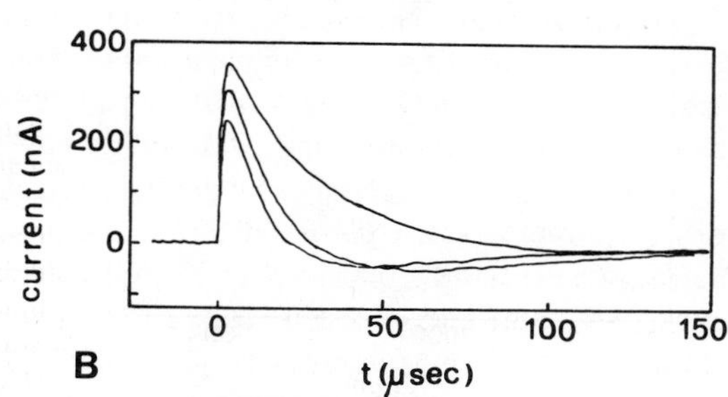

Figure 21. (A) Fast photoresponses specific to a membrane–water interface. All records are from the same membrane containing mesoporphyrin esters of magnesium octaethylporphyrin (MgOEP). The aqueous redox composition (with 0.1 *M* NaCl) is indicated, e.g., 10:0/0:10 means 10 m*M* ferricyanide and 0 m*M* ferrocyanide on the acceptor side and 0 m*M* ferricyanide and 10 m*M* ferrocyanide on the donor side. The membrane is voltage-clamped at $V = 0$. The wavelength is 590 nm. (B) Time courses of transient photocurrents and the effect of varying the redox composition of the aqueous phase. The records are from a membrane containing diethyl ester of Mg^{2+} mesoporphyrin, voltage-clamped at $V = 0$ mV. The aqueous redox composition (with 100 m*M* KCl), in the order of diminishing responses is 10:0/0:0; 10:0.5/0:0; 10:1/0:0. The wavelength is 590 nm. (Modified from Hong, 1976.)

Recently, bacterial reaction centers have been incorporated in planar lipid bilayers (Schönfeld *et al.*, 1979; Packham *et al.*, 1982). Upon illumination, this system displays transient and steady-state voltages and currents. Figure 22 shows the photoresponses and the associated action spectrum. The results are interpreted as associated with light-driven electron transfer across the reaction centers, which are thought to span the entire thickness of the membrane. This approach appears to serve as a basis for understanding structural and functional aspects of the photosynthetic apparatus.

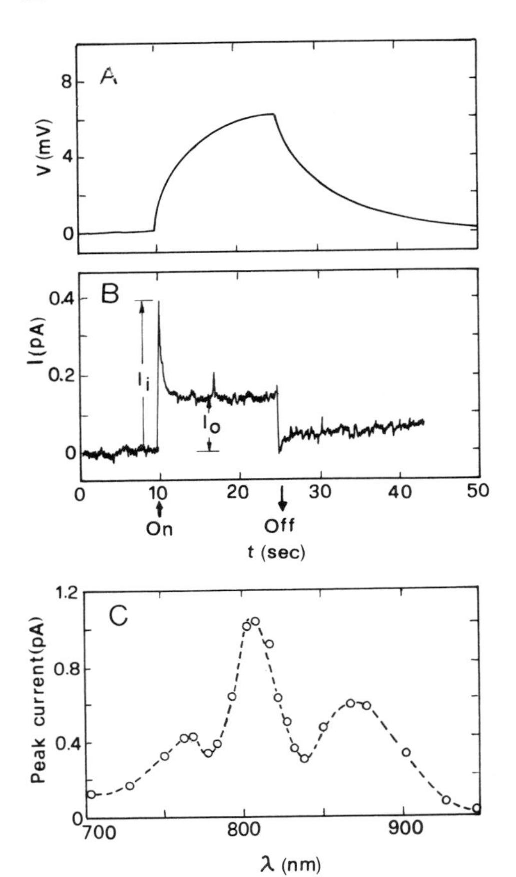

Figure 22. Photovoltages (A) and photocurrents (B) of planar lipid bilayers containing reaction centers (RC) from *Rhodopseudomonas sphaeroides* R-26. The wavelength of the incident light is 800 nm. The aqueous compartments contain 30 μM cytochrome *c* on one side and 0.25 mM quinones on the opposite side. (C) (○), action spectrum of the peak current I_i; (---) optical absorbance in the RC–lipid complex in hexane used to form the membrane. (Modified from Schönfeld *et al.*, 1979.)

4.3. Bacteriorhodopsin and the Purple Membrane

The cell membrane of *Halobacterium halobium* exhibits characteristic patches containing the chromoprotein bacteriorhodopsin, a unique protein. These specialized regions, known as the purple membrane, are made up of approximatively 75% bacteriorhodopsin and 25% lipids by weight. X-ray diffraction of a purple membrane demonstrates the presence of a hexagonal lattice. Bacteriorhodopsin is chemically similar to the visual pigment rhodopsin of animal eyes. In both molecules, retinol is attached via a Schiff base to a lysine residue of the apoprotein opsin. The protonated Schiff base absorbs maximally at 570 nm. In light absorption, bacteriorhodopsin proceeds through a cyclical sequence of at least five spectral intermediates. Only the first step of this photochemical cycle requires light. The reversion to the original pigment (bR_{570}) is accompanied by a reprotonation. The rate of these transitions is a function of temperature. At room temperature and physiological pH, the half-time of photoreaction is ~10 msec.

As a result of the photochemical cycle, protons are translocated across the membrane, thereby generating a substantial electrochemical gradient. The cell uses this energy to drive ATP synthesis and other energy-requiring functions. Bacteriorhodopsin thus acts as a light-driven electron pump (cf., e.g., Stoeckenius *et al.,* 1979).

Reconstitution of the purple membrane functions in an artificial system under controlled experimental conditions seems an appealing task. The intrinsic simplicity of the purple membrane and the ease of in isolating it by simple osmotic shocks make it the ideal system to study in a reconstitution approach.

4.3.1. Bacteriorhodopsin in Vesicles

A successful reconstitution of the purple membrane functions was made by Racker and Stoeckenius (1974), who succeeded in incorporating purple membrane fractions in phospholipid vesicles. A measure of the external pH showed that, upon illumination, the reconstituted vesicles took up protons, which were released in the dark. The direction of proton flow, from outside to inside, is opposite to that observed in intact bacteria. This fact suggests that the purple membrane has a definite sidedness and, during reconstitution, it has been turned inside out, as shown by freeze-fracture electron micrographs. Uncouplers of oxidative phosphorylation, such as 1799 (bis-hexafluoracetonyl-acetone), which increase the permeability of protons and thus dissipate by backdiffusion the electrochemical gradient, reduce the response to a lower level.

This system provides a brilliant experimental model of the Mitchell chemiosmotic theory of energy transduction. In fact, incorporation of beef heart mitochondrial ATPase into the same liposomes made the system capable of catalyzing light-dependent ATP formation from adenosine diphosphosphate (ADP) and inorganic phosphorus.

A direct measure of a pH gradient and of a consequent electric potential difference was provided by Kayushin and Skulachev (1974), using, as a molecular probe, a fluorescent dye and a hydrophobic anion, respectively.

4.3.2. Bacteriorhodopsin in Lipid–Water Systems

Purple membrane characteristics were also studied in lipid–water systems. A mixture of purple membrane fragments and phospholipids was spread at an air–water interface, forming an insoluble film that, according to Hwang *et al.*

(1977*a*), consists of nonoverlapping membrane fragments connected by a lipid monolayer. Bacteriorhodopsin seems to maintain its properties, showing spectral change characteristics in darkness and upon flash illumination, similar to those observed in aqueous suspensions. In addition, electron microscopy of shadow-cast replicas indicates that the membrane fragments are highly oriented, with their intracellular surface toward the aqueous phase.

The photopotentials generated in these films upon illumination were observed by spreading over the film a thin layer (0.3 mm) of decane. Since the purple fragments are oriented, protons are expected to be transported from the aqeuous to the organic phase. However, owing to the relative insolubility of protons, only a small photopotential may be recorded by an ionizing electrode. These photovoltages are relatively slow, having a rising time of the order of seconds.

An increase in photopotential was observed by adding proton carriers such as carbonylcyanide-*p*-trifluoromethoxyphenylhydrazone (FCCP) and dinitrophenol (DNP) (Boguslavsky *et al.*, 1976; Hwang *et al.*, 1977*b*). This result was interpreted considering that the interface-linked bacteriorhodopsin was able to pump H^+ from water to the lipid phase. The uncouplers were postulated to act as lipid-soluble H^+ acceptors. This interpretation was recently questioned by Drachev *et al.* (1982). Several reasons were adduced by these investigators, among these, the fact that phenylcarbaundecarborane anion (PCB^-) was shown to be even more effective than the potent protonophore, CCCP. Therefore, if H^+ acceptance were a function of uncouplers, PCB^-, which is a strong acid, should not substitute for uncouplers. According to Drachev *et al.*, there is a water cavity between the bacteriorhodopsin membrane and the bulk lipid phase. Thus, the system cannot be considered a simple biphasic system with interface-linked bacteriorhodopsin. They suggest that the primary electrogenic event is indeed bacteriorhodopsin-mediated transmembrane H^+ movement; however, it occurs from the bulk aqueous phase to the water space in the cavity between the bacteriorhodopsin membrane and the bulk lipid. Consequently, the slow photoelectric responses cannot be defined in the simple terms exposed earlier. These responses are analyzed in terms of the equivalent electrical circuit. The molecular events are not yet completely clear; a contribution to the photoresponse may be derived from a light-dependent conversion of retinal released by bacteriorhodopsin under illumination in the decane–water system.

Similar results were obtained by the same investigators in a lipid–water system, in which a decane solution of asolectin was used as the lipid phase.

4.3.3. Bacteriorhodopsin in Planar Layers

Fast photovoltages have been found in interfacial films containing bacteriorhodopsin apposed to a vertically mounted Teflon septum, separating two aqueous compartments (Trissl and Montal, 1977). Analysis with a quasi-short circuit (tunable voltage clamp) method with microsecond time resolution has shown that there are two components in the displacement photocurrent (Hong and Montal, 1979). A fast component, which has no detectable latency (with an instrumental time constant of 1.5 μsec), and a slower component, in the millisecond range, of opposite polarity, inhibited by low temperature (5 °C) and low pH (3.0). These capacitive photovoltages are the electrical correlates of the molecular charge displacements associated with the photochemical cycle.

With multilayered films of dry fragments of purple membrane and lipid sandwiched between two metal electrodes, Hwang *et al.* (1978) found photovoltages arising from bacteriorhodopsin intermediates in the photoreaction cycle. Unlike the bR_{570} previously described, the latter photovoltages made the extracellular surface of purple membrane negative with respect to the intracellular one. These workers suggest that these signals arise from a charge displacement in the bacteriorhodopsin molecule in response to the absorption of light. In turn, this charge displacement might reflect a transfer of charged particles along the bacteriorhodopsin, so that the events underlying the fast photovoltages may be linked to the molecular mechanism that pumps protons through the purple membrane.

Many attempts to study bacteriorhodopsin properties using planar lipid bilayers were recently made. Two primary methods have been used. In the first one, vesicles containing purple membrane fractions were added to one aqueous phase of a BLM (Herrmann and Rayfield, 1978). The photocurrents observed in this model system were enhanced dramatically only in the presence of proton carriers; this finding suggests that the vesicle fusion leaves an aqueous compartment between the purple membrane sheet and the planar bilayer, as shown in Fig. 23A. In the second method, purple membrane fragments were added directly to one side of a positively charged lipid bilayer. Also in this case a considerable increase in the steady-state photocurrent was induced by the addition of gramicidin, which acts as an ionic carrier. It is therefore suggested that the purple membrane sheets might be bound to the surface of the bilayer forming a sandwichlike structure, as shown in Fig. 23B. Figure 23C,D shows short-circuit photocurrents after addition of purple membrane fragments, in the absence and in the presence of gramicidin, respectively. In both models, photoresponses are analyzed in terms of an equivalent electrical circuit. They are consistent with a proton flux toward the bacteriorhodopsin-free solution, in close agreement with the absorbance spectrum of the latter.

Association of bacteriorhodopsin with lipid-impregnated filters has also been studied (Blok and Van Dam, 1979). Bacteriorhodopsin vesicles were associated with cellulose nitrate filters impregnated with a solution of phospholipids in hexadecane. The generation of photopotentials upon illumination of the filter is suggested to arise from protons driven into aqueous compartments located

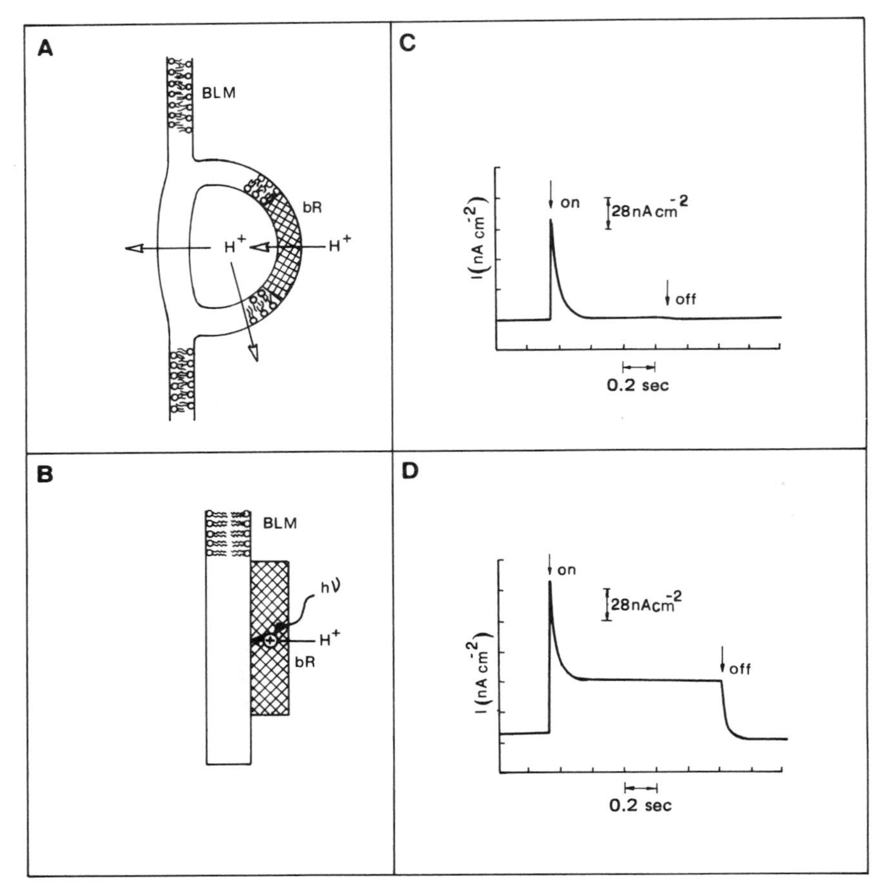

Figure 23. (A) Schematic drawing of a vesicle containing bacteriorhodopsin (bR), fused with a black lipid membrane (BLM). (After Hermann and Rayfield, 1978.) (B) Proposed model for purple membrane interaction with a positively charged membrane and short-circuit photocurrents recorded from the same system in the absence (C) and in the presence (D) of gramicidin A. λ_{max} = 550 nm. (From Bamberg *et al.*, 1979.)

in the filter. These compartments are analogous to those present in the model of Drachev *et al.* (1982) and in that of Herrmann and Rayfield (1978), which describes the fusion of bacteriorhodopsin vesicles with BLM (cf. Fig. 23A). It is suggested that fusion in this artificial system occurs between vesicles and the lipids of the system.

4.4. Rhodopsin in Model Membranes

The primary process leading to a visual excitation is still considered the photoisomerization of retinaldehyde, the chromophore of the visual pigment rhodopsin, from the 11-*cis* to the all-*trans* configuration. The photoisomerization initiates a series of light-induced conformational changes in the protein. In the case of vertebrate animals, the sites of visual transduction are the cones and the rod-shaped outer segment (ROS). There is general agreement that the light-sensitive permeability of the ROS plasma membrane must be regulated by some internal transmitters that mediate between the site of photon absorption in the disk membrane and the conductance mechanism of the plasma membrane (Baylor and Fuortes, 1970; Hagins, 1972; Cone, 1973). This is probably the primary process leading to cellular excitation. In the case of vertebrate animals, two general mechanisms are currently under consideration. According to the first, the calcium ions would act as transmitters connecting the disk membrane, where rhodopsin is located, to the plasma membrane. Calcium would close the Na channels in the plasma membrane, producing a hyperpolarization capable of triggering the electrical signal in the optic nerve (Hagins, 1972). No specific role for rhodopsin was proposed, but one possibility is that rhodopsin itself forms a transmembrane Ca channel when photolyzed.

The second model involves amplification by enzymatic reactions (Hubbel and Bownds, 1979). A specific version of this model (Polans *et al.*, 1979) suggests that the plasma Na^+ conductance depends on the intracellular levels of cyclic GMP. Rhodopsin bleaching regulates GMP levels through activation of GTPase and a phosphodiesterase.

These two alternative models have stimulated many investigators to study light-induced transmembrane ion movements in rhodopsin-containing membranes. Hong and Hubbel (1972) showed that highly purified rhodopsin could be incorporated into phospholipid vesicles without denaturation. Subsequent experiments have shown that phospholipid–rhodopsin vesicles, containing Ca^{2+} inside, release Ca^{2+} ions when illuminated. Bleaching of rhodopsin would thus increase ionic permeability; this finding plays in favor of the model in which rhodopsin itself would act as a channel that opens and closes with light. These permeability changes are, however, not selective, because Na^+, Cs^+, glucose, and glycerol are also released upon illumination (Darszon *et al.*, 1977).

The electrical properties of planar lipid bilayers (obtained by hydrophobic apposition of two monlayers) containing bovine rhodopsin showed light- and voltage-dependent conductance changes occurring with a pronounced latency of 3–200 sec, depending on rhodopsin concentration in the membrane (Montal *et al.*, 1977). The observed results are interpreted in terms of formation of a transmembrane channel, ~10 Å in diameter. However, the long latency time after light application (explained as attributable to aggregation of subunits) and the nonselective permeability to Ca^{2+} after illumination do not permit consideration of experiments on artificial systems as definite evidence for channel formation.

Current fluctuations similar to those obtained with bovine rhodopsin, indicating the formation of transmembrane ionic channels, were also found by incorporating a photopigment from the honeybee eye in BLM that does not seem to be directly identifiable as a rhodopsin (Gambale *et al.,* 1980). This finding indicates that caution must be exercised in interpreting results from single bilayer experiments.

In fact, the use of nonnative phospholipids and possible structural changes of the protein in the reconstituted membrane might give rise to physical phenomena that have little to do with the phototransduction process *in vivo.* On the other hand, it is now apparent that cyclic GMP changes stimulated by illumination are sufficiently rapid to be involved in membrane permeability changes. Further progress in reconstitution techniques appears to be important in understanding the events underlying phototransduction.

ACKNOWLEDGMENTS. We thank E. Carbone, A. Chiabrera, and I. M. Pepe for many helpful discussions and suggestions and T. Gulik and V. Luzzati for giving us the electron micrographs of the freeze-fractured membrane. The expert secretarial assistance of Miss O. Graffigna and Mrs. S. Ostric is also greatly appreciated.

REFERENCES

Almers, W., 1978, Gating currents and charge movements in excitable membranes, *Rev. Physiol. Biochem. Pharmacol.* **82**:96–190.

Alvarez, O., Diaz, E., and Latorre, R., 1975, Voltage-dependent conductance induced by hemocyanin in black lipid films, *Biochim. Biophys. Acta* **389**:444–448.

Anderson, C. R., and Stevens, C. F. 1973, Voltage clamp analysis of acetylcholine produced endplate current fluctuations at the frog neuromuscular junction, *J. Physiol. (Lond.)* **235**:655–691.

Armstrong, C. M., 1974, Ionic pores, gates and gating currents, *Q. Rev. Biophys.* **7**(2):179–210.

Armstrong, C. M., and Bezanilla, F., 1973, Currents related to the gating particles of the sodium channels, *Nature (Lond.)* **242**:459–461.

Armstrong, C. M., and Bezanilla, F., 1974, Charge movement associated with the opening and closing of the activation gates of Na channels, *J. Gen. Physiol.* **63**:533–552.

Bamberg, E., Apell, H. J., Dencher, N. A., Spelling, W., Stieve, H., Läuger, P., 1979, Photocurrents generated by bacteriorhodopsin in planar bilayer membranes, *Biophys. Struct. Mech.* **5**:277–292.

Baumann, G., and Mueller, P., 1974, A molecular model of membrane excitability, *J. Supramol. Struct.* **2**:538–557.

Baylor, D. A., and Fuortes, M. G. F., 1970, Electrical responses of single cones in the retina of the turtle, *J. Physiol. (Lond.)* **207**:77–92.

Bean, R. C., Shepherd, W. C., Chan, H., and Eichner, J., 1969, Discrete conductance fluctuations in lipid bilayer protein membranes, *J. Gen. Physiol.* **53**:741–757.

Begenisich, T., and Stevens, C. F., 1975, How many conductance states do potassium channels have? *Biophys. J.* **15**:843–846.

Benz, R., and Läuger, P., 1976, Kinetic analysis of carrier-mediated ion transport by the charge-pulse technique, *J. Membrane Biol.* **27**:171–191.

Benz, R., Fröhlich, O., Läuger, P., and Montal, M., 1975, Electrical capacity of black lipid films and of lipid bilayers made from monolayers, *Biochim. Biophys. Acta* **394**:323–334.

Benz, R., Fröhlich, O., and Läuger, P., 1977, Influence of membrane structure on the kinetics of carrier-mediated ion transport through lipid bilayers, *Biochim. Biophys. Acta* **464**:465–481.

Berns, D. S., 1976, Photosensitive bilayer membranes as model systems for photobiological processes, *Photochem. Photobiol.* **24**:117–139.

Bernstein, J., 1902, Untersuchungen zur thermodynamik der bioelektrischen Ströme, *Arch. Ges. Physiol.* **92**:521–62.

Bezanilla, F., and Armstrong, C. M., 1972, Negative conductance caused by the entry of sodium and cesium ions into the potassium channel of squid axon, *J. Gen. Physiol.* **60**:588–608.

Blok, M. C., and Van Dam, D., 1979, Association of bacteriorhodopsin with lipid-impregnated filters, *Biochim. Biophys. Acta* **550**:527–542.

Boguslavsky, L. I., Boystsov, V. G., Volkov, A. G., Kozlov, I. A., Metelsky, S. T., 1976, Light-dependent translocation of H^+ from water to octane by bacteriorhodopsin, *Bioorg. Khimi* **2**:1125–1130.

Boheim, G., 1974, Statistical analysis of alamethicin channels in black lipid membranes, *J. Membr. Biol.* **19**:277–303.

Bullock, J. O., and Schauf, C. L., 1978, Combined voltage-clamp and dialysis of *Myxicola* axons: behaviour of membrane asymmetry currents, *J. Physiol. (Lond.)* **278**:309–324.

Bu'Lock, J. D., De Rosa, M., and Gambacorta, A., 1982, Isoprenoid biosynthesis in archaebacteria, in: *Polyisoprenoid Biosynthesis* (J. N. Porte, ed.), pp. 159–189 Wiley, London.

Carbone, E., Wanke, E., Prestipino, G., Possani, L., and Maelicke, A., 1982, Selective blockage of voltage-dependent K^+ channels by a novel scorpion toxin, *Nature (Lond.)* **296**:90–91.

Changeux, J. P., Thiery, J., Tung, Y., and Kittel, X., 1967, On the cooperativity of biological membranes, *Proc. Natl. Acad. Sci. U.S.A.* **57**:335–341.

Chen, C. H., and Berns, D. S., 1976, Photosensitivity of artificial bilayer membranes: lipid–chlorophyll interaction, *Photochem. Photobiol.* **24**:255–260.

Ciani, S., Eisenman, G., Laprade, R., and Szabo, G., 1973, Theoretical analysis of carrier-mediated electrical properties of bilayer membrane, in: *Membranes—A Series of Advances* Vol. 2 (G. Eisenman, ed.), p. 61, Marcel Dekker, New York.

Ciani, S., Gambale, F., Gliozzi, A., and Rolandi, R., 1975, Effects of unstirred layers on the steady state zero-current conductance of bilayer membranes mediated by neutral carriers of ions, *J. Membr. Biol.* **24**:1–34.

Cone, R. A., 1973, The internal transmitter model for visual excitation: some quantitative implications, in: *Biochemistry and Physiology of Visual Pigments* (H. Läuger, ed.), pp. 275–284, Springer-Verlag, New York.

Conti, F., 1984, Noise analysis and single channel recordings, in: *Current Topics in Membrane and Transport, Vol. 22: The Squid Axon* (P. F. Baker, ed.), pp. 371–405, Academic-Press, New York.

Conti, F., and Neher, E., 1980, Single channels recording of K^+ currents in squid axons, *Nature (Lond.)* **255**:140–143.

Conti, F., and Wanke, E., 1975, Channel noise in nerve membranes and lipid bilayers, *Q. Rev. Biophys.* **8**:451–506.

Conti, F., DeFelice, L. J., and Wanke, E., 1975, Potassium and sodium ion current noise in the membrane of the squid giant axon, *J. Physiol. (Lond.)* **248**:45–82.

Conti, F., Neumcke, B., Nonner, W., and Stämpfli, R., 1980, Conductance fluctuations from the inactivation process of sodium channels in myelinated nerve fibres, *J. Physiol. (Lond.)* **308**:217–238.

Danielli, J. F., 1975, The bilayer hypothesis of membrane structure, in: *Cell Membranes* (G. Weissmann and R. Claiborne, eds.), pp. 3–11, HP Publishing, New York.

Danielli, J. F., and Davson, H., 1935, A contribution to the theory of permeability of thin films, *J. Cell Comp. Physiol.* **7**(3):495.

Darszon, A., Montal, M., and Zarco, J., 1977, Light increases the ion and non-electrolyte permeability of rhodopsin-phospholipid vesicles, *Biochem. Biophys. Res. Commun.* **76**:820–827.

De Rosa, M., Gambacorta, A., Nicolaus, B., and Bu'Lock, J. D., 1980, Complex lipids of *Caldariella acidophila,* a thermoacidophile archaebacterium, *Phytochemistry* **19**:821–825.

Drachev, L. A., Kaulen, A. D., Skulachev, V. P., and Voytsitsky, V. M., 1982, Bacteriorhodopsin-mediated photoelectric responses in lipid/water systems, *J. Membr. Biol.* **65**:1–12.

Ehrenstein, G., and Lecar, H., 1977, Electrically gated ionic channels in lipid bilayers, *Q. Rev. Biophys.* **10**(1):1–34.

Ehrenstein, G., Lecar, H., and Nossal, R., 1970, The nature of the negative resistance in bimolecular lipid membranes containing excitability-inducing material, *J. Gen. Physiol.* **55**:119–133.

Eisenberg, M., Hall, J. E., and Mead, C. A., 1973, The nature of the voltage-dependent conductance induced by alamethicin in black lipid membranes, *J. Membr. Biol.* **14**:143–176.

Eyring, H., 1935, The activated complex in chemical reactions, *J. Chem. Phys.* **3**:107.

Eyring, H., Lumry, R., and Woodbury, J. W., 1949, Some applications of modern rate theory to physiological systems, *Record. Chem. Progr.* **10**:100.

Fatt, P., and Katz, B., 1952, Spontaneous subthershold activity at motor nerve endings, *J. Physiol. (Lond.)* **117**:107–128.

Finkelstein, A., and Mauro, A., 1974, Physical principles and formalisms of electrical excitability, in: *Handbook of Physiology: The Nervous System. I.* (E. R. Kandel, ed.), pp. 161–214, American Physiological Society, Bethesda, Md.

Fishman, H. M., 1972, Excess noise from small patches of squid axon membrane, *Biophys. Soc. Annu. Meet. Abstr.* 12p, 119a.

Fishman, H. M., Moore, L. E., and Poussart, D. J. M., 1975, Potassium-ion conduction noise in squid axon membrane, *J. Membr. Biol.* **24**:305–328.

Foster, M., and McLaughlin, S., 1974, Complexes between uncouplers of oxidative phosphorylation, *J. Membr. Biol.* **17**:155–180.

Frehland, E., 1980, Current fluctuations in discrete transport systems far from equilibrium. Breakdown of the fluctuation dissipation theorem, *Biophys. Chem.* **12**:63–71.

Frye, C. D., and Edidin, M., 1970, The rapid intermixing of cell surface antigens after formation of mouse–human heterokaryons, *J. Cell Sci.* **7**:319.

Gambale, F., Gliozzi, A., and Robello, M., 1973, Determination of rate constants in carrier-mediated diffusion through lipid bilayers, *Biochim. Biophys. Acta* **330**:325–334.

Gambale, F., Gliozzi, A., Pepe, M., Robello, M., and Rolandi, R., 1980, Photopigment inducing pores in lipid bilayer membranes, in: *Developments in Biophysical Research* (A. Borsellino, P. Omodeo, R. Strom, A. Vecli, and E. Wanke, eds.), pp. 93–107, Plenum Press, New York.

Gilly, W. F., and Armstrong, C. M., 1980, Gating current and potassium channels in the giant axon of the squid, *Biophys. J.* **29**:485–492.

Gliozzi, A., 1980*a*, The lipid bilayer: a model system for biological membranes, in: *Bioenergetics and Thermodynamics: Model Systems* (A. Braibanti, ed.), pp. 377–390, D. Reidel Publishing Co., Dordrecht, Holland.

Gliozzi, A., 1980*b*, Carriers and channels in artificial and biological membranes, in: *Bioenergetics and Thermodynamics: Model systems* (A. Braibanti, ed.), pp. 339–353, D. Reidel Publishing Co., Dordrecht, Holland.

Gliozzi, A., Rolandi, R., De Rosa, M., and Gambacorta, A., 1982*a*, Artificial black membranes from bipolar lipids of thermophilic Archaebacteria, *Biophys. J.* **37**:563–566.

Gliozzi, A., Rolandi, R., De Rosa, M., Gambacorta, A., and Nicolaus, B., 1982*b*. Membrane models in Archaebacteria, in: *Transport in Biomembranes: Model Systems and Reconstitution* (R. Antolini, A. Gliozzi, and A. Gorio, eds.), pp. 39–47, Raven Press, New York.

Gliozzi, A., Rolandi, R., De Rosa, M., and Gambacorta, A., 1983, Monolayer black membranes from bipolar lipids of Archaebacteria and their temperature-induced structural changes, *J. Membrane Biol.* **75**:45–56.

Goldman, D. E., 1943, Potential impedance and rectification in membranes, *J. Gen. Physiol.* **27**:37–60.

Gordon, L. G. M., and Haydon, D. A., 1972, The unit conductance channel of alamethicin, *Biochim. Biophys. Acta* **255**:1014–1018.

Gordon, L. G. M., and Haydon, D. A., 1976, Kinetics and stability of alamethicin conducting channels in lipid bilayers, *Biochim. Biophys. Acta* **436**:541–556.

Gorter, E., and Grendel, F., 1925, On bimolecular layers of lipids on the chromocytes of the blood, *J. Exp. Med.* **41**:439–443.

Grell, E., Funck, T., and Eggers, F., 1975, Structure and dynamics properties of ion-specific antibiotics, in: *Membranes—A Seriesof Advances,* Vol. 3 (G. Eisenman, ed.), p. 1, Marcel Dekker, New York.

Griffith, O. H., Dehlinger, P. J., and Van, S. P., 1974, Shape of the hydrophobic barrier of phospholipid bilayers: evidence for water penetration in biological membranes, *J. Membr. Biol.* **15**:159–192.

Hagins, W. A., 1972, The visual process: excitatory mechanisms in the primary receptor cells, *Annu. Rev. Bioeng.* **1**:131–158.

Hall, D. O., and Rao, K. K., 1977, *Photosynthesis,* Edward Arnold, London.

Hall, J. E., and Latorre, R., 1976, Nonactin–K^+ complex as a probe for membrane asymmetry, *Biophys. J.* **16**:99–103.

Herrmann, T. R., and Rayfield, G. W., 1978, The electrical response to light of bacteriorhodopsin in planar membranes, *Biophys. J.* **21**:111–125.

Hill, T. L., and Chen, Y. D., 1972, On the theory of ion transport across the nerve membrane. IV. Noise from the open–close kinetics of K-channels, *Biophys. J.* **12**:948–959.

Hille, B., 1970, Ionic channels in nerve membranes, in: *Progress in Biophysics and Molecular Biology,* Vol. 21 (J. A. V. Butler and D. Noble, eds.), pp. 1–32, Pergamon Press, Oxford.

Hille, B., 1973, Potassium channels in myelinated nerve. Selective permeability to small cations, *J. Gen. Physiol.* **61**:669–686.

Hille, B., 1975*a*, Ionic selectivity, saturation and block in sodium channels. A four barrier model, *J. Gen. Physiol.* **66**:535–560.

Hille, B., 1975*b*, Ionic selectivity of Na and K channels in nerve membranes, in: *Membranes—A series of Advances vol. 3 Dynamic Properties of Lipid Bilayers and Biological Membranes* (G. Eisenmann, ed.), pp. 255–323, Marcel Dekker, New York.

Hille, B., and Schwartz, W., 1978, Potassium channels as multiion single-file pores, *J. Gen. Physiol.* **72**:409–442.

Hladky, S. B., 1972, The steady state theory of carrier transport of ions, *J. Membr. Biol.* **10**:67.

Hladky, S. B., and Haydon, D. A., 1972, Ion transfer across lipid membranes in the presence of Gramicidin A. I. Studies of the unit conductance channel, *Biochim. Biophys. Acta* **279**:244–312.

Hladky, S. B., and Haydon, D. A., 1973, Membrane conductance and surface potential, *Biochim. Biophys. Acta* **318**:464–468.

Hodgkin, A. L., and Huxley, A. F., 1952*a*, Currents carried by sodium and potassium ions through the membrane of giant axon of *Loligo, J. Physiol. (Lond.)* **116**:449–472.

Hodgkin, H., and Huxley, A. F., 1952*b*, A quantitative description of membrane current and its application to conduction and excitation in nerve, *J. Physiol. (Lond.)* **117**:500–544.

Hodgkin, A. L., and Katz, B., 1949, The effect of sodium ions on the electrical activity of the giant axon of the squid, *J. Physiol. (Lond.)* **108**:37–77.

Hong, F. T., 1976, Charge transfer across pigmented bilayer lipid membrane and its interfaces, *Photochem. Photobiol.* **24**:155–189.

Hong, F. T., and Montal, M., 1979, Bacteriorhodopsin in model membranes. A new component in the displacement photocurrent in the microsecond time scale, *Biophys. J.* **25**:465–472.

Hong, K., and Hubbel, W. L., 1972, Preparation and properties of phospholipid bilayers containing rhodopsin, *Proc. Natl. Acad. Sci. USA* **69**:2617–2621.

Hubbel, W. L., and Bownds, M. D., 1979, Visual transduction in vertebrate photoreceptors, *Annu. Rev. Neurosci.* **2**:17–34.

Hwang, S., Korenbrot, J. I., and Stoeckenius, W., 1977*a*, A structural and spectroscopic characteristics of bacteriorhodopsin at an air–water interface film, *J. Membr. Biol.* **36**:115–135.

Hwang, S., Korenbrot, J. I., and Stoeckenius, W., 1977*b*, Proton transport by bacteriorhodopsin through an interface film, *J. Membr. Biol.* **36**:137–158.

Hwang, F., Korenbrot, J. I., and Stoeckenius, W., 1978, Transient photovoltages in purple membrane multilayers, *Biochim. Biophys. Acta* **509**:300–317.

Katz, B., 1966, *Nerve, Muscle and Synapses,* McGraw-Hill, New York.

Katz, B., and Miledi, R., 1970, Membrane noise produced by acetylcholine, *Nature (Lond.)* **225**:962–963.

Katz, B., and Miledi, R., 1972, The statistical nature of acetylcholine potential and its molecular components, *J. Physiol. (Lond.)* **224**:665–699.

Kayushin, L. P., and Skulachev, V. P., 1974, Bacteriorhodopsin as an electrogenic proton pump: reconstitution of bacteriorhodopsin proteoliposomes generating $\Delta\psi$ and ΔpH, *FEBS Lett.* **39**:39–42.

Keynes, R. D., Malachowsky, G. C., Van Helden, D. F., and Greef, N. G., 1980, Components of the asymmetry current in the squid giant axon, in: *Twenty-eighth International Congress of Physiological Sciences, Budapest, 1980,* pp. 26–29, Pergamon Press–Akademiai Kiado, Oxford.

Keynes, R. D., and Rojas, E., 1976, The temporal and steady state relationships between activation of the sodium conductance and movement of the gating particles in the squid giant axon, *J. Physiol. (Lond.)* **255**:157–189.

Keynes, R. D., Greeff, N. G., and van Helden, D. F., 1982, The relationship between the inactivating fraction of the asymmetry current and gating of the sodium channel in the squid giant axon, *Proc. R. Soc. Lond. B* **215**:391–409.

Kolb, H. A., and Läuger, P., 1978, Spectral analysis of current noise generated by carrier-mediated ion transport, *J. Membr. Biol.* **41**:167–187.

Kubo, R., 1957, Statistical mechanical theory of irreversible processes. General theory and simple applications to magnetic and conduction processes, *Nippon Seirigaku Zasshi* **12**:570.

Latorre, R., and Alvarez, O., 1981, Voltage-dependent channels in planar lipid membranes, *Physiol. Rev.* **61**:77–150.

Latorre, R., Ehrenstein, G., and Lecar, H., 1972, Ion transport through excitability inducing material (EIM) channels in lipid bilayer membranes, *J. Gen. Physiol.* **60**:72–85.

Laüger, P., 1975, Shot noise in ion channels, *Biochim. Biophys. Acta* **413**:1–10.

Laüger, P., 1980, Kinetic properties of ion carriers and channels, *J. Membr. Biol.* **57**:163–178.

Le Blanc, O. H., Jr., 1971, The effect of uncouplers of oxidative phosphorylation on lipid bilayer membranes: Carbonylcyanide *m*-chlorophenylhydrazone, *J. Membr. Biol.* **4**:227.

Liberman, E. A., Topaly, V. P., Silberstein, A., and Okhlobistin, O., 1971, Mobile ion-carriers and the negative resistance of membranes. 1. Uncoupling agents of oxidative phosphorylation-proton carriers, *Biophysics* **16**:637–639.

MacDonald, D. K., 1962, *Noise and Fluctuations, An Introduction,* Wiley, New York.

McLaughlin, S., 1977, Electrostatic potentials at membrane–solution interfaces, in: *Current Topics in Membrane Transport,* vol. 9 (F. Bronner and A. Kleinzeller, eds.), pp. 71–139, Academic Press, New York.

Menestrina, G. F., and Antolini, R., 1981, Ion transport through hemocyanin channels in oxidized cholesterol artificial bilayer membranes, *Biochim. Biophys. Acta* **643**:616–625.

Montal, M., 1979, Rhodopsin in model membranes, *Biochim. Biophys. Acta* **559**:231–257.

Montal, M., and Mueller, P., 1972, Formation of bimolecular membranes from lipid monolayers and a study of their electrical properties. *Proc. Natl. Acad. Sci. USA* **69**:3561–3566.

Montal, M., Darszon, A., and Trissl, H. W., 1977, Transmembrane channel formation in rhodopsin-containing bilayer membranes, *Nature (Lond.)* **267**:221–225.

Montal, M., Darszon, A., and Schindler, A., 1981, Functional reassembly of membrane proteins in planar lipid bilayers, *Q. Rev. Biophys.* **14**(1):1–79.

Moore, L. E., and Neher, E., 1976, Fluctuation and relaxation analysis of monazomycin-induced conductance in black lipid membranes, *J. Membr. Biol.* **27**:347–362.

Mueller, P., and Rudin, D. O., 1963, Induced excitability in reconstituted cell structure, *J. Theor. Biol.* **4**:268–280.

Mueller, P., and Rudin, D. O., 1968, Action potentials induced in bimolecular lipid membranes, *Nature (Lond.)* **217**:713–719.

Mueller, P., Rudin, D. O., Ti Tien, H., and Wescott, W. C., 1962, Reconstitution of excitable membrane structure in vitro, *Circulation* **26**:1167.

Mueller, R. V., and Finkelstein, A., 1972, The effect of surface charge on the voltage-dependent conductance induced in thin lipid membranes by monazomycin, *J. Gen. Physiol.* **60**:285–306.

Neher, E., and Sackman, B., 1976, Single-channel currents from membrane of denervated frog muscle fibers, *Nature (Lond.)* **260**:799–802.

Neher, E., and Zingsheim, H. P., 1974, The properties of ionic channels measured by noise analysis in thin lipid membranes, *Pfluegers Arch.* **351**:61–67.

Neumcke, B., 1978, 1/f noise in membranes, *Biophys. Struct. Mech.* **4**:179–199.

Neumcke, B., and Bamberg, E., 1975, The action of uncouplers on lipid bilayer membranes, in: *Membranes—A Series of Advances* Vol. 3 (G. Eisenman, ed.), p. 215, Marcel Dekker, New York.

Nicholson, G. L., 1976, Transmembrane control of the receptors on normal and tumor cells. I. Cytoplasmatic influence over cell surface components, *Biochim. Biophys. Acta* **457**:57–108.

Nonner, W., Rojas, E., and Stämpfli, R., 1975, Displacement currents in the node of Ranvier. Voltage and time dependence, *Pfluegers Arch.* **354**:1–18.

Packham, N. K., Dutton, P. L., and Mueller, P., 1982, Photoelectric currents across planar bilayer membranes containing bacterial reaction centers, *Biophys. J.* **37**:465–473.

Parling, B., and Eyring, H., 1954, Membrane permeability and electrical potential, in: *Ion Transport across Membranes* (H. T. Clark, ed.), pp. 103–118, Academic Press, New York.

Pepe, I. M., and Gliozzi, A., 1983, Model photoresponsive membranes, in: *Molecular Models of Photoresponsiveness* (G. Montagnoli and B. F. Erlanger, eds.), pp 337–354, Plenum Press, New York.

Polans, A. S., Hermolin, J., and Bownds, D., 1979, Light-induced dephosphorylation of two proteins in frog rod outer segments. Influence of cyclic nucleotides and calcium, *J. Gen. Physiol.* **74**:595–613.

Poussart, D. J. M., 1971, Membrane current noise in lobster axon under voltage clamp, *Biophys. J.* **11**:211–234.

Racker, E., and Stoeckenius, W., 1974, Reconstitution of purple membrane vesicles catalyzing light-driven proton uptake and adenosine triphosphate formation, *J. Biol. Chem.* **249**:662–663.

Schauf, C. L., and Bullock, J. O., 1979, Ion channels in membranes: the physical basis of excitability in nerve and muscle, *Sci. Prog.* **66**:231–248.

Schindler, H., 1980, Formation of planar bilayers from artificial or native membrane vesicles, *FEBS Lett.* **104**:157–160.

Schindler, H., and Quast, U., 1980, Functional acetylcholine receptor from *Torpedo marmorata* in planar membranes, *Proc. Natl. Acad. Sci. USA* **77**:3052–3056.

Schoch, P., Sargent, D. F., and Swyzer, R., 1979, Capacitance and conductance as tools for the measurement of asymmetric surface potentials and energy barriers of lipid bilayer membranes, *J. Membr. Biol.* **46**:71–89.

Schönfeld, M., Montal, M., and Feher, G., 1979, Functional reconstitution of photosynthetic reaction centers in planar lipid bilayers, *Proc. Natl. Acad. Sci. USA* **76**:6351–6355.

Schwartz, T. L., 1971, The thermodynamic foundations of membrane physiology, in: *Biophysics and Physiology of Excitable Membranes* (W. J. Adelman, Jr., ed.), pp. 47–95, Van Nostrand–Reinhold, New York.

Siebenga, E. and Verveen, A. A., 1972, Membrane noise and ion transport in the node of Ranvier, *Biomembranes 3*:473–482.

Sigworth, F. J., and Neher, E., 1980, Single Na^+ channel currents observed in cultured rat muscle cells, *Nature (Lond.)* **287**:447–449.

Singer, S. J., and Nicolson, G. L., 1972, The fluid mosaic model of the structure of cell membranes, *Science* **175**:720–731.

Stark, G., Ketterer, B., Benz, R., and Läuger, P., 1971, The rate constants of valinomycin-mediated ion transport through thin lipid membranes, *Biophys. J.* **11**:981–994.

Stevens, C. F., 1972, Inferences about membrane properties from electrical noise measurements, *Biophys. J.* **12**:1028–1047.

Stevens, C. F., 1977, Study of membrane permeability changes by fluctuation analysis, *Nature (Lond.)* **270**:391–396.

Stoeckenius, W., Lozier, R. H., and Bogomolni, R. A., 1979, Bacteriorhodopsin in the purple membrane of halobacteria, *Biochim. Biophys. Acta* **505**:215–278.

Tien, H. Ti, 1968, Light induced phenomena in black lipid membranes constituted from photosynthetic pigments, *Nature (Lond.)* **219**:272–274.

Tien, H. Ti, 1976, Electronic processes and photoelectric aspects of bilayer lipid membranes, *Photochem. Photobiol.* **24**:97–116.

Tien, H. Ti., 1979, Photoeffects in pigmented bilayer lipid membranes, in: *Photosynthesis in Relation to Model Systems* (J. Barber, ed.), pp. 116–173, Elsevier North-Holland Biomedical Press, New York.

Träuble, H., and Sackmann, E., 1972, Studies on the crystalline-liquid crystalline phase transition of lipid model membranes. III. Structure of a steroid–lecithin system below and above the lipid phase transition, *J. Am. Chem. Soc.* **94**(13):4499–4510.

Träuble, H., and Sackmann, E., 1973, Lipid motion and rhodopsin rotation, *Nature (Lond.)* **245**:209–211.

Trissl, H. W., and Läuger, P., 1970, Photoelectric effects in thin chlorophyll films, *Z. Naturforsch.* **25b**:1059.

Trissl, H. W., and Montal, M., 1977, Electrical demonstration of rapid light-induced conformational changes in bacteriorhodopsin *Nature (Lond.)* **266**:655–657.

Ulbricht, W., 1977, Ionic channels and gating currents in excitable membranes, *Annu. Rev. Biophys. Bioeng.* **6**:7.

Ullrich, H. M., and Kuhn, H., 1972, Photoelectric effects in biomolecular lipid-dye membranes, *Biochim. Biophys. Acta* **266**:584–596.

Verveen, A. A., and De Felice, L. J., 1974, Membrane noise, in: *Progress in Biophysics and Molecular Biology,* Vol. 28 (J. A. V. Butler and D. Noble, eds.), pp. 189–265, Pergamon Press, Oxford.

Verveen, A. A., and Derksen, H. E., 1965, Fluctuations in membrane potential of axons and the problem of coding, *Kybernetik* **2**:152–160.

White, S., 1978, Formation of "solvent-free" black lipid bilayer membranes from glyceryl monooleate dispersed in squalene, *Biophys. J.* **23**:337–347.

Woodbury, J. W., 1971, Eyring rate theory model of the current–voltage relationships of ion channels in excitable membranes, in: *Advances in Chemical Physics* (J. Hirshfelder, ed.), Vol. XXI, pp. 601–617, Wiley (Interscience), New York.

Chapter 2

Biochemistry of Chemosensory Behavior in Prokaryotes and Unicellular Eukaryotes

Barry L. Taylor and Sharon M. Panasenko

1. DIVERSITY AND UNITY IN CHEMOTAXIS

1.1. Diverse Roles of Chemosensory Responses

Chemotaxis, a migratory response to a chemical gradient, serves a variety of purposes among microorganisms by means of a corresponding diversity of chemotactic mechanisms. Even when chemotaxis serves the same end, such as migration toward sources of food, the mechanism and stimulus specificity reflect a given organism's unique needs. Bacteria such as *Escherichia coli* and *Salmonella typhimurium* have a refined chemotactic response to a variety of compounds signaling the presence of sources of carbon and nitrogen (Adler, 1975; Koshland, 1980*a;* Taylor and Laszlo, 1981). Representative amino acids and sugars are strong attractants for these bacteria (Mesibov and Adler, 1972; Adler *et al.,* 1973). Likewise, *Paramecium,* which feeds on bacteria, is attracted to various excretion products of bacterial metabolism, such as lactose, acetate, folate, and ammonium ion (for a review of chemotaxis in protozoa, see Van Houten *et al.,* 1981).

In addition to providing a means of acquiring food, chemotaxis can enable an organism to escape dangerous or harmful environments. The bacteria

Barry L. Taylor • Department of Biochemistry, School of Medicine, Loma Linda University, Loma Linda, California 92350. **Sharon M. Panasenko** • Chemistry Department, Pomona College, Claremont, California 91711.

described above are able to sense certain compounds, such as acetate and indole, as repellents and will move away from regions in which these compounds are present at high concentrations (Tsang *et al.*, 1973; Tso and Adler, 1974). It is likely that acetate and some other repellents are indicators of overcrowding in a bacterial population and that their role as repellents is to achieve dispersal of the cells before nutrients are depleted. *Paramecium* is also able to avoid unfavorable growth conditions as indicated by extremes of pH, guanidine-HCl, and tetraethylammonium (Van Houten, 1977).

Multicellular organisms have homeostatic mechanisms that obviate the need of individual cells to seek a favorable microenvironment. Instead, chemotactic behavior is displayed by specialized cells for defense or differentiation. For example, leukocytes are not attracted toward nutrients, but rather toward compounds that signal infection or damage of the host, such as formylated peptides and products of the complement cascade (for reviews see Zigmond, 1978; Snyderman and Goetzl, 1981; Schiffman, 1982).

Yet another purpose served by chemotaxis is to provide and maintain contact among cells, as required during mating or aggregation. A variety of microorganisms respond chemotactically to mating pheromones (Mascarenhas, 1978; O'Day, 1979); and several, such as *Dictyostelium discoideum* and perhaps *Myxobacteria,* undergo a chemotactically stimulated aggregation phase before multicellular development (for reviews see Loomis, 1975; Kaiser *et al.*, 1979; Gerisch, 1982). The attractants that lead to aggregation are called acrasins. Aggregation in *D. discoideum* is mediated by waves of cyclic adenosine monophosphate (cAMP) (Konijn *et al.*, 1967; Robertson *et al.*, 1972). An aggregate of 105 amebae becomes delimited by a slime sheath (O'Day, 1979). This constitutes a slug capable of coordinated taxes that promote migration to the most favorable environment. The taxes may also be mediated by waves of cAMP released by cells in the tip of the slug to guide the movement of the other cells (Durston and Vork, 1979) and by a slug-turning factor (STF) that is antagonistic to cAMP (Fisher *et al.*, 1981; Williams, 1982).

The distinction between the various roles of chemotaxis just discussed is not simply a formal one, as can be seen in organisms such as *Dictyostelium* and *Myxobacteria* that exhibit more than one of the various chemotactic behaviors. For example, *D. discoideum* is attracted by folic acid derived from bacteria on which it preys (Pan *et al.*, 1972); in addition, it is attracted by cAMP, an acrasin (Konijn *et al.*, 1967). The cells can synthesize cAMP periodically so that alternating cycles of cAMP release and chemotactic response to cAMP result in a generation of pulses of cAMP that synchronizes the activities of the cells (Alcantara and Monk, 1974; Tomchik and Devreotes, 1981; Gerisch, 1982). Aggregating *D. discoideum* cells frequently form a spiral wave connecting all cells that are releasing cAMP at the same time (Fig. 1). The specificity of chemoreception varies depending on the growth phase of the

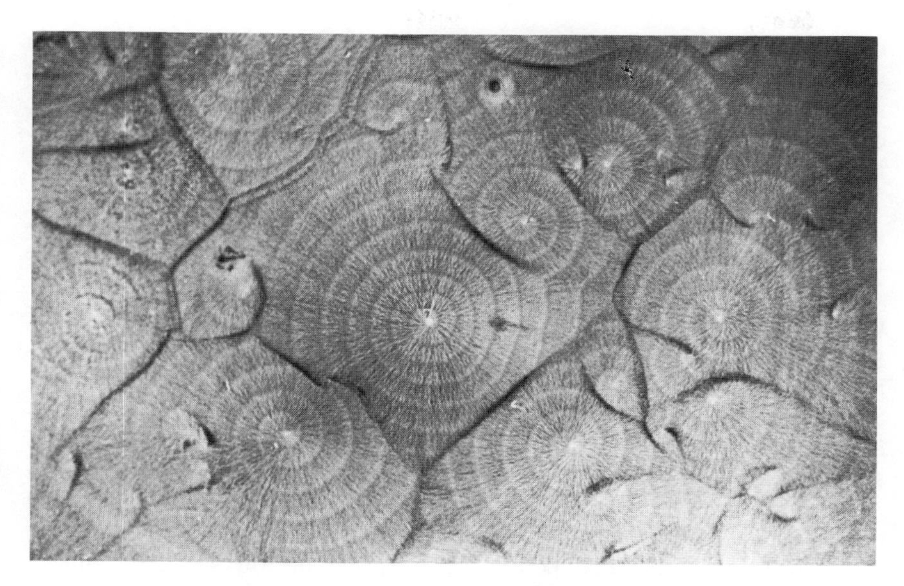

Figure 1. Organized waves of cell movement during aggregation of *Dictylostelium discoideum*. Territories are ~1 cm in diameter. (Photograph by P. N. Devreotes.)

organism so that, during vegetative growth when the cells are seeking food, the response to folate is most acute (Varnum and Soll, 1981). When the cells enter the multicellular phase of their life cycle, their responsiveness to folate declines and the number of cAMP binding sites increases dramatically (Henderson, 1975). Thus the organism suits its behavior to its needs.

In *Myxococcus xanthus,* motility is governed by two genetic systems (Hodgkin and Kaiser, 1979*a,b*). Wild-type cells can move singly or in groups. Mutants in one class are able to move only singly, whereas mutants in the other class are able to move only in groups. Because single cell movement is likely to be involved in food seeking, and movement in groups is appropriate for aggregation, dual genetic control facilitates maintenance, by selection pressure, of the two types of motility.

Just as the roles of chemotactic behavior are quite diverse, there is also great variety in the means by which the behavior is expressed. Locomotion can be the result of the rotation of helical flagella, beating of cilia, extension of pseudopods, or gliding over surfaces. Furthermore, as commonly used, the term chemotaxis encompasses a variety of behaviors consisting of quite different mechanisms for the modulation of movement. For example, chemotaxis, in its strict definition, refers to the modulation of the direction of motion or angle of turn by chemoeffectors (Diehn *et al.,* 1977). In chemoklinokinesis, the frequency of turning is altered, but the direction and angle of the turn remain random. Chemorthokinesis refers to the modulation of speed of locomotion without regard to direction. Each of these types of behavior has been observed; occasionally a given organism may display several. We use the term chemotaxis in the general sense and apply precise terminology where necessary.

Because of the diversity that exists in chemotactic responses, it is not practical to undertake an exhaustive treatment of the range of chemotactic behavior observed in nature. This review explores the common mechanistic elements in diverse responses and identifies principles of sensory transduction that are common, not only to microbial chemotaxis, but to more complex systems as well. The biochemistry of the chemosensory response has been intensively studied in bacteria (for additional reviews see Hazelbauer and Parkinson, 1977; M. S. Springer *et al.*, 1979; Koshland, 1980*a*, 1981; Taylor and Laszlo, 1981; Parkinson, 1981; Macnab, 1982; Boyd and Simon, 1982), but our knowledge of sensory transduction at the molecular level in other microorganisms is limited. As a result, the prokaryotic mechanism is usually discussed in more detail, but each section includes relevant observations regarding other microbial systems. It is hoped that this juxtaposition of information will stimulate further research.

1.2. Response–Regulator Model

The variety and complexity of chemotactic behavior may be variations on a relatively simple theme. The observed responses can be explained in terms of changes in the level of a response-regulator parameter that acts like a thermostat in controlling behavior (Koshland, 1977). The response regulator could be a small molecule, an ion gradient, or a particular conformation of a regulatory protein; the important concepts are that (1) the regulator turns the response on or off, and (2) the regulator is responsive to sensory stimuli. This is not a new concept, but Koshland (1977, 1980*a,b,* 1981) and Macnab and Koshland (1972) have been most successful in presenting this model as a unifying element in sensory systems.

Bacterial chemotaxis is readily described in terms of the response–regulator model. Swimming bacteria may assume one of two states: tumbling or smooth swimming. Net migration is governed by controlling the frequency of tumbling. When the swimming bacteria experience a favorable stimulus, such as an increase in attractant or a decrease in repellent, they suppress tumbling for a time and continue swimming in the favorable direction. Eventually tumbling is restored to the normal frequency, unless there is a further increase in stimulus level. A response regulator X is commonly designated as a suppressor of tumbling, although with appropriate changes the model would work equally well if the regulator promoted tumbling (Macnab and Koshland, 1972). Under steady-state conditions the level of response regulator (X_{ss}) fluctuates spontaneously about a critical value (X_{crit}) (Fig. 2). When X_{ss} rises above X_{crit}, tumbling is suppressed; when X_{ss} falls below X_{crit}, the bacterium tumbles. Thus fluctuation of X_{ss} accounts for the runs (i.e., smooth swimming) and tumbles in the normal swimming pattern observed in unstimulated cells. A favorable stimulus causes a transient increase in X, which then suppresses tumbling transiently; an unfavorable stimulus causes a transient decrease in X and an interval of tumbling.

The level of tumble regulation is controlled by the relative rates of for-

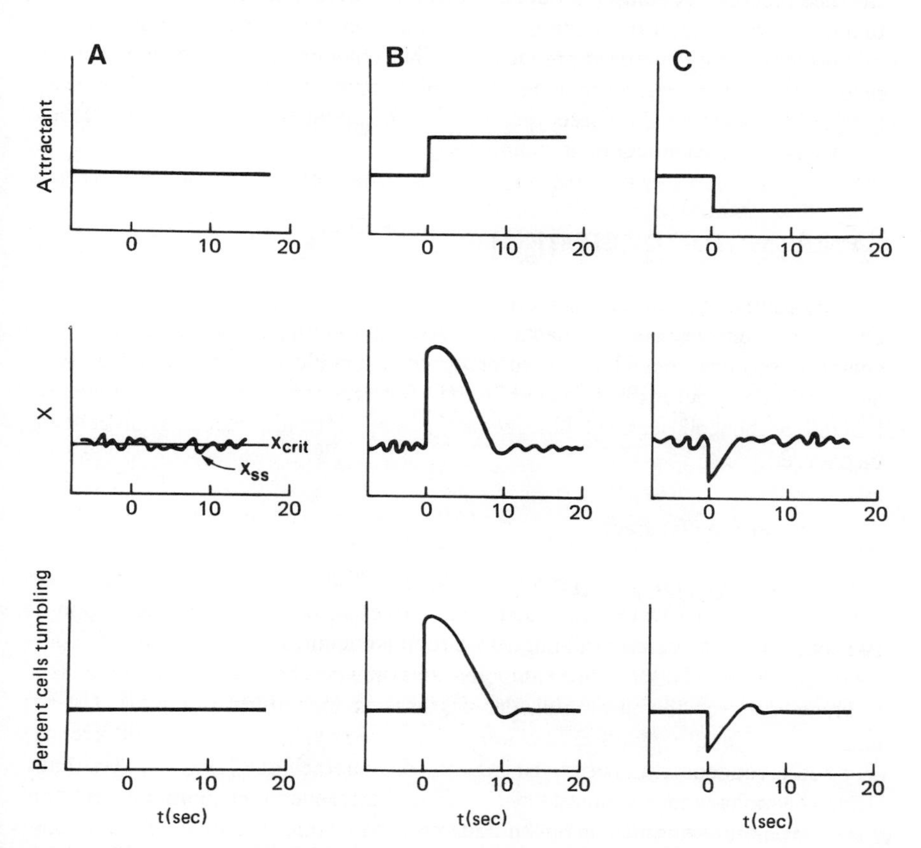

Figure 2. Transient response of bacteria to step changes in attractant concentration as explained by a response regulator model. (A) Cells preadapted to a moderate concentration of attractant. (B) Step increase in attractant concentration and transient smooth swimming response. (C) Step decrease in attractant concentration and brief tumbling response. X, Response regulator. The steady-state (X_{ss}) concentration of X in adapted cells fluctuates relative to a critical level (X_{crit}).

mation and destruction of *X*. A favorable stimulus increases the rate of both the formation and degradation of tumble regulator, but formation is stimulated faster than decomposition, accounting for the transient increase in the level of regulator (Macnab and Koshland, 1972). This is similar to a class of two-process models previously advanced to explain behavioral responses, which involve competing processes to initiate and terminate the response (Rashevsky, 1933; Hill, 1936; Delbruck and Reichardt, 1956).

Complex chemosensory behavior is explained in this model by the state of a two-way switch. In subsequent sections it will become evident to the reader that the response-regulator model has been a powerful concept in studies of the biochemistry of bacterial chemotaxis. It is also evident that this model will be invaluable in suggesting approaches to the biochemistry of chemosensory responses in other microorganisms. We will return at the end of this chapter to consider more sophisticated versions of the model that reflect the present knowledge of the molecular mechanisms.

2. PERCEPTION OF CHEMICALS

Gradients of chemicals in the environment of microorganisms are perceived by receptor proteins that have a specificity comparable to the substrate specificity of enzymes. That is, the receptor is specific for one ligand or for a small class of structurally similar ligands. The receptor proteins are generally located on the cell surface, but recently some internal receptors have been discovered.

2.1. Sensing the Gradient

Bacteria respond to a gradient of chemoeffector rather than to the absolute concentration. Over a limited range in effector concentration, the response is proportional to relative change in effector concentration (Dahlquist *et al.*, 1972; Koshland, 1980*a*). An analogous relationship, the Weber–Fechner law, is known to describe the magnitude of sensation in higher organisms. In the shallow gradients to which *E. coli* and *S. typhimurium* respond, the difference in effector concentration over the length of a bacterium (2 μm) is less than 0.01% (Macnab and Koshland, 1972). This difference is insignificant relative to the statistical fluctuations in the distribution of effector molecules at a concentration of 10^{-6} *M*. Instead of relying on comparison of stimuli detected simultaneously (e.g., spatial sensing), a bacterium compares the present with the immediate past (e.g., temporal sensing). The swimming of the cell is essential in this transposition from a gradient in space to a gradient in time, because it results in an ability to compare effector concentrations at sites separated by many times the cell's length. The cell's discrimination problem is thus reduced considerably, and the problem posed by inhomogeneities in the gradient is diminished.

Paramecium has a well-documented avoidance reaction in which the ciliate backs away from a stimulus and then swims forward in a direction selected at random (Jennings, 1906). This may be considered a form of spatial sensing, because an avoidance response is triggered by mechanical stimulation of the anterior end but not of the posterior end (Naitoh and Eckart, 1969). However, *Paramecium* also shows characteristics of a temporal sensing mechanism: When subjected to a step increase in attractant it responds and then adapts to the new conditions (Dryl, 1973; Van Houten, 1978). This appears to be the dominant mechanism for sensing chemicals in *Paramecium,* as it responds to chemoeffectors by altering the frequency of turns (klinokinesis) and swimming speed (orthokinesis) and travels many body lengths before adjusting its path. Temporal sensing has also been demonstrated in another ciliate, *Tetrahymena thermophilia* (Almagor *et al.*, 1981).

True chemotaxis with oriented movement is more efficient than kinesis in microorganisms that are large relative to the spatial fluctuations of chemoeffector. Kinesis is efficient when the size of the organism is in the range in which spatial fluctuations of the effector concentration are significant. For example, chemokinesis is more energy efficient in *E. coli,* but chemotaxis is more efficient in leukocytes.

Because leukocytes have distinct morphological features (Fig. 3), it is possible to determine the direction of their locomotion from their morphology. Analysis of leukocyte orientation and movement in concentration gradients indicates that leukocytes sense a gradient before moving and thus may use a spatial sensing mechanism (Zigmond, 1977). There are, however, conditions under which a response to a temporal gradient can be observed (Zigmond and Sullivan, 1979). Leukocytes, when subjected to a temporal gradient by rapid mixing, undergo a marked change in cellular morphology that decays over time, in a manner reminiscent of the temporal response of chemotactic bacteria. The significance of this phenomenon is not yet understood, but it would be prudent to have some reservations about the generally accepted view that leukocytes employ a spatial sensing mechanism.

It is likely that *D. discoideum* has a spatial sensing mechanism because application of cAMP causes these cells to extend pseudopods in the direction of the source and then to move toward it (Gerisch *et al.*, 1975). However, other interpretations are possible; in fact, *D. discoideum* shows several transient components in its response to cAMP that are characteristic of adaptation and may indicate that temporal sensing also occurs (Rossier *et al.*, 1980).

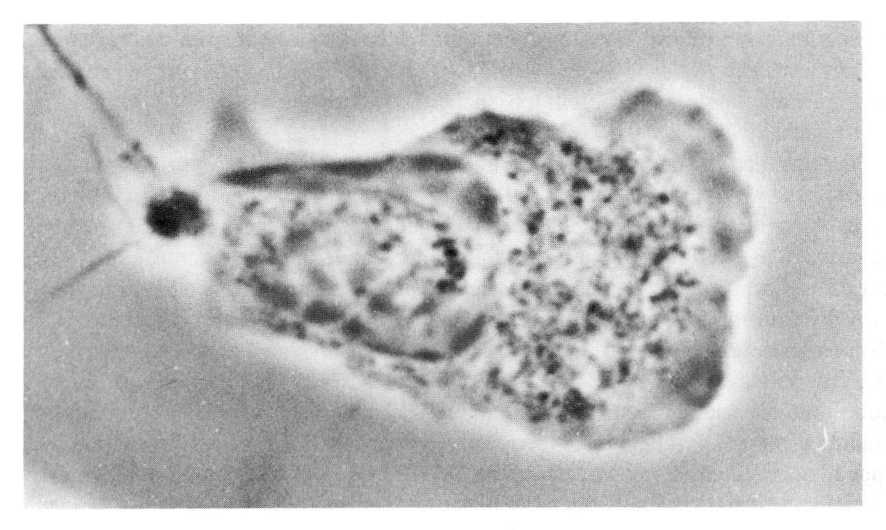

Figure 3. Migrating horse neutrophil in a spatial gradient of effector. The cell was moving from left to right. Note the broad advancing cell border with underlying zone of clear cytoplasm. Granules are numerous in the central part of the cell. The nuclear lobes are seen toward the rear of the cell. At the trailing margin is seen a knoblike tail from which strands are radiating. Phase-contrast micrography (×1600). (From Zigmond and Hirsch, 1972.)

2.2. Chemoreceptor Proteins on the Surface of Bacteria

In the envelope of the gram-negative bacterium, the plasma membrane is surrounded by the cell wall or by peptidoglycan, a thin, rigid layer that maintains cell shape and prevents lysis in a hypotonic environment. The outer membrane adheres to peptidoglycan and is the first permeability barrier encountered by molecules entering the cell. Hydrophilic pores in the outer membrane freely admit polar molecules smaller than ~600 daltons (Nikaido and Nakae, 1979). Between the inner membrane and the cell wall is an aqueous layer of uncertain, and perhaps variable, dimensions known as the periplasm. Two classes of chemoreceptors are found on the bacterial cell surface: soluble receptors in the periplasm and receptors that are integral proteins in the cytoplasmic membrane.

In *E. coli* there are about 25 receptors for attractants and an undetermined, but smaller, number of receptors for repellents (Adler, 1975; Hazelbauer and Parkinson, 1977; Koshland, 1980*a*). Some of these receptors have been isolated and characterized, while others have been identified by competition studies measured with a behavioral assay; thus the evidence for each is not equally strong. Definitive identification of a chemoreceptor is best accomplished by genetic techniques. This has been a powerful tool in bacterial studies, and there is likely to be rapid progress in the investigation of sensory phenomena in eukaryotes with increasing understanding of the genetics of eukaryotic microorganisms.

The best characterized receptors, i.e., the D-ribose-, D-galactose-, and maltose-binding proteins, are all periplasmic, are readily released from the bacterium by osmotic shock, and function in both active transport and chemotaxis (Anraku, 1968; Heppel, 1969; Aksamit and Koshland, 1972; Willis and Furlong, 1974; Kellerman and Szmelcman, 1974; Hazelbauer, 1975). In each case the purified receptor has a molecular weight of 30,000–40,000 and is a monomer with one sugar-binding site (Zukin *et al.*, 1977*b;* Hazelbauer and Parkinson, 1977). There may be specific structural similarities between these sugar receptors and other nonchemotactic soluble binding proteins for sugar transport (Parsons and Hogg, 1973; Quiocho *et al.*, 1977). Definitive structural information from X-ray crystallography will soon be available for direct comparison of receptors for sugar transport and chemotaxis in *E. coli* (Quiocho *et al.*, 1979; Alber *et al.*, 1981).

Some of the earlier competition studies (Mesibov and Adler, 1972; Adler *et al.*, 1973) suggested a broad specificity for many chemoreceptors, but definitive studies have now determined that receptors are highly specific for ligands. D-Glucose and D-galactose, which differ only by inversion at the C-4 position, bind with high affinity to the galactose receptor with dissociation constants of 10^{-7}. Sugars with slightly greater differences in structure, i.e., D-arabinose, lactose, and D-fucose, have affinities 1000-fold less (Zukin *et al.*, 1977*b,* Koshland, 1980*a*). Other sugars do not bind appreciably to the galactose receptor. The serine receptor shows nearly absolute specificity for L-serine (Clarke and Koshland, 1979). Even L-alanine and L-homoserine bind so weakly to the serine receptor that these interactions are of no physiological significance. Such specificity is typical of receptors that have been studied in detail, including eukaryotic receptors. For this reason, it will be surprising if chemosensory (e.g., olfactory and gustatory) receptors in humans are not similarly specific (Taylor and Laszlo, 1981).

All known periplasmic chemoreceptors are sugar receptors and have a dual role both as chemoreceptors and as receptors for the active transport of the sugars (Hazelbauer and Adler, 1971; Aksamit and Koshland, 1974; Hazelbauer, 1975). Periplasmic binding proteins for amino acid transport appear not to function as chemoreceptors (Ordal *et al.*, 1978; Schellenberg, 1978). In fully induced bacteria about 10^4 copies of each periplasmic chemoreceptor are present (Strange and Koshland, 1976). This is an appropriate number of receptors to balance the cell's need for efficient detection against the space requirements for 40 different receptors (Berg and Purcell, 1977; Koshland, 1980*a*). The periplasmic location ensures protection of these receptors as well as retention inside the cell wall. The relatively high (10^{-3}–10^{-4} M) concentration of receptors is also effective in retaining attractant in the periplasm when the attractant concentration in the external environment is very low (Willis and Furlong, 1974; Silhavy *et al.,* 1975). At attractant concentrations of 10^{-6} M, which bacteria readily detect, there would be, on the average, about one molecule of free attractant in the periplasm of each bacterium, if the receptors were absent. If there were only a few receptors in the periplasm, perception of effectors would obviously be a hit-or-miss proposition; the response would be drastically affected by local fluctuations. The retention of attractant by the high concentration of receptors buffers against these local fluctuations.

Other surface chemoreceptors are integral membrane proteins. In *S. typhimurium* and *E. coli* these include the high-affinity serine and aspartate chemoreceptors, the Tsr and Tar proteins, which also function in transduction of the signal from periplasmic receptors, transmitting the signal from the outside to the inside of the cell (Clarke and Koshland, 1979; Hedblom and Adler, 1980; Wang and Koshland, 1980). The glucose and mannitol chemoreceptors are representative of another class of sugar receptors that are hydrophilic but attached to the surface of the plasma membrane (Adler and Epstein, 1974). These chemoreceptors also have a dual role and are the sugar-specific receptor (enzyme II) in the phosphotransferase active-transport system.

2.3. Internal Receptors in Bacteria

No chemoreceptors for repellents have been isolated or characterized, and it now appears that many repellents have an indirect effect on the chemotactic system. Membrane-permeant organic acids, such as acetate, are potent repellents of *S. typhimurium* and *E. coli* and act by decreasing the cytoplasmic pH level in the bacteria (Kihara and Macnab, 1981). The response is mediated by an internal H^+ receptor, either in the cytoplasm or on the cytoplasmic side of the inner membrane. In addition, there appears to be a surface H^+ chemoreceptor (Slonczewski *et al.,* 1982).

Bacteria detect and respond to changes in the proton motive force (pmf), which is the sum of ΔpH and $\Delta\psi$, the electrical potential across the membrane. Uncouplers, respiratory inhibitors, and ionophores that decrease the membrane pmf in *B. subtilis* and *S. typhimurium* also induce a tumbling response similar to that elicited by repellents (Ordal and Goldman, 1975, 1976; Miller and Koshland, 1977, 1980; Taylor *et al.,* 1979). An increase in pmf induces smooth swimming similar to the response to an attractant. This pmf-mediated response is involved in the physiological response of bacteria to proline and to oxygen and to other electron acceptors for the respiratory chain, such as nitrate and fumarate (Taylor *et al.,* 1979, Laszlo and Taylor, 1981; Clancy *et al.,* 1981).

The best documented pmf-mediated response is aerotaxis, i.e., chemotaxis to oxygen. Increased binding of oxygen to the terminal oxidase of the respiratory chain stimulates electron transport, thereby increasing the pmf (Laszlo and Taylor, 1981; Fandrich and Laszlo, 1981; Laszlo, 1981). The bacteria detect the change in pmf and respond by decreasing the probability of tumbling. The responses of *S. typhimurium* to oxygen can be mediated by either ΔpH or $\Delta\psi$ as predicted for a pmf-mediated response (Shioi *et al.*, 1982*a*), and aerotaxis is inhibited by a voltage clamp across the membrane (Taylor and Shioi, 1982). The identity of the pmf sensor is unknown, but it is likely that it spans the membrane, permitting comparison of the electrochemical potential of the proton on the inside and outside of the cell.

2.4. Chemoreceptors in Eukaryotes

2.4.1. Leukocytes

Leukocytes respond to a number of chemoeffectors, including formylated peptides derived from bacteria (Schiffman *et al.*, 1975), a complement cleavage product, C5a (Fernandez and Hugli, 1978), lymphokines produced by lymphocytes that have been stimulated by antigens and mitogens, and derivatives of the lipoxygenation of arachidonic acid known as leukotrienes (Goetzl and Pickett, 1980). Leukotrienes released from neutrophils in response to external attractants, such as formylated peptides, may serve to amplify the chemotactic response to the initial stimulus (Snyderman and Goetzl, 1981; Schiffmann, 1982). Studies comparing the structure–activity relationships of several synthetic formylated peptides indicate that the response of leukocytes is mediated by highly specific receptor proteins (Freer *et al.*, 1980). Membrane-bound receptors have been identified for formylated peptides (Williams *et al.*, 1977; Aswanikumar *et al.*, 1977; Snyderman and Fudman, 1980) and for C5a (Chenoweth and Hugli, 1978). Partial characterization of the neutrophil receptor for formylated peptides (Niedel *et al.*, 1980; Schiffmann *et al.*, 1980) has been followed by the isolation of three polypeptides with fMet-Leu-Phe-binding activity (Goetzl *et al.*, 1981). One polypeptide (MP2) has the same specificity as the neutrophil receptor *in vivo*. It would be interesting to ascertain whether leukocyte chemoreceptors resemble the serine and asparate receptors in bacteria and serve as focusing elements mediating the response to additional stimuli.

The number of receptors for serine and aspartate in *E. coli* and *Salmonella* is influenced by neither prior exposure nor adaptation to the attractant. However, in leukocytes, receptor–chemoeffector complexes may be internalized by endocytosis (Niedel *et al.*, 1979; Sullivan and Zigmond, 1980; Vitkauskas *et al.*, 1980). In this way, high concentrations of attractants may actually decrease the availability of binding sites, and this down regulation of chemoreception may be, in part, responsible for the densitization of leukocytes to further stimulation (Donabedian and Gallin, 1981). No strict correlation has been found between the number of receptors and the ability of cells to respond, however, and desensitization is likely to be a complicated process. Increased chemoreceptor activity has also been observed after treatment of leukocytes with agents that affect the properties of the membrane, such as aliphatic alcohols (Liao and Freer, 1980). The physiological significance of up- and downregulation by leukocyte chemoeffectors has not been determined.

2.4.2. *Dictyostelium discoideum*

Slime molds have a common food receptor, the folate receptor, and possibly a common environmental sensor, the receptor for slug-turning factor, but the receptors that enable the cells to find each other may be different in different genera. For example, the acrasin is cAMP in *D. discoideum* (Konijn *et al.*, 1967) and *N*-propionyl-γ-L-glutamyl-L-ornithine-δ-lactam ethyl ester (glorin) in *Polysphondylium violaceum* (Shimomura *et al.*, 1982). Specific receptors for cAMP and folate have been identified in *Dictyostelium discoideum*, even though the study of chemoreception was complicated by degradative enzymes on the surface that break down both cAMP and folate as part of the chemosensory process (Malchow and Gerisch, 1974; Green and Newell, 1975). Use of dithiothreitol, which inhibits phosphodiesterase, has permitted investigators to characterize the structural requirements for cAMP binding and to determine that these requirements differ from those of cAMP-binding components of protein kinases (Mato *et al.*, 1978). However, cAMP binding to solubilized receptors has not yet been achieved, and the receptor has not been purified.

The number of cAMP receptors increases by one order of magnitude during development, as a result of the interaction of cells with external cAMP (Henderson, 1975; Roos *et al.*, 1977). Downregulation of the number of cAMP-binding sites also occurs when cells are exposed to high concentrations of cAMP (Klein and Juliani, 1977). As in the case of leukocytes, this may be a form of desensitization. The responsiveness of cells to folate is reduced considerably during aggregation (Pan *et al.*, 1972).

2.4.3. *Paramecium*

Behavioral studies have investigated effects of one attractant on the ability of *Paramecium* to respond to a second attractant (see Van Houten, 1981). The results indicate that chemoreception in these cells is highly specific and is probably initiated via binding of chemoeffector to receptor proteins. Chemoeffectors generate a transient receptor potential—a change in membrane potential graded to the magnitude of the stimulus—that triggers an action potential (see Kung and Saimi, 1982). As yet, however, individual receptors have not been characterized.

3. SIGNAL TRANSDUCTION

The binding of attractant to the appropriate chemoreceptors produces a signal that is somehow transmitted across the membrane into the cell. Within the cell, the signal is processed so that it alters the level of response regulator, and thereby behavior. This sequence of reactions is known as excitation.

3.1. Focusing the Signals in Bacteria

The first step in signal transduction for chemotaxis is assumed to involve a conformational change in the receptor. The clearest evidence for ligand-induced conformational change in chemoreceptors is that from physical studies by Zukin and co-workers (Zukin *et al.*, 1977*a*, 1979; Zukin, 1979). Binding of galactose to its receptor perturbed both the fluorescence of a tryptophan residue presumed to be near the binding site and an extrinsic fluorophore separated from the tryptophan residue by 40 Å. The conformational change induced by binding the 5-Å-diameter galactose must be propagated at least 30 Å through the receptor molecule. Fluorescence and circular dichroism studies indicate a similar conformational change in the ribose and maltose receptors (Zukin *et al.*, 1979; Zukin, 1979). Furthermore, the maltose receptor binds to the Tar signaling protein in the presence, but not in the absence, of maltose (Koiwai and Hayashi, 1979; Richarme, 1982). Interestingly, a membrane potential is apparently required for receptor binding (Richarme, 1982).

With more than 25 types of receptors sampling the environment, bacteria must process and integrate information from diverse, and sometimes conflicting, stimuli. This is achieved, in part, by focusing the sensory pathways, in a series of stages, so that they converge to elicit a net response to the algebraic sum of the stimuli (Spudich and Koshland, 1975). Focusing begins at the receptor level, where two substrates may compete for binding to the receptor. For example, the galactose receptor can bind either glucose or galactose, but not ribose, as described earlier. Further focusing occurs in competition between the galactose receptor and ribose receptor for binding to the Trg signaling protein that transmits the signal to the inside of the cell (Fig. 4). Evidence suggesting such focusing was first obtained from competition studies by Mesibov and Adler (1972), subsequently confirmed by Strange and Koshland (1976), who demonstrated that ribose inhibited chemotaxis toward galactose, but only

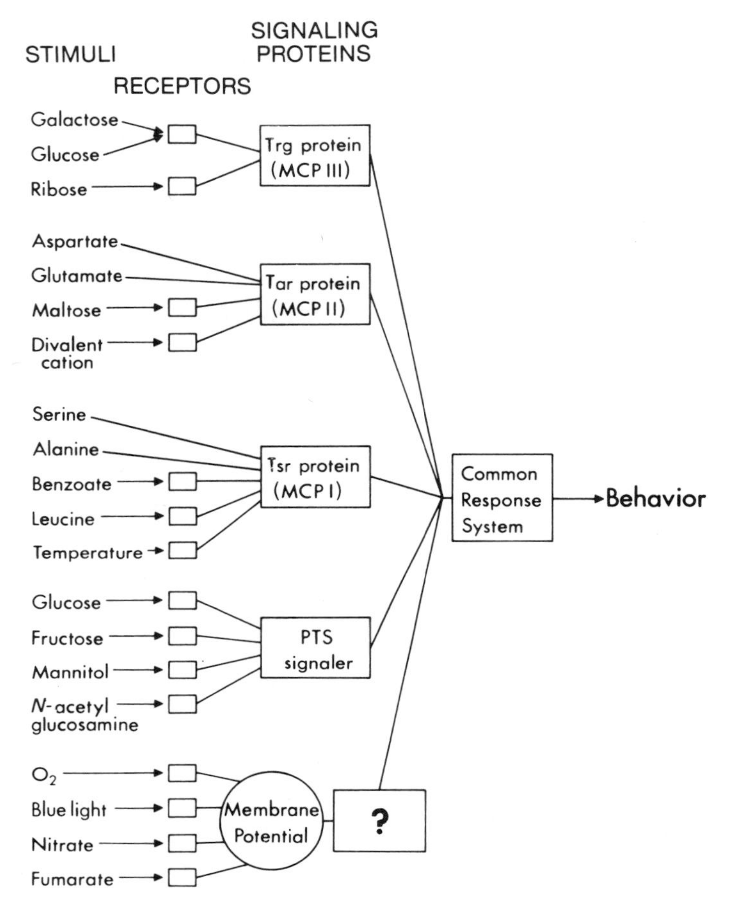

Figure 4. Pathways for focusing selected chemotactic stimuli into a common response system. PTS, Phosphotransferase system.

if the ribose receptor was present. A mutation in *trg* abolishes taxis to both ribose and galactose (Ordal and Adler, 1974; Fahnestock and Koshland, 1979; Hazelbauer and Harayama, 1979).

In addition to the Trg protein, at least four other signaling proteins are involved in focusing, of which the tsr and tar proteins are the best characterized (Fig. 4). The latter are multifunctional proteins involved in perception, signal transmission, and adaptation. For example, the Tsr protein binds serine and alanine at separate active sites and somehow mediates sensory signals from the internal H^+ receptor and from the thermoreceptor (Clarke and Koshland,

1979; Maeda and Imae, 1979; Hedblom and Adler, 1980; Kihara and Macnab, 1981). The Tar protein binds aspartate, the maltose–receptor complex, and mediates the response to metal cationic repellents (M. S. Springer *et al.*, 1977*b;* Silverman and Simon, 1977). The product of the *tap* gene is similar to the Tar protein and is believed to be a signaling protein, but the physiological role of tap is undetermined (Boyd *et al.*, 1981; Wang *et al.*, 1982). Additional signaling proteins are required for chemotaxis to oxygen and to sugars transported by the phosphotransferase system (Hazelbauer and Engström, 1980; Niwano and Taylor, 1982*a*). The latter signaling proteins have not yet been identified, however. A scheme illustrating focusing through the signaling proteins and convergence to a common tumble-regulating system is shown in Fig. 4. The role of the signaling proteins in adaptation is discussed in Section 5.

The excitation process by which the signal from the signaling proteins elicits a behavioral response remains a mystery. The signaling proteins appear to be distributed independently over the cell surface and are not physically associated with the basal bodies of the flagella (Engström and Hazelbauer, 1982). On the other hand, the CheC protein, which can function as a switch on the flagellar motor, is part of the basal body (Silverman and Simon, 1973; Warrick *et al.*, 1977; Rubik and Koshland, 1978; DeFranco and Koshland, 1982). In transmitting the signal to the common tumble-regulating system, the signaling protein could induce a conformational change in another protein, catalyze the formation of a metabolite, or open an ion channel. One clue as to the mechanism of excitation is provided by the elegant work of Berg and associates, in which attractant was applied iontophoretically to tethered *E. coli* (Segall *et al.*, 1982). A normal latency of 200 msec was observed between the addition of attractant and the commencement of a response as evidenced by a reversal of the flagellar motors. In *che*Z strains the latency was about 2 sec, so it is possible that the CheZ product is involved in excitation.

3.2. Mechanism of Excitation in *Paramecium*

In contrast to the situation in other chemotactic systems, excitation and signal transduction is rather well understood in *Paramecium* (for a detailed review, see Doughty and Dryl, 1981; Kung and Saimi, 1982). The response to touch is initiated by a transient receptor potential that depolarizes the membrane and triggers a second depolarization of the ciliary membrane (the action potential) (Fig. 5). The action potential is the result of the opening of a Ca channel, which permits influx of Ca^{2+}, followed by a delayed opening of a K^+ channel and efflux of K^+ (Naitoh *et al.*, 1972). The Ca^{2+} channel has been shown to be located only in the ciliary membrane (Ogura and Takahashi, 1976; Dunlap, 1977), whereas the channels responsible for the initial receptor potential and the delayed K^+ current are thought to be located in the body mem-

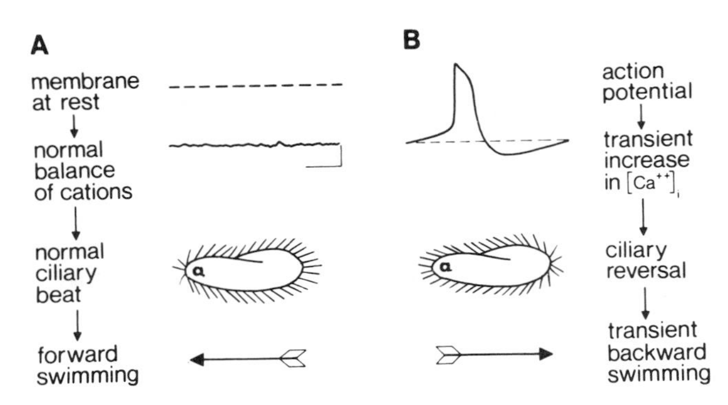

Figure 5. Relationship of membrane potential and locomotor behavior of *Paramecium*. (A) Cell at rest. Tracing represents a membrane potential of ~ -25 mV. (B) Cell stimulated with barium. Note the relationship of action potential with reversed beating of cilia, (a) anterior. (From Kung *et al.*, 1975.)

brane of the cell (Machemer and Ogura, 1979). In this manner the electrical signal occurs throughout the membrane, but the increase in Ca^{2+} concentration is limited to the cilia where, in fact, the Ca^{2+} is apparently sequestered. The rise in Ca^{2+} concentration during the action potential has been shown to cause the cilia to beat in reverse with increased frequency (Machemer, 1974*a*). The mechanism of this effect of Ca^{2+} on motility is not yet known, although several proposed mechanisms will be discussed briefly in Section 4.

The action potential is not an all-or-none phenomenon, but is graded. As a result, different degrees of membrane depolarization or hyperpolarization can be achieved; it has been suggested that these degrees of change in membrane potential determine the balance that will be struck between the various behavioral responses, e.g., increased or decreased speed and increased or decreased frequency of avoiding reactions (Van Houten, 1978, 1979, 1981). In *Tetrahymena pyriformis,* another protozoan, there is also evidence that depolarization of the membrane controls negative chemotaxis (Tanabe *et al.,* 1980).

3.3. Signal Transduction in Leukocytes and *Dictyostelium*

3.3.1. Leukocytes

The role of Ca^{2+} during signal transduction in leukocyte chemotaxis has been extensively investigated, and it is clear that chemotactic stimuli may have multiple effects on Ca^{2+} movement. Studies of $^{45}Ca^{2+}$ uptake indicate attractant-stimulated transport of Ca^{2+} (Boucek and Snyderman, 1976; Naccache *et al.,* 1977). However, increased levels of cytoplasmic Ca^{2+} are also the result of release of internally sequestered pools (Naccache *et al.,* 1979). A model (Fig. 6) has been proposed in which the release of internal Ca^{2+} is an initial event in signal transduction (Becker and Stossel, 1980; Schiffman, 1982). The transient rise in $[Ca^{2+}]$ generated in this way is then the trigger for a variety of reactions involving phospholipid metabolism leading to the influx of external Ca^{2+}. These findings are consistent with a role of Ca^{2+} in excitation and signal transduction similar to that just discussed for *Paramecium.* Indeed, Ca^{2+} may have a direct influence on motility in leukocytes.

In leukocytes, intracellular levels of cAMP are affected by the presence of chemotactic stimuli (Simchowitz *et al.,* 1980). Furthermore, a Ca^{2+}-dependent kinase is activated by the increased levels of cAMP and phosphorylates a specific 90,000-dalton protein (Wedner *et al.,* 1980). It is therefore tempting to suggest that cyclic nucleotides are part of the signal-transduction mechanism. However, the significance of these events has not yet been demonstrated.

Depolarization or hyperpolarization of the cytoplasmic membrane repre-

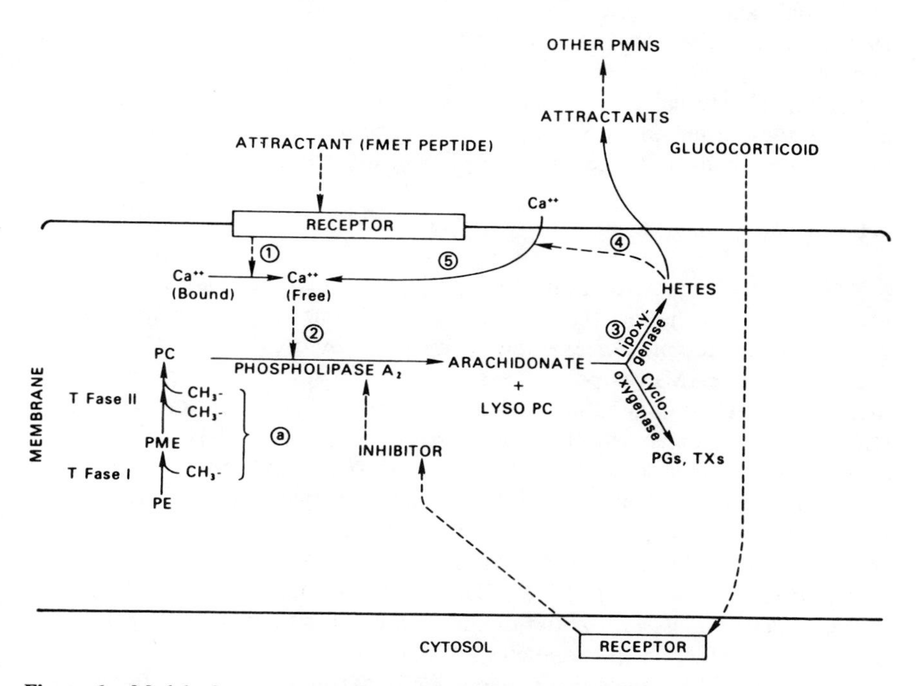

Figure 6. Model of sensory transduction in leukocytes. FMET, formylmethionine; HETEs, hydroxyeicosatetraenoic acids; PC, phosphatidylcholine; PE, phosphatidylethanolamine; PGs, prostaglandins; PME, phosphatidylmonoethylethanolamine; PMNs, TFase, methyltransferase; TXs, thromboxanes. (From Schiffmann, 1982.)

sent other possible means of signal transduction, and a brief depolarization has been observed after exposure of leukocytes to C5a and formylated peptides (Gallin *et al.,* 1978). The change is apparently the result of Ca^{2+} and Na^{+} influx followed by a Ca^{2+}-triggered K^{+} efflux. Aside from the effects of Ca^{2+}, no direct biochemical linkages have been documented, however, between these changes in membrane potential and the motility apparatus.

Changes in phospholipid metabolism have been observed during chemotaxis in leukocytes (Pike *et al.,* 1978; Schiffman *et al.,* 1979). These changes are apparently initiated by the effects of chemoattractants on the methylation of phosphatidylethanolamine to phosphatidylcholine and the release of arachidonic acid. Arachidonic acid can then be converted to a variety of products including hydroxyeicosatetraenoic (HETE) acids, and these may then be released from the cell, where they may serve as chemoattractants. Alternatively, they may be incorporated into cellular phospholipids or proteins and thus modulate membrane properties or enzymatic activities required for chemotaxis. For example, the finding that inhibition of lipooxygenation of arachidonic acid suppressed Ca^{2+} uptake during chemotaxis (Naccache *et al.,* 1979), suggests that these reactions may serve a vital role during signal transduction as the link between chemoreceptor binding, release of internal Ca^{2+}, and influx of external Ca^{2+}. There are a number of complicated relationships between these events, however, and the detailed picture of the effects of phospholipid metabolism on chemotactic behavior is far from clear.

3.3.2. *Dictyostelium*

Study of the excitation phase of the response of *D. discoideum* to cAMP is complicated by the fact that chemotaxis is only one of several responses elicited by the attractant. For example, stimulation by cAMP causes production and release of cAMP as part of a signal relay system (Shaffer, 1975). The increase in cAMP can be blocked by caffeine with no effect on chemotaxis (Brenner and Thoms, 1984). Thus increases in adenylcyclase activity after stimulation are caused by the signal relay response and are not involved in chemotaxis itself. However, several reactions have been observed that do display the properties expected of an excitatory system of signal transduction and that fit into a unified scheme. For example, stimulation with cAMP causes an influx of Ca^{2+} (Wick *et al.,* 1978). Calmodulin has been identified in *Dictyostelium* (Clarke *et al.,* 1980), and the interaction of calmodulin with Ca^{2+} may have a direct effect on the organelles of motility.

Internal cGMP levels also rise transiently during the chemotactic response of *Dictyostelium* (Mato *et al.,* 1977) due to increased activity of guanyl cyclase (Mato and Malchow, 1978). A cGMP phosphodiesterase-deficient mutant shown to have prolonged elevation of cGMP and a prolonged chemotactic response has been reported (Van Haastert *et al.*, 1982*b*). These results are promising, and it is apparent that cGMP may act via a cGMP-sensitive protein. The presence of a cGMP-binding protein has been demonstrated, and the kinetics of cGMP-binding suggest that the protein may be involved in chemotaxis (Van Haastert *et al.*, 1982*a*).

4. BEHAVIORAL RESPONSES

The chemotactic response consists of net migration of an organism toward or away from stimuli. The mechanisms by which chemotactic organisms move may differ enormously, however. Some motility mechanisms, such as flagellar rotation in *E. coli* and *Salmonella* and the beating of cilia in *Paramecium*, are understood at the molecular level, but others, for example, gliding in myxobacteria, remain obscure. The various mechanisms are highlighted in this discussion by way of illustrating our theme of diversity and unity in chemotaxis. For in-depth treatments of the various motility types, the reader is referred to several excellent reviews on *Dictyostelium* (Clarke and Spudich, 1977; Poff and Whitaker, 1979), on *E. coli* and *Salmonella* (Macnab, 1979), on gliding (Burchard, 1980), and on leukocytes (Stossel, 1978; Snyderman and Goetzl, 1981).

4.1. Klinokinesis in Bacteria

Peritrichous bacteria such as *E. coli* and *S. typhimurium* propel themselves through liquid media by rotating flagella that protrude from the surface of the cell (Berg and Anderson, 1973; Silverman and Simon, 1974). The flagellum is composed of a long, helical filament attached to the inner and outer membranes via a specialized structure resembling a hook (DePamphilis and Adler, 1971). The individual subunits of the filament may assemble into one of several different helical conformations, the most probable conformation having a left-handed twist. Movement is achieved when the individual flagella form a bundle at the "rear" of the bacterium and rotate within the bundle to propel the bacterium forward (Macnab, 1977; Macnab and Ornston, 1977). Normally, the flagella rotate in a counterclockwise direction, which promotes formation of a stable bundle and enables the bacterium to swim forward smoothly. However, periodic reversal of the direction of rotation causes the bundle to fly apart, and the bacterium is seen to tumble briefly until the bundle can be reorganized. This reorganization occurs when normal counterclockwise rotation is resumed. If, however, clockwise rotation is prolonged, the flagella are forced into a second, right-handed helical formation, which promotes formation of a bundle once again along with restoration of a slower, smooth swimming pattern. In either event, smooth, swimming and tumbling behaviors are seen to be the result of continuous rotation interrupted by brief reversals in direction of rotation.

The energy required to drive the flagellar rotation is derived from the membrane proton motive force (Larsen *et al.*, 1974; Manson *et al.*, 1977, 1980; Matsuura *et al.*, 1977; Khan and Macnab, 1980*b;* Shioi *et al.*, 1980). Neither the nature of the motor itself nor the identity of the molecule(s) responsible for regulating tumble frequency is known.

How does the swimming behavior result in the directed movement required for chemotaxis? Tumbling enables the bacterium to change its direction, although the direction assumed at the end of a tumble is random with respect to the initial direction. During the response to chemoeffectors, the frequency of tumbling is altered (Berg and Brown, 1972). When traveling in a favorable direction, the frequency of tumbling is reduced and the bacterium continues to travel in the favorable direction. If the direction of travel is unfavorable, the frequency of tumbling is slightly increased. Since each tumble provides an opportunity for the bacterium to resume swimming in a more favorable direction, this regulation of tumble frequency causes net migration by *biased random walk* (Macnab and Koshland, 1972; Berg and Brown, 1972). Thus it can be seen that a rather complex behavioral response can be elicited by means of a mechanism involving a simple switch to determine the sense of rotation of the flagella and a regulator to control the probability of switching. Note that this type of behavior is well suited to a temporal gradient-sensing mechanism whereby the organism cannot predict which direction to take.

4.2. Klinokinesis and Orthokinesis in Ciliates

Ciliated organisms such as *Paramecium* display a more complicated behavioral response. The organisms propel themselves by beating their cilia in a coordinated manner similar to the stroking of oars. The speed of swimming is determined by the frequency and angle of beating which, as we have seen, can vary as a function of membrane potential (Eckert, 1972). In addition, the beat can be completely reversed, again as a result of changes in membrane potential, which causes the organism to stop forward movement, back up, and swim off in a new direction. This phenomenon has been called the avoidance reaction. Thus a remarkable similarity exists between the avoidance reaction and tumbling in bacteria.

One might therefore predict that oriented movement could be achieved in *Paramecium* by a biased random walk mechanism; in fact, under certain conditions, such a mechanism has been demonstrated. For example, a class of attractants and repellents has been identified that cause modulation of the frequency of the avoidance reaction (Van Houten, 1977). The result is analogous to the situation in bacteria in that increased frequency of tumbling caused by repellents promotes movement away from the source, whereas decreased frequency caused by attractants promotes movement toward the source. Mutants that are unable to reverse the beat of their cilia do not respond to these compounds. However, in addition to this klinokinesis, *Paramecium* displays a second type of behavior, orthokinesis, based on the principle that compounds that decrease its swimming speed will cause accumulation, whereas those that increase its speed will promote dispersal (Van Houten, 1978).

The picture is complicated by the fact that both speed of swimming and direction of beat are controlled by the same factor: changes in potential. A compound may cause a decrease in frequency of the avoidance reaction and a simultaneous increase in speed. Whether the organism responds to the compound as an attractant or repellent will depend on a delicate balance of these effects (Van Houten, 1979). Apparently, as discussed in Section 3.2, the extent of depolarization or hyperpolarization of the membrane is the governing factor in this balance.

The cilia of *Paramecium* are similar to the cilia of other eukaryotic cells in that they consist of an axoneme containing a 9 + 2 arrangement of microtubules (Warner, 1974). Although the ciliary proteins in *Paramecium* have not been well studied, they have been assumed to be similar to those of *Tetrahymena,* which have been thoroughly investigated. The microtubules are connected by "arms" composed of high-molecular-weight protein complexes called dyneins. The dynein arms have an associated Mg^{2+}-dependent ATPase activity; presumably ciliary beating is caused by sliding of the microtubules relative to each other and the driving force is provided by the hydrolysis of ATP by the dyneins (for review, see Blum and Hines, 1979).

The beat cycle of a cilium consists of a power stroke and a recovery stroke. The orientation of the beat has been shown to be controlled by the internal concentration of Ca^{2+} by using Triton-treated *Paramecium* with membranes that are permeable to Ca^{2+} and varying concentrations of external Ca^{2+} (Naitoh *et al.,* 1972). The power stroke is normally toward the posterior right, but an increase in intraciliary Ca^{2+} concentration changes the orientation of the beat to the anterior right, as well as increasing the beat frequency (Machemer, 1974*b*). The mechanism by which the axoneme responds to the rise in intraciliary Ca^{2+} concentration is not understood.

4.3. Tactic Behavior

Cellular slime molds such as *Dictyostelium discoideum* display ameboid movement and true taxis. It is now widely accepted that the driving force for this type of motility is generated by contraction of filaments containing actin and myosin (Clarke and Spudich, 1977). Phosphorylation of myosin heavy chains has recently been shown to inhibit the aggregation of monomers into thick filaments (Kuczmarski and Spudich, 1980). Stimulation of *Dictyostelium* with the chemoattractant cAMP causes dephosphorylation of the heavy chains of myosin (Malchow *et al.,* 1981). Reversible phosphorylation or dephosphorylation may therefore be involved in the regulation of motility of this organism. It also seems likely that the phosphorylation–dephosphorylation reactions will be seen to be linked to the changes in Ca^{2+} or cGMP, or both, that occur during excitation. Such a link has not yet been demonstrated, however, and the details of the regulation of the tactic response are unknown.

Leukocytes also display ameboid motion and migrate toward attractants by orientation toward the source, rather than by altering the frequency with which they stop or turn (Nossal and Zigmond, 1976). In addition, uniform concentrations of chemotactic factors may increase the rate of movement of leukocytes (Keller and Sorkin, 1966). It is interesting to note that whereas stimulation of swimming speed in *Paramecium* resulted in dispersal or repulsion, increased speed when coupled with directed orientation in leukocytes resulted in attraction.

Leukocytes contain actin and myosin as well as microtubules. Chemoattractants were recently shown to stimulate tyrosylation of tubulin (Nath *et al.,* 1981), but the role of microtubules in leukocyte motility is unclear. Stendahl and Stossel (1980) have proposed that actin interacting with myosin is the motor for leukocyte chemotaxis and that the state of the actin lattice is governed by a number of cofactors, including an actin-binding protein that holds the lattice in a gel. These investigators suggest that the ATPase activity of the myosin is stimulated during transition to the gel state and that this results in contraction. Another factor, gelsolin, a Ca^{2+}-binding protein, binds to the actin filaments and apparently leads to breaking of the links made by actin-binding protein and to conversion of the lattice to the sol state. Thus entry of Ca^{2+} into the cell could stimulate dissolution of the lattice, while the contractions occurring in the crosslinked gel areas would produce movement. Although a number of questions remain to be answered, the present model is one that explains the linkage between excitation and motility.

Many organisms, such as myxobacteria, exhibit gliding behavior, perhaps the least understood form of motility. Gliding, in this case, refers to movement of the organism over a solid surface without accompanying changes in cell shape. No obvious organelles of motility such as flagella have been observed. Several models have been proposed to account for this mysterious form of movement. One such model suggests that contraction of subsurface filaments may generate translocation across a surface to which the bacterial envelope may adhere (Burchard, 1980). Another proposal involves the observation that some gliding bacteria appear to have small rotary assemblies at the surface and that these spin to drive the bacterium forward (Pate and Chang, 1979). A third model suggests that adhesion sites may run along moving tracks fixed to the cell wall (Lapidus and Berg, 1982).

Myxobacteria are presumed to be chemotactic because they have been observed to locate and move toward other bacteria, which they then devour (Kuhlwein and Reichenbach, 1968). In addition, under appropriate conditions, myxobacteria undergo a pattern of aggregation and sporulation similar to that found in cellular slime molds (Kaiser *et al.,* 1979). The aggregation phase of this developmental program may also involve a chemotactic response to some as yet unidentified acrasin. However, it is possible to detect chemotaxis to isolated chemicals only under very special circumstances (Schimkets *et al.,* 1979; Ho and McCurdy, 1979); there is also some doubt as to whether chemotaxis is the guiding force in aggregation (Dworkin and Eide, 1983). Numerous mutants are available that are defective in motility and sporulation (Hagen *et al.,* 1978; Hodgkin and Kaiser 1979*a,b*), and changes in patterns of protein synthesis and modification during development are under intense investigation (Inouye *et al.,* 1979; Komano *et al.,* 1980, 1982; Panasenko, 1983). Myxobacteria have many advantages as a model system; future work should reveal interesting features of signal transduction during both chemotaxis and development in this organism.

5. ADAPTATION AND SIGNAL PROCESSING

5.1. Role of Adaptation in Sensory Responses

By adapting to continuous stimulation, an organism avoids saturation of the sensory pathways and optimizes its response to change in the immediate environment. Just as background noise from a motor may be perceived by a shift worker entering a factory at the start of a shift but is soon forgotten, until the motor stops, comparable adaptation to chemosensory stimuli has been observed in bacteria. Bacteria that experience a step increase in attractant concentration suppress tumbling and swim smoothly for a time interval proportional to the change in chemoreceptor occupancy (Macnab and Koshland, 1972; Spudich and Koshland, 1975; Berg and Tedesco, 1975). The bacteria then adapt to the new concentration of attractant and return to the random swimming pattern characteristic of unstimulated cells. If the attractant concentration is then returned to the former concentration, the bacteria undergo a period of constant tumbling similar to the response to a repellent. Thus increased occupancy of the attractant receptors was found to induce a favorable response at first, but then became the norm to which future changes were compared. As a result, after adaptation to a higher occupancy level, a decrease in receptor occupancy was interpreted as an unfavorable stimulus. An increase in repellent is also an unfavorable stimulus, and a decrease in repellent is a favorable stimulus (Tsang *et al.,* 1973).

Adaptation of eukaryotic microorganisms to chemoeffectors is not so well understood. There is evidence that leukocytes and *Paramecium* have adaptive mechanisms, but their importance in chemotaxis remains to be established. Eukaryotes frequently employ an alternative to cellular adaptation: they destroy the chemical signal.

5.2. Biochemistry of Adaptation in Bacteria

Recently, remarkable progress has been made in our understanding of the biochemistry of adaptation in *E. coli* and *S. typhimurium.* This work has involved a combination of genetic, biochemical, and behavioral approaches to the problem. Although the mechanism of adaptation is more complicated than was thought at first, it appears to be effected by a covalent modification of the signaling proteins. Adaptation of *E. coli* and *S. typhimurium* to most chemoeffectors involves methylation of a signaling protein, but adaptation to some attractants is a methylation-independent mechanism. These two adaptation mechanisms are discussed separately.

5.2.1. Methylation-Dependent Adaptation

The signals from most chemotactic stimuli impinging on *E. coli* and *S. typhimurium* are focused through one of three signaling proteins that are the products of *tsr*, *tar*, and *trg* (Fig. 5), as pointed out in Section 3. Increased binding of an attractant or periplasmic receptor to a signaling protein not only initiates the behavioral response, but activates selective methylation of that signaling protein as well. As a result, these signaling proteins are also known as methyl-accepting chemotaxis proteins (MCPs).

The MCPs were discovered when Kort *et al.* (1975) incubated *E. coli* with radioactive-labeled methionine and identified radioactive-labeled cellular proteins by electrophoresis and liquid scintillation counting of gel slices. A methionine requirement for chemotaxis had been demonstrated previously (Adler and Dahl, 1967; Aswad and Koshland, 1974), and *S*-adenosylmethionine had been implicated as the active form of methionine (Armstrong, 1972; Aswad and Koshland, 1975). The *tsr*, *tar*, and *trg* genes are responsible for at least 10 bands of MCP that appear in the 55,000- to 65,000-dalton region of polyacrylamide gels run in the presence of sodium dodecylsulfate; the Tap protein is a similar MCP of unknown function (Silverman and Simon, 1977; M. S. Springer *et al.,* 1977*b;* Kondoh *et al.,* 1979; DeFranco and Koshland, 1980; Chelsky and Dahlquist, 1980; Koiwai *et al.,* 1980; Boyd *et al.,* 1981; Wang *et al.,* 1982). The presence of multiple protein bands of different apparent molecular weights coded by a few genes was puzzling until it was realized that about four methyl groups can be attached to each molecule of MCP. The protein distribution among the bands is changed by methylation and demethylation; highly methylated forms show the greatest electrophoretic mobility (DeFranco and Koshland, 1980; Chelsky and Dahlquist, 1980; Boyd and Simon, 1980; Engström and Hazelbauer, 1980). The methyl moiety is attached to the MCP at glutamyl residues, forming a γ-carboxyl methyl ester (Kleene *et al.*, 1977; Van der Werf and Koshland, 1977).

There is a continuous turnover of methyl groups attached to MCPs. A methyltransferase encoded by the *cheR* gene catalyzes the methylation of MCP using *S*-adenosylmethionine as substrate (W. R. Springer and Koshland, 1977; Clarke *et al.*, 1980), and the CheB protein, methylesterase, hydrolyzes the methyl ester, releasing methanol (Stock and Koshland, 1978; Toews and Adler, 1979; Hayashi *et al.*, 1979). These opposing processes maintain a steady-state level of methylated signaling protein in unstimulated cells (M. S. Springer *et al.*, 1979). Although the Tsr, Tar, and Trg proteins are all substrates for the same enzymes, the methylation level of a given signaling protein can vary independently in response to a stimulus (Toews *et al.*, 1979; Kleene *et al.*, 1979). Binding of an attractant–receptor complex to the signaling protein apparently induces a conformational change that enhances the protein as a substrate for the transferase and produces an unfavorable configuration for esterase activity. Demethylation is blocked, and net incorporation of methyl groups increases for about as long as the behavioral response lasts. Demethylation then resumes at the same rate as occurs in unstimulated cells (Toews *et al.*, 1979). The reactivity of the methylatable γ-carboxyl residues on the Tar and Tsr MCPs is not uniform, and the residues appear to be methylated and demethylated in a preferred order (Stock and Koshland, 1981; M. S. Springer *et al.*, 1982).

The net result of these reciprocal effects of an attractant stimulus on the methyltransferase and methylesterase reactions is an increase in methylated MCP to a new steady-state level (Goy *et al.*, 1977). For a given attractant, the new plateau is a function of the fraction of receptors occupied by attractant; behavioral adaptation is complete when the new plateau level is attained (Goy *et al.*, 1977). The new methylation level persists until another change in affector concentration occurs. Studies of an *in vitro* methylation system by Kleene *et al.* (1979) demonstrated that addition of repellent rapidly reduces the methylation level by slowing incorporation and stimulating hydrolysis.

The signaling proteins are essential in initiating a chemotactic response (M. S. Springer *et al.*, 1977*b;* Silverman and Simon, 1977). But initiation appears to be independent of methylation. Bacteria that exhibit a defective methylation reaction respond appropriately to a step increase in attractant or repellent (Aswad and Koshland, 1974; M. S. Springer *et al.*, 1977*a;* Parkinson and Revello, 1978; Goy *et al.*, 1978). This means that methylation does not mediate transmission of the signal from the receptors to the flagella. Cells that exhibit impaired methylation are defective in adapting to the stimulus. If starved of methionine or deficient in the methyltransferase (CheR), they fail to adapt to an increase in effector concentration and continue responding indefinitely (Aswad and Koshland, 1974; M. S. Springer *et al.*, 1975; Goy *et al.*, 1978). Inability to adapt prevents the cells from responding to spatial assays in a measurable fashion (Adler and Dahl, 1967).

In addition to methylation–demethylation, MCPs can undergo another covalent modification. In *cheR cheB* strains of *E. coli,* MCP is synthesized in a form, designated 2*, that is unmethylated but that has several properties of a methylated MCP. The CheB methylesterase processes 2* to 1*, which no longer resembles methylated MCP (Sherris and Parkinson, 1981; Rollins and Dahlquist, 1981). The rate of conversion of 2* to 1* is modulated by chemotactic stimuli. The physiological role of the conversion is unknown, but it is evident that covalent modifications, other than methylation of the signaling proteins, can mediate adaptation. In *Bacillus subtilis* adaptation to an attractant appears to be effected by demethylation rather than by methylation (Goldman *et al.*, 1982).

5.2.2. Methylation-Independent Adaptation

Taxis to oxygen (aerotaxis) and taxis to sugars transported by the phosphotransferase active transport system (PTS) use signaling proteins that are not MCPs (Fig. 5). In *E. coli* AW660 (*tsr*, *tar*, *trg*), which lacks the known MCPs, chemotaxis is normal to oxygen and to D-mannose, D-glucose, and *N*-acetyl-D-glucosamine, which are substrates for the PTS (Niwano and Taylor, 1982*a*). Adaptation to these attractants is independent of protein methylation and is unaffected by depletion of *S*-adenosylmethionine or deletion of *cheR*, the structural gene for the protein methyltransferase required for methylation-dependent chemotaxis (Niwano and Taylor, 1982*b*).

The mechanism of methylation-independent adaptation remains to be determined. It seems likely that the sensory adaptation mechanism for chemotaxis to PTS substrates is different from the adaptation mechanism for taxis mediated by the pmf. The membrane-bound enzymes IIA in the PTS, or a nearby transmembrane protein, may be chemically modified in response to an increased concentration of PTS substrate. Phosphorylation by phosphoenolpyruvate is an attractive mechanism for signal processing in this system. For aerotaxis, the pmf sensor is presumably a transmembrane protein and may be chemically modified in adapation. In this respect it is of interest that adenosine triphosphate (ATP) is required for chemotaxis in addition to the requirements for *S*-adenosylmethionine and pmf (Kondoh, 1980; Shioi *et al.*, 1982*b*). It is possible that the pmf sensor is phosphorylated or adenylylated. Alternatively, the pmf sensor and adaptation system might be associated with the flagellar basal body, because the flagellar motor is energized by the pmf, which also influences tumbling frequency in unstimulated bacteria (Khan and Macnab, 1980*a;* Laszlo and Taylor, 1981).

In the absence of protein methyltransferase activity, an auxiliary adaptation system may replace some functions of the methylation-dependent adaptation mechanism in *S. typhimurium.* A *cheR* strain of *S. typhimurium* (ST1038) was able to adapt to a mixture of the repellents phenol, acetate, and leucine, although adaptation was somewhat impaired (Stock *et al.,* 1981). A pseudorevertant of ST1038 was chemotactic to serine and aspartate, even though methyltransferase activity was not restored.

Although there are many signaling proteins and a variety of mechanisms for adaptation, all chemotactic pathways appear to converge to a common tumble-regulating system (Fig. 5). The same flagella and flagellar motors are involved in all behavioral responses, implying the presence of a common response regulator modulated by diverse signals.

5.3. Adaptation in Eukaryotes

Do eukaryotic microorganisms adapt to chemotactic stimuli? This is harder to determine than in *Salmonella* and *E. coli,* especially in the organisms that respond to spatial rather than temporal gradients. One indication that adaptation may be occurring is the transient nature of components of the response in the presence of persistent stimuli. These types of transient effects have been observed in *Paramecium, Dictyostelium,* and leukocytes.

The klinokinetic behavior of *Paramecium* is suitable for direct observation of adaptation. The response to a sudden temporal increase in attractants or repellents eventually decays, and the turning frequency and speed of the organism return to near basal levels. This adaptation to the new environment reflects the establishment of a new resting membrane potential, as well as a corresponding change in the threshold for action potentials (Machemer, 1976). The adapted cell is therefore capable of responding to further slight changes in chemoeffector concentration. The mechanism establishing the new resting potential and threshold has not yet been identified.

Leukocytes apparently respond to a spatial gradient by true chemotaxis; however, they also display transient behavioral changes when subjected to a temporal gradient (Zigmond and Sullivan, 1979). Increasing attractant causes leukocytes to stop moving and form ruffles; decreasing attractant causes them to become round. After a delay the cells regain their normal shape and resume normal motility. Efforts to understand the molecular details of this adaptation have focused on methylation reactions. Protein methylation coincident with the chemotactic response has been observed in rabbit polymorphonuclear leukocytes (O'Dea *et al.*, 1978). However, human polymorphonuclear leukocytes are chemotactic without a comparable change in protein methylation, hence it appears that protein methylation is not the mechanism of adaptation in leukocytes (Pike *et al.*, 1978). Phospholipid methylation in leukocytes may be directly involved in signal transduction, as discussed in Section 3.3.1, but it may not participate in adaptative mechanisms. Another mechanism for creating a transient response to chemoeffector may be the control of chemoreceptor activity during up- and down-regulation of chemoreceptor binding. It should be noted, however, that such a reduction in chemoeffector binding causes a *desensitization,* but may not restore the ability of the cells to respond to prestimulus levels. In contrast, adaptation in bacteria and *Paramecium* restores the ability to respond to further changes.

Increased formation of cyclic guanyl monophosphate (cGMP) in *Dictyostelium discoideum* is transient even in the continuous presence of nondegradable analogues of cAMP (Mato and Malchow, 1978). Methylation of proteins and phospholipids occurs as a result of stimulation with cAMP, and the latter may be partly influenced by cGMP (Mato and Marin-Cao, 1979; Alemany *et al.*, 1980), but methylation is not directly mediated by cGMP (Van Haastert *et al.*, 1982). It should be noted that caution must be applied in equating transient changes with adaptation when it is not known whether the ability of the cells to respond normally has been fully restored. The adaptation in *D. discoideum* of the cAMP-signaling response during cAMP stimulation has been demonstrated and the mechanism investigated. Extracellular cAMP activates adenylate cyclase, which leads to an increase in intracellular cAMP and the rate of cAMP secretion (Dinauer *et al.*, 1980*a*). Cells adapt to any constant cAMP stimulus after several minutes and can then respond to an increase in the concentration of the stimulus. The observed changes in the cells are consistent with a two-process model (Dinauer *et al.*, 1980*b*) similar to the model used to explain adaptation to chemoattractants in bacteria (Koshland, 1977). Dinauer *et al.* (1980*c*) proposed that occupancy of surface cAMP receptors leads to changes in two opposing cellular processes—excitation and adaptation—that control the activity of adenylate cyclase in *D. discoideum.*

6. MODELS FOR CHEMOTAXIS

Although a response regulator model underlies most current research in chemotaxis, the specific details of the model assume different forms when applied to specific organisms. In *Paramecium,* migration is controlled by the speed of swimming and by a stochastic process that triggers the avoidance response. Membrane potential is the regulating parameter: A depolarization increases the avoidance frequency and decreases speed, whereas a hyperpolarization decreases the avoidance frequency and increases speed (Van Houten, 1979). The chemotactic response—attraction or repulsion—has been correlated with changes in membrane potential (Van Houten, 1978, 1981; Van Houten *et al.,* 1981). But, as discussed in Section 3.2., it is likely that the response regulator is the intraciliary concentration of Ca^{2+}, rather than membrane potential. The basic schema of excitation in *Paramecium* chemotaxis can be summarized as follows: Chemoeffector binds to a receptor protein → change in receptor conformation and membrane conductance → receptor potential → action potential → change in intraciliary [Ca^{2+}] → alteration in speed and probability of reversal (Beidler, 1954, 1971; Kung and Saimi, 1982). Not all steps in this sequence have been demonstrated satisfactorily. For example, there is no proof that the increase in beat frequency caused by hyperpolarization is mediated by an increase in internal Ca^{2+} concentration, but in view of the supporting evidence it is likely that the schema for excitation is essentially correct. There is insufficient evidence to construct a model for adaptation in *Paramecium,* although Kung and Saimi (1982) have suggested that more than one mechanism is involved.

It would be premature to propose a response regulator for chemotaxis in leukocytes or in *D. discoideum,* even though Ca^{2+} is known to modulate the cytoskeletal interactions, because the relative importance of other components of the chemotactic mechanism is undetermined. For example, phospholipase A_2, metabolism of arachidonate, phospholipid methylation and cyclic nucleotides in leukocytes, and cyclic nucleotides and phospholipid methylation in *D. discoideum* may be involved in regulating chemotactic behavior.

As in leukocytes and *D. discoideum,* the response regulator has not been identified in bacterial chemotaxis. An earlier report (Ordal, 1977; Ordal and Fields, 1977) that Ca^{2+} might be the response regulator in *B. subtilis* has not yet been confirmed in other laboratories. It is unlikely that Ca^{2+} mediates chemotaxis in *S. typhimurium,* because chemotaxis is unaffected by growth in the presence of a calcium chelator (Snyder *et al.,* 1981). Identification of the response regulator should be facilitated by the development of a procedure to prepare bacterial cell envelopes with functional flagella (Eisenbach and Adler, 1981).

For methylation-dependent bacterial chemotaxis, models have been proposed in which each MCP exists in two interconvertible conformations, the active and passive states (Fig. 7). A tumble regulator that binds to the motor switch is proportional to the fraction of MCP that is in an active conformation (Taylor and Laszlo, 1981; Springer *et al.,* 1982). The motor, which can rotate spontaneously in either direction, is biased in the counterclockwise (smooth swimming) sense, and the tumble regulator increases the probability of clockwise rotation and tumbling. Interconversion of the active and passive forms is

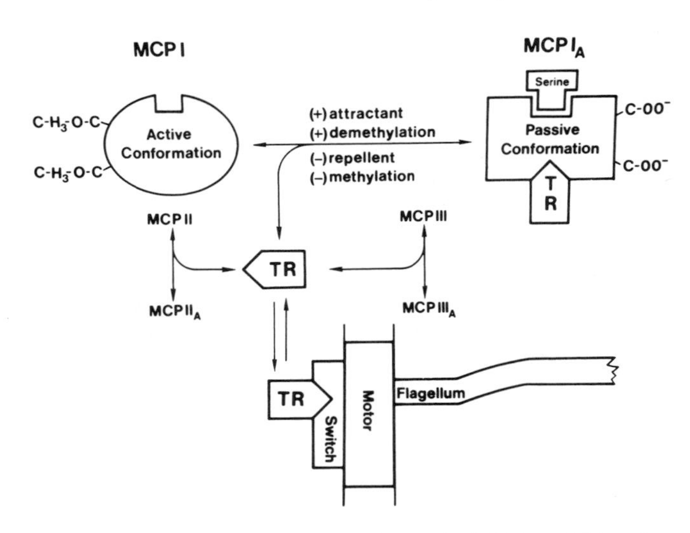

Figure 7. Regulatory methyl-accepting chemotaxis protein (MCP) model for chemotaxis. TR, Tumble regulator. Subscript A designates the active conformation of a MCP, and (+) and (−) indicate activators and inhibitors, respectively, of the formation of the passive conformation.

influenced by methylation of the MCP and by binding of effectors or receptor complexes to the MCP. Methylation by the methyltransferase converts the passive to the active form, which is the substrate of the methylesterase. The methylesterase converts the active form to the passive demethylated form such that each class of MCP is in equilibrium between active and passive conformations. The active form of each MCP has a similar effect on the level of tumble regulator, and their combined effect in unstimulated cells is appropriate in maintaining random motility.

The addition of attractant to the medium induces the formation of chemoreceptor complexes that bind to both passive and active forms of a specific MCP class. This induces a conformational change that stabilizes the passive conformation. Other classes of MCP are unchanged. The abrupt decrease in regulator level permits smooth swimming. The methyltransferase acts on the passive form to reverse the effects of the attractant by further increasing methylation, which favors formation of the active form. Adaptation is complete when the forms of the MCP reach an equilibrium in which the regulator level is similar to the level in unstimulated cells. At that point, the level of methylation is a function of attractant bound to the chemoreceptor. When attractant is removed, dissociation of the chemoreceptor removes a constraint favoring the passive form, and the active form increases as a result. Constant tumbling ensues, until the methylesterase completes the process of deadaptation by reducing the extent of methylation to the prestimulus level. Repellent chemoreceptors are postulated to bind at an additional site. The sequence of events is the reverse of those shown for the attractant response cycle. Repellent stimulation stabilizes the active form and causes tumbling; demethylation restores the active form to the basal level.

In contrast to most previous models, we have postulated that the regulator is a promoter, rather than an inhibitor, of tumbling. This convention is consistent with the apparent counterclockwise bias of the motor (Khan and Macnab, 1980*a;* Eisenbach and Adler, 1981) and with the predominance of smooth over tumbling phenotype in *che* mutants (Parkinson, 1977).

7. CONCLUSION

Even though chemoresponses take different forms in different organisms, there are obvious similarities in the basic mechanisms. Most chemoreceptors are proteins located on the outside of the cell. The signal generated by effector molecules binding to a receptor is transmitted to the inside of the cell and is processed to generate an appropriate behavioral response. Processing includes focusing of the signals into a common response system and adaptation of the response to a step change in the level of effector. In the final analysis, it is likely that these complex behavioral responses are achieved by the interaction of relatively simple regulatory processes, such as two-position switches controlled by the level of a response regulator.

ACKNOWLEDGMENTS. We thank Daniel Laszlo and Jun-ichi Shioi for helpful discussions and P. Devreotes, K. Williams, and S. Zigmond for critical reading of the manuscript. This work was supported by grants (to B.L.T.) from the American Heart Association and from the National Institute of General Medical Sciences (GM29481).

REFERENCES

Adler, J., 1975, Chemotaxis in bacteria, *Annu. Rev. Biochem.* **44**:341–356.

Adler, J., and Dahl, M. M., 1967, A method for measuring the motility of bacteria and for comparing random and nonrandom motility, *J. Gen. Microbiol.* **46**:161–173.

Adler, J., and Epstein, W., 1974, Phosphotransferase-system enzymes as chemoreceptors for certain sugars in *Escherichia coli* chemotaxis, *Proc. Natl. Acad. Sci. USA* **71**:2895–2899.

Adler, J., Hazelbauer, G. L., and Dahl, M. M., 1973, Chemotaxis toward sugars in *Escherichia coli, J. Bacteriol.* **115**:824–847.

Aksamit, R. R., and Koshland, D. E., Jr., 1972, A ribose binding protein of *Salmonella typhimurium, Biochem. Biophys. Res. Commun.* **48**:1348–1353.

Aksamit, R. R., and Koshland, D. E., Jr., 1974, Identification of the ribose binding protein as the receptor for ribose chemotaxis in *Salmonella typhimurium, Biochemistry* **13**:4473–4478.

Alber, T., Fahnestock, M., Mowbray, S. L., and Petsko, G. A., 1981, Preliminary X-ray data for the galactose binding protein from *Salmonella typhimurium, J. Mol. Biol.* **147**:471–474.

Alcantara, F., and Monk, M., 1974, Signal propagation during aggregation in the slime mould *Dictyostelium discoideum, J. Gen. Microbiol.* **85**:321–334.

Alemany, S., Gil, M. G., and Mato, J. M., 1980, Regulation by guanosine 3′:5′-cyclic monophosphate of phospholipid methylation during chemotaxis in *Dictyostelium discoideum, Proc. Natl. Acad. Sci. USA* **77**:6996–6999.

Almagor, M., Ron, A., and Bar-Tana, J., 1981, Chemotaxis in *Tetrahymena thermophila, Cell Motil.* **1**:261–268.

Anraku, Y., 1968, Transport of sugars and amino acids in bacteria. I. Purification and specificity of the galactose- and leucine-binding proteins, *J. Biol. Chem.* **243**:3116–3122.

Armstrong, J. B., 1972, An S-adenosylmethionine requirement for chemotaxis in *Escherichia coli, Can. J. Microbiol.* **18**:1695–1701.

Aswad, D., and Koshland, D. E., Jr., 1974, Role of methionine in bacterial chemotaxis, *J. Bacteriol.* **118**:640–645.

Aswad, D. W., and Koshland, D. E., Jr., 1975, Evidence for an S-adenosylmethionine requirement in the chemotactic behavior of *Salmonella typhimurium, J. Mol. Biol.* **97**:207–223.

Aswanikumar, S., Corcoran, B. A., Schiffman, E., Day, A. R., Freer, R. J., Showell, H. J., Becker, E. L., and Pert, C. B., 1977, Demonstration of a receptor on rabbit neutrophils for chemotactic peptides, *Biochem. Biophys. Res. Commun.* **74**:810–817.

Becker, E. L., and Stossel, T. P., 1980, Chemotaxis, *Fed. Proc.* **39**:2949–2952.

Beidler, L. M., 1954, A theory of taste stimulation, *J. Gen. Physiol.* **38**:133–139.

Beidler, L. M., 1971, Taste receptor stimulation with salts and acids, in: *Handbook of Sensory Physiology,* Vol. 4, Pt. 2, (L. M. Beidler, ed.), pp. 200–220, Springer-Verlag, New York.

Berg, H. C., and Anderson, R. A., 1973, Bacteria swim by rotating their flagellar filaments, *Nature (Lond.)* **239**:380–382.

Berg, H. C., and Brown, A. D., 1972, Chemotaxis in *E. coli* analyzed by three-dimensional tracking, *Nature (Lond.)* **239**:500–504.

Berg, H. C., and Purcell, E. M., 1977, Physics of chemoreception, *Biophys. J.* **20**:193–219.

Berg, H. C., and Tedesco, P. M., 1975, Transient response to chemotactic stimuli in *Escherichia coli, Proc. Natl. Acad. Sci. USA* **72**:3235–3239.

Blum, J. J., and Hines, M., 1979, Biophysics of flagellar motility, *Q. Rev. Biophys.* **12**:103–180.

Boucek, M. M., and Snyderman, R., 1976, Calcium influx requirement for human neutrophil chemotaxis: inhibition by lanthanum chloride, *Science* **193**:905–907.

Boyd, A., and Simon, M. I., 1980, Multiple electrophoretic forms of methyl-accepting chemotaxis proteins generated by stimulus-elicited methylation in *Escherichia coli, J. Bacteriol.* **143**:809–815.

Boyd, A., and Simon, M., 1982, Bacterial chemotaxis, *Annu. Rev. Physiol.* **44**:501–517.

Boyd, A., Krikos, A., and Simon, M., 1981, Sensory transducers of *E. coli* are encoded by homologous genes, *Cell* **26**:333–343.

Brenner, M., and Thoms, S., 1984, Caffeine blocks activation of cyclic AMP synthesis in *Dictyostelium discoideum, Dev. Biol.* **101:**136–146.

Burchard, R. P., 1980, Gliding motility of bacteria, *Bioscience* **30**:157–162.

Chelsky, D., and Dahlquist, F. W., 1980, Structural studies of methyl-accepting chemotaxis proteins of *Escherichia coli:* evidence for 5 multiple methylation sites, *Proc. Natl. Acad. Sci. USA* **77**:2434–2438.

Chenoweth, D. E., and Hugli, T. E., 1978, Demonstration of specific C5a receptor on intact human polymorphonuclear leukocytes, *Proc. Natl. Acad. Sci. USA* **75**:3943–3947.

Clancy, M., Madill, K. A., and Wood, J. M., 1981, Genetic and biochemical requirements for chemotaxis to L-proline in *Escherichia coli, J. Bacteriol.* **146**:902–906.

Clarke, M., and Spudich, J. A., 1977, Nonmuscle contractile proteins: The role of actin and myosin in cell motility and shape determination, *Annu. Rev. Biochem.* **46**:797–822.

Clarke, M., Bazari, W. L., and Kayman, S. C., 1980, Isolation and properties of calmodulin from *Dictyostelium discoideum, J. Bacteriol.* **141**:397–400.

Clarke, S., and Koshland, D. E., Jr., 1979, Membrane receptors for aspartate and serine in bacterial chemotaxis, *J. Biol. Chem.* **254**:9695–9702.

Clarke, S., Sparrow, K., Panasenko, S., and Koshland, D. E., Jr., 1980, *In vitro* methylation of bacterial chemotaxis proteins: characterization of protein methyltransferase activity in crude extracts of *Salmonella typhimurium, J. Supramol. Struct.* **13**:315–328.

Dahlquist, F. W., Lovely, P., and Koshland, D. E., Jr., 1972, Quantitative analysis of bacterial migration in chemotaxis, *Nature New Biol.* **236**:120–123.

DeFranco, A. L., and Koshland, D. E., Jr., 1980, Multiple methylation in the processing of sensory signals during bacterial chemotaxis, *Proc. Natl. Acad. Sci. USA* **77**:2429–2433.

DeFranco, A. L., and Koshland, D. E., Jr., 1982, Construction and behavior of strains with mutations in two chemotaxis genes, *J. Bacteriol.* **150**:1297–1301.

Delbrück, M., and Reichardt, W., 1956, System analysis for the light growth reactions of *Phycomyces,* in: *Cellular Mechanisms in Differentiation and Growth* (D. Rudnick, ed.), pp. 3–44, Princeton University Press, Princeton, N.J.

DePamphilis, M. L., and Adler, J., 1971, Fine structure and isolation of the hook-basal body complex of flagella from *E. coli* and *B. subtilis, J. Bacteriol.* **105**:396–407.

Diehn, B., Feinleib, M., Haupt, W., Hildebrand, E., Lenci, F., and Nultsch, W., 1977, Terminology of behavioral responses in motile microorganisms, *Photochem. Photobiol.* **26**:559–560.

Dinauer, M. C., MacKay, S. A., and Devreotes, P. N., 1980*a,* Cyclic 3′,5′-AMP relay in *Dictyostelium discoideum.* III. The relationship of cAMP synthesis and secretion during the cAMP signalling response, *J. Cell. Biol.* **86**:537–544.

Dinauer, M. C., Steck, T. L., and Devreotes, P. N., 1980*b,* Cyclic 3′,5′-AMP relay in *Dictyostelium discoideum.* IV. Recovery of the cAMP signalling response after adaptation to cAMP, *J. Cell. Biol.* **86**:545–553.

Dinauer, M. C., Steck, T. L., and Devreotes, P. N., 1980*c,* Cyclic 3′,5′-AMP relay in *Dictyostelium discoideum.* V. Adaptation of the cAMP signalling response during cAMP stimulation, *J. Cell. Biol.* **86**:554–561.

Donabedian, H., and Gallin, J. I., 1981, Deactivation of human neutrophil chemotaxis by chemoattractants: effect on receptors for the chemotactic factors fMet-Leu-Phe, *J. Immunol.* **127**:839–844.

Doughty, M. J., and Dryl, S., 1981, Control of ciliary activity in *Paramecium:* an analysis of chemosensory transduction in a eukaryotic unicellular organism, *Prog. Neurobiol.* **16**:1–115.

Dryl, S., 1973, Chemotaxis in ciliate protozoa, in: *Behavior of Microorganisms* (A. Perez-Miravete, ed.), pp. 16–30, Plenum Press, New York.

Dunlap, K., 1977, Localization of calcium channels in *Paramecium caudatum, J. Physiol. (Lond.)* **271**:119–134.

Durston, A. J., and Vork, F., 1979, A cinematographical study of the development of vitally stained *Dictyostelium discoideum, J. Cell Sci.* **36**:261-279.

Dworkin, M., and Eide, D., 1983, *Myxococcus xanthus* does not respond chemotactically to moderate concentration gradients, *J. Bacteriol.* **154**:437–442.

Eckert, R., 1972, Bioelectric control of ciliary activity, *Science* **176**:473–481.

Eisenbach, M., and Adler, J., 1981, Bacterial cell envelopes with functional flagella, *J. Biol. Chem.* **256**:8807–8814.

Engström, P., and Hazelbauer, G. L., 1980, Multiple methylation of methyl-accepting chemotaxis proteins during adaptation of *E. coli* to chemical stimuli, *Cell* **20**:165–171.

Engström, P., and Hazelbauer, G. L., 1982, Methyl-accepting chemotaxis proteins are distributed in the membrane independently from basal ends of bacterial flagella, *Biochim. Biophys. Acta* **686**:19–26.

Fahnestock, M., and Koshland, D. E., Jr., 1979, Control of the receptor for galactose taxis in *Salmonella typhimurium, J. Bacteriol.* **137**:758–763.

Fandrich, B., and Laszlo, D. J., 1981, Cytochrome *o* as the receptor for aerotaxis in *Salmonella typhimurium, Fed. Proc.* **40**:1637.

Fernandez, H. N., and Hugli, T. E., 1978, Primary structural analysis of the polypeptide portion of human C5a anaphylatoxin, *J. Biol. Chem.* **253**:6955–6964.

Fisher, P. R., Smith, E., and Williams, K. L., 1981, An extracellular chemical signal controlling phototactic behavior by *D. discoideum* slugs, *Cell* **23**:799–807.

Freer, R. J., Day, A. R., Radding, J. A., Schiffman, E., Aswanikumar, S., Showell, H. J., and Becker, E. L., 1980, Further studies on the structural requirements for synthetic peptide chemoattractants, *Biochemistry* **19**:2404–2410.

Gallin, J. I., Gallin, E. K., Malech, H. L., and Cramer, E. B., 1978, Structural and ionic events during leukocyte chemotaxis, in: *Leukocyte Chemotaxis* (J. I. Gallin and P. G. Quie, eds.), pp. 123–141, Raven Press, New York.

Gerisch, G., 1982, Chemotaxis in *Dictyostelium, Annu. Rev. Physiol.* **44**:535–552.

Gerisch, G., Fromm, H., Huesgen, A., and Wick, U., 1975, Control of cell-contact sites by cyclic AMP pulses in differentiating *Dictyostelium* cells, *Nature (Lond.)* **255**:547–549.

Goetzl, E. J., and Pickett, W. C., 1980, Human PMN leukocyte chemotactic activity of complex hydroxy-eicosatetraenoic acids (HETEs), *J. Immunol.* **125**:1789–1791.

Goetzl, E. J., Foster, D. W., and Goldman, D. W., 1981, Isolation and partial characterization of membrane protein constituents of human neutrophil receptors for chemotactic formylmethionyl peptides, *Biochemistry* **20**:5712–5722.

Goldman, D. J., Worobec, S. W., Siegel, R. B., Hecker, R. V., and Ordal, G. W., 1982, Chemotaxis in *Bacillus subtilis:* effects of attractants on the level of methylation of methyl-accepting chemotaxis proteins and the role of demethylation in the adaptation process, *Biochemistry* **21**:915–920.

Goy, M. F., Springer, M. S., and Adler, J., 1977, Sensory transduction in *Escherichia coli:* role of a protein methylation reaction in sensory adaptation, *Proc. Natl. Acad. Sci. USA* **74**:4964–4968.

Goy, M. F., Springer, M. S., and Adler, J., 1978, Failure of sensory adaptation in bacterial mutants that are defective in a protein methylation reaction, *Cell* **15**:1231–1240.

Green, A. A., and Newell, P. C., 1975, Evidence for the existence of two types of cAMP binding sites in aggregating cells of *Dictyostelium discoideum, Cell* **6**:129–136.

Hagen, D. C., Bretcher, A. P., and Kaiser, D., 1978, Synergism between morphogenic mutants of *Myxococcus xanthus, Dev. Biol.* **64**:284–296.

Hayashi, H., Koiwai, O., and Kozuka, M., 1979, Studies on bacterial chemotaxis. II. Effect of *cheB* and *cheZ* mutations on the methylation of methyl-accepting chemotaxis protein of *Escherichia coli, J. Biochem.* **85**:1213–1223.

Hazelbauer, G. L., 1975, The maltose chemoreceptor of *Escherichia coli, J. Bacteriol.* **122**:206–214.

Hazelbauer, G. L., and Adler, J., 1971, Role of the galactose binding protein in chemotaxis of *Escherichia coli* toward galactose, *Nature (Lond.)* **230**:101–104.

Hazelbauer, G. L., and Engström, P., 1980, Parallel pathways for transduction of chemotactic signals in *Escherichia coli, Nature (Lond.)* **283**:98–100.

Hazelbauer, G. L., and Harayama, S., 1979, Mutants in transmission of chemotactic signals from two independent receptors of *Escherichia coli, Cell* **16**:617–625.

Hazelbauer, G. L., and Parkinson, J. S., 1977, Bacterial chemotaxis, in: *Microbial Interaction (Receptors and Recognition)* (J. Reissig, ed.), pp. 59–98, Chapman and Hall, London.

Hedblom, M. L., and Adler, J., 1980, Genetic and biochemical properties of *Escherichia coli* mutants with defect in serine chemotaxis, *J. Bacteriol.* **144**:1048–1060.

Henderson, E. J., 1975, The cyclic adenosine 3′,5′-monophosphate receptor of *Dictyostelium discoideum, J. Biol. Chem.* **250**:4730–4736.

Heppel, L. A., 1969, The effect of osmotic shock on release of bacterial proteins and active transport, *J. Gen. Physiol.* **54**:95s–109s.

Hill, A. V., 1936, Excitation and accommodation in nerve, *Proc. R. Soc. (Lond.)* [*Biol.*] **B119**:305–354.

Ho, J., and McCurdy, H. D., 1979, Demonstration of positive chemotaxis to cyclic GMP and 5′-AMP in *Myxococcus xanthus* by means of a simple apparatus for generating stable concentration gradients, *Can. J. Microbiol.* **25**:1214–1218.

Hodgkin, J., and Kaiser, D., 1979*a*, Genetics of gliding motility in *Myxococcus xanthus* (Myxobacteriales): Two gene systems control movement, *Mol. Gen. Genet.* **171**:177–191.

Hodgkin, J., and Kaiser, D., 1979*b*, Genetics of gliding in *Myococcus xanthus* (Myxobacteriales): Genes controlling movement of single cells, *Mol. Gen. Genet.* **171**:167–176.

Inouye, M., Inouye, S., and Zusman, D., 1979, Biosynthesis and self-assembly of protein S, a development-specific protein of *Myxococcus xanthus, Proc. Natl. Acad. Sci. USA* **76**:209–213.

Jennings, H. S., 1906, *Behavior of Lower Animals,* Indiana University Press, Bloomington, Indiana.

Kaiser, D., Manoil, C., and Dworkin, M., 1979, Myxobacteria: cell interactions, genetics and development, *Annu. Rev. Microbiol.* **33**:595–639.

Keller, H. U., and Sorkin, E., 1966, Studies on chemotaxis. IV. The influence of serum factors on granulocyte locomotion, *Immunology* **10**:409–416.

Kellerman, O., and Szmelcman, S., 1974, Active transport of maltose in *Escherichia coli* K12. Involvement of a "periplasmic" maltose binding protein, *Eur. J. Biochem.* **47**:139–149.

Khan, S., and Macnab, R. M., 1980*a*, The steady-state counterclockwise/clockwise ratio of bacterial flagellar motors is regulated by protonmotive force, *J. Mol. Biol.* **138**:563–597.

Khan, S., and Macnab, R. M., 1980*b*, Proton chemical potential, proton electrical potential, and bacterial motility, *J. Mol. Biol.* **138**:599–614.

Kihara, M., and Macnab, R. M., 1981, Cytoplasmic pH, pH taxis, and weak-acid repellent taxis in bacteria, *J. Bacteriol.* **148**:1209–1221.

Kleene, S. J., Toews, M. L., and Adler, J., 1977, Isolation of glutamic acid methyl ester from an *Escherichia coli* membrane protein involved in chemotaxis, *J. Biol. Chem.* **252**:3214–3218.

Kleene, S. J., Hobson, A. C., and Adler, J., 1979, Attractants and repellents influence methylation and demethylation of methyl-accepting chemotaxis proteins in an extract of *Escherichia coli, Proc. Natl. Acad. Sci. USA* **76**:6309–6313.

Klein, C., and Juliani, M. H., 1977, cAMP-induced changes in cAMP-binding sites on *D. discoideum* amoebae, *Cell* **10**:329–335.

Koiwai, O., and Hayashi, H., 1979, Studies on bacterial chemotaxis. IV. Interaction of maltose receptor with a membrane-bound chemosensing component, *J. Biochem.* **86**:27–34.

Koiwai, O., Minoshima, S., and Hayashi, H., 1980, Studies on bacterial chemotaxis. V. Possible involvement of four species of the methyl-accepting chemotaxis protein in chemotaxis of *Escherichia coli, J. Biochem.* **87**:1365–1370.

Komano, T., Inouye, S., and Inouye, M., 1980, Patterns of protein production in *Myxococcus xanthus* during spore formation induced by glycerol, dimethyl sulfoxide and phenethyl alcohol, *J. Bacteriol.* **144**:1076–1082.

Komano, T., Brown, N., Inouye, S., and Inouye, M., 1982, Phosphorylation and methylation of protein during *Myxococcus xanthus* spore formation, *J. Bacteriol.* **151**:114–118.

Kondoh, H., 1980, Tumbling chemotaxis mutants of *Escherichia coli*: possible gene-dependent effect of methionine starvation, *J. Bacteriol.* **142**:527–534.

Kondoh, H., Ball, C. B., and Adler, J., 1979, Identification of a methyl-accepting chemotaxis protein for the ribose and galactose chemoreceptors of *Escherichia coli, Proc. Natl. Acad. Sci. USA* **76**:260–264.

Konijn, T. M., van der Meene, J. G. C., Bonner, J. T., and Barkley, D. S., 1967, The acrasin activity of adenosine-3′,5′-cyclic phosphate, *Proc. Natl. Acad. Sci. USA* **58**:1152–1154.

Kort, E. N., Goy, M. F., Larsen, S. H., and Adler, J., 1975, Methylation of a membrane protein involved in bacterial chemotaxis, *Proc. Natl. Acad. Sci. USA* **72**:3939–3943.

Koshland, D. E., Jr., 1977, A response regulator model in a simple sensory system, *Science* **196**:1055–1063.

Koshland, D. E., Jr., 1980*a*, *Bacterial Chemotaxis as a Model Behavioral System. Distinguished Lecture Series of the Society of General Physiologists,* Vol. 2, Raven Press, New York.

Koshland, D. E., Jr., 1980*b*, Bacterial chemotaxis in relation to neurobiology, *Annu. Rev. Neuroscience* **3**:43–76.

Koshland, D. E., Jr., 1981, Biochemistry of sensing and adaptation in a simple bacterial system, *Annu. Rev. Biochem.* **50**:765–782.

Kuczmarski, E. R., and Spudich, J. A., 1980, Regulation of myosin self-assembly: phosphorylation of *Dictyostelium* heavy chain inhibits formation of thick filaments, *Proc. Natl. Acad. Sci. USA* **77**:7292–7296.

Kuhlwein, H., and Reichenbach, H., 1981, *Encyclopedia od Cinematography and Film,* C893/1965, pp. 335–359, Inst. Wiss. Film, Gottingen.

Kung, C., and Saimi, Y. 1982, The physiological basis of taxes in *Paramecium, Annu. Rev. Physiol.* **44**:519–534.

Kung, C., Chang, S-Y., Satow, Y., Van Houten, J., and Hansma, H., 1975, Genetic dissection of behavior in *Paramecium, Annu. Rev. Physiol.* **44**:519–534.

Lapidus, I. R., and Berg, H. C., 1982, Gliding motility of *Cytophaga* sp strain U67, *J. Bacteriol.* **151**:384–398.

Larsen, S. H., Adler, J., Gargus, J. J., and Hogg, R. W., 1974, Chemomechanical coupling without ATP. The source of energy for motility and chemotaxis in bacteria, *Proc. Natl. Acad. Sci. USA* **71**:1239–1243.

Laszlo, D. J., 1981, "The mechanism of aerotaxis in *Salmonella typhimurium*," Ph.D. thesis, Loma Linda University, Loma Linda, Calif.

Laszlo, D. J., and Taylor, B. L., 1981, Aerotaxis in *Salmonella typhimurium:* the role of electron transport, *J. Bacteriol.* **145**:990–1001.

Liao, C. S., and Freer, R. J., 1980, Cryptic receptors for chemotactic peptides in rabbit neutrophils, *Biochem. Biophys. Res. Commun.* **93**:566–571.

Loomis, W. F., 1975, *Dictyostelium discoideum: A Developmental System,* Academic Press, New York.

Machemer, H., 1974*a*, Frequency and directional response of cilia to membrane potential changes in *Paramecium, J. Comp. Physiol.* **92**:293–316.

Machemer, H., 1974*b*, Ciliary activity and metachronism in protozoa, in: *Cilia and Flagella* (M. A. Sleigh, ed.), pp. 199–286, Academic Press, New York.

Machemer, H., 1976, Interactions of membrane potential and cations in regulation of ciliary activity in *Paramecium, J. Exp. Biol.* **65**:427–448.

Machemer, H., and Ogura, A., 1979, Ionic conductances of membranes in ciliated and deciliated *Paramecium, J. Physiol. (Lond.)* **296**:49–60.

Macnab, R. M., 1977, Bacterial flagella rotating in bundles: a study in helical geometry, *Proc. Natl. Acad. Sci. USA* **74**:221–225.

Macnab, R. M., 1979, Locomotion in microbial plants, in: *Encyclopedia of Plant Physiology*, Vol. 7, (W. Haupt and M. E. Feinleib, eds.), pp. 207–233, Springer-Verlag, New York.

Macnab, R. M., 1982, Sensory reception in bacteria, in: *Prokaryotic and Eukaryotic Flagella, Society for Experimental Biology Symposium No. XXXV* (W. B. Amos and J. G. Duckett, eds.), pp. 77–104, Cambridge University Press, London.

Macnab, R. M., and Koshland, D. E., Jr., 1972, The gradient sensing mechanism in bacterial chemotaxis, *Proc. Natl. Acad. Sci. USA* **69**:2509–2512.

Macnab, R. W., and Ornston, M. K., 1977, Normal-to-curly flagellar transitions and their role in bacterial tumbling. Stabilization of an alternative quaternary structure by mechanical force, *J. Mol. Biol.* **112**:1–30.

Maeda, K., and Imae, Y., 1979, Thermosensory transduction in *Escherichia coli:* inhibition of the thermoresponse by L-serine, *Proc. Natl. Acad. Sci. USA* **76**:91–95.

Malchow, D., and Gerisch, G., 1974, Short-term binding and hydrolysis of cyclic 3′,5′-adenosine monophosphate by aggregating *Dictyostelium* cells, *Proc. Natl. Acad. USA* **71**:2423–2427.

Malchow, D., Böhme, R., and Rahmsdorf, H. J., 1981, Regulation of phosphorylation of myosin heavy chain during the chemotactic response of *Dictyostelium* cells, *Eur. J. Biochem.* **117**:213–218.

Manson, M. D., Tedesco, P., Berg, H. C., Harold, F. M., and Van der Drift, C., 1977, A protonmotive force drives bacterial flagella, *Proc. Natl. Acad. Sci. USA* **74**:3060–3064.

Manson, M. D., Tedesco, P. M., and Berg, H. C., 1980, Energetics of flagellar rotation in bacteria, *J. Mol. Biol.* **138**:541–561.

Mascarenhas, J. P., 1978, Sexual chemotaxis and chemotropism in plants, in: *Taxis and Behavior* (G. L. Hazelbauer, ed.), pp. 169–203, Halsted Press, New York.

Mato, J. M., and Malchow, D., 1978, Guanylate cyclase activation in response to chemotactic stimulation in *Dictyostelium discoideum, FEBS Lett.* **90**:119–122.

Mato, J. M., and Marin-Cao, D., 1979, Protein and phospholipid methylation during chemotaxis in *Dictyostelum discoideum* and its relationship to calcium movement, *Proc. Natl. Acad. Sci. USA* **76**:6106–6109.

Mato, J. M., Krens, F. A., van Haastert, P. J. M., and Konijn, T. M., 1977, 3′:5′-cyclic AMP-dependent 3′:5′-cyclic GMP accumulation in *Dictyostelium discoideum, Proc. Natl. Acad. Sci. USA* **74**:2348–2351.

Mato, J. M., Jastorff, B., Morr, M., and Konijn, T. M., 1978, A model for cyclic AMP–chemoreceptor interaction in *Dictyostelium discoideum, Biochim. Biophys. Acta* **544**:309–314.

Matsuura, S., Shioi, J., and Imae, Y., 1977, Motility in *Bacillus subtilis* driven by an artificial protonmotive force, *FEBS Lett.* **82**:187–190.

Mesibov, R., and Adler, J., 1972, Chemotaxis toward amino acids in *Escherichia coli, J. Bacteriol.* **112**:315–326.

Miller, J. B., and Koshland, D. E., Jr., 1977, Sensory electrophysiology of bacterial: relationship of the membrane potential to motility and chemotaxis in *Bacillus subtilis, Proc. Natl. Acad. Sci. USA* **74**:4752–4756.

Miller, J. B., and Koshland, D. E., Jr., 1980, Protonmotive force and bacterial sensing, *J. Bacteriol.* **141**:26–32.

Naccache, P. H., Showell, H. J., Becker, E. L., and Sha'afi, R. I., 1977, Transport of sodium, potassium, and calcium across rabbit polymorphonuclear leucocyte membranes, *J. Cell Biol.* **73**:428–444.

Naccache, P. H., Volpi, M., Showell, H. J., Becker, E. L., and Sha'afi, R. I., 1979, Chemotactic factor-induced release of membrane calcium in rabbit neutrophils, *Science* **203**:461–463.

Naitoh, Y., and Eckert, R., 1969, Ionic mechanisms controlling behavioral responses of *Paramecium* to mechanical stimulation, *Science* **164**:963–965.

Naitoh, Y., Eckert, R., and Friedman, K., 1972, A regenerative calcium response in *Paramecium, J. Exp. Biol.* **56**:667–681.

Nath, J., Flavin, M., and Schiffmann, E., 1981, Stimulation of tubulin tyrosylation in rabbit leukocytes evoked by the chemoattractant formyl-methionyl-leucyl-phenylalanine, *J. Cell Biol.* **91**:232–239.

Niedel, J. E., Kahane, I., and Cuatrecasas, P., 1979, Receptor-mediated internalization of fluorescent chemotactic peptide by human neutrophils, *Science* **205**:1412–1414.

Niedel, J., Davis, J., and Cuatrecasas, P., 1980, Covalent affinity labeling of the formyl peptide chemotactic receptor, *J. Biol. Chem.* **255**:7063–7066.

Nikaido, H., and Nakae, T., 1979, The outer membrane of gram negative bacteria, *Adv. Microb. Physiol.* **20**:163–250.

Niwano, M., and Taylor, B. L., 1982*a*, Novel sensory adaptation mechanism in bacterial chemotaxis to oxygen and phosphotransferase substrates, *Proc. Natl. Acad. Sci. USA* **79**:11–15.

Niwano, M., and Taylor, B. L., 1982*b*, Requirement of *CheB* product for methylation-independent chemotaxis to oxygen in bacteria, *Fed. Proc.* **41**:759.

Nossal, R., and Zigmond, S. H., 1976, Chemotropism indices for polymorphonuclear leukocytes, *Biophys, J.* **16**:1171–1182.

O'Day, D. H., 1979, Aggregation during sexual development in *Dictyostelium discoideum, Can. J. Microbiol.* **25**:1416–1426.

O'Dea, R. F., Viveros, O. H., Axelrod, J., Aswanikumar, S., Schiffman, E., and Corcoran, B. A., 1978, Rapid stimulation of protein carboxymethylation in leukocytes by a chemotactic peptide, *Nature (Lond.)* **272**:462–464.

Ogura, A., and Takahashi, K., 1976, Artificial deciliation causes loss of calcium-dependent responses in *Paramecium, Nature (Lond.)* **264**:170–172.

Ordal, G. W., 1977, Calcium ion regulates chemotactic behavior in bacteria, *Nature (Lond.)* **270**:66–67.

Ordal, G. W., and Adler, J., 1974, Isolation and complementation of mutants in galactose taxis and transport, *J. Bacteriol.* **117**:509–516.

Ordal, G. W., and Fields, R. B., 1977, A biochemical mechanism for bacterial chemotaxis, *J. Theor. Biol.* **68**:491–500.

Ordal, G. W. and Goldman, D. J., 1975, Chemotaxis away from uncouplers of oxidative phosphorylation in *Bacillus subtilis, Science* **189**:802–805.

Ordal, G. W., and Goldman, D. J., 1976, Chemotactic repellents of *Bacillus subtilis, J. Mol. Biol.* **100**:103–108.

Ordal, G. W., Villani, D. P., Nicholas, R. A., and Hamel, F. G., 1978, Independence of proline chemotaxis and transport in *Bacillus subtilis, J. Biol. Chem.* **253**:4916–4919.

Pan, P., Hall, E. M., and Bonner, J. T., 1972, Folic acid as second chemotactic substance in the cellular slime moulds, *Nature (Lond.)* **237**:181–182.

Panasenko, S. M., 1983, Protein and lipid methylation by methionine and s-adenosylmethionine in *Myxococcus xanthus, Can. J. Microbiol.* **29**:1224–1228.

Parkinson, J. S., 1977, Behavioral genetics in bacteria, *Annu. Rev. Genet.* **11**:397–414.

Parkinson, J. S., 1981, Genetics of bacterial chemotaxis, in: *Genetics as a Tool in Microbiology* (S. W. Glover and D. A. Hopwood, eds.), pp. 265–290, Cambridge University Press, London.

Parkinson, J. S., and Revello, P. T., 1978, Sensory adaptation mutants of *E. coli, Cell* **15**:1221–1230.

Parsons, R. G., and Hogg, R. W., 1973, A comparison of the L-arabinose and D-galactose-binding proteins of *Escherichia coli* B/r, *J. Biol. Chem.* **249**:3608–3614.

Pate, J. L., and Chang, L. Y. E., 1979, Evidence that gliding motility in prokaryotic cells is driven by rotary assemblies in the cell envelopes, *Curr. Microbiol.* **2**:59–64.

Pike, M. C., Kredich, N. M., and Snyderman, R., 1978, Requirement of S-adenosylmethionine-mediated methylation for human monocyte chemotaxis, *Proc. Natl. Acad. Sci. USA* **75**:3928–3932.

Poff, K. L., and Whitaker, B. D., 1979, Movement in slime molds, in: *Encyclopedia of Plant Physiology,* Vol. 7, pp. 355–382, Springer-Verlag, New York.

Quiocho, F. A., Gilliland, G. L., Miller, D. M., and Newcomer, M. E., 1977, Crystallographic and chemical studies of L-arabinose-binding protein from *E. coli, J. Supramol. Struct.* **6**:503–518.

Quiocho, F. A., Meador, W. E., and Pflugrath, J. W., 1979, Preliminary crystallographic data of receptors for transport and chemotaxis in *Escherichia coli:* D-galactose and maltose binding proteins, *J. Mol. Biol.* **133**:181–184.

Rashevsky, N., 1933, Outline of a physico-mathematical theory of excitation and inhibition, *Protoplasma* **20**:42–56.

Richarme, G., 1982, Interaction of maltose-binding protein with membrane vesicles of *Escherichia coli, J. Bacteriol.* **149**:662–667.

Robertson, A., Drage, D. J., and Cohen, M. H., 1972, Control of aggregation in *Dictyostelium discoideum* by an external periodic pulse of cyclic adenosine monophosphate, *Science* **175**:333–335.

Rollins, C., and Dahlquist, F. W., 1981, The methyl-accepting chemotaxis proteins of *E. coli:* a repellent-stimulated, covalent modification, distinct from methylation, *Cell* **25**:333–340.

Roos, W., Malchow, D., and Gerisch, G., 1977, Adenylate cyclase and the control of cell differentiation in *Dictyostelium discoideum, Cell Diff.* **6**:229–239.

Rossier, C., Eitle, E., van Driel, R., and Gerisch, G., 1980, Biochemical regulation of cell development and aggregation in *Dictyostelium discoideum,* in: *The Eukaryotic Microbial Cell* (G. W. Gooday, D. Lloyd, and A. P. J. Trinci, eds.), pp. 405–424, Cambridge University Press, London.

Rubik, B. A., and Koshland, D. E., Jr., 1978, Potentiation, desensitization, and inversion of response in bacterial sensing of chemical stimuli, *Proc. Natl. Acad. Sci. USA* **75**:2820–2824.

Schellenberg, G. D., 1978, "The multiplicity of glutamate and aspartate transport systems in *Escherichia coli,*" Ph.D. Thesis, University of California, Riverside.

Schiffmann, E., 1982, Leukocyte chemotaxis, *Annu. Rev. Physiol.* **44**:553–568.

Schiffmann, E., Corcoran, B. A., and Wahl, S. M., 1975, N-Formylmethionyl peptides as chemoattractants for leucocytes, *Proc. Natl. Acad. Sci. USA* **72**:1059–1062.

Schiffmann, E., O'Dea, R. F., Chiang, P. K., Venkatasubramanian, K., Corcoran, B., Hirata, F., and Axelrod, J., 1979, Role for methylation-leukocyte chemotaxis, in: *Modulation of Protein Function, ICN–UCLA Symposium,* Vol. 13 (D. E. Atkinson and C. F. Fox, eds.), pp. 299–313, Academic Press, New York.

Schiffmann, E., Aswanikumar, S., Venkatasubramanian, K., Corcoran, B. A., Pert, C. B., Brown, J., Gross, E., Day, A. R., Freer, R. J., Showell, H. J., and Becker, E. L., 1980, Some characteristics of the neutrophil receptor for the chemotactic peptides, *FEBS Lett.* **117**:1–7.

Schimkets, L. J., Dworkin, M., and Keller, K. H., 1979, A method for establishing stable concentration gradients in agar suitable for studying chemotaxis on a solid surface, *Can. J. Microbiol.* **25**:1460–1467.

Segall, J. E., Manson, M. D., and Berg, H. C., 1982, Signal processing times in bacterial chemotaxis, *Nature (Lond.)* **296**:855–857.

Shaffer, B. M., 1975, Secretion of cyclic AMP induced by cyclic AMP in the cellular slime mold *Dictyostelium discoideum, Nature (Lond.)* **255**:549–552.

Sherris, D., and Parkinson, J. S., 1981, Posttranslational processing of methyl-accepting chemotaxis proteins in *Escherichia coli, Proc. Natl. Acad. Sci. USA* **78**:6051–6055.

Shimomura, O., Suthers, H. L. B., and Bonner, J. T., 1982, Chemical identity of the acrasin of the cellular slime mold *Polysphondylium violaceum, Proc. Natl. Acad. Sci. USA* **79**:7376–7379.

Shioi, J., Shusuke, M., and Imae, Y., 1980, Quantitative measurements of proton motive force and motility in *Bacillus subtilis, J. Bacteriol.* **144**:891–897.

Shioi, J., Thomsen, G. E., Rowsell, E. H., and Taylor, B. L., 1982*a*, Protonmotive force in bacterial chemotaxis to oxygen, *Fed. Proc.* **41**:759.

Shioi, J., Galloway, R. J., Niwano, M., Chinnock, R. E., and Taylor, B. L., 1982*b*, Requirement of ATP in bacterial chemotaxis, *J. Biol. Chem.* **257**:7969–7975.

Silhavy, T. J., Szmelcman, S., Boos, W., and Schwartz, M., 1975, On the significance of the retention of ligand by protein, *Proc. Natl. Acad. Sci. USA* **72**:2120–2124.

Silverman, M., and Simon, M., 1973, Genetic analysis of bacteriophage Mu-induced flagellar mutants in *Escherichia coli, J. Bacteriol.* **116**:114–122.

Silverman, M., and Simon, M., 1974, Flagellar rotation and the mechanism of bacterial motility, *Nature (Lond.)* **249**:73–74.

Silverman, M., and Simon, M., 1977, Chemotaxis in *Escherichia coli:* Methylation of *che* gene products, *Proc. Natl. Acad. Sci. USA* **74**:3317–3321.

Simchowitz, L., Fischbein, L. C., Spilberg, I., and Atkinson, J. P., 1980, Transient elevation in intracellular cyclic AMP by chemotactic factors, *J. Immunol.* **124**:1482–1491.

Slonczewski, J. L., Macnab, R. M., Alger, J. R., and Castle, A. M., 1982, Effects of pH and repellent tactic stimuli on protein methylation levels in *Escherichia coli, J. Bacteriol.* **152**:384–399.

Snyder, M., Stock, J. B., and Koshland, D. E., Jr., 1981, Role of membrane potential and calcium in chemotactic sensing by bacteria, *J. Mol. Biol.* **149**:241–257.

Snyderman, R., and Fudman, E. J., 1980, Demonstration of a chemotactic factor receptor on macrophages, *J. Immunol.* **124**:2754–2757.

Snyderman, R., and Goetzl, E. J., 1981, Molecular and cellular mechanisms of leukocyte chemotaxis, *Science* **213**:830–837.

Springer, M.S., Kort, E. N., Larsen, S. H., Ordal, G.W., Reader, R.W., and Adler, J., 1975, Role of methionine in bacterial chemotaxis: requirement for tumbling and involvement in information processing, *Proc. Natl. Acad. Sci. USA* **72**:4640–4644.

Springer, M. S., Goy, M. F., and Adler, J., 1977*a*, Sensory transduction in *Escherichia coli:* a requirement for methionine in sensory adaptation, *Proc. Natl. Acad. Sci. USA* **74**:183–187.

Springer, M. S., Goy, M. F., and Adler, J., 1977*b*, Sensory transduction in *Escherichia coli:* two complementary pathways of information processing that involve methylated proteins, *Proc. Natl. Acad. Sci. USA* **74**:3312–3316.

Springer, M. S., Goy, M. F., and Adler, J., 1979, Protein methylation in behavioral control mechanism and in signal transduction, *Nature (Lond.)* **280**:279–284.

Springer, M. S., Zanolari, B., and Pierzchala, P. A., 1982, Ordered methylation of the methyl-accepting chemotaxis proteins of *Escherichia coli, J. Biol. Chem.* **257**:6861–6866.

Springer, W. R., and Koshland, D. E., Jr., 1977, Identification of a protein methyltransferase as the *cheR* gene product in the bacterial sensing system, *Proc. Natl. Acad. Sci. USA* **74**:533–537.

Spudich, J. L., and Koshland, D. E., Jr., 1975, Quantitation of the sensory response in bacterial chemotaxis, *Proc. Natl. Acad. Sci. USA* **72**:710–713.

Stendahl, O. I., and Stossel, T. P., 1980, Actin-binding protein amplifies actomyosin concentration, and gelsolin confers calcium control on direction of contraction, *Biochem. Biophys. Res. Commun.* **92**:675–681.

Stock, J. B., and Koshland, D. E., Jr., 1978, A protein methylesterase involved in bacterial sensing, *Proc. Natl. Acad. Sci. USA* **75**:3659–3663.

Stock, J. B., and Koshland, D. E., Jr., 1981, Changing reactivity of receptor carboxyl groups during bacterial sensing, *J. Biol. Chem.* **256**:10826–10833.

Stock, J. B., Maderis, A. M., and Koshland, D. E., Jr., 1981, Bacterial chemotaxis in the absence of receptor carboxyl-methylation, *Cell* **27**:37–44.

Stossel, T. P., 1978, The mechanism of leukocyte locomotion, in: *Leukocyte Chemotaxis* (J. I. Gallin and P. G. Quie, eds.), pp. 143–160, Raven Press, New York.

Strange, P. G., and Koshland, D. E., Jr., 1976, Receptor interactions in a signaling system: Competition between ribose receptor and galactose receptor in the chemotactic response, *Proc. Natl. Acad. Sci. USA* **73**:762–766.

Sullivan, S. J., and Zigmond, S. H., 1980, Chemotactic peptide receptor modulation in polymorphonuclear leukocytes, *J. Cell. Biol.* **85**:703–711.

Tanabe, H., Kurihara, K., and Kobatake, Y., 1980, Changes in membrane potential and membrane fluidity in *Tetrahymena pyriformis* in association with chemoreception of hydrophobic stimuli: fluorescence studies, *Biochemistry* **19**:5339–5344.

Taylor, B. L., and Laszlo, D. J., 1981, The role of proteins in chemical perception in bacteria, in: *Perception of Behavioral Chemicals* (D. M. Norris, ed.), pp. 1–27, Elsevier, Amsterdam.

Taylor, B. L., and Shioi, J., 1982, Protonmotive force as a signal in bacterial sensory transduction, *Twelfth International Congress of Biochemistry, Perth.*

Taylor, B. L., Miller, J. B., Warrick, H. M., and Koshland, D. E., Jr., 1979, Electron acceptor taxis and blue light effect on bacterial chemotaxis, *J. Bacteriol.* **140**:567–573.

Toews, M. L., and Adler, J., 1979, Methanol formation *in vivo* from methylated chemotaxis proteins in *Escherichia coli, J. Biol. Chem.* **254**:1761–1764.

Toews, M. L., Goy, M. F., Springer, M. S., and Adler, J., 1979, Attractants and repellents control demethylation of methylated chemotaxis proteins in *Escherichia coli, Proc. Natl. Acad. Sci. USA* **76**:5544–5548.

Tomchik, K. J., and Devreotes, P. N., 1981, Adenosine 3′-5′-monophosphate waves in *Dictyostelium discoideum:* a demonstration by isotope dilution–fluorography, *Science* **212**:443–446.

Tsang, N., Macnab, R. M., and Koshland, D. E., Jr., 1973, Common mechanism for repellents and attractants in bacterial chemotaxis, *Science* **181**:60–63.

Tso, W.-W., and Adler, J., 1974, Negative chemotaxis in *Escherichia coli, J. Bacteriol.* **118**:560–576.

Van der Werf, P., and Koshland, D. E., Jr., 1977, Identification of a gamma-glutamyl methyl ester in bacterial membrane protein involved in chemotaxis, *J. Biol. Chem.* **252**:2793–2795.

Van Driel, R., 1981, Binding of the chemoattractant folic acid by *Dictyostelium discoideum* cells, *Eur. J. Biochem.* **115**:391–396.

Van Haastert, P. J. M., Van Walsum, H., and Pasveer, F. J., 1982*a*, Nonequilibrium kinetics of a cyclic GMP-binding protein in *Dictyostelium discoideum, J. Cell. Biol.* **94**:271–278.

Van Haastert, P. J. M., van Lookeren Campagne, M. M., and Ross, F. M., 1982*b*, Altered cGMP-phosphodiesterase activity in chemotactic mutants of *Dictyostelium discoideum, FEBS Lett.* **147**:149–152.

Van Houten, J., 1977, A mutant of *Paramecium* defective in chemotaxis, *Science* **198**:746–748.

Van Houten, J., 1978, Two mechanisms of chemotaxis in *Paramecium, J. Comp. Physiol.* **127**:167–174.

Van Houten, J., 1979, Membrane potential changes during chemokinesis in *Paramecium, Science* **204**:1100–1103.

Van Houten, J., 1981, Chemosensory transduction in *Paramecium:* role of membrane potential, *Olfact. Taste* **7**:53–56.

Van Houten, J., Hauser, D. C. R., and Levandowsky, M., 1981, Chemosensory behavior in protozoa, in: *Biochemistry and Physiology of Protozoa,* Vol. 4 (M. Levandowsky and S. H. Hunter, eds.), pp. 67–124, Academic Press, New York.

Varnum, B., and Soll, D. R., 1981, Chemoresponsiveness to cAMP and folic acid during growth, development and dedifferentiation in *Dictyostelium discoideum, Cell Diff.* **18**:151–160.

Vitkauskas, G., Showell, H. J., and Becker, E. L., 1980, Specific binding of synthetic chemotactic peptides to rabbit peritoneal neutrophils: effects on dissociability of bound peptide, receptor activity and subsequent biologic responsiveness (deactivation), *Mol. Immunol.* **17**:171–180.

Wang, E. A., and Koshland, D. E., Jr., 1980, Receptor structure in the bacterial sensing system, *Proc. Natl. Acad. Sci. USA* **77**:7157–7161.

Wang, E. A., Mowry, K. L., Clegg, D. O., and Koshland, D. E., Jr., 1982, Tandem duplication and multiple functions of a receptor gene in bacterial chemotaxis, *J. Biol. Chem.* **257**:4673–4676.

Warner, F. D., 1974, The fine structure of ciliary and flagellar axonemes, in: *Cilia and Flagella* (M. A. Sleigh, ed.), pp. 11–37, Academic Press, London.

Warrick, H. M., Taylor, B. L., and Koshland, D. E., Jr., 1977, Chemotactic mechanism of *Salmonella typhimurium:* preliminary mapping and characterization of mutants, *J. Bacteriol.* **130**:223–231.

Wedner, H. J., Sinchowitz, L., Atkinson, J., and Stenson, W., 1980, Chemotactic factors induce rapid phosphorylation of a 90,000 dalton protein in human PMN leukocytes, *Fed. Proc.* **39**:1950 (abst).

Wick, U., Malchow, D., and Gerisch, G., 1978, Cyclic AMP stimulated calcium influx into aggregating cells of *Dictyostelium discoideum, Cell Biol. Int. Rep.* **2**:71–79.

Williams, K. L., 1982, Molecules involved in morphogenesis in the multicellular stage of *Dictyostelium discoideum,* Thirty-third Mosbach Colloquium, in: *Biochemistry of Differentiation and Morphogenesis* (Z. Jaenicke, ed.), pp. 231–246, Springer-Verlag, Berlin.

Williams, L. T., Snyderman, R., Pike, M. C., and Lefkowitz, R. J., 1977, Specific receptor sites for chemotactic peptides on human polymorphonuclear leukocytes, *Proc. Natl. Acad. Sci. USA* **74**:1204–1208.

Willis, R. C., and Furlong, C. E., 1974, Purification and properties of a ribose-binding protein from *Escherichia coli, J. Biol. Chem.* **249**:6926–6929.

Wurster, B., and Butz, U., 1980, Reversible binding of the chemoattractant folic acid to cells of *Dictyostelium discoideum, Eur. J. Biochem.* **109**:613–618.

Zigmond, S. H., 1977, The ability of polymorphonuclear leukocytes to orient in gradients of chemotactic factors, *J. Cell Biol.* **75**:606–616.

Zigmond, S. H., 1978, Chemotaxis by polymorphonuclear leukocytes, *J. Cell Biol.* **77**:269–287.

Zigmond, S. H., and Hirsch, J. G., 1972, Effects of cytochalasin B on polymorphonuclear leukocyte locomotion, phagocytosis and glycolysis, *Exp. Cell Res.* **73**:383–393.

Zigmond, S. H., and Sullivan, S. J., 1979, Sensory adaptation of leukocytes to chemotactic peptides, *J. Cell Biol.* **82**:517–527.

Zukin, R. S., 1979, Evidence for a conformational change in the *Escherichia coli:* maltose receptor by excited-state fluorescence lifetime data, *Biochemistry* **18**:2139–2145.

Zukin, R. S., Hartig, P. R., and Koshland, D. E., Jr., 1977*a,* Use of a distant reporter group as evidence for a conformational change in a sensory receptor, *Proc. Natl. Acad. Sci. USA* **74**:1932–1936.

Zukin, R. S., Strange, P. G., Heavy, L. R., and Koshland, D. E., Jr., 1977*b,* Properties of the galactose-binding protein of *Salmonella typhimurium* and *Escherichia coli, Biochemistry* **16**:381–386.

Zukin, R. S., Hartig, P. R., and Koshland, D. E., Jr., 1979, Effect of an induced conformational change on the physical properties of two chemotactic receptor molecules, *Biochemistry* **18**:5599–5605.

Chapter 3

Mechanosensory Transduction in Protozoa

Yutaka Naitoh

1. INTRODUCTION

When a forward-moving protozoan collides with a mechanical obstacle at its anterior end, it moves backward for a while, then resumes forward movement in a direction different from that before the collision. Thus protozoans avoid mechanical obstacles standing in their way in order to continue forward locomotion. The transient locomotor response has been termed an *avoiding reaction* (Jennings, 1906).

Another type of locomotor response commonly seen in protozoans is a transient acceleration of the forward movement in response to mechanical disturbances made around their posterior region. Thus protozoans can escape an unfavorable condition behind them. This transient locomotor response has been termed an *escape reaction* (Naitoh, 1974).

The dual responsiveness of protozoans to mechanical stimulation depending on the site of stimulation was first demonstrated experimentally by Jennings (1906). When he poked an arrested specimen of *Paramecium* in its anterior region with a glass needle, the specimen moved backward for a short distance; in contrast, it moved forward when poked in the posterior region.

Doroszewski (1961, 1965, 1970) studied locomotor responses to mechanical touch in a large, slowly moving ciliate *Dileptus,* stimulating various portions on the cell surface with the tip of a glass needle. Doroszewski demonstrated that anteroposterior differentiation in the membrane function

Yutaka Naitoh • Institute of Biological Sciences, University of Tsukuba, Ibaraki 305, Japan.

produces two different kinds of locomotor response to mechanical stimulation (Fig. 1).

Employing conventional electrophysiological techniques, Naitoh and Eckert (1969*a*) demonstrated that mechanical stimulation of the anterior membrane of *Paramecium* produces a depolarizing response, while stimulation of the posterior membrane produces a hyperpolarizing response (Fig. 2A). The depolarizing response is always accompanied by a transient ciliary reversal,

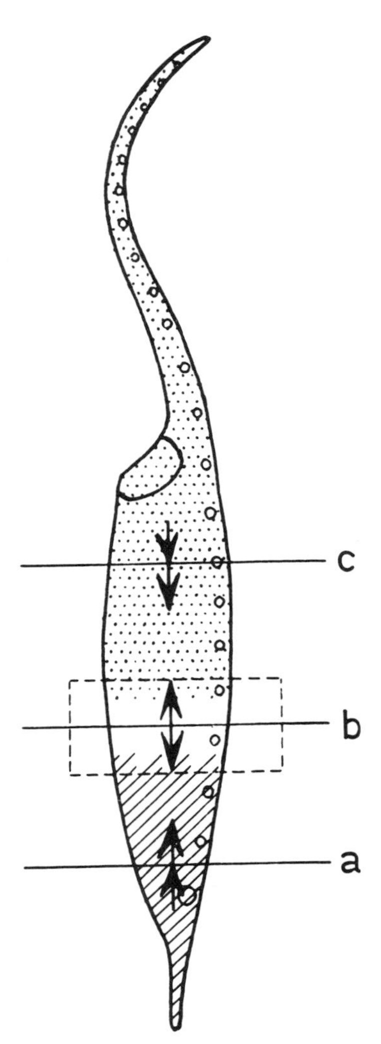

Figure 1. Topographical differentiation in the mechanosensitivity of the surface membrane in a ciliate, *Dileptus*. (a, diagonally-lined area) Produces forward locomotion when incised or mechanically stimulated. (b, dashed portion) Transition area produces sometimes forward locomotion, sometimes backward locomotion. (c, dotted area) Produces backward locomotion. (From Doroszewski, 1965.)

and the hyperpolarizing response is accompanied by a transient increase in the beat frequency of cilia in the normal direction.

Some ciliates, such as *Euplotes* (Fig. 2B) (Naitoh and Eckert, 1969*b*), *Stylonychia* (Fig. 2C) (de Peyer and Machemer, 1978*a*), and *Tetrahymena* (Fig. 2D) (Takahashi *et al.*, 1980), show a topographical differentiation in the membrane function in a single cell that is able to produce two different kinds of mechanoelectric response.

A motor response of protozoan cells to a mechanical stimulation other than the ciliary response is a protoplasmic contraction. *Spirostomum, Stentor, Zoothamnium, Carcesium,* and *Vorticella* are well-known examples of contractile ciliates. These protozoans have well-differentiated subcellular contractile organelles such as myonemes and spasmonemes.

Electrophysiological examinations of a ciliate in relationship to its protoplasmic contraction were carried out by Wood (1970, 1975) in *Stentor.* He recorded a mechanoreceptor potential through a microelectrode inserted into the intracellular vacuole. The amplitude of the response increased to its saturated level with increasing intensity of mechanical stimulation (Fig. 3A).

Some ciliates that have no well-differentiated contractile organelles, such as *Paramecium* and *Euplotes,* also show a contraction of their ectoplasm in response to a mechanical stimulation of the posterior membrane (Naitoh, 1982).

Amoebae are highly sensitive to mechanical stimulation in pseudopod formation relating to prey-capturing behavior. Employing a sensitive electric current detector, Nuccitelli *et al.* (1977) measured a minute electric current around an amoeba generated in association with its forward locomotion (Fig. 3B). Mechanical stimulation produced a local eddy current—an inward current at the stimulation site surrounded by an outward current area—which seemed to trigger the formation of a new pseudopod. It has generally been accepted that the pseudopod formation occurs in close association with protoplasmic contraction, although the contraction sites in the cell are still disputed.

A most peculiar and fascinating cellular response to a mechanical stimulation is the emission of light in *Noctiluca.* An important series of electrophysiologic examinations of *Noctiluca* in relationship to light emission were done by Eckert (1965*a,b*) (Fig. 3C).

Negative geotactic behavior seen in many ciliates and flagellates has long been believed to be dependent on predominant upward orientation of the cells caused by some active responses of cilia or flagella to a mechanical distortion of the cell produced by gravitational as well as viscous forces (see Chapter 5). No electrophysiological examinations on the effect of the gravitational field on the electrical activities of the protozoan cells have been made.

This chapter deals mainly with electrophysiological aspects of the mechanosensory transduction in some ciliate protozoans based on the data pre-

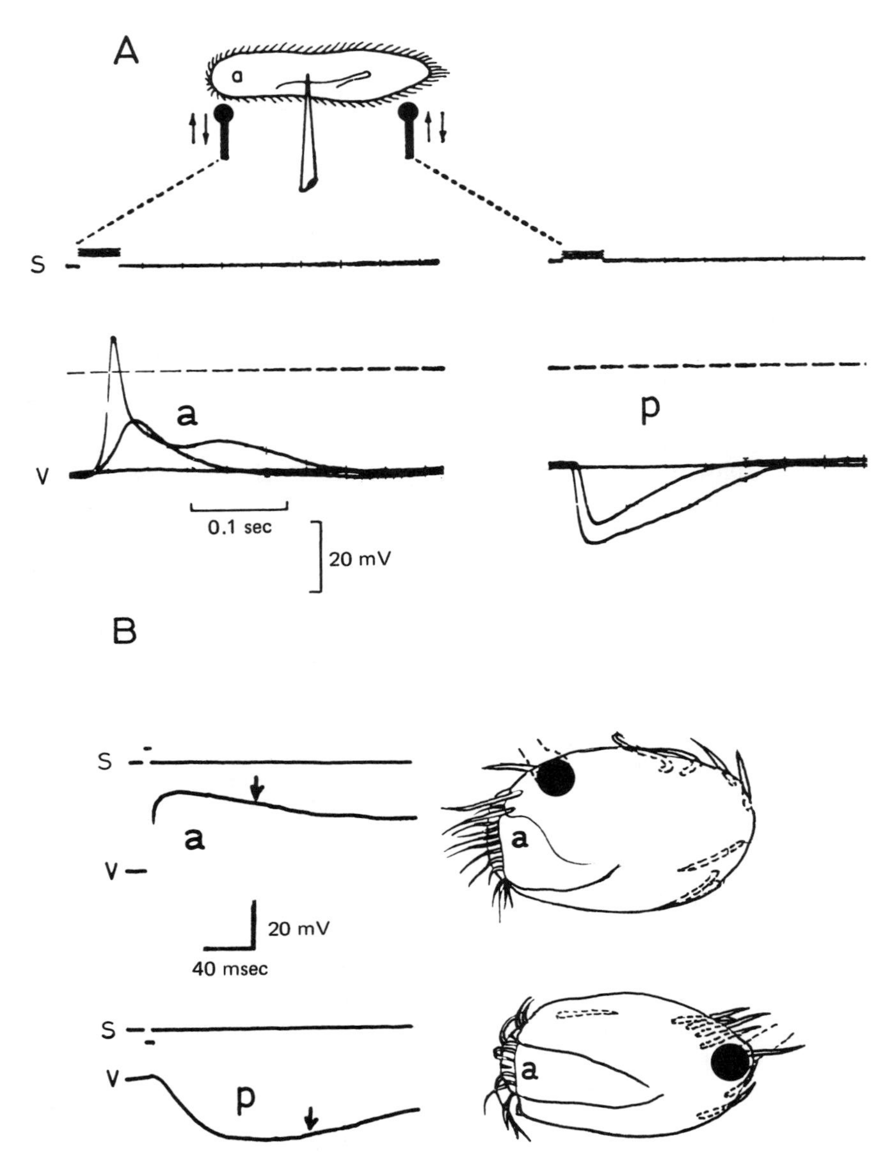

Figure 2. Electrical responses (V) of the anterior (a) and posterior (p) membrane to a mechanical stimulation(s) in four species of ciliates. (A) *Paramecium*, electrical responses of three different amplitudes to mechanical stimulations of three different intensities are superimposed in each potential trace (a and p), (From Naitoh and Eckert, 1969*a*.) (B) *Euplotes*, small arrows in potential traces indicate the moment when the specimen was photographed to examine the ciliary orientation. Traces of the photographs are shown at the right of each potential trace. Note the anteriorly pointing (reversed) cirri during the depolarizing response to the anterior stimulation. Stimulation site is indicated by a black dot in each cell trace (From Naitoh and Eckert, 1969*b*.) (C) *Stylonychia*. (From de Peyer and Machemer, 1978*a*.) (D) *Tetrahymena*. (From Takahashi *et al.*, 1980.)

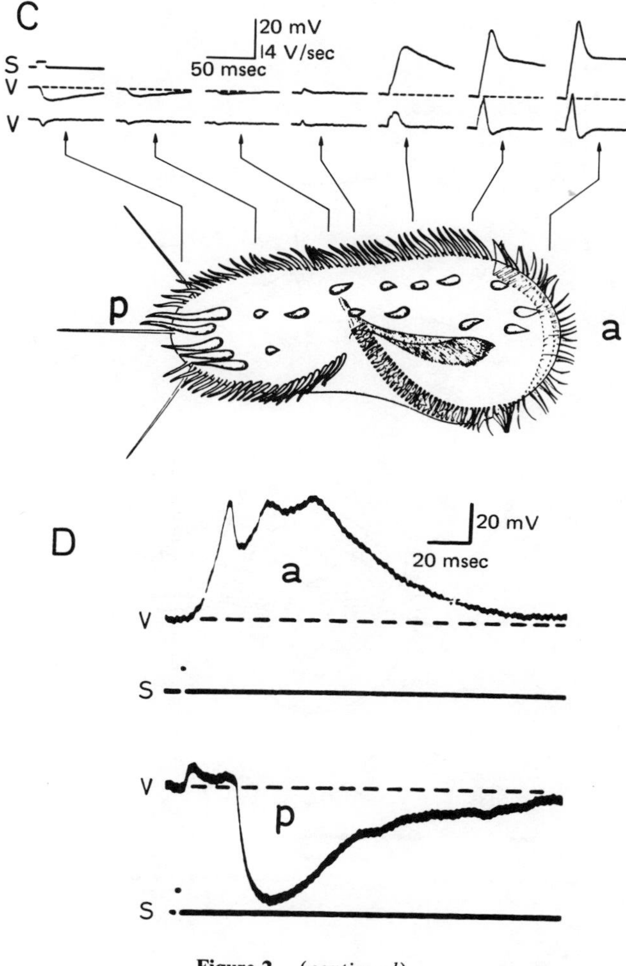

Figure 2. (*continued*)

sented by the research groups of Naitoh, Eckert, and Machemer during the past decade (Eckert, 1972; Eckert and Naitoh, 1972; Naitoh and Eckert, 1974; Naitoh, 1974; Machemer and de Payer, 1977; Eckert *et al.*, 1976; Eckert and Brehm, 1979; Naitoh, 1982).

2. DEPOLARIZING AND HYPERPOLARIZING MEMBRANE RESPONSES TO MECHANICAL STIMULATION

A head-on collision of a free-swimming protozoan with a mechanical obstacle can be mimicked in an immobilized specimen impaled with micro-

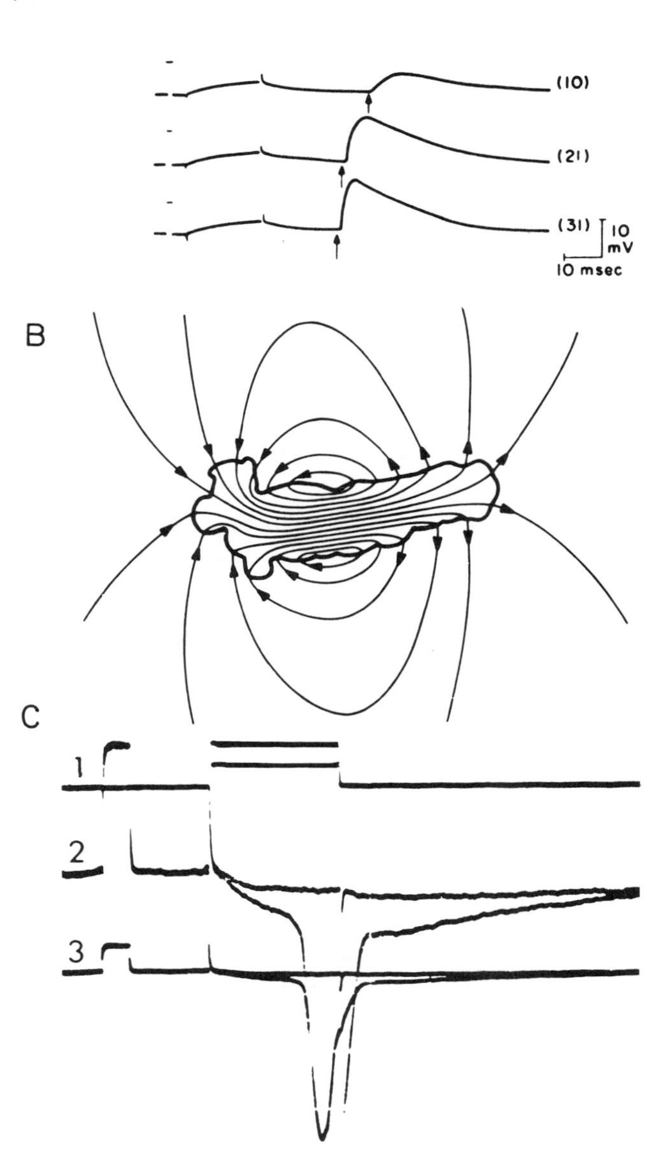
A
(10)
(21)
(31)
10 mV
10 msec
B
C
1
2
3

electrodes by hitting the anterior end of the specimen with the tip of a glass microneedle (Naitoh and Eckert, 1969*a*). The needle is mounted on a piezoelectric element and driven vertically to the cell surface by activating the element with an electrical pulse. Rate, extent, and duration of displacement of the needle can be controlled by changing the rate of rise, final voltage, and duration of the electrical pulse (Eckert *et al.*, 1972; Naitoh and Eckert, 1973; de Peyer and Machemer, 1978*a*).

When the anterior membrane of *Paramecium* is deformed by the needle, it produces a transient depolarizing response in association with a transient ciliary reversal, triggering an avoiding reaction in a free-swimming specimen. In contrast, when the posterior membrane is deformed, it produces a transient hyperpolarizing response accompanied by a transient increase in the beating frequency of cilia in normal forward-swimming direction, causing an escape reaction in a free-swimming specimen (Fig. 2A).

A depolarizing response to a mechanical stimulation of the anterior membrane consists of two components: the first slow mechanoreceptor potential and the second fast spikelike action potential activated by the mechanoreceptor potential. The mechanoreceptor potential is enhanced, while the action potential is suppressed when the membrane is hyperpolarized by an injection of inward current (Naitoh *et al.*, 1972; Eckert *et al.*, 1972). If the membrane potential is kept clamped at its resting level to prevent activation of the action potential, a mechanical stimulation of the anterior membrane produces only a transient inward current responsible for the receptor potential in a nonclamped specimen (Fig. 4) (Ogura and Machemer, 1980).

A hyperpolarizing response to a mechanical stimulation of the posterior membrane consists only of a mechanoreceptor potential, since hyperpolarization does not produce an active potential response in the membrane in normal external ionic conditions. In a voltage-clamped specimen, the posterior

Figure 3 (A) Electrical response of the ciliate *Stentor* to a mechanical stimulation. Numbers at the right of each trace indicate the relative intensity of the mechanical stimulus. (From Wood, 1975.) (B) Electrical current generated by a monopodial amoeba during its forward locomotion. The arrow on each current line indicates the direction of the net transmembrane current. Relative line densities indicate relative current densities. (From Nuccitelli *et al.*, 1977.) (C) Electrical responses of the dinoflagellate *Noctiluca* to mechanical stimulations: (1) superimposed traces of two electrical pulses corresponding to mechanical stimulations of two different intensities; (2) higher amplification; (3) superimposed traces of two membrane potential responses to the mechanical stimulations. A weak stimulus produced only a small, slow mechanoreceptor potential. A stronger stimulus produced a larger receptor potential followed by an action potential, which triggered a bioluminescent flash. Calibration pulses in (2) and (3), 10 mV and 10 msec. (From Eckert, 1965*a*.)

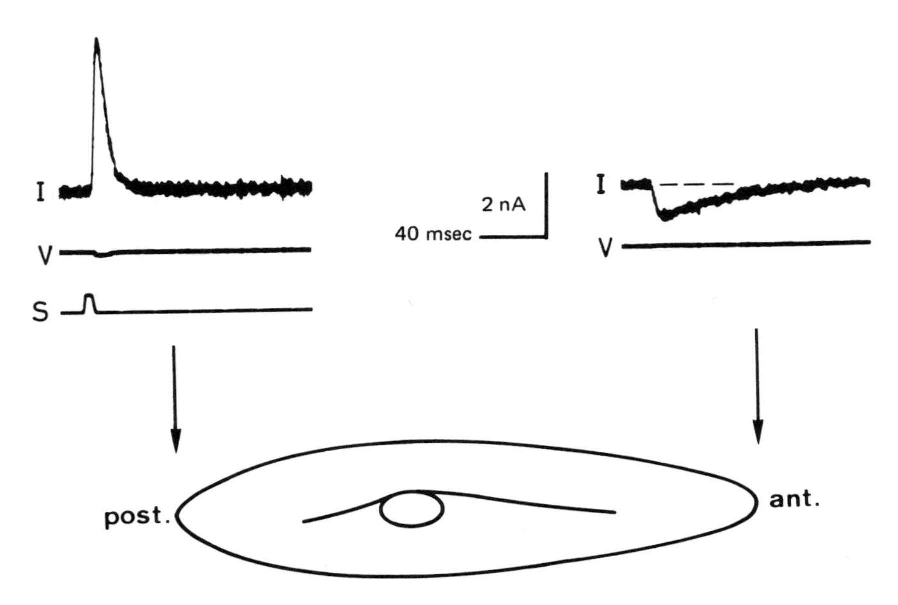

Figure 4. The anterior inward (right) and the posterior outward (left) mechanoreceptor currents in a voltage-clamped specimen of *Paramecium*. (I) Membrane current; (V) membrane potential; (S) voltage pulse applied to a piezoelectric driver of the stimulation needle. (From Ogura and Machemer, 1980.)

mechanoreceptor response appears as a transient outward current (Fig. 4) (Ogura and Machemer, 1980).

Both anterior and posterior mechanoreceptor responses increase in their amplitude in a graded fashion, reaching saturated levels as the rate of displacement of the stimulation needle is increased while the extent of the displacement is kept constant (Fig. 5A,C). The amplitude also increases when the extent of displacement of the needle is increased (Fig. 5B,D) (Eckert *et al.,* 1972; Naitoh and Eckert, 1973; de Peyer and Machemer, 1978*a*).

No refractory period is detectable in the mechanoreceptor responses. Summation of two mechanoreceptor potentials produced by two successive stimulations of a given portion of the cell surface is consistently observed, when the amplitude of each response is smaller than its saturated level (Naitoh and Eckert, 1973; de Peyer and Machemer, 1978*a*).

3. IONIC MECHANISMS OF THE MECHANORECEPTOR RESPONSES

The anterior mechanoreceptor potential decreases in amplitude as the membrane is depolarized by an injection of outward current into the cell (Fig.

6A) (Eckert *et al.*, 1972; de Peyer and Machemer, 1978*a*). When the depolarization exceeds a certain potential level, the depolarizing mechanoreceptor potential reverses its polarity to become a hyperpolarizing response. The potential level is termed a *reversal potential* for the anterior mechanoreceptor response. In a voltage-clamped specimen, the inward anterior receptor current becomes outward, when the clamped voltage is more positive (depolarized) than the reversal potential (Ogura and Machemer, 1980).

The reversal potential approximates the equilibrium potential for that species of ion by which most of the receptor current is carried through the mechanically activated ion channels in the membrane. The reversal potential is therefore very sensitive to a change in the external concentration of the ion.

The reversal potential for the anterior mechanoreceptor response shifts toward the depolarizing direction with increasing external Ca^{2+} concentration ($[Ca^{2+}]_o$). The slope of the plot of the reversal potential against log $[Ca^{2+}]_o$ is ~22 mV in both *Stylonychia* (de Peyer and Machemer, 1978*a*) and *Paramecium* (Fig. 7A) (Ogura and Machemer, 1980). The value approximates the ideal Nernst slope of 29 mV for a divalent cation battery. Change in the external K^+ concentration ($[K^+]_o$) is less effective than $[Ca^{2+}]_o$ in shifting the rever-

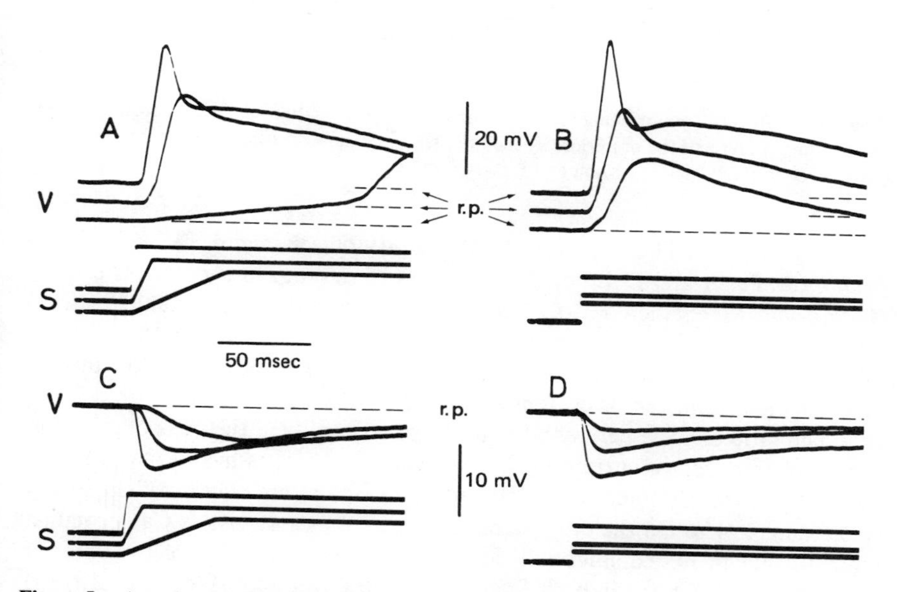

Figure 5. Anterior (A, B) and posterior (C, D) membrane responses to mechanical stimulation in *Stylonychia*. The rate of the displacement of the stimulation needle is varied while the extent of the displacement is kept constant in (A) and (C) and varied in (B) and (D). (V) Membrane potential; (S) voltage applied to a piezoelectric driver of the stimulation needle. (From de Peyer and Machemer, 1978*a*.)

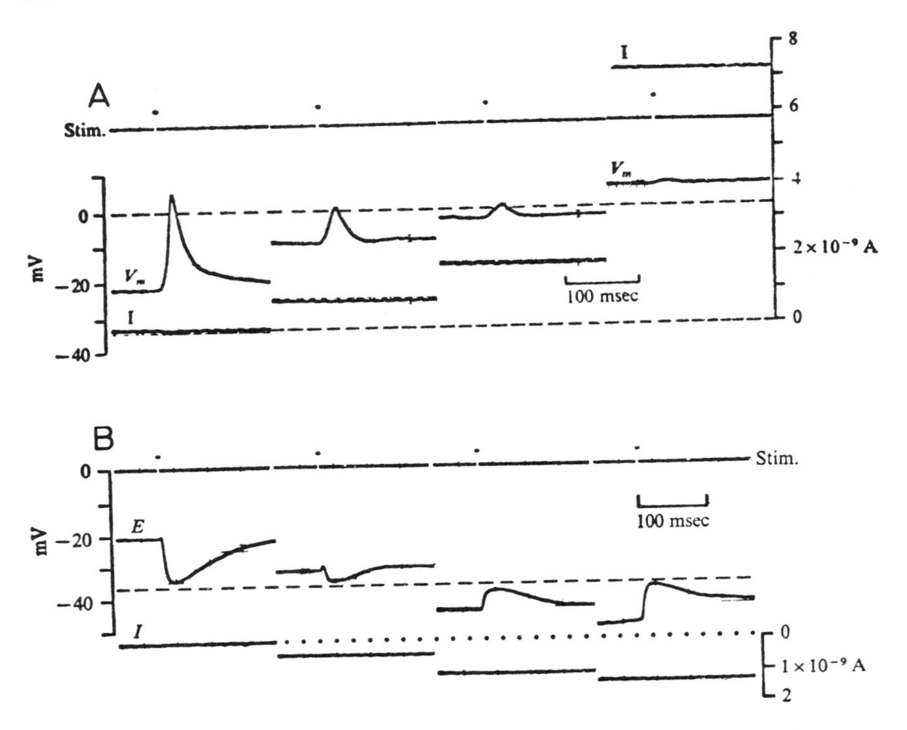

Figure 6. Effects of conditioning electrical current on the amplitude of the electrical responses of the anterior (A) and the posterior (B) membranes to mechanical stimulation in *Paramecium*. (A, Eckert *et al.*, 1972; B, Naitoh and Eckert, 1973.)

sal potential level. The slope is ~5 mV per 10-fold change in $[K^+]_o$ in *Stylonychia* (de Peyer and Deitmer, 1980) and ~15 mV in *Paramecium* (Fig. 7B) (Ogura and Machemer, 1980). These facts indicate that a mechanical stimulation of the anterior membrane produces a transient increase in the permeability of the membrane predominantly to Ca^{2+} ions. In other words, the inward anterior mechanoreceptor current is carried mostly by Ca^{2+} ions.

The ratio of Ca^{2+} permeability to K^+ permeability at the peak of the anterior mechanoreceptor current in *Stylonychia* is ~18:1, as has been calculated by de Peyer and Deitmer (1980) according to Goldman's constant-field theory (Goldman, 1943). On the other hand, the ratio of an increase in Ca^{2+} conductance to that in K^+ conductance on a mechanical stimulation of the anterior membrane is about 3:1 in both *Stylonychia* (de Peyer and Machemer, 1978*a*) and *Paramecium* (Ogura and Machemer, 1980). This value has been calculated according to Hodgkin and Horowicz's equation (Hodgkin and Horowicz, 1959).

Alkaline-earth cations, such as Mg^{2+}, Sr^{2+}, and Ba^{2+}, can carry the mechanoreceptor current through the mechanically activated Ca channels in the anterior membrane (Fig. 8). Mn^{2+} in the external solution inhibits the mechanoreceptor current carried by alkaline-earth cations (de Peyer and Deitmer, 1980).

The saturated peak value of the posterior mechanoreceptor potential in *Paramecium* shifts with a positive slope of ~50 mV per 10-fold increase in $[K^+]_o$. The value is comparable to the ideal Nernst's slope of 58 mV for a monovalent cation battery. $[Ca^{2+}]_o$ is far less effective in the shift of the peak potential (Fig. 9) (Naitoh and Eckert, 1969*a,* 1973). The addition of Na^+, Mg^{2+}, and Mn^{2+} to the external solution never produces significant change in the peak potential value (Naitoh and Eckert, 1973; Deitmer, 1982). These facts indicate that mechanical stimulation of the posterior membrane produces a transient increase in the permeability of the membrane almost exclusively to K^+ ions.

The posterior hyperpolarizing mechanoreceptor potential decreases in amplitude as the membrane is hyperpolarized by an injection of inward current into the cell; it then reverses its polarity to a depolarizing response when the hyperpolarization exceeds the reversal potential level (Fig. 6B) (Naitoh and Eckert, 1973; de Peyer and Machemer, 1978*a*). In a voltage-clamped specimen, the outward mechanoreceptor current becomes inward when the clamped voltage is more negative (hyperpolarized) than the reversal potential.

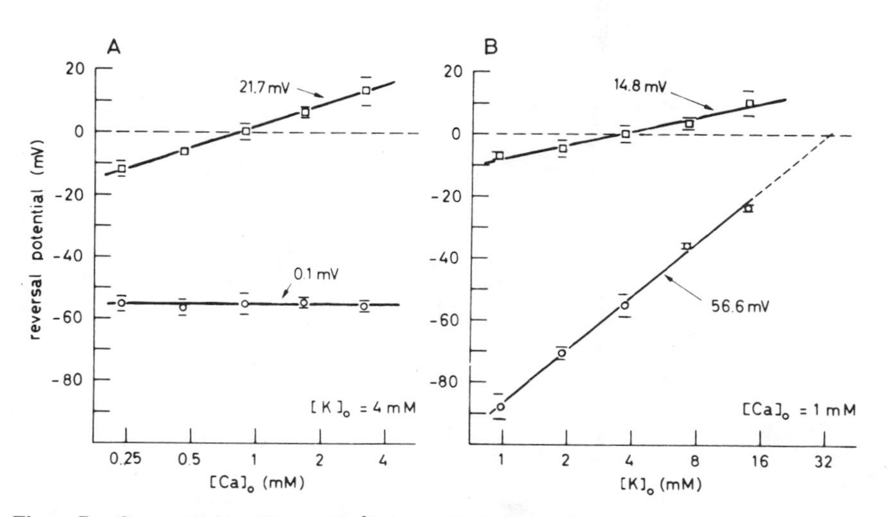

Figure 7. Concentration effects of Ca^{2+} (A) and K^+ (B) in the external solution on the reversal potentials for the anterior mechanoreceptor current (□) and the posterior mechanoreceptor current (○) in a voltage-clamped *Paramecium*. (From Ogura and Machemer, 1980.)

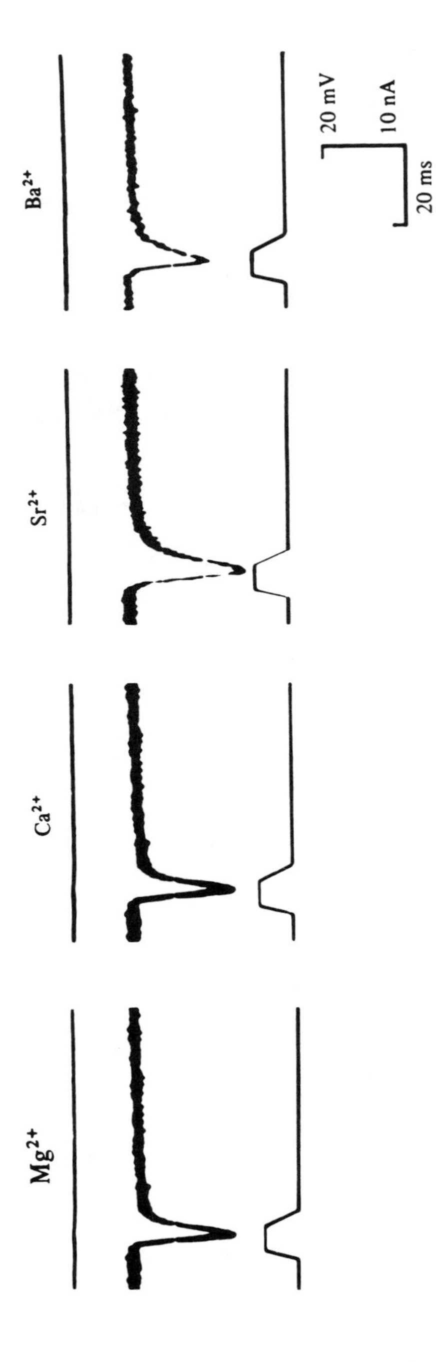

Figure 8. The anterior mechanoreceptor current (middle traces) in a voltage-clamped *Stylonychia* in different solutions containing four different kinds of alkaline-earth cations: Mg^{2+}, Ca^{2+}, Sr^{2+}, and Ba^{2+}. Upper traces: membrane potential. Lower traces: voltage pulse applied to a piezoelectric drive of the stimulation needle. (From de Peyer and Deitmer, 1980.)

The reversal potential is a linear function of log $[K^+]_o$ with a positive slope of 57 mV (Fig. 7B) (Ogura and Machemer, 1980). This indicates that the reversal potential for the posterior mechanoreceptor response is very close to the equilibrium potential for K^+ ions.

The amplitude of the hyperpolarizing mechanoreceptor potential in *Paramecium* caused by a mechanical stimulation of the posterior membrane decreases if tetraethylammonium (TEA), which is an effective blocking reagent of K channels in various kinds of excitable cells, is added to the external solution. When the concentration of TEA is 4 m*M*, the posterior membrane fails to produce the hyperpolarizing response; instead, it produces a depolarizing response to a mechanical stimulation, as does the anterior membrane. This indicates that blocking the outward K^+ current by TEA unmasks an inward Ca^{2+} current through the mechanically activated Ca channels coexisting with the mechanosensitive K channels in the posterior membrane. This inward Ca^{2+} current produces the depolarizing membrane response (Naitoh and Eckert, 1973). The presence of both kinds of mechanosensitive ion channels in the posterior region of *Paramecium* in a mosaic pattern has been dem-

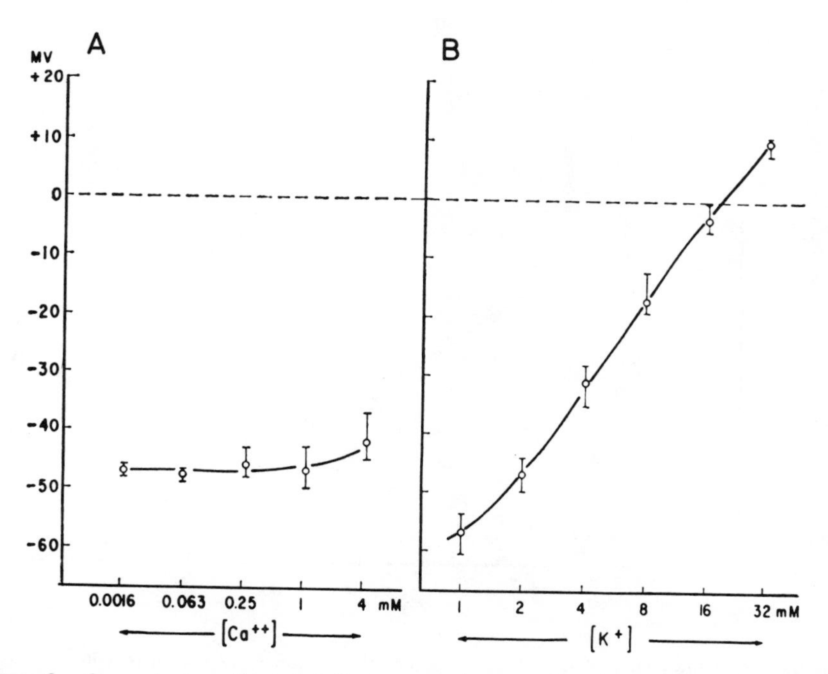

Figure 9. Concentration effects of Ca^{2+} (A) and K^+ (B) in the external solution on the peak level of the saturated posterior mechanoreceptor potential in *Paramecium*. (From Naitoh and Eckert, 1969*a*.)

onstrated by Ogura and Machemer (1980) (cf. Fig. 13), although the tip of the posterior region possesses the K channels only.

The posterior membrane also starts to produce a depolarizing response when $[K^+]_o$ is as high as 32 m*M* (Naitoh and Eckert, 1973). Since the resting potential level is close to the K^+ equilibrium potential in such a high K^+ solution, the outward K^+ current evoked by a mechanical stimulation is negligible, thereby unmasking the inward Ca^{2+} current.

In *Stylonychia* the inward Ca^{2+} current is never observed in response to mechanical stimulation of the posterior membrane in the presence of TEA (Deitmer, 1982). This fact indicates the absence of mechanosensitive Ca channels in the posterior region of the cell in this species.

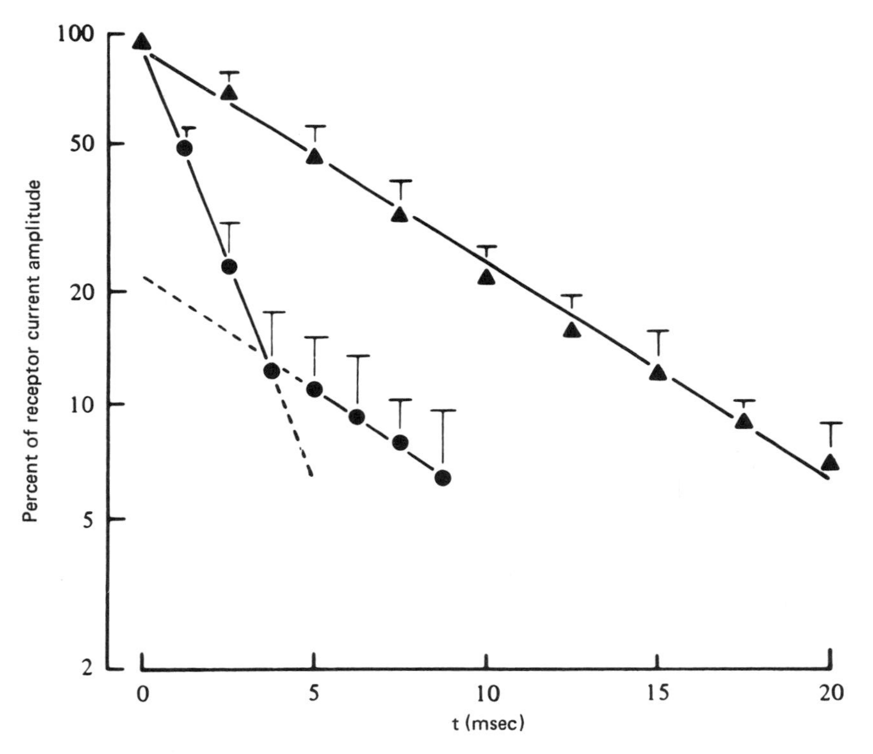

Figure 10. Time course of the decay in the anterior mechanoreceptor current in a voltage-clamped *Stylonychia* in the Ca solution (●) and in the Mg solution (▲). Logarithm of the current amplitude (percentage of the peak value) is plotted against time. (From de Peyer and Deitmer, 1980.)

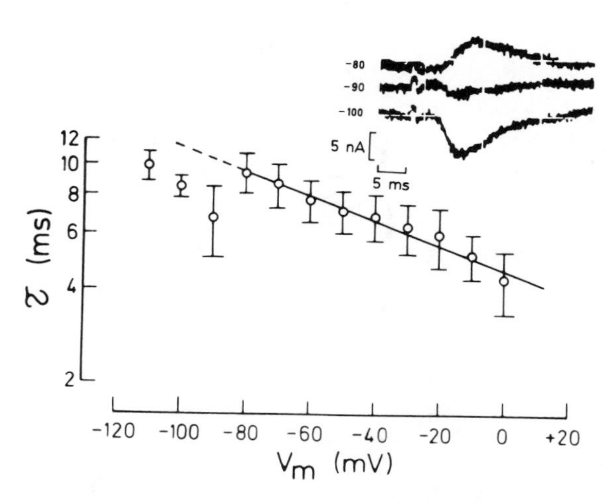

Figure 11. Logarithmic plot of the decay time constant (τ) in the posterior mechanoreceptor current against the membrane potential (V_m) in a voltage-clamped *Stylonychia*. Three mechanoreceptor currents produced under three different holding membrane potentials are shown at the upper right portion of the figure. (From Deitmer, 1981.)

4. KINETIC ANALYSIS OF THE MECHANORECEPTOR CURRENTS

Recently, de Peyer and Deitmer (1980) showed that the time course of the decay in the anterior mechanoreceptor current in a voltage-clamped *Stylonychia* consists of two single exponential components, one with a shorter time constant of ~1.8 msec followed by another with a longer time constant of ~7.2 msec, when the specimen is placed in a Ca^{2+}-containing solution. On the other hand, in a Ca^{2+}-free Mg^{2+} solution, in which the receptor current can only be carried by Mg^{2+} ions, the time constant of the decay is ~7.6 msec, which is comparable to that for the slow component in a Ca^{2+}-containing solution (Fig. 10). de Peyer and Deitmer have attributed the fast decay to Ca^{2+}-dependent inactivation of the mechanosensitive Ca channels and the slow decay to Ca^{2+}-independent inactivation of the channels.

Deitmer (1981) examined the effect of the membrane potential level on the decay kinetics of the posterior outward receptor current in *Stylonychia.* The time course of the decay is exponential with a time constant of ~7.3 msec. The time constant increases with increasing membrane hyperpolarization up to the reversal potential level (Fig. 11). A membrane hyperpolarization beyond the reversal potential produces a decrease in the time constant (Fig. 12).

Kinetic analysis of the mechanoreceptor current is essential to an understanding of the molecular mechanisms underlying mechanosensory transduc-

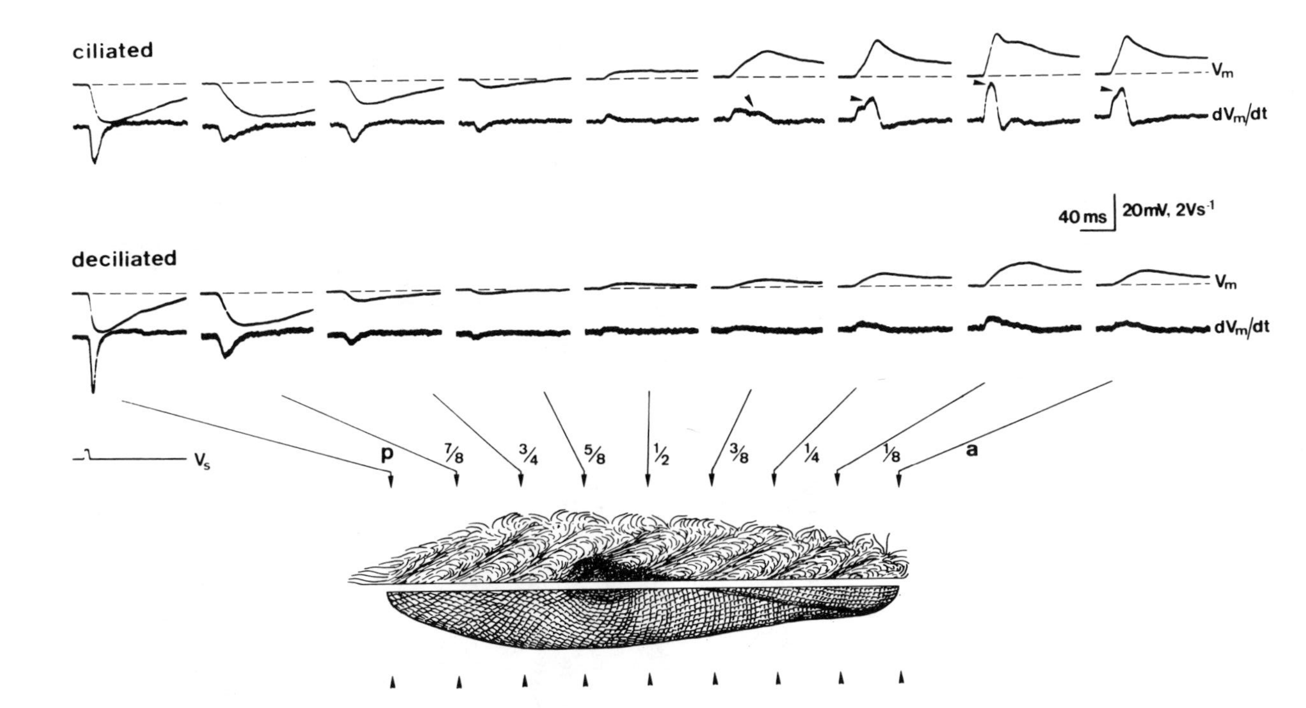

Figure 12. Membrane potential responses to a mechanical stimulation in both normal (ciliated) and deciliated *Paramecium* as a function of the stimulation site along the longitudinal axis of the specimens. V_m, Membrane potential; dV_m/dt, time derivative of V_m; V_s, voltage applied to the piezoelectric driver of the stimulation needle. (From Ogura and Machemer, 1980.)

tion in the cell membrane, although we are just beginning to take the first steps toward the solution of the problem.

5. TOPOGRAPHICAL DISTRIBUTION OF THE MECHANORECEPTOR CHANNELS

5.1. Ciliary Membrane or Somatic Membrane?

When most of the cilia of *Paramecium* are removed by combined treatment of the cells with various agents (chloralhydrate solution or an alcohol–Ca^{2+} mixture) and mechanical agitation, *Paramecium* loses its ability to produce a Ca action potential in response to a depolarizing stimulus (Ogura and Takahashi, 1976; Dunlap, 1977). The deciliated *Paramecium* can, however, produce depolarizing and hyperpolarizing mechanoreceptor potentials (Machemer and Ogura, 1978; Ogura and Machemer, 1980). These facts clearly indicate that the depolarization-sensitive Ca channels reside in the ciliary membrane, while both mechanosensitive K and Ca channels reside in the somatic membrane. The fact that a mechanical stimulation of nonciliated area of *Stylonychia* produces a mechanoreceptor potential also supports this hypothesis (de Peyer and Machemer, 1978*a*).

Machemer and Machemer-Röhnish (1982) showed the latent time for the receptor response to be longer when a mechanical stimulation is applied to the caudal cilia than to the soma membrane from which those cilia arise. If the ciliary membrane produces the mechanoreceptor response, as does the soma membrane, prolongation of the latent time is not expected, because the potential response evoked at any region of the ciliary membrane spreads instantaneously over the cell because of the electric isopotential nature of the cilia and the cell body (Eckert and Naitoh, 1970; Dunlap, 1977). It is conceivable that a mechanical distortion produced in a portion of a cilium is mechanically transmitted down the ciliary shaft toward the somatic membrane around the cilium, activating the mechanosensitive ion channels.

5.2. Anteroposterior Distribution

The amplitude of the depolarizing response in *Paramecium* to a mechanical stimulation of a given intensity decreases progressively from the anterior end toward the mid-region of the cell. Similarly, the amplitude of the hyperpolarizing response decreases as the mechanical stimulus is applied off the posterior region. The mid-region is apparently insensitive to mechanical stimulus (Naitoh and Eckert, 1969*a*, 1973; Eckert *et al.*, 1972).

Recently Ogura and Machemer (1980) found a series of successive changes in the amplitude and polarity of the mechanoreceptor response to mechanical stimulation of a given intensity, changing the stimulation site on the cell surface along the anteroposterior axis in both normal and deciliated *Paramecium*. Only a very small potential response can be seen in the mid-region (Fig. 12). The absence of the mechanoreceptor response in the mid-region of *Paramecium,* however, does not imply absence of mechanoreceptor channels in the region. Precise examinations of the membrane current in a voltage-clamped *Paramecium* (Ogura and Machemer, 1980) have revealed that mechanical stimulation of the mid-region produces a small current response, which is the sum of an inward Ca^{2+} current and an outward K^+ current. Since both currents are almost equal in size and opposite in direction, the net membrane current becomes virtually zero. Therefore, a membrane potential response as well as a ciliary response cannot be produced by a mechanical stimulation of the mid-region.

Increases in both K^+ and Ca^{2+} conductance in response to a mechanical stimulation of a given intensity in various regions on the cell surface of *Paramecium* can be estimated from the reversal potential for the mechanoreceptor response evoked in each region, according to Hodgkin and Horowicz (1959)

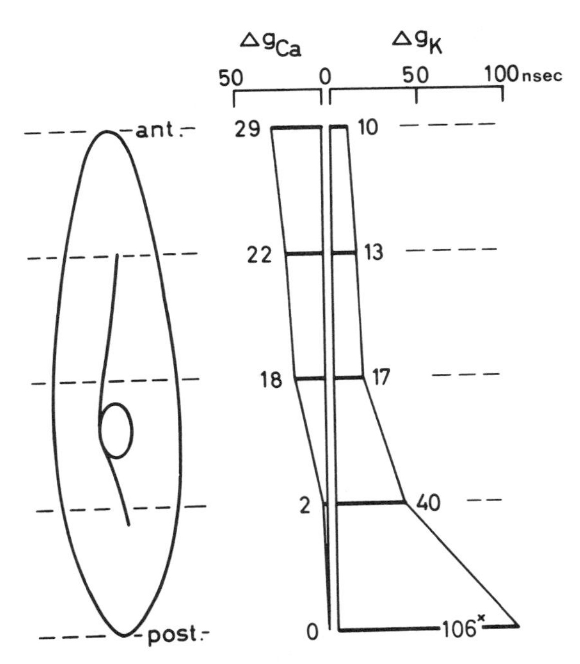

Figure 13. Increase in both Ca^{2+} and K^+ conductance (Δg_{Ca} and Δg_K) in response to mechanical stimulation of five different portions of *Paramecium*. (From Ogura and Machemer, 1980.)

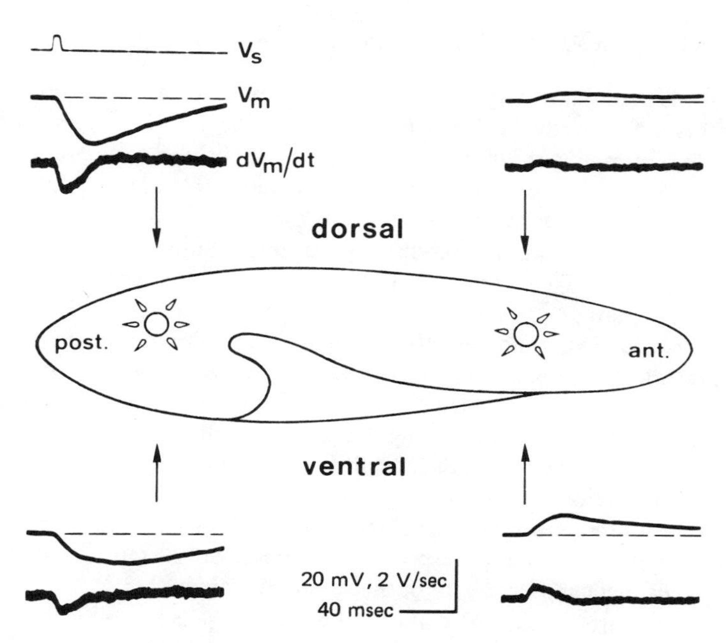

Figure 14. Dorsoventral difference in mechanosensitivity of *Paramecium*. V_m; membrane potential. dV_m/dt, time derivative of V_m; V_s, voltage applied to the piezoelectric driver of the stimulation needle. (From Ogura and Machemer, 1980.)

(Fig. 13) (Ogura and Machemer, 1980). On the basis of the conductance data, we can assume that the density of the mechanosensitive Ca channels is highest at the anterior end of the cell, decreasing progressively toward the posterior region, while the density of the mechanosensitive K channels is highest at the posterior end, decreasing progressively toward the anterior region. The two kinds of mechanoreceptor channels, can therefore be said to be arranged over the cell membrane in a more or less mosaic pattern.

Anteroposterior changes in the mechanoreceptor response essentially similar in their basic characteristics to *Paramecium* are also seen in *Stylonychia* (Fig. 2C) (de Peyer and Machemer, 1978*a*).

5.3. Dorsoventral Distribution

The amplitude of the depolarizing response to a mechanical stimulation of the anterior region of *Paramecium* is larger when the stimulation is applied to the ventral surface rather than to the dorsal surface. In contrast, the posterior hyperpolarizing response is more prominent in the dorsal membrane than in the ventral membrane (Fig. 14) (Ogura and Machemer, 1980).

5.4. Species-Dependent Differences in Mechanosensitivity

The mechanosensitivity of *Paramecium* in producing a mechanoreceptor potential is higher in the posterior region than in the anterior region. A barely detectable posterior mechanoreceptor potential appears in response to a stimulus far smaller than one producing a barely detectable anterior mechanoreceptor potential. Furthermore, the posterior response is saturated by a mechanical stimulus smaller than that producing a saturated anterior response (Naitoh and Eckert, 1973). Therefore, a weak mechanical agitation of the specimens, such as a light tap on the culture vessel, always produces a transient increase in their forward swimming velocity in response to the predominant activation of the posterior hyperpolarizing response. By contrast, *Stylonychia* produces an anterior depolarizing response larger in amplitude than the posterior hyperpolarizing response. Therefore, a light tap on the culture vessel always produces an avoiding reaction of the specimens caused by the predominant activation of the anterior depolarizing response (de Peyer and Machemer, 1978*b*). Some hypotrich ciliates other than *Stylonychia, Urostyla,* and *Euplotes* also show the avoiding reaction in response to a tapping on the culture vessel, indicating higher mechanosensitivity in their anterior regions.

The carnivorous ciliate *Didinium* produces a depolarizing mechanoreceptor potential whenever the cell surface is stimulated mechanically, although the mechanosensitivity in producing the receptor potential is higher in the anterior region than in the posterior region (Hara *et al.,* 1981).

6. COUPLING OF THE MECHANORECEPTOR RESPONSES WITH THE BEHAVIORAL RESPONSES

Mechanical stimulation of the anterior region of a ciliate protozoan, such as *Paramecium* and *Stylonychia,* activates the mechanosensitive Ca channels residing predominantly in the anterior somatic membrane. External Ca^{2+} ions move down their electrochemical gradient into the cell through the activated channels (the anterior receptor current), producing a depolarization of the membrane (the anterior mechanoreceptor potential). The receptor potential spreads electrotonically over the cell owing to the cable property of the cell body and cilia (Eckert and Naitoh, 1970; Dunlap, 1977), depolarizing the membrane of all the cilia. Depolarization of the ciliary membrane activates the depolarization-sensitive Ca channels, which reside exclusively in the ciliary membrane (Ogura and Takahashi, 1976; Dunlap, 1977; Machemer and Ogura, 1978), producing a regenerative Ca action potential accompanied by an inflow of external Ca^{2+} ions into the cilia (Naitoh *et al.,* 1972). The inflow produces an increase in Ca^{2+} concentration in the cilia, which in turn activates a Ca^{2+}-sensitive mechanism of the ciliary motile machine (Naitoh, 1969; Naitoh and Kaneko, 1972, 1973). The activation of this mechanism produces a counterclockwise shift in the direction of the effective power stroke of the cilia (if we look down on the cilia) together with an increase in the beating frequency of cilia (Machemer and Eckert, 1973; Machemer, 1974, 1976). The ciliary response has conventionally been termed *ciliary reversal*. Ciliary reversal produces a backward locomotion of the ciliate.

The mechanically activated Ca channels as well as the depolarization-activated Ca channels inactivate within a short time. The entry of Ca^{2+} into the cilia therefore soon stops. The Ca^{2+} concentration in the cilia decreases rapidly to its prestimulus lower level ($<10^{-7}$ *M*) as Ca^{2+} ions are pumped out by an active Ca^{2+} transport system in the ciliary membrane. Thus, the ciliary motile activity resumes its prestimulus state. The ciliate consequently resumes forward locomotion.

Mechanical stimulation of the posterior region activates the mechanosensitive K channels, which reside predominantly in the somatic membrane of the region, producing an outflow of the intracellular K^+ ions through the channels according to their electrochemical gradient (the posterior receptor current). An outflow of K^+ ions results in a hyperpolarization of the membrane (the posterior mechanoreceptor potential). The receptor potential spreads electrotonically over the cell and hyperpolarizes the ciliary membrane. A hyperpolarization of the ciliary membrane produces a clockwise shift in the direction of the effective power stroke of cilia together with an increase in the beating frequency of cilia through mechanisms not yet clarified. The ciliary response has been conventionally called *ciliary augmentation;* it is responsible for the behavioral response, i.e., escape reaction, of the ciliate to mechanical agitation of its posterior region.

The possibility that the receptor current might modify the motile activity of cilia directly has been excluded by the observation that both anterior and posterior receptor currents do not produce any kind of ciliary responses in *Stylonychia* when its membrane potential is kept clamped at the resting potential level (de Peyer and Machemer, 1978*b*). Modification or control of the ciliary motile activity is mediated exclusively by the membrane potential shift or by the accompanying ionic currents across the ciliary membrane, or by both mechanisms.

ACKNOWLEDGMENT. This work was supported by the Mitsubishi Foundation, Ministry of Education of Japan (grants 411802, 510902, 50105002) and by the Deutsche Forschungsgemeinschaft, grant SFB 114, TPA 5. I thank Dr. H. Machemer for his critical reading of the manuscript.

REFERENCES

Deitmer, J. W., 1981, Voltage and time characteristics of the potassium mechanoreceptor current in the ciliate *Stylonychia, J. Comp. Physiol.* **141**:173–182.

Deitmer, J. W., 1982, The effects of tetraethylammonium and other agents on the potassium mechanoreceptor current in the ciliate *Stylonychia, J. Exp. Biol.* **96**:239–249.

de Peyer, J. E., and Deitmer, J. W., 1980, Divalent cations as charge carriers during two functionally different membrane currents in the ciliate *Stylonychia, J. Exp. Biol.* **88**:73–89.

de Peyer, J. E., and Machemer, H., 1978*a,* Hyperpolarizing and depolarizing mechanoreceptor potentials in *Stylonychia, J. Comp. Physiol.* **127**:255–266.

de Peyer, J. E., and Machemer, H., 1978*b,* Are receptor-activated ciliary motor responses mediated through voltage or current? *Nature (Lond.)* **276**:285–287.

Doroszewski, M., 1961, Reception areas and polarization of ciliary movement in ciliate *Dileptus, Acta Biol. Exp.* **21**:15–34.

Doroszewski, M., 1965, The response of *Dileptus cygnus* to the bisection, *Acta Protozool.* **3**:175–182.

Doroszewski, M., 1970, Responses of the ciliate *Dileptus* to mechanical stimuli, *Acta Protozool.* **7**:353–362.

Dunlap, K., 1977, Localization of calcium channels in *Paramecium caudatum, J. Physiol. (Lond.)* **271**:119–133.

Eckert, R., 1965*a,* Bioelectric control of bioluminescence in the dinoflagellate *Noctiluca.* I. Specific nature of triggering events, *Science* **147**:1140–1142.

Eckert, R., 1965*b,* Bioelectric control of bioluminescence in the dinoflagellate *Noctiluca.* II. Asynchronous flash initiation by a propagated triggering potential, *Science* **147**:1142–1145.

Eckert, R., 1972, Bioelectric control of ciliary activity, *Science* **176**:473–481.

Eckert, R., and Brehm, P., 1979, Ionic mechanisms of excitation in *Paramecium, Annu. Rev. Biophys. Bioeng.* **8**:353–383.

Eckert, R., and Naitoh, Y., 1970, Passive electrical properties of *Paramecium* and problems of ciliary coordination, *J. Gen. Physiol.* **55**:467–483.

Eckert, R., and Naitoh, Y., 1972, Bioelectric control of locomotion in the ciliates, *J. Protozool.* **19**:237–243.

Eckert, R., Naitoh, Y., and Friedman, K., 1972, Sensory mechanisms in *Paramecium.* I. Two components of the electric response to mechanical stimulation of the anterior surface, *J. Exp. Biol.* **56**:683–694.

Eckert, R., Naitoh, Y., and Machemer, H., 1976, Calcium in the bioelectric and motor functions of *Paramecium, Symp. Soc. Exp. Biol.* **30**:233–255.

Goldman, D. E., 1943, Potential, impedance and rectification in membranes, *J. Gen. Physiol.* **27**:37–60.

Hara, R., and Naitoh, Y., 1980, Electrophysiological responses of *Didinium nasutum* to mechanical and electric stimulations, *Dobutsugaku Zasshi* [*Zool. Mag., Tokyo*] **89**:450.

Hodgkin, A. L., and Horowicz, P., 1959, The influence of potassium and chloride ions on the membrane potential of single muscle fibres, *J. Physiol. (Lond.)* **148**:127–160.

Jennings, H. S., 1906, *Behavior of the Lower Organisms,* Columbia University Press, New York.

Machemer, H., 1974, Frequency and directional responses of cilia to membrane potential changes in *Paramecium, J. Comp. Physiol.* **92**:293–316.

Machemer, H., 1976, Interaction of membrane potential and cations in reguration of ciliary activity in *Paramecium, J. Exp. Biol.* **65**:427–448.

Machemer, H., and Eckert, R., 1973, Electrophysiological control of reversed ciliary beating in *Paramecium, J. Gen. Physiol.* **61**:572–587.

Machemer, H., and Machemer-Röhnisch, S., 1982, Tail cilia of *Paramecium* passively transmit mechanical stimuli to the cell soma, *J. Submicrosc. Cytol.* **15**:281–288.

Machemer, H., and Ogura, A., 1978, Ionic conductances of membrane in ciliated and deciliated *Paramecium, J. Physiol. (Lond.)* **296**:49–60.

Machemer, H., and de Peyer, J., 1977, Swimming sensory cells: electrical membrane parameters, receptor properties and motor control in ciliate protozoa, *Verh. Dtsch. Zool. Ges.* **1977**:86–110.

Naitoh, Y., 1969, Control of the orientation of cilia by adenosine-triphosphate, calcium and zinc in glycerol extracted *Paramecium caudatum, J. Gen. Physiol.* **53**:517–529.

Naitoh, Y., 1974, Bioelectric basis of behavior in protozoa, *Amer. Zool.* **14**:883–893.

Naitoh, Y., 1982, Protozoa, in: *Conduction and Behavior in 'Simple' Invertebrates* (G. A. B. Shelton, ed.), pp. 1–48, Clarendon Press, Oxford.

Naitoh, Y., and Eckert, R., 1969*a*, Ionic mechanisms controlling behavioral responses of *Paramecium* to mechanical stimulation, *Science* **164**:963–965.

Naitoh, Y., and Eckert, R., 1969*b*, Ciliary orientation: Controlled by cell membrane or by intracellular fibrills? *Science* **166**:1633–1635.

Naitoh, Y., and Eckert, R., 1973, Sensory mechanisms in *Paramecium*. II. Ionic basis of the hyperpolarizing mechanoreceptor potential, *J. Exp. Biol.* **59**:53–65.

Naitoh, Y., and Eckert, R., 1974, The control of ciliary activity in protozoa, in: *Cilia and Flagella* (M. A. Sleigh, ed.), pp. 305–352, Academic Press, London and New York.

Naitoh, Y., and Kaneko, H., 1972, Reactivated Triton-extracted models of *Paramecium:* modification of ciliary movement by calcium ions, *Science* **176**:523–524.

Naitoh, Y., and Kaneko, H., 1973, Control of ciliary activities by adenosinetriphosphate and divalent cations in Triton-extracted models of *Paramecium caudatum, J. Exp. Biol.* **58**:657–676.

Naitoh, Y., Eckert, R., and Friedman, K., 1972, A regenerative calcium response in *Paramecium, J. Exp. Biol.* **56**:667–681.

Nuccitelli, P., Poo, M. M., and Jaffe, L. F., 1977, Relations between ameboid movement and membrane-controlled electrical currents, *J. Gen. Physiol.* **69**:745–763.

Ogura, A., and Machemer, H., 1980, Distribution of mechanoreceptor channels in the *Paramecium* surface membrane, *J. Comp. Physiol.* **135**:233–242.

Ogura, A., and Takahashi, K., 1976, Artificial deciliation causes loss of calcium-dependent response in *Paramecium, Nature (Lond.)* **264**:170–172.

Takahashi, M., Onimaru, H., and Naitoh, Y., 1980, A mutant of *Tetrahymena* with non-excitable membrane, *Proc. Jpn. Acad.* **56B**:585–590.

Wood, D. C., 1970, Electrophysiological studies of the protozoan, *Stentor coeruleus, J. Neurobiol.* **1**:363–377.

Wood, D. C., 1975, Protozoa as models of stimulus transduction, in: *Aneural Organisms in Neurobiology, Advances in Behavioral Biology,* Vol. 13 (E. M. Eisenstein, ed.), pp. 5–23, Plenum Press, New York.

Chapter 4

Temperature Sensing in Microorganisms

Kenneth L. Poff, Donna R. Fontana, and Bruce D. Whitaker

1. INTRODUCTION

1.1. Overview

Many microorganisms are known to sense and respond to their thermal environment with a behavior modification. The objective of this chapter is to review the current state of knowledge concerning this thermosensing phenomenon. Toward this end, we discuss some of the difficulties encountered in a study of thermosensing mechanisms as well as the types of elements that might be involved in the initial sensing steps of the transduction process. Finally, we conclude with a survey of specific biological examples of thermosensing.

1.2. Difficulties Peculiar to Thermosensing

Although thermosensing in microorganisms was first described almost a century ago (Schenk, 1893), progress in this area of sensory physiology has been particularly slow until very recently. We believe this slow progress to be

Kenneth L. Poff and Donna R. Fontana • MSU-DOE Plant Research Laboratory, Michigan State University, East Lansing, Michigan 48824. **Bruce D. Whitaker** • Department of Biochemistry, University of Wisconsin, Madison, Wisconsin 53706. Present address for Dr. Fontana: Department of Physiological Chemistry, Johns Hopkins University, Baltimore, Maryland 21205.

in part the direct result of specific identifiable conceptual difficulties. It is the intention of this section to identify some of these difficulties with the hope that this will lead toward possible approaches to their solution.

Thermosensing has been used in the literature by those in the field to imply a considerable level of specificity. For example, the terminology of sensing has not been applied to the fact that an organism has an optimum growth temperature—a clear case of thermosphysiology but not of thermosensing. Thus, the literature leans toward the idea that a response that is a general outgrowth of thermal effects on metabolism probably does not constitute sensing. The term thermoreceptor should be reserved for the molecule(s) present at the beginning of the thermosensory reaction chain and directly perturbed by the thermal energy.

It is assumed in most instances of sensory transduction that there is a specific sensor and transduction pathway leading from a specific stimulus to a specific behavioral responses. For example in chemosensing and in photosensing, the stimulus is highly specific (a particular molecule such as cAMP or a specific wavelength of light such as 450 nm light). The comparable assumptions concerning specificity generally have not been made with thermosensing. Heat energy must inherently affect all cellular components and metabolic reactions. Thus, there is little if any specificity in the interaction of the stimulus with its receptor—the stimulus is interacting with all molecules. In spite of this there is an accumulation of evidence (based on mutants, specific inhibitors, specific ion effects, and extreme sensitivity) in several systems strongly supporting the conclusion that there is indeed a specific sensor and transduction pathway for thermosensing, i.e., in *E. coli* (Section 3.1), *Paramecium* (Section 3.2), *Dictyostelium discoidem* (Section 3.4), and *Caenorhabditis elegans* (Section 3.5).

The arguments above support the contention that a specific transduction pathway mediates thermosensing. The element of this pathway which interacts with heat energy and which limits the pathway must then be the thermosensor or biothermometer. Unfortunately, since all components interact with the heat energy, the mere interaction of the stimulus with a molecule cannot be used as evidence that particular molecule is the thermosensor. Thus, the identification of the thermosensor becomes considerably more difficult than the identification of a photoreceptor pigment or chemoreceptor, since each of these receptors is one of very few molecules interacting with the stimulus.

One can imagine an organism detecting the absolute temperature T, a change in temperature over time dT/dt, or a change in temperature over some distance dT/dx. In microorganisms, the known examples of thermosensing which result in thermotaxis appear to involve the detection of a change in temperature with time or distance. In contrast, the temperature itself is probably the stimulus in the thermal-adaptation responses.

It is impossible to remove all heat stimulus from a biological system. Thus, if an organism senses temperature, one cannot study the response in the absence of any stimulus. Such an insurmountable barrier is not encountered in any other sensing system. For example, light can easily be completely removed as can specific chemicals to which an organism might respond. Even the forces of gravity can be removed under special conditions. In contrast, the complete removal of heat is by definition nonphysiological.

The methods available for measuring temperature are relatively insensitive, such that one must routinely contend with a situation in which the organism is far more sensitive to temperature than the nonbiological sensors. This is much less a problem with other stimuli, but offers a challenge for understanding the methods whereby such extreme sensitivity is achieved in biological systems.

Few techniques are currently available for studying the interactions of molecules with heat and these are of limited value in thermosensing. For example, a calorimeter may be used to measure large molecule–heat interactions as spectrophotometry may be used for studying molecule–light interactions. However, the sensitivity of calorimetry is limited. Moreover, the molecule of interest—the sensor—is surrounded by other molecules that also interact with heat. Thus, *in vivo* calorimetry is of minimal value as a probe for the thermosensor, whereas *in vivo* spectrophotometry may be of considerable value as a probe for the photoreceptor pigment likely to be one of only a few molecules absorbing one particular wavelength of light.

There are several general although largely circumstantial reasons for believing that the primary steps in thermosensing in some way involve cellular membranes:

1. The evidence provided by Maeda and Imae (1979) with *Escherichia coli:* that the membrane-bound Tsr chemoreceptor is also the thermoreceptor or at least is required for thermoreception.
2. Every known sensory receptor for any stimulus is located on a membrane or involved with a membrane early in the transduction process.
3. Given the sensitivty of many organisms to temperature, it seems likely that an amplification step is involved early in thermosensing. From strictly theoretical considerations, a membrane is an ideal site for such an amplification step.
4. Many thermosensing systems involve the phenomenon of adaptation wherein the characteristics of the temperature response curve are dependent upon the temperature history of the organism. The cellular processes shown to demonstrate adaptation in the short period of time in which adaptation occurs are those regulating membrane lipid composition.

2. CELLULAR COMPONENTS OR PROCESSES AFFECTED BY TEMPERATURE

2.1. Enzymes

A change in temperature can affect an enzyme in various ways. Temperature is a factor in determining enzyme stability, enzyme–substrate affinity, the amount of substrate in instances in which substrate dissolution is necessary (e.g., if the substrate is O_2), enzyme affinity for activators or inhibitors, and the rate of breakdown of the enzyme–product complex (Lee, 1977*a,b*; Overath and Thilo, 1978). Temperature is also an important component in determining the apparent p_K of ionizable groups (Dixon and Webb, 1964), which may affect enzyme activity as well.

The effect of temperature on enzyme stability and affinity to substrates or regulators can be explained in terms of the weak bonds controlling these functions (Somero, 1975; Somero and Hochachka, 1976). Hydrogen bonding, electrostatic interactions, and van der Waals interactions are weakly exothermic and hydrophobic interactions are weakly endothermic; the small enthalpies involved make these interactions temperature sensitive. A reduction in temperature will stabilize the exothermic bonds and weaken the endothermic ones. The converse occurs with a rise in temperature. Depending on which weak interaction dominates when determining the stability of a protein or the binding of a molecule, the effect of a change in temperature will differ. In tightly coupled enzymatic systems, the K_m, used as a measure of enzyme–substrate affinity, has been found not to vary over twofold throughout the physiologically relevant temperature region (Somero and Hochachka, 1976). However, this may not be true of reactions that are not tightly coupled. This suggests that when substrate affinity is critical, adaptation has made it temperature invariant.

The dissolution of substrate, the effect of p_K, and the increase in rate of enzyme–product complex breakdown are understandable when one realizes that the energy of the system is related to temperature—i.e., an increase in temperature adds energy to the system. Also, because the energy of the system determines what fraction of collisions between enzyme and substrate will yield the product, the reaction velocity and reaction equilibria are also dependent on temperature (Segel, 1975; Somero and Hochachka, 1976). However, the activation energy and enthalpy of enzymatic reactions are not temperature dependent (Somero and Hochachka, 1976). Thus when one examines a plot of reaction rate versus temperature, or $1/T$, the plot shape is the result of varying the energy in the system and the effect of this energy on weak interactions, p_K, possibly substrate availability, and the probability of a reaction on enzyme–substrate collision.

Even though so many facets of an enzymatic reaction are temperature dependent, the rate of an enzymatic reaction usually doubles (or less) with a 10° increase in temperature (Dixon and Webb, 1964; Segel, 1975). This lack of sensitivity to environmental changes suggests that if proteins are involved in thermosensing, they are acting in conjunction with another component, such as a membrane, the functional and/or structural properties of which depend much more strongly on temperature.

2.2. Membranes

It is well established that membranes are affected by the ambient temperature. The temperature can determine the physical state of lipids, alter the activity of membrane-bound enzymes, and in living organisms be an important factor in determining membrane composition. In order to understand the interaction between membranes and temperature on a biophysical level, it is necessary to examine the individual components of a membrane and determine how these are altered by a change in temperature. This discussion begins with the dependence of the physical state of lipids on temperature, and continues with a brief examination of protein–lipid interactions as they may be modulated by temperature changes. After the discussion of physical effects of temperature changes, we examine the role of growth temperature in determining membrane composition.

Perhaps the best studied effect of temperature on lipids, particularly phospholipids, is the phase transition. At lower temperatures, lipids in bilayers exist in a gel or ordered state. In this state, the hydrocarbon chains are in an all-trans-conformation and the molecules are closely packed in hexagonal arrays. As the temperature is raised, the number of defects in the packing increases. Near a characteristic temperature called the transition temperature *trans-gauche* isomerizations occur in the hydrocarbon chains at packing defects, leading to an expansion of the bilayer. This rotation about the C—C bond makes it easier for similar rotations to occur in adjacent molecules. In this way, the defects spread rapidly as the temperature is increased until the bilayer is in a fluid or liquid crystalline state. This transition may be accompanied by a change in the orientation of the fatty acids and polar headgroups with respect to the bilayer surface (reviewed in Träuble and Eibl, 1975; Lee, 1977*a*).

This type of phase transition will occur in mixtures of phospholipids as long as the lipids are completely miscible in all physical states. However, complete miscibility is not the case in biological membranes because of the diversity of hydrocarbon chains and polar headgroups. What typically occurs during a biological transformation is that one component of the membrane begins C—C bond rotation before the others. This fluid component will tend to separate from the other components that are still in the gel state. This segregation is called lateral phase separation. As the temperature is raised, more components become fluid and the fluid domains grow at the expense of the ordered patches. At an even greater temperature, all the lipid components are in a fluid state (reviewed by Lee, 1977*b*). This process usually occurs at temperatures well below the physiological range, but there are examples in which this transition is physiologically relevant (Overath and Thilo, 1978). Phase transitions, if physiologically relevant, can act as transducers by altering membrane permeability (DeGier *et al.*, 1968; Murata and Fork, 1976), the rate of lateral diffusion of membrane components (Wu *et al.*, 1977), osmotic behavior (Haest, *et al.*, 1972; Van Zoelen *et al.*, 1975), and the activity of membrane-bound enzymes (Overath and Thilo, 1978).

Other lipid transitions can occur as a result of a temperature change. One such transition is the transition from a lamellar to hexagonal phase. When lipids are in the hexagonal phase, their head groups form an aqueous channel, a water cylinder, which traverses the membrane. This lipid arrangement can only occur in the presence of certain lipids such as phosphatidylethanolamine and cardiolipin and can only occur at temperatures above the gel to liquid-crystalline phase transition (Rand and Sengupta, 1972 Cullis and DeKruijff, 1978*a*). In lipid mixture, this transition to a hexagonal phase is often accompanied by an increase in isotropic motion as seen with 3′ *P*-NMR (Cullis and DeKruijff, 1978*a,b*). This isotropic component, which is not the result of a hexagonal phase, has been seen in mitochondrial and microsomal membranes at physiological temperatures (Stier *et al.*, 1978; DeKruijff *et al.*, 1978; Cullis *et al.*, 1980; DeKruijff *et al.*, 1980) and suggests a nonlamellar lipid arrangement. The existence of these phases (lamellar, hexagonal, or the isotropic phase) in membranes depends on membrane composition, the ionic strength of the medium, the pH level, and the temperature of the environment (Rand and Sengupta, 1972; Stier *et al.*, 1978; Cullis and DeKruijff, 1978*a,b;* DeKruijff *et al.*, 1980; Cullis *et al.*, 1980; Boggs *et al.*, 1981), the phase relationships may be altered by changing any of these factors.

Even in the absence of a phase transition, the biophysical properties of membranes are not temperature independent. Raising the temperature increases the probability of a *trans-gauche* isomerization in the hydrocarbon chains, thereby making this region of the membrane more fluid. This change in fluidity will affect the permeability of small ions and molecules (Lee, 1975) and also the rate of lipid lateral diffusion (Wu *et al.*, 1977). Polar head group movement is also temperature dependent (Caldwell and Haug, 1981). The existence of lipid clusters well above the gel to liquid-crystalline phase transition has been demonstrated (Lee *et al.*, 1974; Ting and Solomon, 1975; Bashford *et al.*, 1976), and their existence is temperature dependent. Thus, a change in environmental temperature can alter the physical properties of membrane lipids in addition to causing phase changes.

2.3. Protein–Lipid Interactions

The fact that lipids are often necessary and able to alter the activity of membrane-bound proteins is well documented (for recent reviews see Capaldi, 1977; Sandermann, 1978; Jain, 1980). The degree of enzyme activation which results from lipid addition may depend on the amount and identity of the added lipid, and may also depend on the physical state of that lipid. The actual interaction between the protein and lipids may be hydrophobic, electrostatic or by means of covalent attachment.

Because of these protein–lipid interactions, a change in membrane fluidity or composition can be reflected in a change in the distribution of a membrane-bound enzyme. The preference of some enzymes for a particular lipid or particular phase can result in the aggregation of proteins into clusters or patches. This aggregation is often observed with the electron microscope when proteins are embedded in a membrane undergoing a lateral phase separation. The proteins seem to prefer the more fluid regions.

Altering membrane fluidity may also cause a change in protein orientation and alter interactions between protein subunits. The effect of fluidity on the vertical placement of proteins has recently been receiving attention (Borochov and Shinitzky, 1976; Heron *et al.*, 1980; Muller and Shinitzky, 1981). It appears that altering the fluidity may change the vertical position of a protein (Somero, 1975; Heron *et al.*, 1980)—i.e., a decrease in fluidity might squeeze a protein out of a membrane (Borochov and Shinitzky, 1976). This squeeze increases the exposure of the protein to its environment, thereby altering protein–lipid interactions and possibly altering substrate binding (Heron *et al.*, 1980) and protein–protein interactions. For example, hyperrigidification of membrane lipids has been shown to induce the shedding of erythrocyte antigens (Muller and Shinitzky, 1981).

If a protein must turn in the membrane in order to act, the membrane fluidity must be very important because it reflects the lateral compressibility of the membrane. This degree of lateral compressibility may also affect the conformation of an enzyme, thereby exerting a more subtle effect on its activity. If the substrate must dissolve into the membrane before it can be acted on, lateral compressibility can again play a role.

In summary, a temperature change may alter the biophysical state of membrane lipids. This state can subsequently affect the activity of membrane-bound enzymes by altering their distribution, orientation, and ability to move within the membrane. The state may also play a role in protein conformation and substrate interaction.

2.4. Membrane Lipid Composition

Studies of the effect of growth temperature on membrane lipid composition have been performed with a variety of microorganisms, including bacteria (Marr and Ingraham, 1962, Sinensky, 1974), yeasts (Kates and Baxter, 1962; Meyer and Bloch, 1963; Kates and Paradis, 1973), algae (Patterson, 1970), and protozoans (Erwin and Block, 1963). The generalization derived from these studies is that microorganisms appear to use the mechanism of homeoviscous adaptation to cope with changing environmental temperature (Marr and Ingraham, 1962; Meyer and Bloch, 1963; Patterson, 1970; Sinensky, 1971; Huang *et al.*, 1974*a;* Pugh and Kates, 1975). Homeoviscous adaptation is a term coined by Sinensky (1974) to describe the process through which bacteria alter the fatty acid composition of membrane phospholipids in response to a change of growth temperature and thereby maintain a constant membrane fluidity.

The change in membrane phospholipids of *E. coli* and other microorganisms brought about by decreasing growth temperature is an increase in the proportions of unsaturated and short-chain fatty acids (Marr and Ingraham, 1962; Meyer and Bloch, 1963; Patterson, 1970; Sinensky, 1974; Huang *et al.*, 1974*b;* Pugh and Kates, 1975). Raising the ratio of unsaturated to saturated fatty acids in a phospholipid bilayer reduces the degree of ordering of the hydrocarbon tails and increases the fluidity (Esfahani, *et al.*, 1971; Keith *et al.*, 1973; Chapman, 1975). This decrease in order and fluidity compensates for the increase in order and fluidity which results from the reduction in growth temperature.

In those microorganisms for which a distinct effect of temperature on fatty acid composition has been demonstrated, one does not observe a fluctuation in the levels of all the fatty acids, but rather a change in the ratios of a few select fatty acids. There is also frequently specificity in the class of phospholipids, which is modified by a change in temperature. Finally, a linear relationship between temperature and the unsaturation of fatty acids is not always observed. Instead, abrupt changes in the ratios of one group of fatty acids can occur at a certain critical temperature (Erwin, 1973). These features of temperature-induced adaptation of lipids in microoorganisms appear to reflect the underlying effects of temperature on the synthesis and activity of specific enzymes involved in phospholipid biosynthesis or restructuring and in fatty acyl desaturation (Erwin, 1973).

Some question remains as to the significance of the temperature-induced changes in membrane lipids of microorganisms with respect to survival and continued growth. In a classical study of the effect of temperature on the fatty acid composition of *E. coli* membranes, Marr and Ingraham (1962) demonstrated that the proportion of unsaturated fatty acids incorporated into membrane phospholipids at a given growth temperature could be varied over a wide range by altering nutritional conditions. Shaw and Ingraham (1965) subsequently showed that if cultures of *E. coli* were glucose starved for several hours before a temperature shift from 37° to 10°C, there was no increase in fatty acid unsaturation at the lower growth temperature, and no change in the 4.5-hr growth lag that normally followed such a shift down in temperature. Later, Gelmann and Cronan (1972) performed similar experiments with unsaturated fatty acid auxotrophs of *E. coli* that required either palmitoleate (16:1) or vaccenate (18:1) for growth. The proportion of unsaturated fatty acids in membranes of the mutants was not altered by lowering the growth temperature from 37° to 15°C, and the growth rate at 15°C remained close to that of wild-type cultures. The fact remains, however, that the phase properties of membrane lipids are of vital importance to *E. coli,* both fluid and nonfluid phospholipids being requisite for growth (Cronan and Gelmann, 1975). If the membrane of *E. coli* is altered to the extent of containing less than roughly one-third the normal proprotion of fluid or nonfluid phospholipids, cell death ensures; yet within these limits, cell growth is insensitive to the fluid:gel ratio of membrane phospholipids. These data suggest that there are few specific interactions between bulk lipid and other membrane components in *E. coli* (Cronan and Gelmann, 1975).

Thus, although the low-temperature-induced increase in the proportion of unsaturated fatty acids in membranes of *E . coli* and other microorganisms may not be essential for growth, the resultant optimization of the physical state of the membrane has been sufficient to ensure evolutionary selection of this thermal adaptation process (Cronan and Gelmann, 1975).

3. BIOLOGICAL EXAMPLES

3.1. Bacteria

The bacterium *Escherichia coli* is the most primitive organism that has been shown to exhibit directed movement in response to a temperature gradient (Maeda *et al.,* 1976). If *E. coli* grown at 34°C are exposed to a thermal gradient of 17°–39°C, the bacteria will accumulate and form a dense band at 34°C; this band will slowly move toward lower temperatures. Thus, the bacteria are initially accumulating at their growth temperature. While it suggests a direct thermal response, this type of experiment, is not conclusive because temperature may affect the rate of oxygen consumption and *E. coli* is known to exhibit aerotaxis (Adler, 1966). Maeda *et al.* (1976) have suggested that the initial accumulation is the result of thermotaxis and the slow movement of the band to lower temperatures is the result of aerotaxis.

To avoid this complication, individual cells were examined as the temperature of the medium was changed. This type of experiment was used successfully in the study of bacterial chemotaxis because *E. coli* senses changing chemical gradients temporally, as opposed to spatially. A change in the concentration of a chemical stimulus alters the ratio of clockwise to counterclockwise motion of the flagella. [For recent reviews see Macnab (1978) and Koshland (1981).] Clockwise flagellar motion results in smooth swimming, while counterclockwise motion causes tumbling. By using a video recorder and examining the swimming tracks of an individual bacterium, the level of sensory stimulation can be determined.

Using this technique, Maeda *et al.* (1976) examined bacteria as the temperature of the medium was dropped from 30° to 20°C at a rate of 0.17°C/sec. These workers found that the tumbling frequency was maximal at 30 sec after the drop was initiated. After 1 min, the tracks resembled those of the controls held at 20°C. If the temperature is raised from 27° to 34°C at a rate of 0.12°C/sec, the tumbling frequency drops reaching a minimum at 30 sec and then rises to the control level. This combination of responses would result in swimming toward higher temperatures and tumbling if the bacterium were to turn toward a cooler temperature. Maeda *et al.* were unable to perform similar experiments above 34°C because these temperatures result in highly curved swimming tracks, which interfere with the analysis, and the high temperature can cause nonreversible changes in tumbling.

Maeda and Imae (1979) attempted to relate the thermotatic response to the extensively studied chemotactic responses of *E. coli*. They found L-serine to be a potent inhibitor of the thermotactic response, with 4 μM providing 50% inhibition. AiBu (α-aminoisobutyrate) and L-alanine, the chemoresponses of which follow the same pathway (Tsr) as the serine response, are only weak inhibitors of the thermoresponse with > 0.01 *M* necessary for 50% inhibition. L-aspartate, a chemoeffector the stimulus of which is transduced through the tar gene product (Tar pathway), is also only a weak inhibitor of the *E. coli* thermoresponse. Mutants blocked in the Tar pathway exhibit wild-type thermoresponses. Tsr mutants show little, if any, response to a sudden decrease in temperature. These results suggest that the Tsr pathway tranduces the thermosensory stimulus as well as the stimuli of various chemoeffectors such as L-serine, L-alanine, AiBu, and glycine. The results also suggest that the interaction between serine and the thermal stimulus occurs before the transduction pathway merges with those of L-alanine and AiBu.

E. coli possesses two chemoreceptor systems for serine, both of which are mediated by the *tsr* gene product (Springer *et al.*, 1977; Maeda and Imae,

1979). The high-affinity chemoreceptor system has a K_m of about 10^{-6}, while the low-affinity system has a K_m of about 10^{-4} *M*. *CheB* mutants, which only respond to serine concentrations of 0.1–10 m*M*, show no thermoresponse. These data, as well as the similarity between the concentration of L-serine needed for 50% inhibition of the thermoresponse and the K_m of the high-affinity receptor for L-Ser, suggest that the thermoresponse is mediated by the high-affinity serine receptor.

Maeda and Imae (1979) have proposed that the *E. coli* high-affinity serine receptor is the thermal receptor. Their model suggests that an increase in temperature causes a conformational change in the high-affinity serine receptor, which is similar to that caused by serine binding. This model is supported by recent data which show that the *tsr* gene product, the methyl-accepting chemotaxis protein I, is also the serine receptor (Clarke and Koshland, 1979; Wang and Koshland, 1980; Hedblom and Adler, 1980). Because other chemoeffectors, such as AiBu and L-Ala, the transduction pathways of which involve this methyl-accepting protein only weakly inhibit the thermoresponse, the L-serine inhibition must occur before this interaction. Therefore, the interaction between the serine and thermal stimuli must occur at or very close to the receptor level. These results clearly indicate that the methyl-accepting chemotaxis protein I is intimately involved in thermosensing. Whether a temperature change affects the protein directly or the change is modulated by membrane lipids has not yet been determined.

3.2. Cilates

Thermosensory behavior in the free-swimming, ciliated protozoan, *Paramecium,* was initially described around the turn of the century by Mendelssohn (1895) and Jennings (1906). When placed in a temperature gradient, the organisms are observed to accumulate at or near some optimum temperature (Mendelssohn, 1895; Jennings, 1906). This temperature optimum correlates well with the growth temperature of *Paramecium* and can be shifted up or down by raising or lowering the growth temperature, respectively (Jennings, 1906; Tawada and Oosawa, 1972). In addition to temperature optimum-seeking behavior, *Paramecium* also exhibits a thermal avoidance response upon entering a region of physiologically harmful high temperature (Jennings, 1906; Hennessey and Nelson, 1979). The avoidance response is characterized by a transient reversal of the direction of the ciliary stroke (ciliary reversal), which results in backward swimming. Ciliary reversal is followed by a reorientation of the organism, which then resumes normal swimming off in a new direction (Jennings, 1906; Eckert *et al.,* 1976; Nakaoka and Oosawa, 1977). Detailed studies of thermosensory behavior in *P. caudatum* indicate that accumulation at an optimum temperature probably arises because of an increase in the frequency of ciliary reversal (and thus direction changes) as the animals in a temperature gradient swim away from the "favored" temperature (Nakaoka and Oosawa, 1977).

The avoidance response of *Paramecium* is elicited by a variety of stimuli other than high temperature, including chemical repellants such as tetraethylammonium (TEA), high concentrations of K^+ or Ba^{2+} ions, contact with a physical barrier, and an electrical current through the suspending medium (Jennings, 1906; Eckert *et al.,* 1976). The ciliary reversal that initiates the avoidance response is associated with a rapid, transient depolarization of the cell membrane (action potential) (Naitoh, 1968; Eckert, 1972; Eckert and Brehm, 1979).

Electrophysiological studies have revealed that membrane depolarization is produced by a sudden increase in Ca^{2+} conductance and a resultant rapid influx of Ca^{2+}, which raises the intracellular $[Ca^{2+}]$ from $< 10^{-7}$ to $\sim 10^{-5}$ *M*. Calcium ions are subsequently pumped back out of the cell, relatively slowly, to reestablish an internal $[Ca^{2+}] < 10^{-7}$ *M* (Naitoh, 1968; Eckert, 1972; Eckert *et al.,* 1976; Eckert and Brehm, 1979).

The sequence of events leading from the rapid rise of the intracellular $[Ca^{2+}]$ to reversal of ciliary beat is unknown. Recently it has been found that the calcium channels responsible for the increased Ca^{2+} conductance upon excitation of the membrane are located predominantly in the ciliary membrane (Naitoh, 1968; Eckert, 1972; Ogura and Takahashi, 1976; Eckert and Brehm, 1979). Deciliated cells do not show action potentials (Ogura and Takahashi, 1976). Mutant strains of *Paramecium,* called *pawns,* which upon stimulation fail to show an action potential, also exhibit no ciliary reversal (Kung, 1971; Kung and Eckert, 1972). These mutants, now known to have defective Ca channels (Kung and Naitoh, 1973), do not exhibit thermal avoidance (Hennessey and Nelson, 1979).

In light of these data, it appears certain that the thermal avoidance response of *Paramecium* is mediated by a thermally induced increase in the Ca^{2+} conductance of the ciliary membrane (Hennessey and Nelson, 1979). The question that remains is: How does temperature bring about this change in Ca^{2+} conductance across the membrane?

On the basis of this information and current concept of membranes, it seem logical to propose that the lipids of the ciliary membrane of *Paramecium* serve as the thermotransducer for the thermal avoidance response. At some critically high temperature, a phase transition of the lipids surrounding the Ca channels of the ciliary membrane could result in an abrupt increase in Ca^{2+} conductance (an "opening" of the Ca channels), thereby setting up an action potential and ciliary reversal.

There are already several lines of evidence in support of this *membrane lipid-as-thermotransducer* hypothesis. The quantitative assay of thermal avoidance by *Paramecium* devised by Hennessey and Nelson (1979) revealed that the response proceeds from the threshold to saturation levels over a narrow temperature range (37°–42°C for cells growth at 28°C). This observation is consistent with the suggestion that a phase transition of membrane lipids triggers the avoidance response, as such lipid-phase transitions are generally complete within a narrow temperature range (Chapman, 1975).

Both the temperature optimum for accumulation and the threshold for thermal avoidance of *Paramecium* are shifted up or down in accordance with the growth temperature of the organism (Jennings, 1906; Tawada and Oosawa, 1972; Hennessey and Nelson, 1979). This process has been demonstrated in the ciliate *Tetrahymena,* a close relative of *Paramecium* (Wunderlich *et al.,* 1973; Kitajima and Thompson, 1977). Perhaps the least complicated interpretation of this adaptation is that it arises from an alteration of the organism's membrane lipid composition.

For the study of temperature-induced changes in the lipid composition of *Tetrahymena* membranes, Thompson and co-workers isolated a strain of *T. pyriformis* (NT-1) from a hot spring (Fukushima *et al.,* 1976; Thompson and Nozawa, 1977). The ability of strain NT-1 to grow well at high temperature permitted a comparison of membrane lipid composition before and after a drastic reduction in growth temperature (39.5° down to 15°C). Acclimation of *Tetrahymena* cultures after such a temperature shift required 8 hr, but increased unsaturation of fatty acids acylated to lipids in the endoplasmic reticulum occurred within 30 min (Thompson and Nozawa, 1977). Associated studies employing the techniques of freeze-fracture electron microscopy (Wunderlich *et al.,* 1973) and electron spin resonance (ESR) spectrometry (Nozawa *et al.,* 1974) have correlated the increased unsaturation of membrane lipid fatty acids at low growth temperature with an increase in membrane lipid fluidity. Conversely, Thompson's group has ben able to demonstrate that fatty acid desaturase activity in *Tetrahymena* is dependent on membrane lipid fluidity (Kitajima and Thompson, 1977; Thompson and Nozawa, 1977). Incorporation of the fatty acid analogue 9-methoxystearate into membrane lipids had a fluidizing effect that resulted in a decrease in fatty acid desaturase activity. Thus, the data indicate that membrane lipid fluidity in the endoplasmic reticulum of *Tetrahymena* is self-regulating. A microsomal fatty acid desaturase likely to be involved in this self-regulation of fluidity was recently described by Kameyama *et al.* (1980). The enzyme mediated the direct desaturation of 2-oleoyl-phosphatidylcholine to 2-linoleoyl-phosphatidylcholine and increased fourfold in activity with a decrease in growth temperature of *T. pyriformis* NT-1 from 39.5° to 15°C.

The changes in lipid composition brought about by a change of growth temperature in *Tetrahymena* and other microorganisms result in a shift of the temperatures at which phase transitions of the lipid bilayer occur (Esfahani *et al.*, 1971; Wunderlich *et al.*, 1973; Chapman, 1975). Thus, the observed shift of thermosensory range in *Paramecium* that follows a change of growth temperature may result from alteration of membrane lipid composition and from the concomitant shift of lipid phase-transition temperatures.

Along these same lines, it is noteworthy that thermal adaptation of thermosensing in *Paramecium* and other microorganisms is rapid. A significant shift in thermosensory range occurs within several hours after a jump in ambient temperature (Tawada and Oosawa, 1972; Hedgecock and Russell, 1975; Hennessey and Nelson, 1979; Whitaker and Poff, 1980; Hennessey, 1981). Although one cannot strictly rule out the involvement of *de novo* protein synthesis in thermal adaptation, it seems more likely that a restructuring of membrane lipid is involved (Hennessey and Nelson, 1979; Whitaker and Poff, 1980; Hennessey, 1981).

Quite recently, Hennessey (1981) investigated rapid changes in fatty acids of membrane phospholipids in *Paramecium* that coincide with the loss of thermal avoidance after a shift of growth tempeature from 25° to 35°C. After 4 hr or more at 35°C, cells no longer avoided a test temperature at 42°C. The fatty acid profile of phospholipids from whole cells was grossly similar at 35°C compared with 25°C, but changes in the proportion of several specific fatty acids were noted. The most dramatic change after the shift to 35°C was a reduction in the proportion of acyl-linked-γ-linolenic acid (18:3). In addition, a threefold reduction in the ratio of palmitate (16:0) to stearate (18:0) was shown for the amide-linked fatty acids of ciliary sphingolipids. Both major changes in fatty acid composition began within 4 hr and were complete after 8 hr after the temperature shift from 25° to 35°C. Equally important, the fatty acid changes were largely reversed within 4–8 hr upon return of the cultures from 35° to 25°C. The ability of cells to avoid 42°C was reacquired with roughly the same time course.

Temperature-induced changes in sterol composition were also shown to be reversible with a rapid time course. Transfer of log-phase cultures from 25° to 35°C for 4 hr resulted in an increased ratio of 7-dehydrocholesterol to 7-dehydrostigmasterol. This change in the proportion of the two major membrane sterols was reversed by returning the cultures to 25°C for 4 hr. In addition to the change in 7-dehydrosterol ratio, an increase in the sterol to phospholipid ratio in the ciliary membrane was observed following the temperature shift from 25° to 35°C.

In summary, the available data are consistent with the model that an increase in the Ca^{2+} conductance of the ciliary membrane mediates the avoidance response in *Paramecium*. Adaptation of the temperature at which this response occurs is based in turn on a slower modification of the membrane lipid composition in response to growth temperature.

3.3. Acellular Slime Molds

The plasmodium of *Physarum polycephalum* will tend to accumulate at an optimum or preferred temperature of 29° ± 1°C regardless of the growth temperature of the organism. If presented with two halves of a Petri dish that differ by at least 3°C, the plasmodium will move toward that half which is closest to the optimum temperature. Tso and Mansor (1975) have suggested that this thermotaxis results from the effects of temperature on several physiological processes, such as migration rate. This proposed mechanism lacks the specificity normally associated with a sensor-mediated response, but is consistent with the insensitive nature of the response.

3.4. Cellular Slime Molds

Thermotaxis in *Dictyostelium discoideum* was first described by Raper (1940) when he reported that *D. discoideum* pseudoplasmodia will crawl toward the warmer side of a Petri dish. Later, Bonner *et al.* (1950) showed that pseudoplasmodia of *Dictyostelium* gave a directed movement response when exposed to a temperature gradient of 0.05/cm, which corresponds to a temperature gradient of 0.0005°C across a 1 mm × 0.1 mm pseudoplasmodium. This extraordinary sensitivity was confirmed by Poff and Skokut (1977), who also showed that the thermotatic response was given only over a rather narrow temperature range that was adaptable: the range was dependent on the temperature at which the amoebae making up the pseudoplasmodia were grown. This finding was later extended by Whitaker and Poff (1980), who demonstrated negative thermotaxis (directed migration toward a cooler temperature) at temperatures several degrees below the growth temperature of the amoebae, as well as positive thermotaxis at higher midpoint temperatures.

Without having identified the entire thermosensory transduction sequence in *Dictyostelium,* it is difficult to exclude unequivocally the possibility that thermotaxis results simply from the effects of temperature on the general metabolic reactions of the organism, e.g., in the acellular molds (see Section 3.4). However, this possibility appears unlikely, since (1) the temperature range for thermotaxis is more narrow than that for migration of the pseudoplasmodia (Poff and Whitaker, 1979), (2) positive thermotaxis and negative thermotaxis may both be observed in the temperature range from 19°C to 28°C over which migration rate is constant (Poff and Skokut, 1977), and (3) the phenomenon of adaptation is unlikely to affect metabolic reactions in general. One may therefore argue that thermotaxis in this organism involves a specific temperature sensor or biothermometer. Based almost solely on adaptation, Poff and

Skokut (1977) suggested that the sensor might be a lipid in a membrane matrix and undergo a phase transition over a narrow temperature range.

Since the temperature at which pseudoplasmodia switched from negative to positive thermotaxis (transition temperature) was dependent on gradient-strength, Whitaker and Poff (1980) suggested that at least three temperature sensors were involved in thermotaxis in *D. discoideum:* one sensor regulating positive thermotaxis, a second sensor regulating negative thermotaxis, and a third sensor detecting the growth temperature and regulating thermal adaptation. These workers conjectured that the third sensor regulating adaptation could be a temperature-controlled enzyme regulating fatty acid desaturation.

Recently, mutants of thermotaxis were described from two separate laboratories. Schneider *et al.* (1982) selected strains of *D. discoideum* with aberrant thermotaxis. The phenotypes fall into three types, one type exhibiting decreased response in both negative and positive thermotaxis, a second type exhibiting normal negative thermotaxis but a decreased positive thermotaxis, and a third type exhibiting only positive thermotaxis over the entire temperature range for migration. Fisher and Williams (1982) tested the thermotactic responses of their *D. discoideum* phototaxis mutants and reported strains which appear to be very similar to the second type having relatively normal negative thermotaxis but an abnormal positive thermotaxis. These strains clearly support the contention that positive and negative thermotaxis have separable pathways (although they could conceivably still share the same sensor and have separable transduction pathways).

The conclusions that the transduction pathways for positive and negative thermotaxis are indeed separable was strengthed by the observation that at temperatures several degrees below the growth temperature, the sign of the thermotactic response depended on the strength of the temperature gradient (Fontana, 1982). This dependence of sign on stimulus strength has frequently been observed with phototactic responses (Lenci and Colombetti, 1978) and when further investigated was found to be the result of two concurrent transduction pathways, but not necessarily separate receptors (Häder, 1974; Hildebrand and Dencher, 1975). In the case of the thermotactic response of *D. discoideum* pseudoplasmodia, it has been proposed that, at least near the transition temperature, the observed response is a combination of a positive thermotactic response and a competing negative response. The dependence of the sign of the observed response on gradient strength suggests that the positive and negative response have a different dependence on stimulus strength.

Throughout the work with *D. discoideum* thermotaxis, the assumption was made that the amoebae must be organized into a pseudoplasmodial unit in order to respond to a temperature gradient. This assumption was made because it had not been possible to detect a thermotactic response by the amoebae. Since amoebal thermotaxis has now been observed (Hong *et al.*, 1983), this assumption should be questioned. In fact, we now know that thermotaxis by individual amoebae of *D. discoideum* is the basis for pseudoplasmodial thermotaxis. The data supporting this conclusion are as follows:

1. The sensitivity of amoebal thermotaxis is at least comparable to that for pseudoplasmodial thermotaxis.
2. Amoebal thermotaxis has the same temperature range as pseudoplasmodial thermotaxis.
3. Amoebae respond either positively or negatively to a thermal gradient similar to the response of pseudoplasmodia.
4. Adaptation in amoebal thermotaxis is similar to adaptation in pseudoplasmodial thermotaxis.
5. Mutant strains with altered pseudoplasmodial thermotaxis show similar alterations in amoebal thermotaxis (Hong *et al.*, 1983).

These observations on amoebal thermotaxis indicate that we probably can ignore the organizational complexities of the pseudoplasmodium in our search for an understanding of thermotaxis.

Hong *et al.* (1983) have also shown that positive and negative thermotaxis have a different temporal dependence in the amoebae, with newly starved amoebae showing only positive thermotaxis and individual amoebae starved for 16 hr (starving initiates the developmental stages leading toward pseudoplasmodia formation) showing both positive and negative thermotaxis. This finding provides additional evidence that positive and negative thermotaxis have separable sensory transduction pathways. This acquisition of negative thermotaxis with time after starvation may provide a significant entry point toward identifying one or more steps in the negative thermotaxis pathway.

An additional consequence of the observation of amoebal thermotaxis will be a required reevaluation of the proposed measurement by *D. discoideum* of a spatial temperature gradient. The best evidence for this conclusion is the observation by Poff and Skokut (1977) that 90% of pseudoplasmodia give their first discernible turn in the correct direction when presented with a thermal gradient. If the measurement had a temporal basis, one would have expected one-half of the pseudoplasmodia to have moved randomly in the wrong direction. However, we know now that the basic temperature response is not given by the pseudoplasmodium, but by the amoebae. It is possible that each amoeba is giving a temporal response and that this results in the appearance of a spatial response on the integrated level of the pseudoplasmodium. Clearly, amoebal thermotaxis needs to be studied specifically to see whether it is a spatial or a temporal measurement.

The possible involvement of a membrane or membrane lipid in temperature sensing or in adaptation has been discussed in several studies (Poff and Skokut, 1977; Whitaker and Poff, 1980; Fisher and Williams, 1982), but to date no concrete evidence has been presented to support such an involvement directly. Poff and Skokut (1977) suggested that the sensor is a lipid in a membrane matrix on the basis of the narrow temperature range for thermotaxis (a phase transition would be expected to occur over a narrow temperature range) and the phenomenon of adaptation. The description by Whitaker and Poff (1980) of negative thermotaxis doubled the known temperature range for thermotaxis. It is most unlikely that the sensor could be a lipid phase transition occurring over the ~15°C range for thermotaxis. Whitaker and Poff (1980) then suggested that perhaps adaptation could be a property of the fluidity of the membrane lipid matrix in which the sensor would reside.

Fontana (1982) has directly tested the hypothesis that adaptation is a property of the membrane lipid fluidity. Using strain Ax 2, she found normal adaptation—the temperature for the transition from positive to negative thermotaxis was dependent on the temperature at which the amoebae were grown and developed into pseudoplasmodia. Lipid fluidity was measured using ESR detection of the probe 5-doxyl stearate (Mohan Das *et al.,* 1980; Fontana, 1982). No major lipid phase changes were found in the physiologically relevant temperature range nor were changes in fluidity found as a result of changing the growth/development temperature.

These results of Fontana (1982) are not compatible with the hypothesis that adaptation is a property of the membrane lipid fluidity. However, a cautionary note must be raised. Bulk fluidity was measured. The data cannot be used to exclude a changed fluidity for the lipids in a specific microenvironment. Thus, the phenomenology points to a membrane lipid involvement in adaptation, but the work of Fontana (1982) means that only a small fraction of the membrane lipid could be so involved, and we are left with a hypothesis that the temperature sensor is located in a specific minor region of a membrane for which the lipid fluidity is dependent on the growth/development temperature.

As a result of mutation analysis, a link in the transduction pathways of *D. discoideum* phototaxis and thermotaxis has been proposed (Fisher and Williams, 1982; Fontana, 1982). All of the phototactic mutants described by Fisher and Williams (1982) showed abnormal positive thermotaxis and some showed abnormal negative thermotaxis. All the mutants of positive thermotaxis isolated by Schneider *et al.* (1982) demonstrated an abnormal phototactic response (Fontana, 1982). The mutant strain deficient in negative thermotaxis, but not positive thermotaxis, exhibited a normal phototactic response (Fontana, 1982) and the temperature dependence of phototaxis was similar to the temperature dependence of positive thermotaxis (Fontana, 1982). The proposal that thermotaxis is the result of a light-generated temperature gradient (Bonner *et al.,* 1950) was also directly tested and rejected (Fontana, 1982). Based on the these results, Fontana proposed that the link between phototaxis and thermotaxis occurs in the transduction chain before the pathway for positive thermotaxis joins that of negative thermotaxis.

3.5. Nematodes

Thermosensory behavior in the phytoparasitic nematode *Ditylenchus dipsaci* was first observed by Wallace (1961) in a study of orientation of the organism to various physical stimuli (moisture, soil particle size, and temperature). *D. dipsaci* was found to accumulate at ~10°C within 1 hr after inoculation at the midpoint (15°C) of a temperature gradient of 2°–30°C. The eccritic temperature (i.e., the temperature at which the organisms accumulated) was 10°C; this temperature did not correlate with the temperature of maximum motility, which occurred over a range of 15°–20°C. Although not reported at the time, the nematodes were stored in dessicated oats at ~10°C for 1 year before being used in the temperature-gradient experiments (Croll, 1967). Thus, the eccritic temperature corresponded to the incubation (storage) temperature.

Croll (1967) followed up the Wallace study with an investigation of acclimatization in the eccritic thermal response of *D. dipsaci*. Bulbs of *Narcissus* infected with the organism were stored at 10°, 20°, and 30°C for 1 month. After extraction from the bulbs, the nematodes were maintained at the incubation temperature (i.e., 10°, 20°, or 30°C) for 3 days before testing on a thermal gradient. On a temperature gradient ranging from 10°C to 40°C, the organisms were routinely inoculated at a midpoint temperature of 20°–25°C. Both the temperature at which *D. dipsaci* accumulated and the temperature range of maximum motility were shifted up or down in accordance with the temperature of storage/incubation before the experiments.

The thermotactic response of *D. dipsaci* and two other phytoparasitic nematodes was studied further by El-Sherif and Mai (1969). Attraction of the nematodes *D. dipsaci* and *Pratylenchus penetrans* to the source of a small thermal gradient (a hot wire, an infrared beam, or germinating seedlings) was tested. The organisms were shown to respond to a temperature gradient as small as 0.033°C/cm in 1% aqueous agar. A number of *P. penetrans* were observed to orient their anterior end toward a heat source 1 cm away and to move directly up gradient within 10 min.

The most elegant, in-depth study of thermosensing and thermal acclimation in nematodes was performed by Hedgecock and Russell (1975) with the soil nematode *Caenorhabditis elegans*.

In this organism, it was demonstrated that when grown at a temperature between 16° and 25°C and placed on a radial thermal gradient, the nematodes sought the growth temperature then tracked isothermally. The temperature of isothermal tracking could be shifted up or down by raising or lowering the incubation temperature, respectively. This behavioral adaptation required several hours at the new temperature, indicating that lipid or protein biosynthesis may be involved (Hedgecock and Russell, 1975).

Another feature of thermosensory behavior in wild-type *C. elegans* reported by Hedgecock and Russell (1975) is the reversal of thermal preference under certain experimental conditions. Nematodes raised at extreme temperatures (16° or 25°C) and then displaced many degrees from the growth temperature showed diminished or even reversed thermal preference (i.e., migrated away from the direction of the growth temperature). Hedgecock and Russell proposed that this paradoxical behavior at extreme temperatures may delimit the working range of the eccritic response, noting that the organism reproduces poorly at temperatures above 25° or below 13°C. Starvation of a population of nematodes (as indicated by clearing of a bacterial lawn) for 4–6 hr also produced a reversal of thermosensory behavior; the worms exhibited growth temperature-leaving rather than growth temperature-seeking behavior. It was suggested that this reversal of behavior resulting from starvation could play a role in survival of the organism in nature by ensuring migration to a different level in the soil at which nutritional conditions might be more favorable.

Hedgecock and Russell (1975) were also to isolate a number of mutants deranged in thermotactic behavior, as well as a variety of mutants defective in chemotaxis. Of the behavioral mutants of *Caenorhabditis* selected for defective chemotaxis, one-half were shown to have defects in thermotaxis as well. A correlation was noted between the extent of chemotactic and thermotactic impairment in these strains. Thus, chemotaxis and thermotaxis in *C. elegans* are dependent on a common gene product(s).

Six additional mutants were selected for defects in thermotaxis. All six were recessive; three strains failed to track isothermally at the growth temperature (20°C), and one mutant was found to be defective in its attraction to low concentrations of NaCl. Those mutants that failed to show isothermal tracking were also found to be cryophilic (i.e., migrated toward cooler temperature on a thermal gradient), whereas those mutants that retained the capacity for both isothermal tracking and thermal acclimation were found to be thermophilic. The existence of separate thermophilic and cryophilic classes of thermotaxis mutants suggested the involvement of two different sensors in the growth temperature-seeking behavior of *Caenorhabditis:* one sensor controlling migration down a thermal gradient and the second sensor controlling migration up gradient toward the growth temperature. Behavioral analysis of all the chemotaxis and thermotaxis mutants indicated that the chemosensory and thermosensory transduction pathways in *C. elegans* converge after the cryophilic/thermophilic comparator comes into play.

More recent research in Russell's laboratory has centered on identifying the genetic lesion for any one of the behavioral mutants of *Caenorhabditis* (Culotti and Russell, 1978). Preliminary results indicate that most chemotaxis and thermotaxis mutants have limited defects in the anterior sensory nervous system, e.g., the simple nervous system of *C. elegans* is composed of not more than 300 neurons, according to Hedgecock and Russell (1975). Several osmotic avoidance mutants (unable to avoid high concentrations of NaCl or fructose) have been isolated and subsequently examined anatomically by electron microscopy. Initial findings show defects affecting the sensory endings of anteriorly located cephalic neurons in at least one mutant. The defect has also been correlated with an altered distribution of dopamine, the presumed neurotransmitter, in these neurons. It is hoped that in the future this sort of intensive anatomical and biochemical investigation will be carried out with mutants in thermotaxis as well.

4. SUMMARY

The sensitivity of most thermotactic response (with the exception of *Physarum*) suggests that there is a specific thermal sensor and that the organisms involved are not responding to a general change in metabolic rate. One can conceive of many different processes in the cell that could be used as temperature sensors. Although the sensor or biothermometer has not been identified for any organism, is clear that most if not all workers in the field believe that a central role is held by membranes in thermosensing. The best evidence for this hypothesis is that described in *E. coli* (Section 3.1).

The challenges for the study of thermosensory transduction are to find biophysical and biochemical methods that can be directed toward the identification of the temperature sensors, to determine the role of the membrane in temperature sensing and in adaptation, and to elucidate the interrelationship between temperature sensing and other sensory pathways. It is our hope that this chapter will stimulate work leading eventually toward an understanding at the molecular level of how an organism detects its thermal environment.

ACKNOWLEDGMENTS. The preparation of this chapter was supported in part by the U.S. Department of Energy under contract DE-AC02-76ERO-1338.

REFERENCES

Adler, J., 1966, Chemotaxis in bacteria, *Science* **153**:708–716.

Bashford, C. L., Morgan, C. G., and Radda, G. K., 1976, Measurement and interpretation of fluorescence polarization in phospholipid dispersions, *Biochim. Biophys. Acta* **426**:157–172.

Boggs, J. M., Stamp, D., Hughes, D. W., and Deber, C. M., 1981, Influence of ether linkage on the lamellar to hexagonal phase transition of ethanolamine phospholipids, *Biochemistry* **20**:5728–5735.

Bonner, J. T., Clarke, W. W., Jr., Neely, C. L., Jr., and Slifkin, M. K., 1950, The orientation to light and the extremely sensitive orientation to temperature gradients in the slime mold *Dictyostelium discoideum, J. Cell Comp. Physiol.* **36**:149–158.

Borochov, H., and Shinitzky, M., 1976, Vertical displacement of membrane proteins mediated by changes in microviscosity, *Proc. Natl. Acad. Sci. USA* **73**:4526–4530.

Caldwell, C. R., and Haug, A., 1981, Temperature dependence of the barley root plasma membrane Ca^{2+}- and Mg^{2+}-dependent ATPase, *Physiol. Plant.* **53**:117–124.

Capaldi, R. A., 1977, *Membrane Proteins and Their Interactions with Lipids,* pp. 260, Marcel Dekker, New York.

Chapman, D., 1975, Phase transitions and fluidity characteristics of lipids and cell membranes, *Q. Rev. Biophys.* **8**:185–235.

Clarke, S., and Koshland, D. E. Jr., 1979, Membrane receptors for aspartate and serine in bacterial chemotaxis, *J. Biol. Chem.* **254**:9695–9702.

Croll, N. A., 1967, Acclimatization in the eccritic thermal response of *Didylenchus dipsaci, Nematologica* **13**:385–389.

Crowman, J. E., and Gelmann, E. P., 1975, Lipid changes affecting phase transitions, *Bacteriol. Rev.* **39**:232–256.

Cullis, P. R., and DeKruijff, B., 1978*a,* The polymorphic phase behavior of phosphatidylethanolamine of natural and synthetic origin, *Biochim. Biophys. Acta* **513**:31–42.

Cullis, P. R., and DeKruijff, B., 1978*b,* Polymorphic phase behavior of lipid mixture as detected by P-NMR. Evidence that cholesterol may destabilize bilayer structure in membrane systems containing phosphatidylethanolamine, *Biochim. Biophys. Acta* **507**:207–218.

Cullis, P. R., DeKruijff, B., Hope, M. J., Nayar, R., Riefveld, A., and Verkleij, A. J., 1980, Structural properties of phospholipids in the rat liver inner mitochondrial membrane, *Biochim. Biophys. Acta* **600**:625–635.

Culotti, J. G., and Russell, R. L., 1978, Osmotic avoidance defective mutants of the nematode *Caenorhabditis elegans, Genetics* **90**:243–256.

DeGier, J., Mandersloot, J. G., and Van Deenen, L. C. M., 1968, Lipid composition and permeability of liposomes, *Biochim. Biophys. Acta* **150**:666–675.

DeKruijff, B., Van Den Besselaar, H. M. H. P., Cullis, P. R., Van Den Bosch, H., and Van Deenen, L. M. M., 1978, Evidence for isotropic motion of phospholipids in liver microsomal membranes, *Biochim. Biophys. Acta* **514**:1–8.

DeKruijff, B., Riefveld, A., and Cullis, P. R., 1980, P-NMR studies on membrane phospholipids in microsomes, rat liver slices and intact perfused rat liver, *Biochim. Biphys. Acta* **600**:343–357.

Dixon, M., and Webb, E. C., 1964, *Enzymes,* pp. 145–166, Academic Press, New York.

Eckert, R., 1972, Bioelectric control of cilia, *Science* **176**:473–481.

Eckert, R., and Brehm, P., 1979, Ionic mechanisms of excitation in *Paramecium, Annu. Rev. Biophys. Bioeng.* **8**:353–383.

Eckert, R., Naitoh, Y., and Machemer, H., 1976, Calcium in the bioelectric and motor functions of *Paramecium,* in: *Calcium in Biological Systems* (C. J. Duncan, ed.,), pp. 233–255, Cambridge University Press, New York.

El-Sherif, M., and Mai, W. F., 1969, Thermotactic response of some plant parasitic nematodes, *J. Nematol.* **1**:43–48.

Erwin, J. A., 1973, Comparative biochemistry of fatty acids in eukaryotic microorganisms, in: *Lipids and Biomembranes of Eukaryotic Microorganisms* (Erwin, J. A., ed.), pp. 76–81, Academic Press, New York.

Erwin, J., and Block, K., 1963, Lipid metabolism of ciliated protozoa, *J. Biol. Chem.* **238**:1618–1624.

Esfahani, M., Limbrick, A. R., Knutton, S., Oka, T., and Wakil, S. J., 1971, The molecular organization of lipids in the membrane of *Escherichia coli:* phase transitions, *Proc. Natl. Acad. Sci. USA* **68**:3180–3184.

Farias, R. N., Bloj, B., Moreno, R. D., Sineriz, F., and Trucco, R. E., 1975, Regulation of allosteric membrane-bound enzymes through changes in membrane lipid composition, *Biochim. Biophys. Acta* **415**:231–251.

Fisher, P. R., and Williams, K. C., 1982, Thermotactic behavior of *Dictyostelium discoideum* slug phototaxis mutants, *J. Gen. Microbiol.* **128:**965–971.

Fontana, D. R., 1982, Thermotactic and photactic responses of *Dictyostelium discoideum* pseudoplasmodia, Ph.D. dissertation, Michigan State University, East Lansing, Mich.

Fukushima, H., Martin, C. E., Iida, H., Kitajima, Y., Thompson, G.A., Jr., and Nozawa, Y., 1976, Changes in membrane lipid composition during temperature adaptation by a thermotolerant strain of *Tetrahymena pyriformis, Biochim. Biophys. Acta* **431**:165–179.

Gelmann, E. P., and Cronan, J. E. Jr., 1972, Mutant of *Escherichia coli* deficient in the synthesis of cis-vaccenic acid, *J. Bacteriol.* **112**:381–387.

Häder, D.-P., 1979, Participation of two photosystems in the photophobotaxis of *Phormidium uncinatum, Arch. Microbiol.* **96:**255–266.

Haest, C. W. M., DeGier, J., Van Es, G. A., Verkleij, A. J., and Van Deenen, L. L. M., 1972, Fragility of the permeability barrier of *Escherichia coli, Biochim. Biophys. Acta* **288**:43–53.

Hedblom, M. L., and Adler, J., 1980, Genetic and biochemical properties of *Escherichia coli* mutants with defects in serine chemotaxis, *J. Bacteriol.* **144**:1048–1060.

Hedgecock, E. M., and Russell, R. L., 1975, Normal and mutant thermotaxis in the nematode *Caenorhabditis elegans, Proc. Natl. Acad. Sci. USA* **72**:4061–4065.

Hennessey, T. M., 1981, Membrane lipids of *Paramecium* in their roles in sensory transduction, doctoral dissertation, University of Wisconsin, Madison, Wisc.

Hennessey, T., and Nelson, D. L., 1979, Thermosensory behavior in *Paramecium tetraurelia:* a quantitative assay and some factors that influence thermal avoidance, *J. Gen. Microbiol.* **112**:337–347.

Heron, D. S., Shimitzky, M., Hershkowitz, M., and Samuel, D., 1980, Lipid fluidity markedly modulates the binding of serotonin to mouse brain membranes, *Proc. Natl. Acad. Sci. USA* **77**:7463–7467.

Hildebrand, E., and Dencher, N., 1975, Two photosystems controlling behavioral responses of *Halobacterium halobium, Nature* **257:**46–48.

Hong, C. B., Fontana, D. R., and Poff, K. L., 1983, Thermotaxis of *Dictyostelium discoideum* amoebae and its possible role in pseudoplasmodial thermotaxis, *Proc. Natl. Acad. Sci. USA* **80**:5646–5649.

Huang, L., Lorch, S. K., Smith G. G., and Haug, A., 1974*a,* Control of membrane lipid fluidity in *Acholeplasma laidlawii, FEBS Lett.* 43:1–5.

Huang, L., Jaquet, D. D., and Haug, A., 1974*b,* Effect of fatty acyl chain length on some structural and functional parameters of *Acholeplasma* membranes, *Can. J. Biochem.* **52**:483–490.

Jain, M. K., 1980, Proteins in lipid bilayers, in: *Introduction to Biological Membranes* (M. K. Jain, and R. C. Wagner, eds.), pp. 143–175, Wiley, New York.

Jennings, H. S., 1906, *Behavior of the Lower Organisms,* pp. 41–109, Indiana University Press, Bloomington, Ind.

Kameyama, Y., Yoshioka, S., and Nozawa, Y., 1980, The occurrence of direct desaturation of phospholipid acyl chain in *Tetrahymena:* thermal adaptation of membrane phospholipid, *Biochim. Biophys. Acta* **618**:214–222.

Kates, M., and Baxter, R. M., 1962, Lipid composition of mesophilic and psychrophilic yeasts (*Candida* species) as influenced by environmental temperature, *Can. J. Biochem. Physiol.* **40**:1213–1227.

Kates, M., and Paradis, M., 1973, Phospholipid desaturation in *Candida lipolytica* as a function of temperature and growth, *Can. J. Biochem.* **51**:184–197.

Keith, A. D., Wisnieski, B. J., Henry, S., and Williams, J. C., 1973, Membranes of yeast and

Neurospora: lipid mutants and physical studies, in: *Lipids and Biomembranes of Eukaryotic Microorganisms* (J. A. Erwin, ed.), pp. 286–289, Academic Press, New York.

Kitajima, Y., and Thompson, G., 1977, Self-regulation of membrane fluidity. The effect of saturated normal and methoxy fatty acid supplementation on *Tetrahymena* membrane physical properties and lipid composition, *Biochim. Biophys. Acta* **468**:73–80.

Koshland, D. E., Jr., 1981, Biochemistry of sensing and adaptation in a simple bacterial system, *Annu. Rev. Biochem.* **50**:765–782.

Kung, C., 1971, Genetic mutants with altered system of excitation in *Paramecium aurelia.* I. Phenotypes of the behavioral mutants, *Z. Vergl. Physiol.* **71**:142–164.

Kung, C., and Eckert, R., 1972, Genetic modification of electric properties in an excitable membrane, *Proc. Natl. Acad. Sci. USA* **69**:93–97.

Kung, C., and Naitoh, Y., 1973, Calcium-induced ciliary reversal in the extracted models of "Pawn," a behavioral mutant of *Paramecium, Science* **179**:195–196.

Lee, A. G., 1975, Functional properties of biological membranes: a physical chemical approach, *Proc. Biophys. Mol. Biol.* **29**:5–56.

Lee, A. G., 1977*a,* Lipid phase transitions and phase diagrams. I. Lipid phase transitions, *Biochim. Biophys. Acta* **472**:237–281.

Lee, A. G., 1977*b,* Lipid phase transitions and phase diagrams. II. Mixtures involving lipids, *Biochim. Biophys. Acta* **472**:285–344.

Lee, A. G., Birdsall, N. J. M., Metcalfe, J. C., Troon, P. A., and Warren, G. B., 1974, Clusters in lipid bilayers and the interpretation of thermal effects in biological membranes, *Biochemistry* **13**:3699–3705.

Lenci, F., and Colombetti, G., 1978, Photobehavior of Microorganisms: A biophysical approach, *Ann. Rev. Biophysics. Bioeng.* **7**:341–361.

Macnab, R. M., 1978, Bacterial motility and chemotaxis: the molecular biology of a behavioral system, *Crit. Rev. Biochem.* **5**:291–341.

Maeda, K., and Imae, Y., 1979, Thermosensory transduction in *Escherichia coli:* inhibition of the thermoresponse by L-serine, *Proc. Natl. Acad. Sci. USA* **76**:91–95.

Maeda, K., Imae, Y., Shioi, J., and Oosawa, F., 1976, Effect of temperature on motility and chemotaxis of *Escherichia coli, J. Bacteriol.* **127**:1039–1046.

Marr, A. G., and Ingraham, J. L., 1962, Effect of temperature on the composition of fatty acids in *Escherichia coli, J. Bacteriol.* **84**:1260–1267.

McElhaney, R., 1974, The effect of alterations in the physical state of the membrane lipids on the ability of *Acholeplasma laidlawaii* B to grow at various temperatures, *J. Mol. Biol.* **84**:145–157.

Melchior, D. L., and Steim, J. M., 1976, Thermotropic transitions in biomembranes, *Annu. Rev. Biophys. Bioeng.* **5**:205–238.

Mendelssohn, M., 1895, Uber den thermotropismus einzelliger Organismen, *Pfluegers Arch. Ges. Physiol.* **60**:1–27.

Meyer, F., and Bloch, K., 1963, Effect of temperature on the enzymatic synthesis of unsaturated fatty acids in *Torulopsis utilis, Biochim. Biophys. Acta* **77**:671–673.

Mohan Das, D. V., Herring, F. G., and Weeks, G., 1980, The effect of growth temperature on the lipid composition and differentiation of *Dictyostelium discoideum, Can. Microbiol.* **26**:796–799.

Muller, C. P., and Shinitzky, M., 1981, Passive shedding of erythrocyte antigens induced by membrane rigidification, *Exp. Cell Res.* **136**:53–62.

Murata, N., and Fork, D. C., 1976, Temperature dependence of the light-induced spectral shift of carotenoids in *Cyanidium caldarium* and higher plant leaves. Evidence for an effect of the physical phase of chloroplast membrane lipids on the permeability of the membrane to ions, *Biochim. Biophys. Acta* **461**:365–378.

Naitoh, Y., 1968, Ionic control of the reversal response of cilia in *Paramecium caudatum*. A calcium hypothesis, *J. Gen. Physiol.* **51**:85–103.

Nakaoka, Y., and Oosawa, F., 1977, Temperature sensitive behavior of *Paramecium caudatum, J. Protozool.* **24**:575–580.

Nozawa, Y., Iida, H., Fukushima, H., Ohki, K., and Ohnishi, S., 1974, Temperature-induced alterations in fatty acid composition of various membrane fractions in *Tetrahymena pyriformis* and its effect on membrane fluidity as inferred by spin-label study, *Biochim. Biophys. Acta* **367**:134–147.

Ogura, A., and Takahashi, K., 1976, Artificial deciliation causes loss of calcium-dependent responses in *Paramecium, Nature* **264**:170–172.

Overath, P., and Thilo, L., 1978, Structural and functional aspects of biological membranes revealed by lipid phase transitions, in: *International Review of Biochemistry of Cell Walls and Membranes.* II, Vol. 19 (J. C. Metcalfe, ed.), pp. 1–44, University Park Press, Baltimore, Md.

Patterson, G. W., 1970, Effect of temperature on fatty acid composition of *Chlorella sorokiniana, Lipid* **5**:597–600.

Poff, K. L., and Skokut, M., 1977, Thermotaxis by pseudoplasmodia of *Dictyostelium discoideum, Proc. Natl. Acad. Sci. USA* **74**:2007–2010.

Poff, K. L., and Whitaker, B. D., 1979, Movement of slime molds, in: *Encyclopedia of Plant Physiology,* Vol. 7, (W. Haupt, and Feinleib, M. E., eds.), pp. 355–382, Springer-Verlag, Berlin.

Pugh, L., and Kates, M., 1975, Characterization of a membrane-bound phospholipid desaturase system of *Candida lipolytica, Biochim. Biophys. Acta* **380**:442–453.

Rand, P. R., and Sengupta, S., 1972, Cardiolipin forms hexagonal structure with divalent cations, *Biochem. Biophys. Acta* **255**:484–492.

Raper, K. B., 1940, Pseudoplasmodium formation and organization in *Dictyostelium discoideum, J. Elisha Mitchell Sci. Soc.* **56**:241–282.

Sandermann, H., Jr., 1978, Regulation of membrane enzymes by lipids. *Biochim. Biophys. Acta* **515**:209–237.

Schenk, S. L., 1893, Die Thermotaxis der Mikvoorganismen und ihre Beziehung zur Erkaltung, *Centralblatt für Bakteriologie, Abteilung I* **14**:33–43.

Schneider, M. J., Fontana, D. R., and Poff, K. L., 1982, Mutants of thermotaxis in *Dictyostelium discoideum, Exp. Cell Res.* **140**:411–416.

Segel, I. H., 1975, Enzyme kinetics, in: *Behavior and Analysis of Rapid Equilibrium and Steady-State Enzyme Systems,* pp. 926–941, Wiley, New York.

Shaw, M., and Ingraham, J. L., 1965, Fatty acid composition of *E. coli* as a possible controlling factor of the minimal growth temperature, *J. Bacteriol.* **90**:141–146.

Shinitzky, M., and Henkart, P., 1979, Fluidity of cell membranes—current concepts and trends, *Int. Rev. Cytol.* **60**:121–147.

Sinensky, M., 1971, Temperature control of phospholipid biosynthesis in *Escherichia coli, J. Bacteriol.* **106**:449–455.

Sinensky, M., 1974, Homeoviscous adaptation—a homeostatic process that regulates the viscosity of membrane lipids in *Escherichia coli, Procl Natl. Acad. Sci. USA* **71**:522–525.

Somero, G. N., 1975, Temperature as a selective factor in protein evolution: the adaptational strategy of "compromise," *J. Exp. Zool.* **194**:175–188.

Somero, G. N., and Hochachka, P. W., 1976, Biochemical adaptations to temperature, in: *Adaptation to Environment: Essays on the Physiology of Marine Animals* (R. C. Newell, ed.), pp. 125–189, Butterworths, London.

Springer, M. S., Goy, M. F., and Adler, J., 1977, Sensory transduction in *Escherichia coli:* Two complementary pathways of information processing that involve methylated proteins, *Proc. Natl. Acad. Sci. USA* **74**:3312–3316.

Stier, A., Finch, S. A. E., and Bosterling, B., 1978, Non-lamellar structure in rabbit liver microsomal membranes, *FEBS Lett.* **91**:109–112.

Tawada, K., and Oosawa, F., 1972, Responses of *Paramecium* to temperature change, *J. Protozool.* **19**:53–57.

Thompson, G. A., and Nozawa, Y., 1977, *Tetrahymena:* system for studying dynamic membrane alterations within the eukaryotic cell, *Biochim. Biophys. Acta* **472**:55–92.

Ting, P., and Solmon, A. K., 1975, Temperature dependence of n-phenyl-l-naphthylamine binding in egg lecithin vesicles, *Biochem. Biophys. Acta* **406**:447–451.

Tourtellotte, M. E., 1972, Mycoplasma membranes: structure and function, in: *Membrane Molecular Biology* (C. F. Fox and A. O. Keith, eds.), pp. 439–470, Sinauer Associates, Stamford, Conn.

Trauble, H., and Eibl, H., 1975, Molecular interactions in lipid bilayers, in: *Functional Linkages in Biomolecular Systems* (F. O. Schmitt, D. M. Schneider, and D. M. Crothers, eds.), pp. 59–61, Raven Press, New York.

Tso, W., and Mansor, T. E., 1975, Thermotaxis in a slime mold, *Physarum polycephalum, J. Behav. Biol.* **14**:499–504.

Van Zoelen, E. J. J., Van Der Neut-kok, E. C. M., De Gier, J., and Van Deenen, L. L. M., 1975, Osmotic behavior of *Acholeplasma laidlawaii* cells with membrane lipids in liquid-crystalline and gel state, *Biochem. Biophys. Acta* **394**:463–469.

Wallace, H. R., 1961, The orientation of *Ditylenchus dipsaci* to physical stimuli, *Nematologica* **6**:222–236.

Wang, E. A., and Koshland, D. E., Jr., 1980, Receptor structure in the bacterial sensing system, *Proc. Natl. Acad. Sci. USA* **77**:7157–7161.

Whitaker, B. D., and Poff, K. L., 1980, Thermal adaptation of thermosensing and negative thermotaxis in *Dictyostelium, Exp. Cell Res.* **128**:87–93.

Wu, E.-S., Jacobson, K., and Papahadjopoulous, D., 1977, Lateral diffusion in phospholipid multilayers measured by fluorescence recovery after photobleaching, *Biochemistry* **16**:3936–3941.

Wunderlich, F., Speth, V., Batz, W., and Kleinig, H., 1973, Membranes of *Tetrahymena.* III. The effect of temperature of membrane core structures and fatty acid composition of *Tetrahymena* cells, *Biochim. Biophys. Acta* **298**:39–49.

Chapter 5

Microbial Geotaxis

Barry Bean

1. THE PHENOMENA OF GEOTAXIS

Many motile microorganisms can distribute themselves asymmetrically in a vertical water column by a number of orientation behaviors, among them *geotaxis*. A flurry of early studies (1880–1920) generated reports on the physical, chemical, and behavioral character of geotaxis, as well as several hypotheses that might explain its occurrence. These observations and hypotheses have been discussed by Davenport (1908), Dryl (1974), Kuźnicki (1968), and Haupt (1962). There is widespread agreement on several features of geotaxis.

1.1. Geotaxis and Geotropism

In general, the term geotaxis is used for all behaviors that result in a nonrandom distribution or directional orientation of independent individuals and are dependent on gravity and the active propulsive movement of the individuals. The term is generally most appropriate for short-term movements of freely motile individuals (unicellular and multicellular), although occasionally it is used for attached organisms and for longer-term orientation. Migration against the acceleration of gravity is called negative geotaxis. Distinctions between taxis, orthokinesis, and klinokinesis are usually not made for gravity-oriented behavior. In the older literature, the term geotropism is sometimes equated

Barry Bean • Department of Biology, Lehigh University, Bethlehem, Pennsylvania 18015.

with geotaxis. The present treatment focuses on the behavior of free-swimming individual ciliated and flagellated cells.

The term geotropism (and recently gravitropism) properly refers to responses to gravity that involve growth rather than locomotion. While the effectors of tropisms (cellular growth or division) may be distinct from those of the taxes (e.g., actions of cilia or flagella), it is possible that important parallels exist at the level of sensory transduction and/or reception. This possibility immediately suggests that fruitful comparative studies might be done on species that possess both geotropism and geotaxis (under different growth conditions or at different stages of the life cycle, e.g., certain fungi). According to current thinking, in gravitropism, dense granules (statoliths) have an asymmetric distribution in the sensitive cells, and by their position relative to specialized membranes (and/or other appropriate structures) appear to be responsible for triggering asymmetric growth. In at least some cases, the growth response is the consequence of differential lateral transport of low-molecular-weight growth regulators. But little has been established about the mechanism by which statolith position results in differential growth. Since detailed reviews of gravitropism have recently appeared (Juniper, 1977; Sievers and Volkmann, 1979; Volkmann and Sievers, 1979; Wilkins, 1979), this subject is treated further in this chapter. Other positioning mechanisms have been reviewed by Carlile (1980*a*).

1.2. Occurrence of Geotaxis

Many species whose life cycles include freely moving individuals exhibit geotaxis. Species of prokaryotes, protozoans, algae, fungi, metazoan animals, and the free-swimming gametes of some plants and animals are included.

1.3. Orientation Direction Provided by Gravity

Gravity can and does serve as a physical force to which geotactic behavior is oriented. It is generally agreed that orientation to gentle centrifugal fields ("centrotaxis" in the older literature) occurs by the same mechanisms, making it possible to study the effects of part of the continuum of intensities of stimulus acceleration forces. Attempts to nullify the influence of gravity have resulted in the use of klinostats (Fig. 1) and in proposals for experiments to be done in orbital satellites (Juniper, 1977). Since the geotactic responses of many organisms are indeed easily masked by responses to other stimuli, rigorous control experiments are always in order (cf. Haupt, 1962). Possible responses to light, temperature, chemical gradients (including gases), convection, flow, and other factors must be addressed and properly distinguished from genuine responses to gravity.

Positive (downward), negative (upward), and transverse (horizontal or

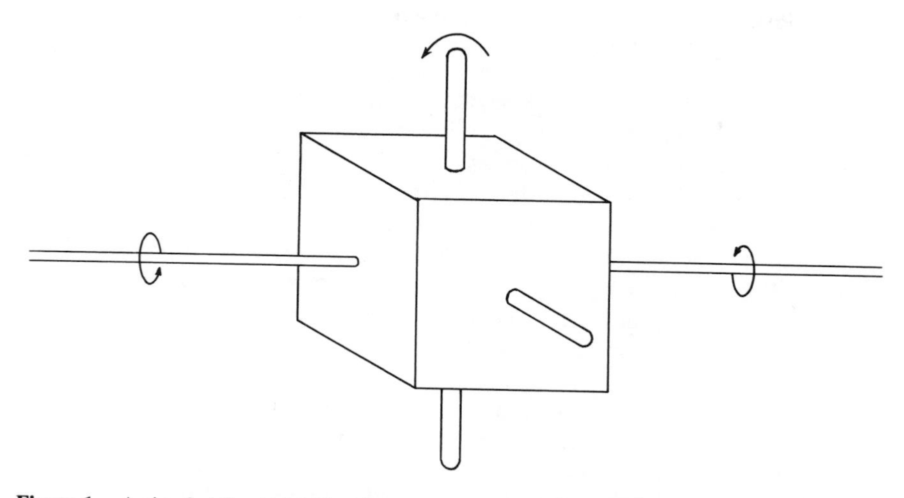

Figure 1. A simple klinostat designed to rotate specimens in a vertical plane. (Redrawn from Davenport, 1908, p. 112.)

inclined) geotactic reactions have been reported. In some cases, the opposite responses may be given by the same species (or even by the same individuals), depending on physical or physiological conditions, e.g., for *Paramecium,* as reported by Moore (1903) and Fox (1925). Examples of conditions in which subpopulations from a single culture migrate in opposite directions are known. Despite these apparent complexities, clear species-typical characterizations are valid; e.g., *Paramecium* and *Chlamydomonas* are negatively geotactic, while *Blepharisma* (Fornshell, 1980) and human spermatozoa are positively geotactic.

1.4. Energetics of Geotaxis

True geotaxis is an energy-dependent process. The nonrandom distribution of individuals is achieved and/or maintained through the expenditure of cellular energy, typically conceived as a small proportion of the adenosine triphosphate (ATP) devoted to propulsive work (e.g., axonemal activity, amoeboid movement). This is manifest in cases in which organisms actively migrate in the direction opposite to that which they would passively accumulate by sedimentation or flotation. Most of the negatively geotactic organisms, in particular, are more dense than their surrounding medium, but migrate toward the surface at a steady rate. By contrast, positively geotaxic cells that are also more dense than their surrounding fluid (e.g., animal sperm, some bacteria) may be expected to sediment at rates faster than those for passive sedimentation (Roberts, 1972) (as should be shown by comparable nonmotile individuals).

1.5. Biological Significance of Geotaxis

There are alternate nonbehavioral mechanisms by which organisms achieve a nonrandom distribution in a water column. Notably, many bacteria, blue-green algae, and other organisms contain gas vacuoles, which keep cells at the surface or distributed within a certain range of depths in their aqueous environment. While vacuole assembly and maintenance are decidedly energy dependent, the actual positioning of cells is a passive process, distinct from geotaxis. Some of the selective forces that lead to the evolution of gas vacuoles, and some of their immediate biological functions, are similar to those of geotaxis. The biology of gas vacuoles has been reviewed in depth by Walsby (1978).

Geotaxis is generally thought to enhance the fitness of microorganisms. The negative geotaxis of microorganisms that are denser than water aids cells in remaining suspended and in areas that improve nutritional opportunities. For *Chlamydomonas,* for instance, the flagella are resorbed during cell division, which usually takes place at night. As the flagella regenerate, geotaxis helps this organism emerge from the sediments toward the surface, thereby optimizing its photosynthetic activity (Bean, 1977). Fungal zoospore geotaxis, as shown by species of *Phytophthora* (Cameron and Carlile, 1977), probably keeps these organisms in soil regions in which plant roots (and their chemoattractants) are available (Carlile, 1980*a*).

Direct contributions of geotaxis to the reproductive success of microbes might derive from spatial concentration of cells at the appropriate time in the life cycle. In those species of *Chlamydomonas* that lack sexual chemoattractants, geotaxis (and phototaxis) by gametes might help create a social environment that optimizes opportunities for sexual reproduction.

According to several hypotheses on the mechanism of geotaxis, including that of Roberts (1970), cell shape is a key factor that determines the character of the geotactic response. If this is true, and if geotaxis does confer a selective advantage, it is clear that the evolution of cell shape as an organismic feature is profoundly influenced by the ability of individuals to perform geotaxis adequately to promote their survival. Thus, for motile species for which geotaxis would be advantageous, selection would favor the evolution of cell shapes that enhance geotaxis (Roberts, 1970, 1981).

1.6. Geotaxis and Other Behaviors

1.6.1. Pattern Swimming or Bioconvection

For many species, geotaxis is but one of several orientation behaviors that may occur. In combination with other responses, geotaxis may be an important component of the complex behavior patterns of some species. For instance, geotaxis is thought to be the driving force of the behavioral phenomenon known as pattern swimming or bioconvection (Wager, 1910, 1911; Gittleson and Jahn, 1968*a,b*; Childress *et al.*, 1975*a,b*; Levandowsky *et al.*, 1975; Plesset *et al.*, 1975; Fornshell, 1978). Dynamic patterns are formed by *Euglena, Tetrahymena, Chlamydomonas, Paramecium, Polytomella,* and other organisms (Fig. 2). Some of the physical, chemical, and physiological variables that influence the formation of these striking patterns have been examined. A comprehensive correlation of the characteristics of geotaxis with those of pattern swimming has not been reported. Several investigators seem to believe that true pattern swimming is shown only by populations of independent, free-swimming unicellular organisms that are denser than water and capable of negative geotaxis (and only within certain limits of cell number, depth, temperature, and so on).

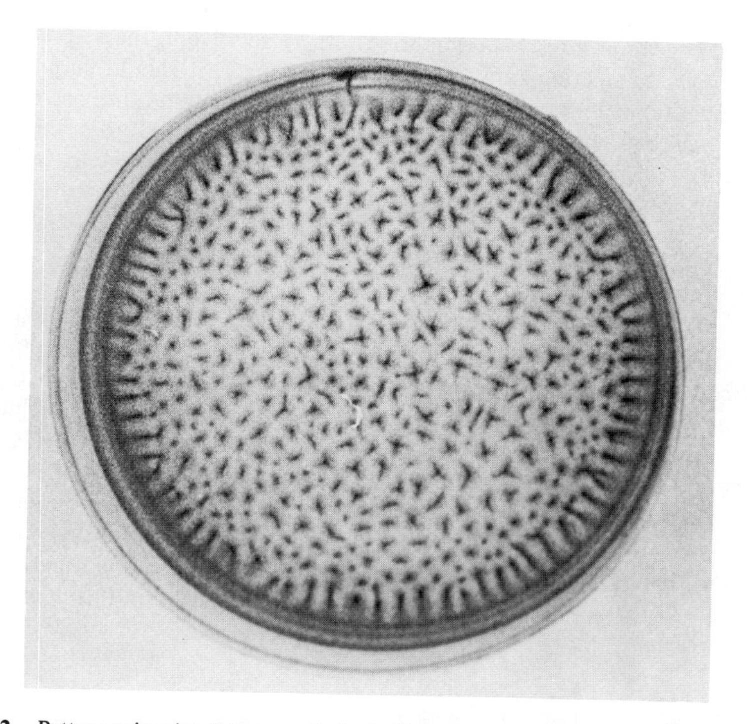

Figure 2. Pattern swimming (bioconvection) of *Chlamydomonas reinhardtii*. The time course for development of this stable pattern parallels that required for strong geotaxis.

1.6.2. Competition among Orientation Behaviors

For most negatively geotactic species, geotaxis is a "slow" response, i.e., it occurs at rates that are small relative to swimming speed. If assayed in competition with "fast" responses of the same cells to light or chemical stimuli, experimental resolution of geotaxis may be lost. Through careful manipulation of stimulus intensity (e.g., actinic light intensity or flash rate for *Chlamydomonas* or *Euglena*) it should be possible to study the competitive influence of the photic and gravitational stimuli. Such competition experiments are done easily for photogeotropic species like *Phycomyces* (Bergman *et al.*, 1973; Medina and Cerda-Olmedo, 1977) and for geotactic species like *Euglena* (Creutz and Diehn, 1976) that are responsive to other stimuli as well. For paramecia, competitive experimental situations have been exploited for the selection of behavioral mutants (Kung, 1971*a*,*b*, 1976; Byrne and Byrne, 1978; Cronkite, 1979; Takahaski, 1979), but many interesting behavioral experiments remain to be done.

The study of competitive experimental situations should make it possible to design relatively simple quantitative (and semiquantitative) bioassays for geotaxis migration rates (or orientation rates?) relative to the responses triggered by the competing stimulus. More importantly, it should be possible to use competition situations to improve selection for mutants that are defective for geotaxis but normal for other selected behaviors (cf. Byrne and Byrne, 1978; Cronkite, 1979). Such mutants might be of critical value in distinction among several proposed mechanisms of geotaxis (see Section 2).

2. ARENAS OF DEBATE

While there is widespread agreement on certain features of geotaxis, as described in Section 1, there remain some basic features for which no wide agreement is evident in the literature. Unfortunately, these areas include the physical mechanics and mechanism(s) of geotaxis, as well as the roles of membranes and sensory transduction events in gravitational orientation behavior. Various aspects of these questions remain inadequately explored or controversial or show unresolved differences among species. A lively controversy on the mechanism(s) of geotaxis surfaced during the 1970s, including serious reconsideration of many of the same hypotheses debated earlier in this century (see Section 4).

2.1. Mechanics versus Mechanism

For purposes of this discussion, an important distinction between the phrases *mechanics of response* and *mechanism of response* should be clarified.

Mechanics refer to the forces on and movements of an organism in the achievement of an orientation response, while *mechanism* refers to the organismic processes (physical or physiological) by which those mechanics are achieved. While it is entirely possible, as several hypotheses purport, that certain systems of mechanics might in fact constitute mechanisms, this is not always a logical necessity. There are other conceivable mechanisms in which biochemical or electrophysiological processes might be responsible for the mechanics that are observed; in such arguments it is essential to distinguish mechanics from mechanism.

For even the most classical of subjects for studies on geotaxis, such as *Paramecium,* there is considerable confusion in the literature about the mechanics of response. Jennings (1904, 1906) and Lyon (1905) stressed the importance of behavioral changes in the achievement of orientation. These behavioral changes included the timing, frequency, and style of avoiding responses, as well as changes in swimming speed. According to current theory (see Chapter 3), such changes are all regulated physiologically. Some recent considerations of the mechanism of geotaxis (Roberts, 1970; Winet and Jahn, 1974; Fukui and Asai, 1980) focus on physical rather than physiological features of the response. In many reports on *Paramecium* and other organisms, investigators are not concerned with the mechanics of response, and in some cases make unwarranted assumptions about the mechanics.

2.2. Mechanisms of Response

Considerations of the mechanics of response are further complicated by failure to recognize (either logically or experimentally) that there may be different mechanisms operating in different phases of the response within a given species: achievement of orientation, maintenance of orientation, and maintenance of distribution. Specifically, it is possible that the mechanism by which geotactic orientation is initially achieved, e.g., by an individual at some depth in a water column, is entirely different from the mechanism that serves in the maintaining orientation after it has been achieved, e.g., during migration toward the surface. Either of these mechanisms, in turn, may be different from a mechanism that serves to keep the individual near its terminal position in asymmetrical accumulation response, in maintaining distribution.

A hypothetical negatively geotactic species might achieve its initial orientation by physiological suppression of avoiding responses attributable to position-dependent stabilization of membrane potential. Its upward orientation might be sustained by passive hydrodynamic forces consequent to cellular morphology and characteristics of locomotion during its upward migration. Finally, maintenance of its position near the surface might be reenforced by physiological factors (decrease in swimming speed) or physical factors (interactions with the surface or other cells), or both.

In order to reach meaningful generalizations about geotaxis, it is essential to consider all possibilities that are not logically excluded, to examine and account for all phases of the response, and to recognize that several mechanisms and physical factors might influence the response. To date, satisfactorily rigorous investigations have not been completed for any species.

2.3. Unproved Existence of Gravireceptors for Geotaxis

For several other well-known microbial motile responses, such as chemotaxis, phototaxis, and thigmotaxis, clear delineations can be drawn, at least at a theoretical level, among three fundamental stages of biological information processing: reception, sensory transduction, and effector action. These are the cellular analogues of the sensory, integrative, and response stages of the stimulus–response systems of higher animals. For the case of cellular geotaxis, only the effector stage is readily accessible. It has not been shown that discrete subcellular gravireceptor structures exist for geotaxis. Whole live and actively motile cells are generally required for the accumulation response to take place, and it may be that whole cells, or substantial portions of cytoplasm, function in an integrated way as the receptor (cf. Kanda, 1918; Haupt, 1962; Kuźnicki, 1968).

2.4. Involvement of Membranes and Sensory Transduction

Furthermore, it is not clear that sensory transduction processes are essential for geotaxis, although it is quite clear that natural or experimental alteration or inhibition of sensory transduction pathways can influence the response. Some investigators (Roberts, 1970; Winet and Jahn, 1974; Fukui and Asai, 1980) have suggested that propulsion is the only essential active cellular ingredient and that regulation of propulsion through sensory transduction pathways is not an active part of the response mechanism. Some arguments and evidence on the roles of membranes and sensory transduction in geotaxis are reviewed in Section 4.

3. METHODS OF ASSESSMENT

In most experimental contexts, geotaxis has been evaluated qualitatively after direct observation and description of the distribution of individuals or of the orientations of individuals within a population. Such methods have been used in both classical (e.g., Massart, 1891; Jensen, 1893; Schwarz, 1884) and modern studies and are fully adequate for many purposes. Capillary tube assays are simple and direct and permit evaluation of many experimental variables. It is important that the experimental conditions—time and conditions of incubation, physical variables of illumination, temperature, chemical composition, and tube length and diameter, as well as the physiological condition of the organisms—must be well explored before standard conditions are defined. Failure to do so can lead to erroneous conclusions.

In most cases, simple bioassay situations can be made more precise or reliable by introduction of photo-, cine-, or videomicrographic techniques. Gręcki and Nowakowska (1977) used simple time-exposure photomacrographs to provide convincing evidence on the effects of differing ciliary reversal conditions on the effectiveness of orientation in populations of paramecia, followed by cinematographic analysis of the movements of individual cells (Nowakowska and Gręcki, 1977). Videomicrographic recording and subsequent analysis are advantageous in some experimental situations for practical and technical reasons, and because it is not intrusive (Bean, 1977).

Fukui and Asai (1980) used interrupted dark-field time-exposure photography to produce records that contained information about direction of movement as well as velocity and angle of orientation. Vectors for 600 cells per experiment were plotted on polar coordinates to provide a quantitative representation of orientation in the population of *Paramecium.*

Other quantitative methods include direct or indirect counting of organisms under defined conditions (e.g., Roberts, 1970). Such data can be used to calculate net rates of geotactic migration in microns per second or other arithmetically related numerical expressions that characterize oriented movements (Bean, 1977).

It should be noted that most quantitative methods characterize the population during or after a period of geotaxic redistribution, and thus do not necessarily relate directly to the rate or mechanism of initial orientation.

4. HYPOTHESES ON THE MECHANISM OF GEOTAXIS

4.1. Old Mechanisms and New Disputes

There has never been consensus on the mechanism(s) of geotaxis. Since the earliest investigations, a variety of hypotheses have been put forward, most of which have been reviewed, discussed, and criticized in some depth (e.g., Verworn, 1899; Jennings, 1904, 1906; Davenport, 1908; Lyon, 1905, 1918; Haupt, 1962; Kuźnicki, 1968; Roberts, 1970, 1981; Jahn and Votta, 1972; Dryl, 1974; Naitoh and Eckert, 1974; Winet and Jahn, 1974; Gręcki and Nowakowska, 1977; Machemer and de Peyer, 1977; Nowakowska and Gręcki, 1977; Fukui and Asai, 1980). Essentially the only area of agreement is that the hypothesis put forward by Jensen (1893)—that cells can perceive differences in pressure across their bodies—is ill conceived. The arguments continued into the 1930s and remained unresolved as studies on geotaxis lost favor through the 1960s. No attempt is made here to present a comprehensive treatment of the older literature or of the details of the various hypotheses. Since 1970, new investigations of geotaxis in several laboratories and the renewed general interest in the behavior of cells have revived several of the better classical hypotheses in the context of present-day scientific sophistication. There are many good observations and good ideas on geotaxis, but little definitive evidence that distinguishes one mechanism from another; there are still several hypotheses, but no theory of the mechanism of geotaxis.

Machemer and de Peyer (1977) reduced the important remaining notions on the mechanism of geotaxis to the four generally different hypotheses summarized in Fig. 3. The four generalized diagrams serve as an introduction to the more detailed treatments of the modern hypotheses of geotaxis that follow.

4.2. Physical Mechanisms

4.2.1. Roberts's "Hydrodynamic" Hypothesis

Roberts (1970) presented a modern theoretical treatment and revision of the mechanical mechanisms of geotaxis in microorganisms. Both differential

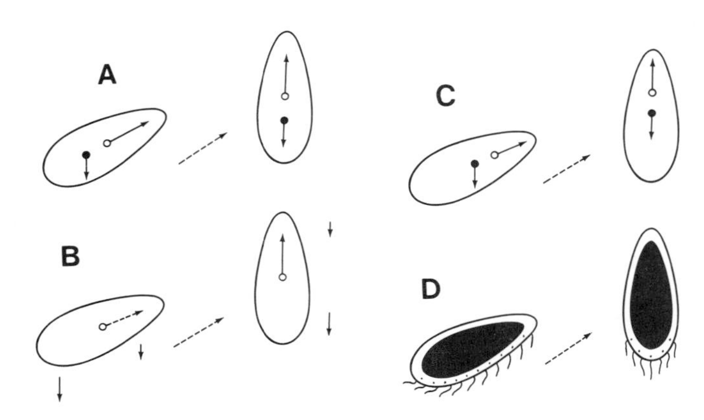

Figure 3. Four concepts of the mechanism of negative geotaxis of *Paramecium.* (A) Static (differential density, gravity–buoyancy) hypothesis: Higher density of the posterior places the center of gravity behind the center of propulsive effort, generating a passive torque (Verworn, 1889). (B) Shape-hydrodynamic (drag–gravity) hypothesis: The bulkier posterior of the cell sediments under gravity with higher velocity than the anterior (Roberts, 1970). (C) Propulsion–gravity hypothesis: The center of effort is anterior to both the center of gravity and the gyration point, and the sedimentation torque unbalances the gyrational torque, resulting in a net increment in upward orientation with each cellular gyration (Winet and Jahn, 1974). (D) Statocyst (physiological) hypothesis: A position-dependent gravity effect on the cytoplasmic substance selectively stimulates the motor apparatus, asymmetrical effector action results in reorientation toward the vertical (Lyon, 1905). (Redrawn from Machemer and de Peyer, 1977.)

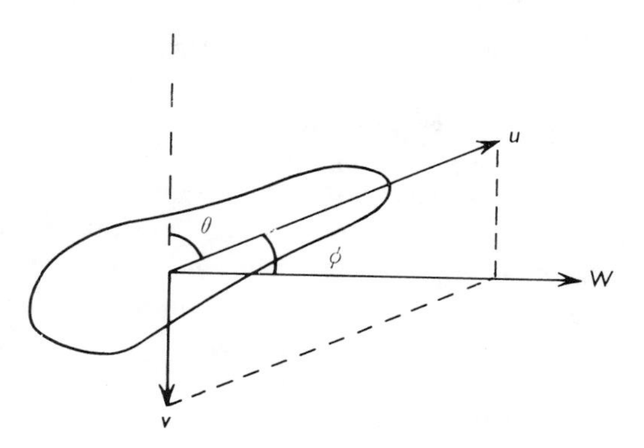

Figure 4. An organism propelled at velocity u by its own motile apparatus, and falling under gravity with velocity v. The resultant velocity is W at angle ϕ to the longitudinal axis. (Redrawn from Roberts, 1970, with permission.)

density within an organism and geometrical asymmetry within an organism were considered possible sources of orienting torque. Practical and theoretical limitations led to the conclusion that density differences within a cell and their consequent effects on cell movement are not of sufficient magnitude to account for geotaxis. The hydrodynamic consequences of cell shape, however, in Roberts's view, are sufficient to account for geotactic orientation rates. By making certain reasonable assumptions and simplifications, Roberts derived a relationship for the rate of orientation for an autonomously motile microorganism (such as that envisioned in Fig. 4) as

$$\frac{d\theta}{dt} = -\beta \sin \theta$$

where β is a constant, the instantaneous orientation rate for a falling nonmotile cell in horizontal position. Importantly, the orientation rate is seen to be the same for motile and nonmotile cells, is constant for a given shape, and is independent of velocity (Roberts, 1970). Relevant values of β lie in the range 0.01–0.08 rad/sec.

Roberts (1970) was careful to distinguish the hydrodynamic consequences of β on the distribution of cells (or physical models of cells) as a separate matter. Here the cellular swimming characteristics take on a critical role in determining the net displacement of cells. Recognizing that in systems of low Reynolds numbers (i.e., with objects of small size) viscosity is very significant, Roberts's approach can be appreciated. For a cell with swimming velocity u, and a mean time between random reorientations of τ, cells with $\beta\tau/2 > v/u$ will move upward (negative geotaxis), while those with $\beta\tau/2 < v/u$ will sediment to the bottom (Roberts, 1970, 1981).

Roberts (1970) also did experiments on the sedimentation (at low Reynolds numbers) of aluminum scale models through glycerol and on the distribution of immobilized *Paramecium caudatum* and reported results that were consistent with his hypothesis. Paramecia immobilized with $NiCl_2$ showed variation in rate of orientation, but for most cells β was in the range of 0.01–0.05 rad/sec with sedimentation velocities in the range 100–200 μm/sec (these data are similar to those of Kuźnicki, 1968).

Roberts (1970) also states that "most organisms fell anterior end upwards . . . ," which is contrary to the observations of Kuźnicki (1968) on the same and three other species of *Paramecium*. However, Gręebecki and Nowakowska (1977) found that chemically induced backward swimming and positively geotactic paramecia did indeed orient with their anteriors upward. Nevertheless, these factors do not distinguish among several possible mechanisms.

Several objections to Roberts's (1970, 1981) hypothesis have been raised (Winet and Jahn, 1974, p. 451). First, over the years there has been no consistent evidence that nonmotile paramecia or other negatively geotactic cells do in fact sediment passively with their anterior parts oriented upward, as would be predicted according to Roberts's hydrodynamic orientation mechanism, called the drag-gravity model and criticized by Winet and Jahn (1974). Notably Kuźnicki (1968) and others (but see Katz and Pedrotti, 1977) find no such propensity in either dead or live immobilized specimens. To such criticism, A. M. Roberts (personal communication, 1978) replies, "Results of sedimentation experiments are equivocal. But all cells show orientation at the correct rate, although many cells don't line up vertically. On the other hand the final orientation is exceedingly shape-dependent, and the fact that cells gyrate must surely average out." In short, striking alignment within a population should not be expected in many experimental circumstances.

Second, Winet and Jahn (1974) object that not all negatively geotactic cells have an anteriad center of surface, particularly flagellates. However, it must be recognized that areas of surface hydration, including those regions involved in the actions of flagella and their mastigonemes, and the orientation-determining actions of protrusive organelles may influence cellular orientation far more greatly than does body shape, particularly when organisms are viewed in a brief time frame.

Third, Winet and Jahn (1974) further object that Roberts's (1970) hypothesis ignores or presumes self-cancelling propulsive asymmetries such as those that result in gyratory swimming paths. These challengers (Jahn and Winet, 1973; Winet and Jahn, 1974) assign a key role to gyration and hypothesize that such "gyration . . . is characteristic of all known negatively geotaxic swimming protozoa." This view is in itself somewhat controversial and was not treated by Roberts (1970) simply because it is not necessary to his argument (Roberts, 1975; personal communication, 1978). To Roberts (1981), "It is not clear how gyration itself can impart directional information to the cell, however, although hydrodynamic orientation must still occur during gyration."

Fourth, Winet and Jahn (1974) also argue, "Some negatively geotaxic species will often display positive geotaxis with no accompanying change in shape." However, it is clear that Roberts (1970, 1981) did consider changes in sign of geotaxis which, according to the hypothesis, result from changes in either swimming velocity and/or the time between random reorientation events (e.g., collisions or spontaneous avoiding reactions). It remains appropriate to challenge Roberts's hypothesis with a rigorous examination of variables that induce such changes. Many lines of evidence to date, e.g., temperature effects on *Paramecium* (Nakaoka and Oosawa, 1977) and *Chlamydomonas* (Majima and Oosawa, 1975), swimming character in *Paramecium* (Grębecki and Nowakowska, 1977), inhibitor effects on *Chlamydomonas* (Bean, 1977; Bean and Harris, 1978), are not inconsistent with the hydrodynamic mechanism suggested by Roberts (1970).

More serious questions about Roberts's (1970) hypothesis grow out of older observations on paramecia with experimentally controlled modification of shape. Koehler (1939) reported that fragments of *Paramecium* lacking their posterior cytoplasm still perform geotaxis in an approximately normal way. Unpublished work of Z. Smiechowska (cited by Kuźnicki, 1968) also documented "typical negative geotactic reactions" in *P. caudatum* specimens during and after division, after injury, and after removal of posterior portions. Further analysis of the behavior of operated animals might prove strong evidence either against or for the approach of Roberts.

It seems safe to conclude that Roberts's (1970) proposed mechanisms play some role in the achievement of initial orientation in geotaxis. The proposed effects are purely physical in nature, are based on reasonable presumptions, and should be expected to contribute to cellular orientation in at least some cases, and to some extent. In particular, the shape–hydrodynamic mechanism is associated with numerical rates of orientation, sedimentation, reorientation frequencies, and swimming velocities that are in the order of magnitude to be significant in contributing to observed geotactic behavior. Under some conditions, viz., when extreme density differences within the organism do exist, the variable-density mechanism, as explicated by Roberts (1970), might be a contributing factor.

On the other hand, it is certainly not safe to conclude that Roberts's (1970) hypothesis constitutes the major mechanism of orientation in even a single species, or that it excludes other mechanisms that may operate in the achievement or maintenance of orientation or in the maintenance of an asymmetrical distribution of cells.

Investigations on the geotaxis of *Paramecium caudatum* (Grębecki and Nowakowska, 1977; Nowakowska and Grębecki, 1977) report that rates of angular orientation in fact reach values in the order of 30°/sec for swimming animals positioned near the horizontal (with average values ~9°/sec) and that rates of orientation by swimming cells are substantially higher than those for passive cells. These observations again suggest that the mechanism proposed by Roberts (1970) is not the primary mechanism that operates in normal paramecia.

4.2.2. The "Propulsion–Gravity Model" of Winet and Jahn

Several workers (Jahn and Votta, 1972; Jahn and Winet, 1973; Winet and Jahn, 1974) advanced as a new hypothesis a propulsion-dependent geotaxis orientation mechanism—the propulsion–gravity model. This model applies to cells that gyrate as they swim (such gyration being attributable to asymmetries in anterior–posterior morphology and/or the distribution or activity of locomotory organelles) and that have a center of propulsive effort E that lies anterior to the centroid (center of mass) C.

According to this hypothesis, orientation is achieved as the consequence of an unbalancing of the vertical components of the gyrational torque by gravity. (Balanced gyration should occur only in the absence of gravity. In low Reynolds number systems, the primary force countering gyration is viscous resistance B, which is uniformly applicable at all points on the cell.) Since gravity acts through C and propulsive forces through E, a torque can be produced. During the downswing of gyration, the vertical components act in the same direction, producing a smaller torque than they do during the upswing, when they act in opposition to each other. These relationships are illustrated more formally in Fig. 5.

The same mechanism can be conceived from the viewpoint of the undulipodia on the cellular surface, and their actions relative to the gyration point G, about which the cell axis gyrates. During the downswing, more propulsive force is dissipated in overcoming viscous resistance, resulting in dampening of gyration (G is located caudally); during the upswing, relatively more propulsive force is used in production of gyrational torque, giving the cell a net increment in its angle with the horizontal plane (θ) with each gyration cycle.

While the propulsion–gravity model of Winet and Jahn (1974) perhaps operates as a mechanism for geotaxis in some species, it is clear that the model is largely based on theoretical considerations, and it has not yet been shown that it constitutes a primary mechanism or that the requirements of the model are met consistently by many negatively geotactic species. The authors understandably did not attempt a comprehensive correlation of their physical requirements and the geotactic behaviors of diverse species. But, given that their arguments (like those of Roberts) derive from first principles of physics, even a single exception would cast serious doubt on the hypothesis (Bean, 1977).

Certain questions immediately arise from the Winet and Jahn (1974) hypothesis. Are there species that do not gyrate with the propensity of *Tetra-*

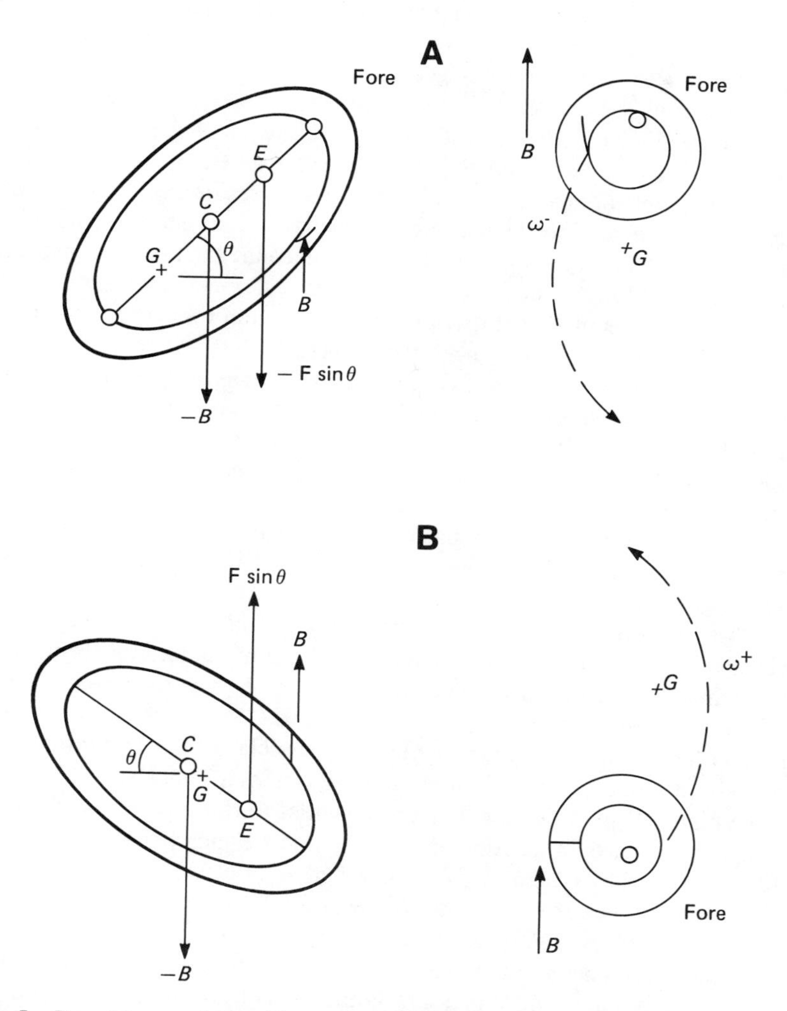

Figure 5. Propulsion–gravity model. (A,B) Left, side views; right, end-on views. In this mixed diagram, both forces and motions are displayed in a nonrigorous manner. The net vertical component of the propulsive force on the body $F \sin \theta$ may be considered to be acting at E. The body is taken to be instantaneously fixed at G about which the body axis gyrates at a ω less than "normal" (i.e., with gravity ignored) during downswing and more than normal during upswing as shown in (A) and (B), respectively. Such a difference is created between the two half-cycles by the countertorque generated by B which increases the viscous energy dissipation of the cilia during downswing, indicated by a bending cilium in (A), and decreases it during upswing, indicated by the unbent cilium in (B). (Redrawn from Winet and Jahn, 1974, with permission.)

hymena or *Paramecium* that nevertheless show strong negative geotaxis? *Chlamydomonas,* for instance, uses a variety of swimming styles and strokes, and propulsion is clearly required for geotactic orientation, but it is by no means clear that cellular gyration or even helical swimming paths are consistently present to a sufficient extent to meet the requirements of the Winet and Jahn hypothesis (Bean, 1977 and unpublished observations; Bean and Harris, 1978). For *Paramecium, Blepharisma,* and other species, examples of shifts among negative, positive, and nongeotactic behavior have been reported, and it is not clear how the Winet and Jahn model would account for them. In particular, Fukui and Asai (1980) specifically "observed that starved *Paramecium* showed gyrating locomotion in almost the same manner as non-starved ciliates, but the starved ciliates did not exhibit negative geotaxis."

Details of the movements of (i.e., gyrating) *Paramecium caudatum* during geotaxis have been studied (Grębecki and Nowakowska, 1977; Nowakowska and Grębecki, 1977). In many respects, the observations of these workers are consistent with a propulsion-gravity mechanism of geotaxis. In several specific aspects, however, their data challenge the Winet and Jahn (1974) hypothesis, and Grębecki and Nowakowska propose an alternate view that emphasizes the translational rather than the gyrational components of propulsion as the key forces in geotaxis.

It was shown (Grębecki and Nowakowska, 1977; Nowakowska and Grębecki, 1977) that ionically induced, backward-swimming paramecia still orient with their anteriors upward, while the direction of their movement is downward. As detailed by Nowakowska and Grębecki (1977), such individuals still gyrate with characteristics that are generally similar to those of forward-swimming cells, although their swimming path describes a somewhat tighter helix, and the plane of gyration is no longer tangential to this helix. Considering the accompanying shifts in location of points *E* and *G,* such behavior is probably consistent with the expectations of Winet and Jahn (1974).

Studies by Nowakowska and Grębecki (1977) on the behavior of specimens that showed normal gyration but no translation (i.e., after treatment with RbCl, during the partial ciliary reversal phase of recovery) raise some important reservations about the Winet and Jahn (1974) hypothesis. Such specimens (1) do not reorient toward the anterior–upward position, (2) show no detectable shifting of the gyration point during the gyration cycle, and (3) show no changes in the angular velocity in the upswing versus downswing phases. Such arguments are open to counterarguments: In particular, the partial ciliary reversal condition, induced by Rb^+ ions, not only reduces the forward translational component of movement, but alters ciliary activity and may shift the center of propulsive effort toward a medial location. It is also expected that such populations are heterogeneous and, while most of the treated cells do migrate toward the surface (Grębecki and Nowakowska, 1977), the data analyzed may be biased in some way toward cells that were not geotactic during filming. H. Winet (personal communication, 1982) responds further that changes in angular velocity are not a necessary expectation of the gravity-propulsion hypothesis.

4.2.3. *Paramecium* Geotaxis Reconsidered (Grębecki and Nowakowska)

An experimental reexamination of the mechanism of geotaxis was undertaken (Grębecki and Nowakowska, 1977; Nowakowska and Grębecki, 1977) through a cinematographic analysis of the movements of *Paramecium caudatum* under several behaviorally interesting conditions (some of which are discussed in the preceding sections). Their data described the mechanics of "rudimentary geotaxic orientation," as reflected in the differences in trajectories (body axis), resolved at 0.02- or 0.04-sec intervals, of cells moving vertically upward as compared with those for the upward and downward gyration phases of cells moving obliquely. Asymmetries present in the movements of nonvertically moving individuals result in progressive reorientation toward the anterior-upward orientation, with instantaneous rates of angular orientation in the range of 4°–30°/sec, increasing in an orderly (but nonlinear) way with average cellular position departing from the vertical. Trajectory analysis further showed that the rate of angular change is essentially constant within each gyration cycle and that the gyration point is fixed within each cell at a point located (on the average) posteriorly at a point 64% of the length of the anteroposterior axis.

In cultures depleted of calcium ion, avoidance responding is eliminated (and average swimming speed is increased); cells performed geotaxis rapidly with striking uniformity of orientation and accentuated rates of angular change. Shortly after treatments with $BaCl_2$ (which generates short periods of forward, alternating with longer periods of backward swimming), or with RbCl (which induces backward swimming with continuous ciliary reversal), cells migrated downward with their anteriors oriented upward. At later times after RbCl treatment, cells showed partial ciliary reversal and made little or no translational progress. During this period net geotactic migration did not take place, and no net angular reorientation of trajectories was detectable. These observations (Grębecki and Nowakowska, 1977; Nowakowska and Grębecki, 1977) provide strong evidence that initial geotactic orientation of paramecia takes place by means of long gradual turns, as it does in *Chlamydomonas* (Bean, 1977) and that avoidance responses, as suggested earlier (Kung, 1971*b*), disrupt this initial orienting process because of their randomizing effect and/or interruptions of forward velocity. Trajectory analysis of Nowakowska and Grębecki (1977) suggested that anteroposterior translational movements, and not gyration, are the important forces in achievement of initial geotactic orientation.

Rates of angular orientation were found (Grębecki and Nowakowska,

1977; Nowakowska and Grębecki, 1977) that were notably higher than those expected under the hypotheses of orientation proposed either by Roberts (1970) or by Winet and Jahn (1974); Grębecki and Nowakowska noted that such rates occurred only in cells that were actively propelled through their medium (in either the forward or backward direction). Preferring a physical mechanism of geotaxis, and perceiving difficulties in justifying their data with the previously discussed hypotheses, Nowakowska and Grębecki (1977) suggested a novel and "purely hypothetical" notion of mechanism, which they called the *lifting force hypothesis*. In contrast to the hypotheses of Roberts (1970) and Winet and Jahn (1974), the generation of geotactically significant

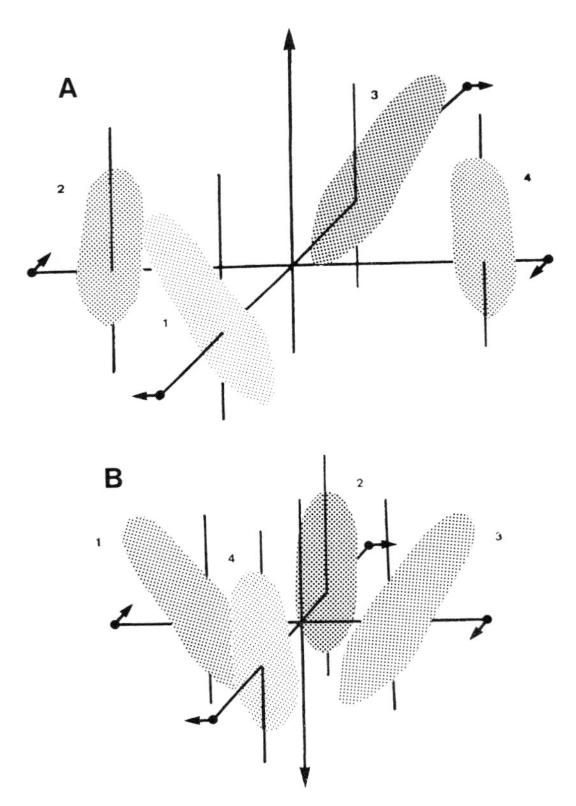

Figure 6. Three-dimensional representation of the body positions of *Paramecium caudatum* during (A) vertical forward-swimming successive images at intervals of 0.04 sec, and (B) vertical backward-swimming successive images at intervals of 0.02 sec. Direction of movement is indicated on the vertical axis, and the sense of the helical trajectory is indicated by arrowheads on the horizontal axes. Instantaneous gyration axes for each body position are given by vertical lines. (Redrawn from Nowakowska and Grębecki, 1977, with permission.)

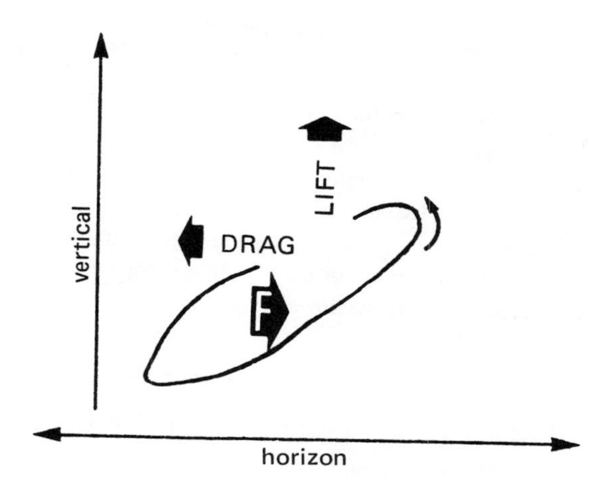

Figure 7. The lifting force concept of Nowakowska and Grębecki. When a gyrating *Paramecium* is oriented obliquely with respect to its net direction of helical movement, interaction with the medium may generate a turning moment. (From Nowakowska and Grębecki, 1977, with permission.)

lifting force is conceived to be tightly dependent on swimming velocity (Nowakowska and Grębecki, 1977; A. Grębecki, personal communication, 1981), and the geometry of the cell as it encounters the viscous resistance of the medium. The torque produced by sedimentation alone would be small, as described by both Roberts (1970) and Kuźnicki (1968), but becomes large when developed against strong propulsive force. In their analysis of trajectories of geotactic versus nongeotactic cells, Nowakowska and Grębecki noted that the cellular axis is somewhat deflected from the direction of locomotion in both forward- and backward-swimming paramecia as a result of the respective gyration patterns illustrated in Fig. 6. Furthermore, as represented in Fig. 7, these workers conceive the viscous resistance to locomotory forces as having two component vectors, drag and lift, which might vary depending on swimming speed and the angle of the cell axis relative to the direction of locomotion. By analogy with an airfoil, lift is generated as a consequence of motion along F, and lift operates through the center of pressure. Since the center of pressure lies anterior to the center of mass, a turning moment is generated. Nowakowska and Grębecki (1977) proposed experimental studies that would test the validity of their hypothesis.

While it may be illustrative of their lifting-force concept, Nowakowska and Grębecki's (1977) comparison with an airfoil is, strictly speaking, a false analogy, which could not directly apply in situations involving low Reynolds numbers. Thus, an actual source of lift remains to be determined.

In interpreting their results, these investigators (Grębecki and Nowakowska, 1977; Nowakowska and Grębecki, 1977) claim to have excluded possible physiological models for geotaxis on the assumption that normal geotaxis takes place at accelerated rates in calcium-depleted cells in which the usual potential-dependent regulation of ciliary action has been disrupted. While such conditions clearly do exclude avoidance responses as an integral part of the mechanism of initial geotactic orientation, they do not rigorously exclude all possible non-calcium-dependent physiological mechanisms. The possibility remains that calcium-indifferent modulation of sensory transduction and ciliary activity are possible in *Paramecium.*

4.2.4. *Paramecium* Geotaxis Reconsidered (Fukui and Asai)

By contrast with other recent studies, Fukui and Asai (1980) interpret their experiments as supporting a physical mechanism of geotaxis in which torque is produced as a consequence of differential density within the cell. These investigators examined the angular distribution (relative to the vertical direction) of cellular positions (but unfortunately not the rates of orientation) of *Paramecium caudatum* populations in relationship to density-relevant experimental variables. Swimming vector distributions for live normal cells showed a strong upward bias, as expected. Cells that had been fed on carbon particles showed an elongated vector distribution on polar coordinates; cells that carried carbon-containing food vacuoles in their posterior parts had a predominantly upward orientation, while those with such vacuoles in their anterior regions were swimming downward (anterior first). Populations that had been starved for 6 days showed significantly slower (but stylistically normal) swimming, were clearly not geotactic, and revealed an essentially random distribution of vectors relative to the vertical. In all these cases, downward swimming speeds were higher than those for cells oriented upward (which in itself would contribute to accumulation of cells near the bottom of the chamber).

Fukui and Asai (1980) also examined the orientation postures of nonmotile (nonionic detergent-extracted) cells after they reached passive sedimentation equilibrium in an 0–10% continuous sucrose density gradient. Models of normal paramecia equilibrated at 5–6% sucrose and were predominantly oriented upward, while models of starved cells equilibrated at 4–5% sucrose and were positioned randomly.

Taken together, the observations of Fukui and Asai (1980) strongly suggest that density biases within a cell can and do influence its ability to orient relative to gravity, presumably by influencing the location of the cellular center of gravity. It is further suggested from their data that such density differences are sufficient to overcome the downward swimming speed advantage that would otherwise generate net downward migration.

While these ideas and observations of Fukui and Asai (1980) are consistent with many reported in the earlier literature (Verworn, 1889, 1899; Harper, 1911, 1912; Dembowski, 1929*a*,*b*, 1931; Merton, 1935), they disagree with other observations, notably with those of Kuźnicki (1968) and others on the sedimentation of immobilized paramecia, those of Kanda (1918), Lyon (1905, 1918), and others on density distribution within paramecia, and of Dembowski (1931) and others on differential swimming velocities. Several of the implied debates cannot be easily resolved on the basis of extant data. Considering the newer theories of initial geotactic orientation, it would be interesting to have a fuller treatment of the experimental approach of Fukui and Asai focusing on parameters that would permit direct relationship to those hypotheses, i.e., rates of directional reorientation, frequencies of randomizing avoidance responding, sedimentation velocities, and so forth. However, even given these additional lines of evidence, it is not clear that definitive cases could be constructed that would rigorously exclude any of the possible theories. It is possible that a density-differential mechanism predominates in density-loaded situations and that other mechanisms predominate under other conditions.

4.3. Physiological Mechanisms

4.3.1. Possible Selective Effects of Centrifugation on Geotaxis

Quite another alternative mechanism was recently invoked by Fornshell (1978), who explained his observations on the effects of centrifugation on the distribution in depths of *Tetrahymena pyriformis, Chlamydomonas moewusii,* and *Polytomella agilis* during bioconvection (pattern swimming), as evidence favoring the statocyst mechanism of geotactic orientation. Fornshell's centrifugation (specifications not given), presumably for extended periods or at relatively high speed, reportedly inhibited geotaxis (as reflected in depth distribution during pattern swimming) for periods that extended well beyond the inhibition of general motility. Fornshell suggested that centrifugation has a disorganizing effect on internal cellular structure and that restoration of structure to reconstitute a functional statocyst requires more time than does restitution of motility. However, Bean (unpublished results) was unable to distinguish any selective effects of centrifugation on the geotaxis of either *C. moewusii* or *C. reinhardtii* using direct assays of geotaxis.

4.3.2. Possible Multiple Mechanisms of Geotaxis in *Euglena*

Creutz and Diehn (1976) tentatively proposed that *Euglena gracilis* uses a full receptor–transducer–effector (physiological) mechanism of geotaxis. In their analysis of the orientation of *Euglena* relative to polarized light stimuli, these investigators found a predominance of specimens that were transversely oriented in the upper regions of their medium (except when bilaterally stimulated with horizontally polarized actinic light). Most of the observations were done in cuvettes of 2 mm × 2 mm × 3 cm, placed on their sides and photographed from below. (Control experiments viewing orientation from the horizontal were not reported.) The effect of depth on transverse orientation was not explored, but similar results were obtained in deeper vessels (Creutz and Diehn, 1976; C. Creutz, personal communication, 1982). Creutz and Diehn recognize that the observed transverse orientation might result from other factors, but if the transverse geotaxis is accepted as a working hypothesis, they would then conclude "that *Euglena* possess a gravity sensing device, and that this sensor interacts with the motor apparatus so that the cell's long axis is maintained at right angles to the force of gravity."

While such an orientation reaction for *Euglena* has not been previously reported, it is quite possible that it does represent a delicate response that might be easily masked by other orientation responses of this responsive protist.

The report of transverse geotaxis in *Euglena* (Creutz and Diehn, 1976) is in apparent conflict with previous reports on this organism. Schwarz (1884), Aderhold (1888), Jensen (1893), Wager (1910, 1911), Thiele (1960), and others have reported that euglenoids are negatively geotactic. Brinkmann (1968*a*) concluded that upward movements are caused by capillary effects and that *E. gracilis* is actually not geotactic. Wager (1910, 1911) and others cite normal negative geotaxis as a key factor in the striking bioconvection patterns formed by *Euglena,* although this argument rests on acceptance of the hypothesized role of geotaxis in pattern formation, which Brinkmann (1968*b*) does not subscribe to. *Euglena* are also centrotactic (Schwarz, 1884), and it would be difficult to explain this observation if they are not negatively geotactic. Further experiments specifically focused on transverse orientation phenomena might prove especially valuable in analysis of the mechanisms of geotaxis. Analysis of the behavior of *Euglena* during achievement of orientation might reveal that transverse orientation is apparent only after establishment of an upwardly asymmetrical distribution of cells. The transverse posture may function in the maintenance of cellular distribution rather than in initial achievement of orientation direction. In such a case, the propensity for cells to assume the transverse orientation after they have approached the surface may indeed involve a physiological suppression of randomizing reorientation maneuvers.

4.3.3. Physiological Mechanisms Reconsidered

Among the venerable ideas on the mechanism of geotaxis is the statocyst hypothesis (Loeb, 1897; Lyon, 1905, 1918; Kanda, 1918; Koehler, 1922, 1930, 1939). The notion that a simple gravity sensor for geotaxis might exist in unicellular eukaryotes has never gained great favor, even though they are commonly discussed for geotropism (e.g., Bentrup, 1979; Volkmann and Sievers, 1979). There is in fact no conclusive evidence for the existence of subcellular gravireceptor structures, nor is there strong evidence for sensory transduction, the transmission of regulated information about orientation relative to gravity into a "primitive" motor apparatus. However, it has become increasingly clear during the past decade alone that what were previously thought of as primitive are now known to be remarkably sophisticated motor apparatuses. Some such apparatuses in themselves consist of an extended systematic array of interacting organelles with the possible complexity of a small telephone system. Should it now be surprising to find that reception and sensory transduction processes are also complex?

Regardless of the possible good sense of nonphysiological explanations, unless one can be shown to be the mechanism that actually does operate in a given organism, the scientist is obliged to maintain some level of skepticism. Of course there are physical mechanics of geotaxis—if rationally describable mechanics were not at work, the cell would not be able to reorient itself. But does the description of mechanics constitute an explanation for them? Which are the causes and which the consequences of the cell's ability to perform geotaxis? If a cell can qualitatively and rapidly reverse the sense of its geotaxis after some shift in a simple physical variable (e.g., temperature, temporary mechanical shock), it is inevitable that a cellular biologist will examine the physiological conditions underlying that cellular capability.

The discussions of Naitoh and Eckert (1974) and of Machemer and de Peyer (1977) reconsidered sensory-based geotactic mechanisms in light of the modern understanding of the electrochemical nature of the regulation of ciliary activity. Naitoh and Eckert (1974) suggested that "certain kinds of behavior of protozoans [e.g., geotaxis and thermosensitive reactions] result from stimulus-dependent modulation of the rate of 'spontaneous' potential fluctuations and thereby the frequency of occurrence of avoiding reactions *in one direction of locomotion* [e.g., with increasing or decreasing temperature] as opposed to the opposite direction." Machemer and de Peyer (1977) suggest that "slight asymmetries in the membrane Ca conductances of upper and lower parts of the cell membrane may alter local ciliary activity in such a way as to establish a negative feedback loop between the stimulus direction and the cellular motor response."

Several generally observed lines of evidence are consistent with the notion of an active physiological mechanism of geotaxis:

1. Forward locomotion is required for negative geotaxis to take place at normal rates.
2. During passive sedimentation, nonmotile cells do not show rates of reorientation toward the vertical sufficient to account for normal geo-

taxis (Kuźnicki, 1968; Grębecki and Nowakowska, 1977; Nowakowska and Grębecki, 1977; Bean, 1977).

3. In paramecia, cells that are swimming upward undergo randomizing avoidance reactions at frequencies that are lower than those of cells on other headings (Jennings, 1906).
4. Stimuli that induce hyperpolarization and accordingly decrease the frequency of avoiding reactions and/or accelerate swimming speed, accelerate negative geotaxis (Naitoh and Eckert, 1974; Grębecki and Nowakowska, 1977).
5. Stimuli that induce depolarization, with accompanying increase in the frequency of avoidance reactions and decrease in swimming speed, inhibit negative geotaxis (Kung, 1971*a,b*; Naitoh and Eckert, 1974; Grębecki and Nowakowska, 1977).
6. Pawn mutants of *Paramecium,* which can only swim forward, perform geotaxis rapidly and with uniform alignment (Kung, 1971*a,b*).
7. Paramecia that are swimming backward or with partial ciliary reversal migrate downward with their anteriors oriented upward (Grębecki and Nowakowska, 1977).
8. Many treatments known to inhibit or alter the sense of orientation, e.g., starvation (Moore, 1903; Merton, 1935; Fukui and Asai, 1980), harsh mechanical, centrifugal, thermal, or chemical treatments (Moore, 1903; Bean, 1975, 1976, 1977; Fornshell, 1978; Bean and Harris, 1978; Bean *et al.,* 1981), and aging or physiological stressing of cells (Moore, 1903; Brinkmann, 1968*a*; Creutz and Diehn, 1976) are also expected to have significant effects on subcellular organization and/or normal coordination of motor activity.
9. Conditions that alter shape but that do not disrupt subcellular organization (Koehler, 1939; Smiechowska, cited by Kuźnicki, 1968) do not generally alter the sense of orientation.
10. The responses of paramecia fed on heavy particles under various conditions are consistent with a statocyst hypothesis (Harper, 1911, 1912; Koehler, 1921, 1922; Machemer and de Peyer, 1977; Fukui and Asai, 1980).
11. Preliminary results on certain inhibitors and mutants for the rate of geotaxis suggest that geotactic response rates can be physiologically uncoupled from effects on general motility of cells (Bean, 1977; Fornshell, 1978).
12. Ciliates are known to have asymmetry around their cell surface with respect to mechanical stimuli and intrinsic ionic properties (Jennings, 1906; Naitoh, 1974; Naitoh and Eckert, 1973, 1974; de Peyer and Machemer, 1978).
13. The motility of artificially reactivated detergent-extracted ciliates

(Naitoh and Kaneko, 1972, 1973; Kung and Naitoh, 1973) and *Chlamydomonas* (Knight and Bean, 1980) is apparently not oriented.

While some of these observations might also be consistent with some purely physical mechanisms of geotactic orientation, all the above observations are consistent with physiological mechanisms of orientation, in which normal regulation of the action of the motor apparatus must be operational.

It is well known from studies on several species that short-term regulation of the axonemal activity of protozoans is governed by membrane-dependent changes in Ca^{2+} concentration (cf. Chapter 3). The specific mechanism of regulation by calcium is unknown, but it seems safe to accept as a working hypothesis on the basis of observations on the physiology of normal, reactivated, and inhibited axonemes (e.g., Machemer and Eckert, 1975; Bean and Harris, 1978; Satir and Ojakian, 1979; Bessen *et al.*, 1980) that a specific protein binds calcium reversibly and differentially affects other axonemal components, depending on the amount of calcium present. It has been reported that the eukaryotic calcium-activated modulator protein, calmodulin, occurs in the ciliary apparatus of *Tetrahymena* (Jamieson *et al.*, 1979; Maihle and Satir, 1979; Satir *et al.*, 1980; Suzuki *et al.*, 1981) and *Paramecium* (Maihle and Satir, 1979; Satir *et al.*, 1980) and in the flagella of *Chlamydomonas* (Gitleman and Witman, 1980). Various motile behaviors of these same organisms are known to be sensitive to calcium-reversible inhibition by phenothiazine drugs (cf. Weiss and Wallace, 1980) and other treatments expected to cause selective inhibition of calmodulin-mediated processes (Bean and Harris, 1978; Satir *et al.*, 1980; Bean and Lawrence, 1981; Suzuki *et al.*, 1981). Dynein ATPases of *Tetrahymena* were found to be sensitive to direct modulation by the calcium–calmodulin complex (Jamieson *et al.*, 1980). It is also possible that the calcium–calmodulin complex acts to stimulate activity of the calcium-activated membrane ATPase (cf. Vincenzi and Hinds, 1980) that functions as a calcium pump to remove calcium ion from the axonemal space. Here again, it should be expected that certain metal ions and drugs should have selective effects on orientation behaviors.

5. DISCUSSION

If there is one primary message that emerges from all the information available on the mechanism(s) of geotaxis, it is that more information is needed. The georesponses of microorganisms are more complex than is often assumed, and various mechanisms might operate in different species or even within a species under different conditions. Several hypothesized mechanisms of geotaxis remain viable. In order to resolve the behavioral complexities involved and to distinguish rigorously among the mechanistic possibilities, new lines of experimental evidence will be required.

In the examination of data, care should be taken to identify the mechanics and the phase of the response addressed. For clear distinction of physical versus physiological mechanisms of response, in particular, previously neglected avenues of analysis will probably be essential.

5.1. Physical and/or Physiological Mechanisms?

For investigators whose interests lie in the areas of biological membranes and sensory transduction, the central question remains: Does geotaxis operate by active physiological mechanisms, or by purely passive physical consequences of cellular propulsion? One of the physical mechanisms previously discussed or some synthetic combination of them may contribute to geotactic behavior. In any circumstance in which the center of effort of a free-swimming cell lies anterior to the center of gravity (and/or, in relevant cases, anterior to the center of buoyancy), some contribution toward an upward torque on the cell should result. In some circumstances, and particularly in gyrating organisms, the resultant torque alone may be sufficient to account for geotaxis.

In order to demonstrate that a physical mechanism is actually operating, and to do so rigorously, requires not only that the natural response fit the predictions of the relevant physical model, but demonstration of evidence against alternate mechanisms. Such evidence might be elusive, but given the fundamental differences between physical and physiological mechanisms, it should be possible to distinguish experimentally between these two broadest types. For instance, if the negative geotaxis of *Paramecium* is determined entirely by a physical mechanism, it should be possible to show that the movements of detergent-extracted models of paramecia (including certain mutants that normally do not show negative geotaxis) after ATP-dependent reactivation of motility, are consistent with the proposed physical mechanism. Likewise, models reactivated to the backward-swimming state should also show movements consistent with the proposed physical mechanism. Such observations would provide compelling evidence that physiological mechanisms are not required.

Furthermore, if physical mechanisms are proposed to be solely responsible for geotactic orientation, this argument should be substantiated with a demonstration that the relevant reorientation rates and energetics of orientation are fully sufficient to account for the observed behavior of cells during the critical phase of the response.

Definitive new data are also required to substantiate the argument for the opposite case, that physical mechanisms alone are not sufficient to account for geotaxis, and that active physiological mechanisms operate in achievement or maintenance of orientation or in the stabilization of an asymmetrically distributed population. To document such arguments convincingly, investigators should be able to produce one or more lines of compelling evidence that active specific physiological processes are required. For instance, it might be possible to isolate specific behavioral mutants that show altered rates or reversed direction of geotaxis without corresponding alteration of cellular shape, surface organization, ultrastructure, density distribution, general motile behavior, and responses to photic or other orientation stimuli.

Alternatively, to support a physiological mechanism it might be possible to demonstrate selective effects of physiological inhibitors on the georesponse relative to other cellular pressures. There are known inhibitors, e.g., Cu^{2+} (Bean *et al.*, 1978; Bean and Yussen, 1979), that are selective for the photoresponses but not for geotaxis, but the reverse has not yet been shown. For *Chlamydomonas*, no agents that inhibit both phototaxis and geotaxis are yet known that do not overlap with inhibition of general motile behavior, e.g., treatment with Ni^{2+} (Bean and Harris, 1978). In particular, if membrane signaling is required for georesponse, inhibitors known to have selective effects on membrane physiology or sensory transduction should be examined in depth, exercising appropriate cautions concerning the specificity of inhibition.

Studies on the mechanics of orientation might also provide strong evidence for a physiological mechanism. If behavior during the achievement of orientation can be characterized in detail, it may be possible to demonstrate rigorously that the actual rate of orientation is faster than the purely physical mechanisms would predict. Preliminary evidence along this line has been given by Nowakowska and Grębecki (see Section 4.2.3) and others.

5.2. Conjectures on Physiological Mechanisms

Given that physiological mechanisms of geotaxis have not been excluded, it is interesting to consider the types of structures and processes that might be exploited by unicellular organisms in detecting orientation relative to gravity, as well as the options by which such information might influence effector coordination. While fundamentally speculative, such considerations may be of some value in the design of future experiments. The basic requirements of a gravireceptor may be generalized as follows: (1) some physical component of the cell (e.g., an organelle or perhaps a specific enzyme) must alter its position or biological activity as a function of the orientation of the cell, and (2) such changes of position or activity must be capable of influencing other cellular activities such that transient asymmetries in motor actions will result. The classical concept of the statocyst may be recast in light of modern notions of cellular ultrastructure to envision several possible receptor mechanisms.

5.2.1. Mechanical Consequences of the Physiology of Effector Coordination

One possibility is that the motor organelles themselves act differentially, depending on their orientation relative to the cytoplasmic center of mass, e.g., by dissipating mechanical energy along different vectors at different points on the cell surface, thereby generating asymmetries of overall effector coordination. Further understanding of this possibility, including its categorization as a mechanical or physiological orientation mechanism, depends on insights into the fundamental nature of effector propulsion and coordination, which at present are rudimentary.

5.2.2. Possible Sensory Functions of the Axoneme

It is further possible that an active physiological reversal of the normal chemomechanical energy interconversion of axonemes (or other organelle systems) may operate, as may be the case for certain receptor functions of animals (cf. Atema, 1973, 1975). For *Chlamydomonas,* flagellar specializations are known to be involved in regulation of adhesiveness, mating interactions, and physiological signalling to the cell body concerning the triggering of specialized events of the mating process; some of these processes are known to involve Ca^{2+} (Solter and Gibor, 1977; Claes, 1980; Goodenough *et al.,* 1982). In this formulation, the major portion of the cytoplasm of the cell body is conceived as a statocyst relative to the motile apparatus, which is asymmetrically situated on the cell surface, serving as both an effector and receptor organelle.

5.2.3. Orientation-Dependent Organelle Actions

An alternative possibility for a physiological gravireceptor is that the juxtaposition of organelles, the distribution of forces within an organelle, or the apposition of organelles with membranes might vary with cellular orientation. In such a case, the key ultrastructural changes might be of very small or undetectable magnitude and duration and might be localized anywhere within the cell. Specific mutants or inhibitor treatments might selectively inactivate or block the physical change or its consequences, and thereby suggest refinements of experimental focus for these lines of investigation. Such concepts are analogous to those suggested by current views on the differential distribution of membrane electrochemical properties on the surface of ciliates and by current thinking on the nature of the gravireceptor for gravitropism.

5.2.4. Membranes and Gravireception

It is further possible that some property of the membrane system of cells is sensitive to cellular orientation. Changes in orientation might have transient selective influence on (1) the distribution of molecules within the fluid mosaic, (2) the relative distribution of membranous materials associated with different organelles, or (3) the site or rates of membrane assembly or disassembly. Such changes might in turn generate slight local perturbation in ion partitioning, within the distances over which membrane-associated processes occur, or in the frequency or rates of membrane-dependent processes. It may be that membrane-localized enzymes, or enzymes that modify membranes, have position-dependent activities. Or, it may be that the dynamics of mosaic flow or associations of membranes with nonmembranous structures are transiently altered with changes in cellular orientation.

One particular variant of this possibility might occur if presumptions about the nature of certain processes of the plasma membrane and about the nature of forces that might be transmitted translationally about the cell surface within the lipid bilayer are made. The structure of the cells of essentially all protists is asymmetrical: asymmetrical anteroposteriad positioning of the effectors of locomotion and asymmetrical distribution of major and minor subcellular organelles. *Chlamydomonas,* for example, has striking anteroposterior organization that is similar in all flagellated motile individuals, while the subcellular organization of the nonmotile phases of the life cycle is neither so regular nor so strongly asymmetrical.

Suppose that one of the organelles were position sensitive in its activity, that it could influence other organelles in a position-dependent way, or that the summed physical or biochemical effects of the position of several organelles could differentially influence the physiological functioning of the plasma membrane system. Most eukaryotic plasma membranes are known to have (and thought to have additional) specific associations or connections with cytoplasmic organelles, including microfilaments, microtubules, intermediate filaments, submembranous fiberous networks, and elements of axonemal basal apparatuses. If these organelles are associated with proteins that are part of the plasma membrane itself, it seems likely that some of these associations might have an influence on one or more of the physiological roles of the plasma membrane. And some of these physiological roles might be differentially affected as the influence of the organelles on different parts of the plasma membrane might be experienced by the cell. Such fluctuations in physiological activities that have short half-lives (such as some found for calmodulin-mediated processes) could be used by the cell as a sensory transduction signal. Such signals might be attributable to differential distribution about the plasma membrane of the sensitive sites or to differential stimulation of uniformly distributed sites during changes in position. The signals might then be propagated mechanically or biochemically, within the membrane system or through the cytoplasm or other organelles.

5.2.5. Speculation on the Gyrosome

In order for membranes to function directly in gravireception, they might contain specialized structures. One can envision a large globular multilipoprotein complex (gyrosome) floating in an oriented way within a local bilayer of phospholipid, its top protruding above the outer surface of the membrane, its bottom protruding through the cytoplasmic surface of the lipid bilayer. When the local membrane surface is calm, the multilipoprotein complex neither bobs nor rolls nor yaws from its vertical pole orientation. The actions of undulipodia, exocytotic events, emptying of contractile vacuoles, or physical encounters of the cell with other objects may bring about symmetrically propagated displacements of the membrane bilayer. When such a perturbation in the calm membrane surface, a single wave, encounters the multilipoprotein globule (being itself a miniscule object in the world of low Reynolds numbers hydrodynamics), the gyrosome is resistant to changing its orientation (as is a gyroscope?), and rides in the wave (as might a human bobbing his head in the surf) and not with the wave (as would a small lipid-soluble molecule) as the wave passes. Or, perhaps the lipoprotein globule is anchored to a rigid three-dimensional object (like a microtubule), forcing it to hold its position regardless of what transpires in the undulations of the membrane surface. As the wave of membrane displacement passes, the globule is tilted or displaced relative to the membrane surface first one way and then the other (always 180° opposite). This exposes to the cytoplasm sequentially, first one portion and then another portion of the globule surface that is not exposed to the cytoplasm when the surface is calm. The wave source may have a particular focus (or multiple but nevertheless fixed foci); waves coming from different directions would be propagated with characteristics (relative to the gyrosome) that depend on the magnitude of the originating force of the perturbation, the physical and chemical properties of the lipid bilayer, possibly the distance from the source, and the direction from which the approach was made relative to the asymmetrical features of the gyrosome. As the newly exposed and possibly enzymatically active globule surface rolls, consequence is "felt" in neighboring molecules, either structurally or enzymatically.

The enzymatic or structural change that occurred may or may not have an immediate effect on its surroundings, but perhaps as more waves arrive in the same region of the membrane, the effect is sufficiently great that it begins itself to spread through neighboring regions of cytoplasm. If attached to a microtubule, it may exert a force, via a covalent bond, on a neighboring tubulin molecule, altering the conformation of a microtubule-tip tubulin subunit into an altered position relative to its neighbors. Such a change might in itself be propagated. As envisioned, such a change might be functionally interpreted at the opposite end of a microtubule by virtue of a miniscule shift in the position of another protein with a bonding partner or substrate molecule. Perhaps such a signal informs the polymerizing region of the microtubule that the position relative to the surface membrane is appropriate for either addition or deletion of a subunit. Or, perhaps the multilipoprotein globule is itself a gate permitting selective influx or efflux of a particular molecule that is normally partitioned differentially by the membrane system. Or, perhaps the consequence of the passage of the physical membrane wave is to activate an ATP-using enzyme, such as an adenylcyclase, an actin-to-myosin interaction, a dynein, or an ion pump.

Thus, by one of many physiological mechanisms, an asymmetrical and possibly propagatable biological signal might be generated, depending on the local physical properties of the globule-containing membrane. The relative distribution of gyrosomes or their ability to influence other structures might vary about the cellular surface.

If gyrosomes can be shown to exist, it might be expected that they are structurally or functionally related to other membrane-associated proteins that act as sensory microreceptors. Such protein complexes might be important in several respects:

1. In the triggering of mitosis or cytokinesis
2. In regulating microtubule length
3. As sites for fusion between apposing membrane surfaces
4. In triggering changes in microtubule-associated proteins (including dynein, ion-binding sites, attachment sites, or associated structural proteins, possibly including basal body tubules, rootlets, or fiber proteins)
5. As mechanoreceptor proteins
6. As sensors of membrane thickness
7. As modulators of membrane-linked processes, possibly including activities of ion gates, membrane ATPases (including pumps), or second-messenger functions

5.3. Conclusion

One hundred years of study have not generated definitive explanations for the geotactic orientation behaviors of free-swimming cells. Recent observations and hypotheses on the mechanism of geotaxis are reviewed in this chapter. Given modern advances in cellular biology, it should now be possible to distinguish rigorously among the hypotheses on mechanism proposed on the basis of sound experimental evidence. However, sufficiently compelling evidence has yet to appear in the literature.

While it is likely that passive hydrodynamic consequences of motility do contribute to orientation relative to gravity, involvement of physiological mechanism(s) remains a significant possibility. Definitive new data will be required to assess the importance of the physiological contribution to orientation. Since physiological mechanisms have not been excluded or treated in detail in recent years, they should be reconsidered seriously. Known complexities of membrane and intracellular signalling systems and reasonable conjectures about the structure and functioning of these systems suggest how physiological mechanisms might operate. In particular, it is possible that a family of membrane-associated microreceptor protein complexes may exist, including a position receptor or gyrosome.

ACKNOWLEDGMENTS. I am grateful to Dr. Bruce R. Hargreaves, Dr. Andrzej Grębecki, Dr. Alun Roberts, Dr. Howard Winet, and Dr. W. Haupt for sharing their perspectives in many ways and to Ms. Marie C. Tracey for excellent secretarial support. This work was supported in part by a grant from the Penrose Fund of the American Philosophical Society.

REFERENCES

Aderhold, R., 1888, Beiträg zur kenntniss richtender kräfte bie der bewegung niederer organismen, *Jen. Z. Naturwiss.* **22**:310–342.

Atema, J., 1973, Microtubule theory of sensory transduction, *J. Theor. Biol.* **38**:181–190.

Atema, J., 1975, Stimulus transmission along microtubules in sensory cells: a hypothesis, in: *Microtubules and Microtubule Inhibitors* (M. Borgers and M. DeBrabander, eds.), pp. 247–257, North-Holland, Elsevier, New York.

Bean, B., 1975, Geotaxis in *Chlamydomonas, J. Cell Biol.* **67**:24a.

Bean, B., 1976, Flagellar coordination in *Chlamydomonas* can be analyzed by studying geotaxis in behavioral mutants, *Genetics* **83**:s5–s6.

Bean, B., 1977, Geotactic behavior of *Chlamydomonas, J. Protozool.* **24**:394–401.

Bean, B., and Harris, A., 1978, Selective inhibition of flagellar activity in *Chlamydomonas* by Ni^{++}, *J. Protozool.* **26**:235–240.

Bean, B., and Lawrence, J. J., 1981, A possible regulatory role of calmodulin in sensory transduction in *Chlamydomonas, J. Cell Biol.* **91**:50a.

Bean, B., and Yussen, P., 1979, Photoresponses of *Chlamydomonas:* differential inhibitions of phototactic and photophobic responses by low concentrations of Cu^{2+}, *J. Cell Biol.* **83**:351a.

Bean, B., Brandt, J., and Prevelige, P., 1978, Selective effects of cupric ion on the motility and photophobic responses of *Chlamydomonas, J. Cell Biol.* **79**:292a.

Bean, B., Savitsky, R., and Grossinger, B., 1981, Strontium ion (Sr^{2+}) induces backward swimming of *Chlamydomonas, J. Protozool.* **29**:296.

Bentrup, F. W., 1979, Reception and transduction of electrical and mechanical stimuli, in: *Physiology of Movements, Encyclopedia of Plant Physiology,* new series, Vol. 7 (W. Haupt and M. E. Feinleib, eds.), pp. 42–70, Springer-Verlag, Berlin.

Bergman, K., 1973, Mutants of *Phycomyces* with abnormal phototropism, *Mol. Gen. Genet.* **123**:1–16.

Bessen, M., Fay, R. B., and Witman, G. B., 1980, Calcium control of waveform in isolated flagellar axonemes of *Chlamydomonas, J. Cell Biol.* **84**:446–455.

Brinkmann, K., 1968*a*, Keine Geotaxis bei *Euglena, Z. Pflanzenphysiol.* **59**:12–16.

Brinkmann, K., 1968*b*, An phasengrenzen induzierte ein- und zweidimensionale kristallmuster in kulturen von *Euglena gracilis, Z. Pflanzenphysiol.* **59**:364–376.

Byrne, B. J., and Byrne, B. C., 1978, Behavior and the excitable membrane in *Paramecium, Crit. Rev. Microbiol.* **6**:53–108.

Cameron, J. N., and Carlile, M. J., 1977, Negative geotaxis of the fungus *Phytophthora, J. Gen. Microbiol.* **98**:599–602.

Carlile, M. J., 1980*a*, Positioning mechanisms—the role of motility, taxis and tropism in the life of microorganisms, in: *Contemporary Microbial Ecology* (D. C. Ellwood, J. N. Hedger, M. J. Latham, J. M. Lynch, and J. H. Slater, eds.), pp. 55–74, Academic Press, New York.

Carlile, M. J., 1980*b*, Sensory transduction in aneural organisms, in: *Photoreception and Sensory Transduction in Aneural Organisms* (F. Lenci and G. Colombetti, eds.), pp. 1–22, Plenum Press, New York.

Childress, W. S., Levandowsky, M., and Spiegel, E. A., 1975*a*, Pattern formation in a suspension of swimming microorganisms: equations and stability theory, *J. Fluid Mech.* **63**:591–613.

Childress, W. S., Levandowsky, M., and Spiegel, E. A., 1975*b*, Non-linear solutions of equations describing bioconvection, in: *Swimming and Flying in Nature,* Vol. 1 (T. Y.-T. Wu, C. J. Brokaw, and C. Brennen, eds.), pp. 361–375, Plenum Press, New York.

Claes, H., 1980, Calcium ionophore-induced stimulation of secretory activity in *Chlamydomonas reinhardii, Arch. Microbiol.* **124**:81–86.

Creutz, C., and Diehn, B., 1976, Motor responses to polarized light and gravity sensing in *Euglena gracilis, J. Protozool.* **23**:552–556.

Cronkite, D. L., 1979, The genetics of swimming and mating behavior in *Paramecium,* in: *Biochemistry and Physiology of Protozoa,* 2d ed., Vol. 2 (M. Levandowsky and S. H. Hutner, eds.), pp. 221–273, Academic Press, New York.

Davenport, C. B., 1908, *Experimental Morphology,* Third Edition, Macmillan, New York.

Dembowski, J., 1929*a*, Die vertikalbewegungen von *Paramecium caudatum.* I. Die Lage des gleichgewichtscentrums im korper des infusors, *Arch. Protistenk.* **66**:104–132.

Dembowski, J., 1929*b*, Die vertikalbewegungen von *Paramecium caudatum.* II. Einfluss einiger assenfaktoren, *Arch. Protistenk.* **74**:153–187.

Dembowski, J., 1931, Die vertikalbewegungen von *Paramecium caudatum.* III. Polemisches und experimentelles, *Arch. Protistenk.* **74**:153–187.

de Peyer, J. E., and Machemer, H., 1978, Hyperpolarizing and depolarizing mechanoreceptor potentials in *Stylonychia, J. Comp. Physiol.* **127**:255–266.

Dryl, S., 1974, Behavior and motor response of *Paramecium,* in: *Paramecium: A Current Survey* (W. J. VanWagtendonk, ed.), pp. 165–218, Elsevier Scientific, Amsterdam.

Fornshell, J. A., 1978, An experimental investigation of bioconvection in three species of microorganisms, *J. Protozool.* **25**:125–133.

Fornshell, J. A., 1980, Positive geotaxis in *Blepharisma persicinum, J. Protozool.* **27**:24A–25A.

Fox, H. M., 1925, The effect of light on the vertical movement of aquatic organisms, *Proc. Camb. Phil. Soc., Biol. Sci.* **1**:219–224.

Fukui, K., and Asai, H., 1980, The most probable mechanism of negative geotaxis of *Paramecium caudatum, Proc. Jpn. Acad.* Ser. B: **56**:172–177.

Gitleman, S. L., and Witman, G. B., 1980, Purification of calmodulin from *Chlamydomonas:* calmodulin occurs in cell bodies and flagella, *J. Cell Biol.* **88**:764–770.

Gittleson, S. M., and Jahn, T. L., 1968*a*, Vertical aggregations of *Polytomella agilis, Exp. Cell Res.* **51**:579–586.

Gittleson, S. M., and Jahn, T. L., 1968*b*, Pattern swimming by *Polytomella agilis, Am. Nat.* **102**:413–425.

Goodenough, U. W., Detmers, P. A., and Hwang, C., 1982, Activation for cell fusion in *Chla-*

mydomonas: analysis of wild-type gametes and non-fusing mutants, *J. Cell Biol.* **92**:378–386.

Gręecki, A., and Nowakowska, G., 1977, On the mechanism of orientation of *Paramecium caudatum* in the gravity field. I. Influence of ciliary reversal and of external Ca deficiency on the geotactic behavior, *Acta Protozool.* **16**:351–358.

Harper, E. H., 1911, The geotropism of *Paramoecium, J. Morphol.* **22**:993–1000.

Harper, E. H., 1912, Magnetic control of geotropism in *Paramoecium, J. Anim. Behav.* **2**:181–189.

Haupt, W., 1962, Geotaxis, in *Encyclopedia of Plant Physiology:* Vol. XVII. *Physiology of Movements* (W. Ruhland, ed.), pp. 390–395, Springer-Verlag, Berlin.

Jahn, T. L., and Votta, J. J., 1972, Locomotion of protozoa, *Annu. Rev. Fluid Mech.* **4**:93–116.

Jahn, T. L., and Winet, H., 1973, Mechanism of negative geotaxis, *Prog. Protozool , Fourth International Congress on Protozoology.,* Clermont-Ferrand, 197.

Jamieson, G. A., Jr., Vanaman, T. C., and Blum, J. J., 1979, Presence of calmodulin in *Tetrahymena, Proc. Natl. Acad. Sci. USA,* **76**:6471–6475.

Jamieson, G. A., Jr., Vanaman, T. C., Hayes, A., and Blum, J. J., 1980, Affinity chromatographic isolation of highly purified Ca–CaM sensitive dynein ATPases from *Tetrahymena* cilia, *Ann. NY Acad. Sci.* **356**:391–392.

Jennings, H. S., 1904, The behavior of *Paramecium.* Additional features and general relations, *J. Comp. Neurol. Psychol.* **XIV**:441–510.

Jennings, H. S., 1906, *Behavior of the Lower Organisms,* Columbia University Press, New York.

Jensen, P., 1893, Ueber den geotropismus niederer organismen, *Pfluegers Arch. Physiol.* **53**:428–480.

Juniper, B. E., 1977, The perception of gravity by a plant, *Proc. R. Soc. Lond. B* **199**:537–550.

Kanda, S., 1918, Further studies on the geotropism of *Paramecium caudatum, Biol. Bull.* **34**:108–119.

Katz, D. F., and Pedrotti, L., 1977, Geotaxis by motile spermatozoa: hydrodynamic reorientation, *J. Theor. Biol.* **67**:723–732.

Knight, W. L., and Bean, B., 1980, Reactivation of motility in detergent-extracted cell models of *Chlamydomonas, Abst. Ann. Meet. Am. Soc. Microbiol.* 1980:81.

Koehler, O., 1921, Über die geotaxis von *Paramecium, Verh. Dtsch. Zool. Ges.* **26**:69–71.

Koehler, O., 1922, Über die geotaxis von *Paramecium, Arch. Protistenk.* **45**:1–94.

Koehler, O., 1930, Über die geotaxis von *Paramecium* II, *Arch. Protistenk.* **70**:297–306.

Koehler, O., 1939, Ein filmprotokoll zum reizverhalten querzertrennter Paramecien, *Zool. Anz. Suppl.* **12**:132–142.

Kung, C., 1971*a*, Genic mutants with altered system of excitation in *Paramecium aurelia,* I. Phenotypes of the behavioral mutants, *Z. Vergl. Physiol.* **71**:142–164.

Kung, C., 1971*b*, Genic mutants with altered system of excitation in *Paramecium aurelia.* II. Mutagenesis, screening and genetic analysis of the mutants, *Genetics* **69**:29–45.

Kung, C., 1976, Membrane control of ciliary motions and its genetic modification, in: *Cell Motility: Book C, Microtubules and Related Proteins* (R. Goldman, T. Pollard, and J. Rosenbaum, eds.), pp. 941–948, Cold Spring Harbor Laboratory, Cold Spring Harbor, N.Y.

Kung, C., and Naitoh, Y., 1973, Calcium-induced ciliary reversal in the extracted models of "Pawn," a behavioral mutant of *Paramecium, Science* **179**:195–196.

Kuźnicki, L., 1968, Behavior of *Paramecium* in gravity fields. I. Sinking of immobilized specimens, *Acta Protozool.* **31**:109–117.

Levandowsky, M., Childress, W. S., Spiegel, E. A., and Hutner, S. H., 1975, A mathematical model of pattern formation by swimming microorganisms, *J. Protozool.* **22**:296–306.

Loeb, J., 1897, Zur theorie der physiologidchen licht und schwerkraftwirkungen, *Pfluegers Arch. Ges. Physiol.* **66**:439–466.

Lyon, E. P., 1905, On the theory of geotropism in *Paramecium, Am. J. Physiol.* **14**:421–432.

Lyon, E. P., 1918, Note on the geotropism of *Paramecium, Biol Bull.* **34**:120.

Machemer, H., and Eckert, R., 1975, Ciliary frequency and orientational responses to clamped voltage steps in *Paramecium, J. Comp. Physiol.* **104**:247–260.

Machemer, H., and de Peyer, J., 1977, Swimming sensory cells: Electrical membrane parameters, receptor properties and motor control in ciliated Protozoa, *Vehr. Dtsch. Zool. Ges.* **1977**:86–110.

Maihle, N. J., and Satir, B. H., 1979, Calmodulin in the ciliates *Paramecium tetraurelia* and *Tetrahymena thermophila, Ann. NY Acad. Sci.* **356**:408–409.

Majima, T., and Oosawa, F., 1975, Response of *Chlamydomonas* to temperature change, *J. Protozool.* **22**:499–501.

Massart, J., 1891, Recherches sur les organisms inferieurs, *Acad. R. Sci. Lett. Beaux-Arts Belg.* **22**:148–167.

Medina, J. R., and Cerda-Olmedo, E., 1977, Allelic interaction in the photogeotropism of *Phycomyces, Exp. Mycol.* **1**:286–292.

Merton, H., 1935, Versche zur Geotaxis von *Paramecium, Arch. Protistenk.* **85**:33–60.

Moore, A., 1903, Some facts concerning geotropic gatherings of paramecia, *Am. J. Physiol.* **9**:238–244.

Naitoh, Y., 1974, Bioelectric basis of behavior in protozoa, *Am. Zool.* **14**:883–893.

Naitoh, Y., and Eckert, R., 1973, Sensory mechanisms in *Paramecium.* II. Ionic basis of the hyperpolarizing mechanoreceptor potential, *J. Exp. Biol.* **59**:53–65.

Naitoh, Y., and Eckert, R., 1974, The control of ciliary activity in protozoa, in: *Cilia and Flagella* (M. A. Sleigh, ed.), pp. 305–352, Academic Press, New York.

Naitoh, Y., and Kaneko, H., 1972, Reactivated triton-extracted models of *Paramecium:* modification of ciliary movement by Ca ions, *Science* **176**:523–524.

Naitoh, Y., and Kaneko, H., 1973, Control of ciliary activities by adenosinetriphosphate and divalent cations in triton-extracted models of *Paramecium caudatum, J. Exp. Biol.* **58**:657–676.

Nakaoka, Y., and Oosawa, F., 1977, Temperature-sensitive behavior of *Paramecium caudatum, J. Protozool.* **24**:575–580.

Nowakowska, G., and Grębecki, A., 1977, On the mechanism of orientation of *Paramecium caudatum* in the gravity field. II. Contributions to a hydrodynamic model of geotaxis, *Acta Protozool.* **16**:359–376.

Plesset, M. S., Whipple, C. G., and Winet, H., 1975, Analysis of the steady state of the bioconvection in swarms of swimming microorganisms, in: *Swimming and Flying in Nature,* Vol. 1 (T. Y-T. Wu, C. J. Brokaw, and C. Brennan, eds.), pp. 339–360, Plenum Press, New York.

Roberts, A. M., 1970, Geotaxis in motile microorganisms, *J. Exp. Biol.* **53**:687–699.

Roberts, A. M., 1972, Gravitational separation of X and Y spermatozoa, *Nature (Lond.)* **238**:223–225.

Roberts, A. M., 1975, The biassed random walk and the analysis of microorganism movement, in: *Swimming and Flying in Nature,* Vol. 1, (T. Y.-T. Wu, C. J. Brokaw, and C. Brennan, eds.), pp. 377–393, Plenum Press, New York.

Roberts, A. M., 1981, Hydrodynamics of protozoan swimming, in *Biochemistry and Physiology of Protozoa,* 2nd ed., Vol. 4 (M. Levandowsky and S. H. Hutner, eds.), pp. 5–66, Academic Press, New York.

Satir, P., and Ojakian, G. K., 1979, Plant cilia, in: *Physiology of Movements, Encyclopedia of Plant Physiology,* new ser., Vol. 7 (W. Haupt and M. E. Feinleib, eds.), pp. 224–249, Springer-Verlag, Berlin.

Satir, B. H., Garofalo, R. S., Gilligan, D. M., Maihle, N. J., 1980, Possible functions of calmodulin in protozoa, *Ann. NY Acad. Sci.* **356**:83–91.

Schwarz, F., 1884, Der einfluss der schwerkraft auf die Bewegunsrichtung von *Chlamidomonas* und *Euglena, Dtsch. Bot. Ges. Ber.* **2**:57–72.

Sievers, A., and Volkmann, D., 1979, Gravitropism in single cells, in: *Physiology of Movements, Encyclopedia of Plant Physiology,* new ser., Vol. 7 (W. Haupt and M. E. Feinleib, eds.) pp. 567–572, Springer-Verlag, Berlin.

Solter, K. M., and Gibor, A., 1977, Evidence for role of flagella as sensory transducers in mating of *Chlamydomonas reinhardi, Nature (Lond.)* **265**:444–445.

Suzuki, Y., Ohnishi, K., Hirabayashi, T., and Watanabe, Y., 1981, Localization of calmodulin in *Tetrahymena* cells and cilia, *J. Cell Biol.* **91**:46a.

Thiele, R., 1960, Uber lichtadaptation und musterbildung bei *Euglena gracilis, Arch. Microbiol.* **37**:379–398.

Verworn, M., 1889, *Psychophysiologisch Protistenstudien,* G. Fischer, Jena.

Verworn, M., 1899, *General Physiology: An outline of the Science of Life,* Macmillan, London.

Vincenzi, F. F., and Hinds, T. R., 1980, Calmodulin and Plasma membrane calcium transport, in: *Calcium and Cell Function,* Vol. I: *Calmodulin* (W. Y. Cheung, ed.), pp. 127–165, Academic Press, New York.

Volkmann, D., and Sievers, A., 1979, Graviperception in multicellular organs, in: *Physiology of Movements, Encyclopedia of Plant Physiology,* new ser., Vol. 7 (W. Haupt and M. E. Feinleib, eds.), pp. 573–600, Springer-Verlag, Berlin.

Wager, H., 1910, The effect of gravity upon the movements and aggregation of *Euglena viridis,* Ehrb., and other micro-organisms, *Proc. Roy. Soc.* **B83**:94–96.

Wager, H., 1911, The effect of gravity upon the movements and aggregation of *Euglena viridis,* Ehrb., and other micro-organisms, *Phil. Trans. R. Soc.* **B201**:333–388.

Walsby, A. E., 1978, The gas vesicles of aquatic prokaryotes, in: *Relations Between Structure and Function in the Prokaryotic Cell, 28th Symposium of the Society for General Microbiology* (R. Y. Stanier, H. J. Rogers, and J. B. Ward, eds.), pp. 327–358, Cambridge University Press, Cambridge.

Weiss, B., and Wallace, T. L., 1980, Mechanisms and pharmacological implications of altering calmodulin activity, in: *Calcium and Cell Function,* Vol. I: *Calmodulin* (W. Y. Cheung, ed.), pp. 329–379, Academic Press, New York.

Wilkins, M. B., 1979, Growth-control mechanisms in gravitropism, in: *Physiology of Movements, Encyclopedia of Plant Physiology,* new ser., Vol. 7 (W. Haupt and M. E. Feinleib, eds.), pp. 601–626, Springer-Verlag, Berlin.

Winet, H., and Jahn, T. L., 1974, Geotaxis in Protozoa I. A propulsion-gravity model for *Tetrahymena* (Ciliata), *J. Theor. Biol.* **46**:449–465.

Chapter 6

Photosensory Responses in Freely Motile Microorganisms

Francesco Lenci, Donat-P. Häder, and Giuliano Colombetti

1. INTRODUCTION

There is growing evidence that membrane-linked and membrane-controlled phenomena can play a fundamental role in photosensory responses of several unicellular organisms. In this chapter we discuss, on the basis of the available experimental results and of theoretical considerations, the different steps of perception and transduction of light stimuli, which induce and control photobehaviors of microorganisms, focusing our attention on the possible role of membrane properties and functions.

For the interested reader, several recent reviews are available on the different aspects of light-induced movements of freely motile microorganisms: from general discussions of photobehavioral responses (Feinleib, 1978; Lenci and Colombetti, 1978; Diehn, 1979; Häder, 1979*a*; Nultsch and Häder, 1979; Colombetti, 1984; Colombetti and Lenci, 1983; Lenci, 1982) to more specific analyses of particular groups of microorganisms, such as prokaryotes (Häder, 1980; Nultsch, 1980; Nultsch and Häder, 1980), flagellates and ciliates (Checcucci, 1976; Doughty and Diehn, 1980; Feinleib, 1980; Foster and Smyth,

Francesco Lenci and Giuliano Colombetti • Istituto di Biofisica, Consiglio Nazionale delle Ricerche, 56100 Pisa, Italy. **Donat-P. Häder** • Fachbereich Biologie Marburg Universität, Lahnberge, D-3350 Marburg/Lahn, Federal Republic of Germany.

1980; Song, 1981; Colombetti *et al.*, 1982*a*), or cellular slime molds (Poff and Whitaker, 1979); from structural, physicochemical, and physiological approaches to the characteristics of photoreceptor pigments (Feinleib and Curry, 1974; Bensasson, 1977, 1980; Colombetti and Lenci, 1980*a*; Hemmerich and Schmidt, 1980; Hildebrand, 1980; Schmidt, 1980; Senger and Briggs, 1981; Song and Walker, 1981; Senger, 1982) to reviews on motility (Blum and Hines, 1979; Lazarides and Revel, 1979) and on experimental techniques and methodologies (Diehn, 1980) to discussions on the evolution of photobehavior (Carlile, 1980*a,b*; Halldal, 1980).

After a brief description of the various types of light-induced motile responses, we discuss the properties of photoreceptor pigments and structures, together with the primary molecular modifications and reactions that can follow the absorption of a photon. The problems of signal transduction and its mechanisms are considered in their different aspects: energetics, ionic and electrical gradients, and enzymatic reactions. We then deal with the question of information transmission, discussing the alternatives of specific transmitter molecules or electrical signals, and concluding with consideration of the final step of the photosensory process, the motor response.

2. PHOTOMOTILE RESPONSES

Light-induced motile reactions in microorganisms can be classified in three principal groups (Diehn *et al.*, 1977):

1. *Phototaxis:* This reaction defines an oriented movement with respect to the light direction. In addition to oriented movements toward the light source (positive phototaxis) or away from it (negative phototaxis) (Fig. 1A), movements at 90° (diaphotaxis) or at another angle with respect to the light direction have been described (Fisher and Williams, 1981). In all these cases, the actual stimulus is the light direction; in other words, the direction of propagation of the light beam, $\mathbf{S}$ ($\mathbf{S} = \mathbf{E} \times \mathbf{H}$ is the Poynting vector, and $\mathbf{E}$ and $\mathbf{H}$ are the electrical and the magnetic field vectors, respectively, of the light wave. $\mathbf{S}$ describes the energy per unit time across the unit surface associated with a plane electromagnetic wave and defines the direction of propagation of this plane electromagnetic wave).
2. *Photokinesis:* This photomotile reaction is independent of the direction of the light stimulus and describes a steady-state dependence of the speed of movement $/\mathbf{v}/$ on light intensity I $[/\mathbf{v}/ = v(I)]$. Photokinesis is called positive when, at a given light intensity, $/\mathbf{v}/$ is higher than in the dark and negative when lower (Fig. 1B). In most studied cases,

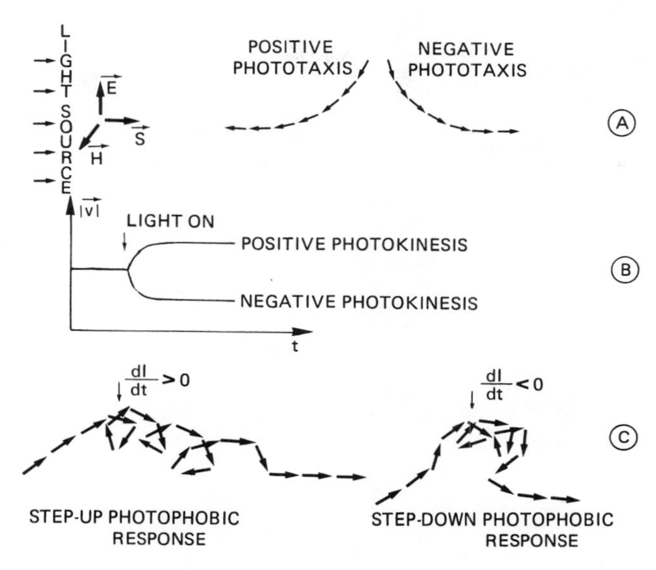

Figure 1. Light-induced motile reactions in microorganisms. (A) Phototaxis; (B) photokinesis; (C) photophobic responses. See text for description of symbols.

however, photokinesis constitutes a typical example of a photocoupling process (through the photosynthetic system (Nultsch, 1980 and references therein)) in which no adaptation to the photic stimulus occurs.

3. *Photophobic responses:* Like photokinesis, these responses are independent of light direction, but they do show adaptation to the stimulus. A sudden change in light intensity (dI/dt) causes a motor response that is generally characteristic of the microorganism under study. A photophobic response in a given organism can be elicited by either an increase (step-up photophobic response) or a decrease (step-down photophobic response) in light intensity (Fig. 1C).

3. PHOTORECEPTOR PROPERTIES

The sensory transduction chain of any photomovement can be dissected in a number of subsequent events (Fig. 2), the first step of which is obviously the interaction of a light quantum with a photoreceptor molecule. The question of which molecules are devoted to the detection of photic stimuli and what physicochemical changes occur from that point is of primary importance to build up reasonable models of sensory transduction chains. In some cases, the

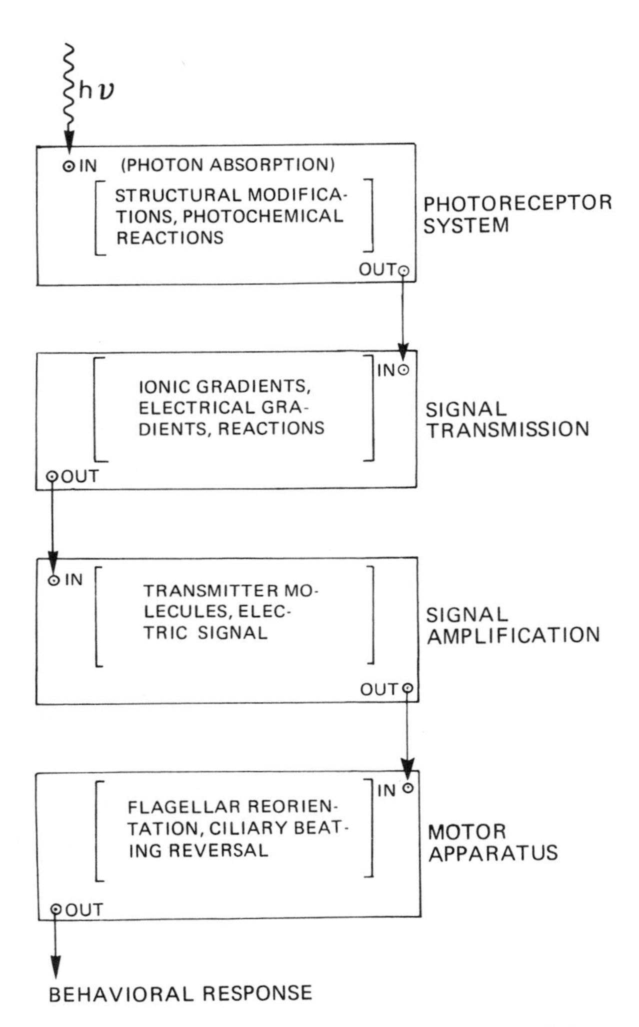

Figure 2. Outline of a photosensory transduction chain.

nature of the photoreceptor pigment has been fairly well established, as in purple bacteria (Throm, 1968), some blue-green algae (Nultsch, 1962; Häder, 1973), desmids (Wenderoth and Häder, 1979), the flagellate *Euglena gracilis* (Diehn, 1969; Checcucci *et al.*, 1976; Benedetti and Lenci, 1977; Barghigiani *et al.*, 1979*a*; Colombetti and Lenci, 1980*b*; Colombetti *et al.*, 1981), and some ciliates (Walker *et al.*, 1979; Song, 1981). In other cases the photopigments involved in sensory responses have not yet been positively identified.

3.1. Microenvironment and Photophysical Characteristics

A photoreceptor pigment suitable for triggering a light-induced sensory response in nature has to absorb wavelengths in the near ultraviolet (UV) and/or the visible region of the solar spectrum. After absorption of a light quantum (or after a decrease of light intensity on the photoreceptor, as in stepdown photophobic reactions), these specialized photoreceptor molecules undergo certain physicochemical modifications (e.g., Song, 1980, and references therein). This will initiate the process of signaling to the organism the transient and steady-state characteristics of the environmental illumination conditions.

As a general feature, almost all photoreceptor molecules seem to be bound to proteins, which in turn are embedded in or associated with membranes. This is ascertained for all cases in which photosynthetically active pigments are also responsible for photomotile responses, as in the case of photosynthetic bacteria, blue-green algae, desmids, and red algae, but there is growing evidence that this is true in other systems as well (Walker *et al.,* 1979; Haupt, 1980; Song and Walker, 1981, and references therein). Also on the basis of theoretical considerations, the complex steps of the sensory transduction chain demand a high degree of structural organization, which can best be achieved with a membrane structure. Actually much experimental evidence indicates an organized texture of the photoreceptor organelle, which is usually closely associated with the plasma membrane. A typical example is the quasicrystalline photoreceptor, the paraflagellar body (PFB) of *Euglena* (Piccinni and Mammi, 1978) (Fig. 3). Also, the presumed photoreceptor of *Ochromonas danica* (Piccinni and Coppellotti, 1979), although apparently less organized, looks like a rather dense organelle inside and is in tight contact with the plasma membrane of one of the flagella (Fig. 4).

In the case of *Chlamydomonas reinhardtii* and *Chlorosarcinopsis,* electron microscopic studies of freeze-fractured membrane have recently shown that the plasmalemma and the outer chloroplast membranes covering the stigma are specialized in carrying a higher density of particles, 8–12 nm in diameter, than any other region of these membranes (Melkonian and Robenek, 1980*a,b*), confirming previous suggestions of Walne and Arnott (1967). These structured localized membrane regions, with the particles they contain, might play the role of photoreceptors in the photobehavior of these microorganisms (Robenek and Melkonian, 1981). A similarly specialized region of the plasma membrane seems to be facing the stigma of *Haematococcus pluvialis* (Ristori *et al.,* 1981) and might also be present in front of the stigma of *Ochromonas.* The structural organization previously sketched can account for the functionality of the photoreceptor system and the closely associated membrane and also influence the photophysical characteristics of the photopigments. In fact the molecular environment in which the chromophores are embedded can have a

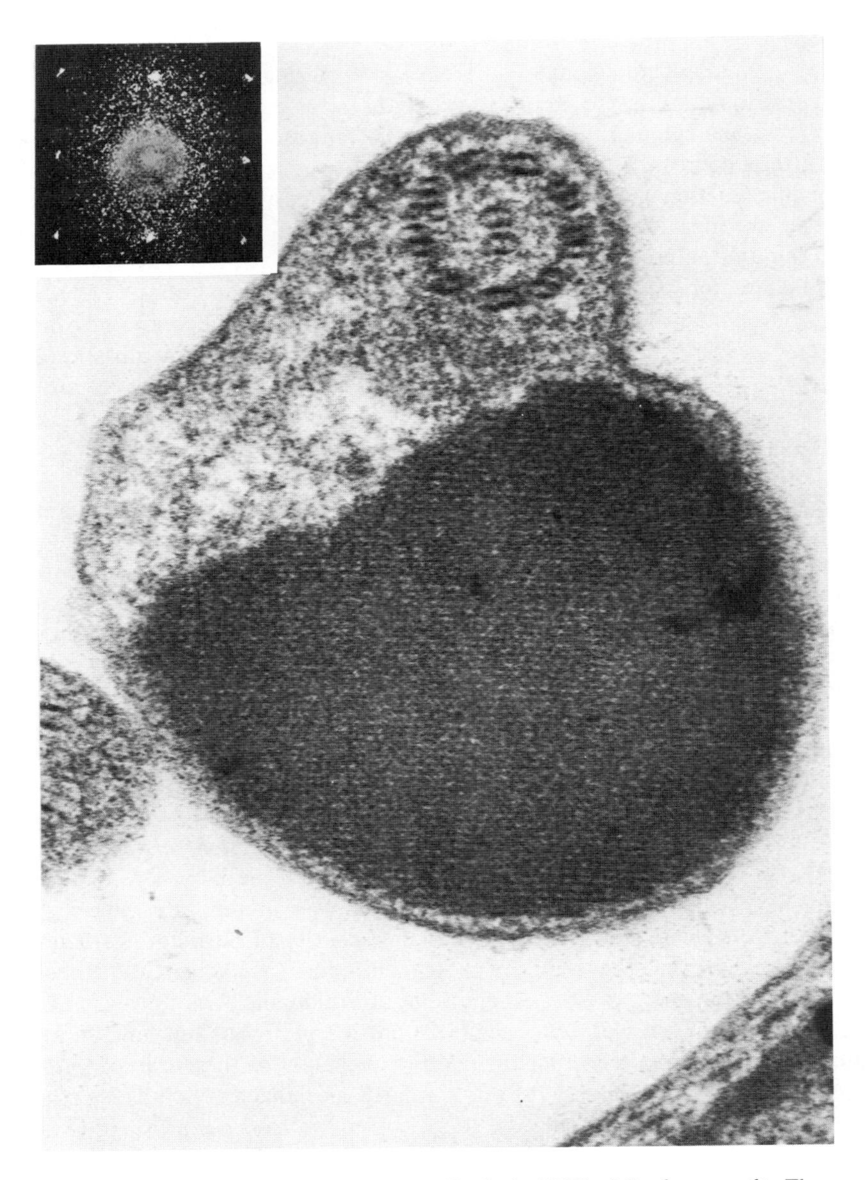

Figure 3. Electron micrograph of the paraflagellar body (PFB) of *Euglena gracilis*. The quasicrystalline structure of the PFB is clearly visible. The inset shows the optical diffraction pattern from the microscopic picture (×120,000). (Courtesy of Dr. E. Piccinni, Padova University.)

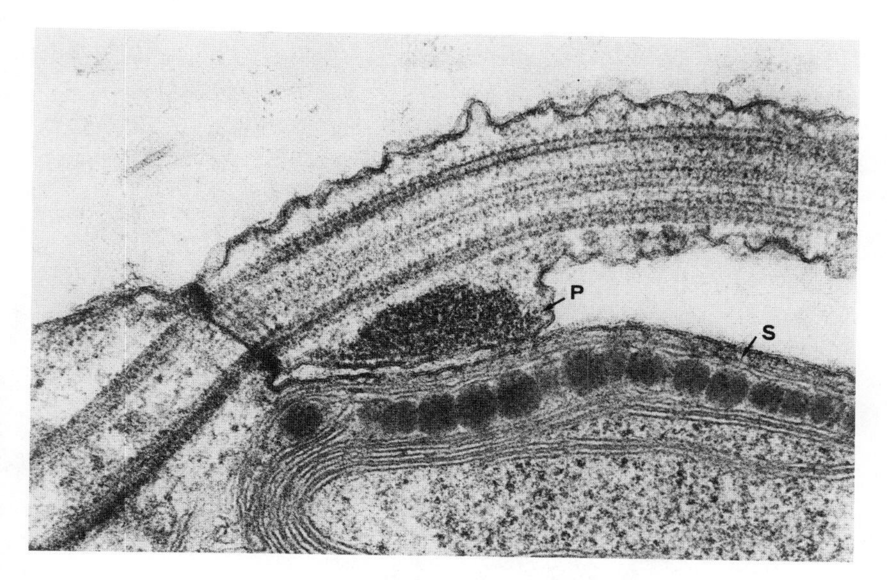

Figure 4. Electron micrograph of the stigma (S) and the paraflagellar swelling (P) of *Ochromonas danica* (×60,200). (Courtesy of Dr. E. Piccinni, Padova University.)

significant effect on their spectroscopic properties, including absorption spectrum, relative quantum yields of the different deexcitation pathways, and reactivity of the transient states.

The interaction of light with photopigments results in a modification in the photoreceptive unit, which triggers and initiates the sensing process. Among the points to be clarified are the physical characteristics of the primary reactive state (Fig. 5). The first excited triplet state T_1 could well be a reasonable candidate, because of its relatively long lifetime, 10^{-6}–10^{-3} sec. In fact during this time the photoreceptor molecules could move by diffusion within the membrane and come into contact with the reactive site. However, in relatively rigid quasicrystalline systems of the PFB type of *Euglena* and the purple membrane of *Halobacterium halobium* (for purple membrane structure see, e.g., Henderson, 1977), for instance, diffusion-controlled processes should not play a relevant role. The short-lived first excited singlet state S_1 (lifetime about 10^{-9} sec) might offer the advantage of being more energetic, and therefore able to initiate reactions that could not start from the less energetic triplet state. In the case of two most typical photoreceptor pigments, flavins and carotenoids, the energy difference between the singlet and triplet state is of the order of 0.2–0.3 eV and about 1.8 eV, respectively. Furthermore, the population of the singlet state is certainly higher than that of the triplet state, which, in an iso-

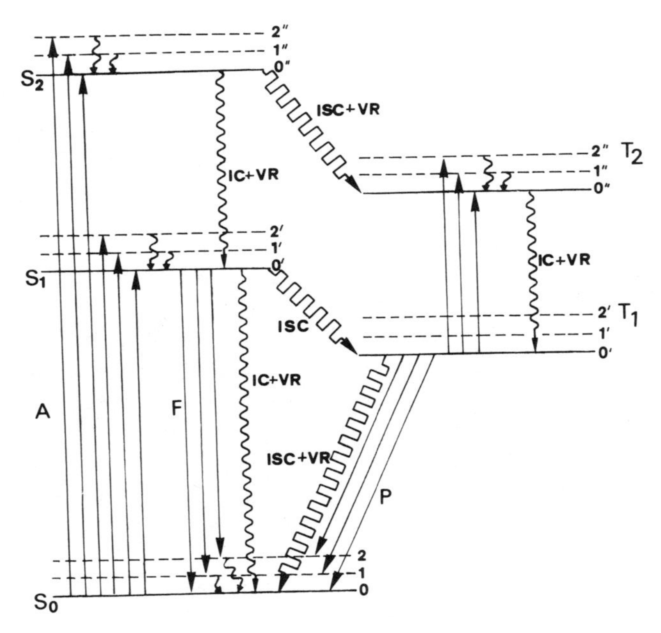

Figure 5. Schematic representation of the energy levels and radiative and non-radiative transitions among them for a typical photoreceptor molecule. (S_0) Ground singlet state; (S_1) first excited singlet state; (S_2) second excited singlet state; (T_1) first excited triplet state; (T_2) second excited triplet state; (A) absorption; (F) fluorescence emission; (P) phosphorescence emission; (IC) internal conversion; (VR) vibrational relaxation; (ISC) intersystem crossing. Dotted lines represent vibrational levels.

lated molecule, is populated from S_1 via intersystem crossing. In the case of carotenoids, for instance, the triplet population is extremely low because of the very inefficient $S_1 \rightsquigarrow T_1$ intersystem crossing, and its energy is as low as 0.8 eV (Bensasson, 1980, and references therein).

In the organized structure of a photoreceptive unit the production of an electronically excited state could in itself be a relevant modification of the photoreceptor pigment fully adequate to initiate the sensing process, without any need for a light-triggered or light-regulated intermolecular or intramolecular reaction. One might, for example, imagine a photoreceptor molecule embedded in a membrane: The difference in electric dipole moments of the ground and of the excited state (S_1, for instance) could change the electrostatic potential in the neighborhood of the receptor site. This could influence a potential-dependent process in the membrane, such as ion permeability, rate of enzymatic reactions, ion binding, and opening and closing of a channel (Lundström and Nylander, 1981). However, although not to be excluded *a priori,* such an electrostatic phenomenon has never been reported in photomotile responses of microorganisms, whereas in the visual process similar events, such as the early receptor potential (ERP), seem not to have any physiological relevance.

Photoreceptor molecules can finally be clustered in such a way that light energy, absorbed anywhere in the cluster, can migrate, during the excitation lifetime, to some kind of reaction center, where this light-produced electronic excitation can be transformed into usable information. These moving excitations, called excitons, play a decisive role in photosynthesis (e.g., Knox, 1973; Pearlstein, 1982) and can be important in other biological processes as well, such as radiation damage of DNA and short-range interactions in protein–chromophore complexes. In principle, excitons can play a role in relatively ordered structures, such as the quasicrystalline photoreceptor of *Euglena,* but no experimental evidence is yet available to indicate such a role in light-sensing processes of any photomotile microorganism.

3.2. Primary Molecular Events

In this section we briefly discuss some of the possible primary reactions that occur after light absorption that can bring about the photosensing process (see also Song, 1980). A relaxation of the excited state in form of intersystem crossing or internal conversion is usually not regarded as the starting point of a photochemical event, but the local heating could alter the microenvironment of the molecule and change diffusion or conductivity characteristics of a membrane.

The pK_a of a ligand on the chromophore or the attached protein can change considerably upon absorption of a quantum, which can induce the uptake or release of a proton. When this is associated with a conformational change of the protein, the proton can be picked up on one side of a membrane and released on the other. The resulting proton gradient can be amplified and used to trigger changes in the membrane potentials and/or ionic gradients (possible mechanisms for amplification are discussed in Section 4.2). Conformational changes and isomerizations are well-known fundamental processes in vision in vertebrates and invertebrates.

Some photopigments have been reported to be photochromic, i.e., they show a shift in their absorption maximum arising from absorption of light with the creation of a biologically active form, as in the case of phycochrome (Björn, 1978, 1979, 1980).

The absorption of a quantum can also cause the dissociation of a photoreceptor molecule (e.g., dissociation of all-*trans*-retinol in the visual process) or, on the other hand, an association that leads to the formation of active molecular complexes. The reorientation of the chromophore induced by light has been discussed, for instance, in the case of phytochrome (Song *et al.,* 1979; Sarkar and Song, 1982).

Other pigments, when absorbing light, can specifically reduce other molecules associated with them in the photoreceptive structure, as has been shown in the flavoprotein-sensitized photoreduction of b-type cytochrome (Poff and Butler, 1975; Senger and Briggs, 1981).

It has been suggested that both in vision (e.g., Maeda and Yoshizawa, 1982, and references therein) and in *Halobacterium* (Hildebrand, 1980, and references therein), the molecules undergo a *cis–trans*-isomerization during the cycle of intermediates, after the absorption of a photon.

4. SIGNAL TRANSDUCTION

Light-mediated reactions in living organisms can be divided into two groups. In one case the energy of the absorbed quanta is converted and stored in the form of chemical energy. These processes, like photosynthesis and at least photokinesis in prokaryotes and desmids, require a relatively high radiation input in order to be effective. In the other case, light energy is used to trigger developmental or motile responses. Many of these reactions function under low light intensities; therefore, in most if not all cases, an amplification step seems to be necessary.

An amplification mechanism is characterized by the fact that for one absorbed photon more than one molecular event can take place (for instance, one photon could lead to the transport of tens of ions or to the activation of several molecules of an enzyme). Thus, through the utilization of previously stored metabolic energy, an energy amplification occurs, the energy involved in the overall phenomenon being higher than the energy of the absorbed photon, as happens in vision, *Phycomyces* phototropism, *Desmid* phototaxis, and so on.

In the case of photomotile responses, it is difficult to estimate the energetic balance involved in the photoresponse itself. In particular, the so-called photoinduced stop reactions (a particular case of photophobic responses) in flagellated microorganisms do not really require any work done on the system. Actually a theoretical analysis of the motion of flagellated microorganisms shows that in this case viscous forces dominate over inertial ones (Berg and Purcell, 1977, and references therein). In fact, an order of magnitude of the ratio of inertial to viscous forces that operate on a system moving steadily through a fluid is given by the Reynolds number, defined as

$$R = \frac{lv\rho}{\eta}$$

where l is a characteristic length of the system that moves with speed v, ρ is the density of the fluid, and η its viscosity. For a fish, R is of the order of 10^6

and this gives a quantitative idea of the well-known fact that fishes are able to propel themselves entirely by means of inertial forces. But for a microorganism such as a bacterium swimming in water, the dynamics must be completely different. In fact, assuming l about 2×10^{-4} cm, v about 3×10^{-3} cm/sec, $\rho = 1$ g/cm^3, and $\eta = 10^{-2}$ dyn sec/cm^2, R is about 5×10^{-5}, hence only viscous forces control the motion. In plain words, when a fish stops swimming, it will coast for several body lengths, whereas a microorganism that stops swimming will coast only for a few tenths of an angstrom. Flagellated cells can therefore be stopped simply by cutting the flux of energy to the motor apparatus, which may require virtually no energy expenditure. Many photoresponsive systems are known to operate over a very wide range of light intensities (usually several decades), as is the case with *Micrasterias* (Neuscheler, 1967*a*,*b*), *Dictyostelium* slugs (K. L. Poff, unpublished data), and *Phycomyces* (Lipson, 1975), which seems difficult to reconcile with a photocoupling process. Moreover, all the adaptation processes that take place in photomotile systems speak against a nonmediated use of the energy of the photon.

As shown in Section 4.1, a number of different mechanisms are used by various microorganisms to transduce the initial photic stimulus.

4.1. Proton Gradients

During the photocycle of bacteriorhodopsin, the protonated Schiff base releases its proton, yielding an intermediate that absorbs at 412 nm. In a subsequent step, the chromophore picks up a proton and regenerates the original state (Fig. 6). One important feature is that the proton is picked up on one side of the membrane and is released on the other side, generating a proton gradient (Takeuchi *et al.*, 1981; Spudich and Spudich, 1982). This could be achieved

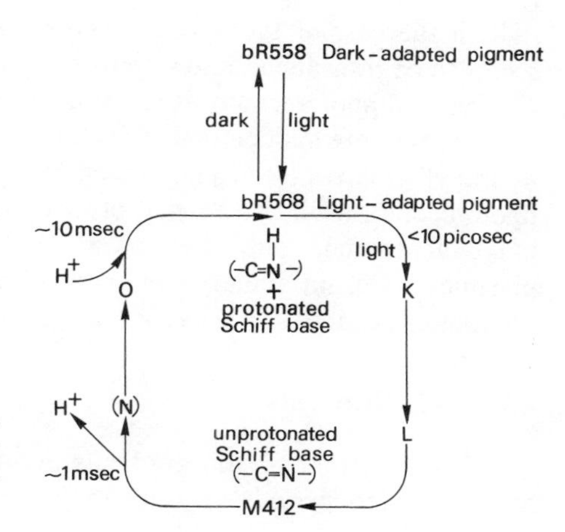

Figure 6. Photocycle of bacteriorhodopsin, the light-harvesting pigment of *Halobacterium halobium* that drives photoinduced ATP synthesis.

by a conformational change in the chromophore-binding protein during the cycle, facilitating the binding on one side and the release on the other side of the membrane.

A similar mechanism has been discussed for the photophobic responses of blue-green algae (Häder, 1979*b*). Absorption of light by the photoreceptor pigments, which are identical to the photosynthetic pigments, causes electrons to be transported through the noncyclic electron-transport chain between the two photosystems, PSII and PSI. Each time plastoquinone, one of the redox components of this chain, cycles between its oxidized and reduced state, it accepts a pair of protons on the outside of the thylakoid membrane, subsequently releasing it into the thylakoid vesicle. This vectorial proton transport generates an electrochemical gradient $\Delta\,\mu_{H^+}$ consisting of a pH and an electrical component ψ:

$$\Delta\,\mu_{H^+} = \Delta\,\text{pH} + \Delta\,\psi$$

In contrast, proton gradients across biomembranes are not very stable unless they are constantly built up; thus, when light intensity decreases, the electrochemical gradient will break down rapidly. This event could be the starting point for a sensory transduction chain, ultimately resulting in a photophobic response, which in blue-green algae consists of a reversal of movement. The light-induced pH gradient across the thylakoid membranes of blue-green algae has been found to be as high as 2–3 pH units (Padan and Schuldinger, 1978). In an attempt to prove that the breakdown of the pH gradient actually elicits the step-down photophobic response, Häder (1981) has studied the effect of pH jumps on the behavior of these systems, finding that a decrease of 2–3 pH units actually caused a reversal of movement similar to a photophobic response (Fig. 7).

In the case of the ciliate *Stentor coeruleus* as well, the photic stimulus seems to be transduced into a transient proton flux from the pigment granules. The excited photoreceptor is supposed to serve as the transient proton source: the excited-state dissociation of protons is caused by a change in the pK_a of the hydroxyl groups of the photoreceptor stentorin upon photoexcitation of the photoreceptor itself. As a result of this phototriggered proton flux, a change in pH is generated, coupled to a Ca^{2+} influx. This transient increase in cytoplasmic and/or intraciliary Ca^{2+} concentration is ultimately responsible for the photomotile reaction of *Stentor* (Song, 1981, and references therein).

4.2. Ionic Gradients

The fact that proton gradients across biomembranes may not be stable over extended periods of time may explain why *Halobacterium* seems to

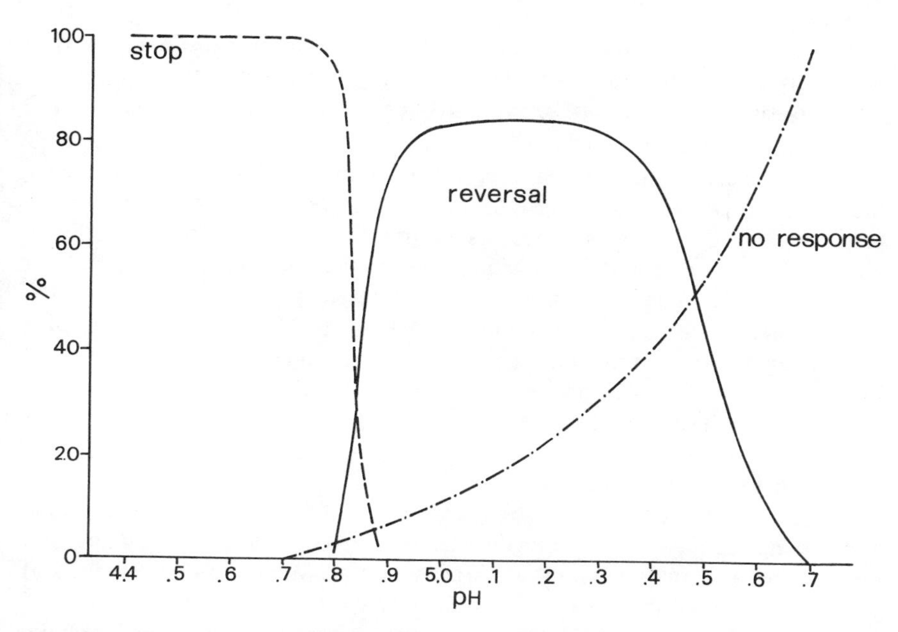

Figure 7. Effect of pH on the motile behavior of a blue-green alga. The ordinate represents percentage of cells affected by the changes in pH value.

exchange the initially generated proton gradient for a potassium or sodium gradient.

Ionic gradients are the molecular basis for electric potentials found in living cells. They are generated by energy-dependent specific ion pumps, which establish a concentration difference across the membrane against a passive leakage (Martonosi, 1980). The existence of ionic gradients can be employed as an amplification mechanism. A prerequisite for this process is the availability of voltage-dependent specific ion gates or channels (Harold, 1977; Hobbs and Albers, 1980).

These channels are believed to respond to small changes in the electrical potential, which can be evoked by a primary receptor potential. If the channels open in response to a small potential change, they permit the ion gradients previously established by active ion pumps to equilibrate rapidly through the open pores, resulting in a dramatic change in the membrane potential. Gating channels open only for a limited time, after which the ion pumps restore the original concentration gradient. Although this amplification and transduction mechanism is well established for behavioral responses in ciliates caused by thermal, mechanical, or chemical stimuli (Eckert, 1977; Hansma, 1979; Eckert and Brehm, 1979; Schusterman *et al.*, 1981), in photoresponsive microorganisms it has been studied only recently and only in a few cases.

Doughty and Diehn (1979) and Doughty *et al.*, (1980) have shown that mono- and divalent cations are involved in photomotile responses of the flagellate *Euglena.* On the basis of experiments with ionophores and ion-current blockers, these workers suggest a model for the step-down photophobic response (Fig. 8). These investigators assume that light, absorbed by a flavin chromophore located in the paraflagellar body (the photoreceptor), activates a Na^+/K^+ pump. When light intensity is decreased (step-down), the Na^+/K^+ pump is inactivated and the subsequent increase of Na^+ in the intraflagellar space triggers the opening of a specific Ca^{2+} gate and causes an increase in the inner Ca^{2+} concentration above the normal value of 10^{-8} M. Access of this gated Ca^{2+} to the axoneme is assumed to induce a flagellar reorientation. After the response, the Ca^{2+} gates are closed again, and a pump restores the normal, lower, inner Ca^{2+} concentration.

Much experimental evidence suggests that in *Chlamydomonas* membrane phenomena mediated by certain ions can be responsible for processing and transducing the photic stimulus (Marbach and Mayer, 1971; Stavis and Hirschberg, 1973; Stavis, 1974; Schmidt and Eckert, 1976; Hyams and Borisy, 1978; Nultsch, 1979; and references therein). Flagellar response and photic stimuli are coupled, in this alga, through Ca^{2+} ions and Ca^{2+}-mediated membrane phenomena: Photostimulation results in an influx of Ca^{2+} ions through the cell membrane, causing a transient increase in the intracellular Ca^{2+} concentration, which in turn causes the backward swimming of the cell. The pho-

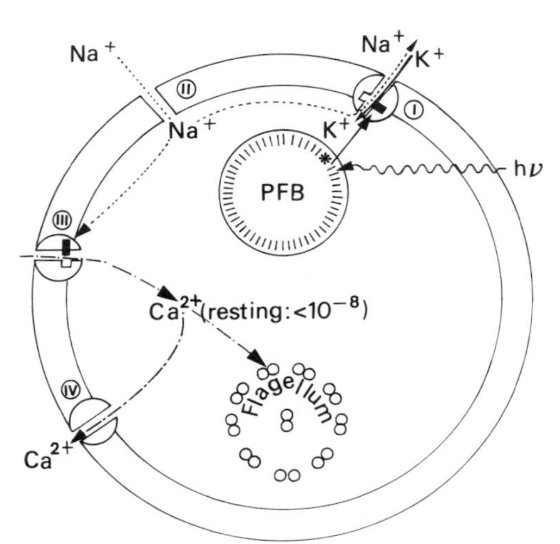

Figure 8. Scheme of the hypothetical ionic model proposed by Doughty and Diehn (1979) and Doughty *et al.* (1980).

toreceptor pigments, implanted in the membrane, might serve as a light-activated gate for the passage of Ca^{2+} ions (see Chapter 1).

4.3. Electrical Gradients

In ciliates it has been shown that changes in ionic gradients are linked with changes in electrical potential (Saimi and Kung, 1980; Hook and Hildebrand, 1980). This is also true for light-induced responses. Song *et al.* (1980) have demonstrated a correlation between step-up photophobic responses of *Stentor coeruleus* and membrane potential changes measured with intracellular microelectrodes.

Similarly, membrane potentials can be evoked by 1-msec light pulses in the flagellate *Haematococcus pluvialis* (Litvin *et al.*, 1978; Ristori *et al.*, 1981). The cell is held by a suction micropipette that electrically insulates a fraction of the cell surface from the remainder. Action potentials can be detected between the inside of the pipette and the rest of the cell when the membrane portion facing the stigma is sucked into the micropipette and illuminated. Illumination of any other part of the cell does not elicit any electrical potential (Ristori *et al.*, 1981).

Membrane potential changes have also been described as means for sensory transduction in the photophobic response of *Phormidium uncinatum* (Häder, 1978*a*,*b*, 1979*b*). The first indication for this mechanism was found in the observation that direct current fields or extremely low-frequency alternating fields of about 1 $V \cdot cm^{-1}$ inhibit photophobic responses at a light–dark border, when the organism moves parallel to the electrical field lines (Häder, 1977). As there is no indication of a galvanotactic effect, it can be assumed that the external field interferes with a light-dependent internal electrical gradient that controls the movement in the filaments. Actually, light-induced potential changes have been detected by means of both extracellular and intracellular microelectrodes (Häder, 1978*a*,*b*). These electrical gradients constitute membrane potentials because they can be decreased by the membrane-penetrating lipophilic cation $TPMP^+$ and have to depend on the photosynthetic machinery, since they are impaired by inhibitors of the photosynthetic electron-transport chain (Häder, 1979*b*).

On the basis of these results, Häder and Burkart (1982) have proposed a computer model that describes photophobic responses in filamentous blue-green algae in terms of an electrical circuit diagram. It is assumed that the sudden breakdown of the proton gradient, caused by the shadowing of the organism when it leaves the light field, triggers the cation channels to open, permitting positive charges to enter the cytoplasm. The intracellular negative potential (usually of $\sim$30–40 mV) is consequently reduced by a few millivolts. In addition to the individual resting potential of each cell, there is a potential gradient between the two ends of the filament established by a constant current through the cross-wall membranes. This electrical gradient determines the direction and speed of movement. During a photophobic response, the gradient reverses, because of the inward current of positive charges in the darkened cells.

These assumptions have been built into a computer program that calculates the actual potential of each cell iteratively with a given cycle time. The potentials and the resulting movements of the trichome are plotted at predefined intervals. Figure 9 shows a simulation of repetitive photophobic responses of a trichome in a light trap. Each time the filament enters the dark area, a phobic potential is elicited and relayed through all cells until the potential gradient between the front and the rear ends is inversed, hence the movement is reversed as well. This model is confirmed by the finding that a reversal of movement can be induced by switching an external dc electrical field on or off, depending on the direction of movement (Häder, 1981).

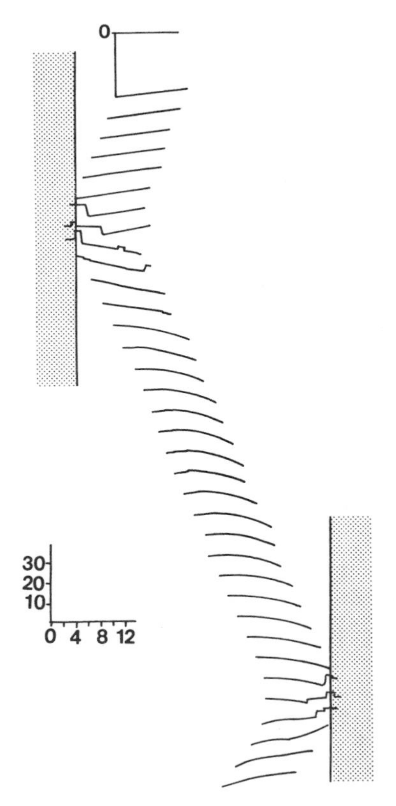

Figure 9. Simulation of repetitive photophobic responses of a trichome in a light trap.

4.4. Enzymatic Reactions

Recent investigations have suggested that a light-activated enzymatic reaction might play an important role in the process of visual transduction in higher organisms. Light could activate a cyclic nucleotide phosphodiesterase in frog rod outer segments (ROS), causing a subsequent decrease in the level of cyclic guanyl monophosphate (cGMP) (for details, see, e.g., Hug *et al.,* 1980, and references therein). The kinetics of this decrease (100–150 msec) would be in good agreement with the time course of the electrical signals recorded from ROS. Moreover, each absorbed photon would activate about 500 molecules of phosphodiesterase. This mechanism can explain both the transduction and the amplification processes that occur in ROS. No result of this type has yet been reported in photosensory processes of microorganisms, but a study of the levels of cGMP and phosphodiesterase in dark-adapted and light-adapted organisms would clearly be of great importance.

A light-induced enzyme system has been characterized in *Neurospora* (Calero *et al.,* 1980; Ninnemann, 1980). In this system, conidiation and nitrate uptake have been suggested to be controlled by blue light via a light-dependent nitrate reductase (Ninnemann and Klemm-Wolfgramm, 1980). However, recent findings on nitrate reductase mutants of *Neurospora* (Paietta and Sargent, 1982) seem to indicate that nitrate reductase is of no or secondary importance as a photoreceptor in blue-light responses of photosuppression and in phase shifting of circadian conidiation.

Transport of ions through membranes does not seem to be a simple process involving only voltage-dependent gates and constant pumping. Recently a calcium-transporting enzymatic system, calmodulin, has been found in a wide variety of eukaryotic organisms. Although first characterized in vertebrates, calmodulin has been analyzed in invertebrates, protozoa, and plants (Blum *et al.,* 1980; Takagi *et al.,* 1980; Vincenzi and Larsen, 1980). The amino acid analysis demonstrated a close similarity among all the calmodulin species studied (Vanaman, 1980).

The Ca^{2+} pump seems to be located in the plasma membrane closely associated with a Ca^{2+}/Mg^{2+}-dependent ATPase. Calmodulin controls a number of other enzymes as the cyclic nucleotide phosphodiesterase (Itano *et al.,* 1980) or as the cGMP-dependent protein kinase (Yamaki and Hidaka, 1980).

In *Neurospora* it could be demonstrated that calmodulin uses a previously generated proton gradient in a $Ca^{2+}/2H^{+}$ antiport reaction that carries Ca^{2+} out of the cell (Stroobant *et al.,* 1980).

Gitelman and Witman (1980) have purified calmodulin from an axoneme fraction of *Chlamydomonas* flagella, which resembles that of vertebrates in a number of properties. The involvement of calmodulin in the light control of flagellar activity may be supported by the fact that chlorpromazine, a drug that binds to calmodulin, causes a reversal of phototaxis of *Chlamydomonas* at moderate light intensities (Hirschberg and Hutchinson, 1980).

5. INFORMATION TRANSMISSION

It is possible that several steps of a photosensory transduction chain, during which the energy used to transmit the signal is changed from one form to another, are interspersed with information transport processes. On the other hand, amplification steps can simultaneously result in a signal transmission over considerable distances, as in the case of electrical potential changes, which can be generated in one region of the cell and sensed in another region. Thus, the formal distinction between amplification and information transport should not be regarded as strict.

5.1. Transmitter Molecules

The hormonal system of vertebrates is based on the production and transport of special transmitter molecules; even in neural processes, which are basically of an electrical nature, neurotransmitters are used. Although there may not be a hormonal system in microorganisms in the strict sense, a number of transmitter molecules have been found and in some cases identified that serve the purpose of information transmission.

Dictyostelium amoebae have developed a system whereby information is exchanged between individual cells, based on pulses of cyclic adenosine monophosphate (cAMP) produced by pacemaker cells, and which other cells respond to by excreting cAMP (Maeda, 1977; Gerisch, 1978). Thus, the information is relayed through a whole population, ensuring aggregation and a simultaneous development of all cells. Recently a transmitter molecule has been found in the phototaxis of *Dictyostelium* pseudoplasmodia (Fisher *et al.*, 1981). Laterally impinging light is focused on the rear side of a pseudoplasmodium, which acts as a cylindrical lens; the photoreceptor molecules that are supposed to be present, at least in all peripheral cells, sense a light gradient with a maximum opposite to the direction of the light source. The light gradient is converted into a chemical gradient by the production and excretion of a slug-turning factor (STF) at the brighter rear side, from which the pseudoplasmodium turns away, heading toward the light source. The STF has not yet been identified; it is not cAMP, but the molecular weight is just below 500 (Fisher *et al.*, 1981).

A rather straightforward mechanism of signal transformation is operative in the photokinesis of photosynthetic prokaryotes. In higher light intensity, the photophosphorylation is increased and the additional ATP is transported to the motor apparatus, where it produces a faster movement (Throm, 1968; Nultsch and Hellmann, 1972). This transport of a transmitter may depend on simple diffusion, which can account for the long times necessary to vary the speed of movement when the light intensity is altered.

5.2. Electrical Signals

In faster reactions a chemical transmitter may be ruled out, since diffusion of even a small molecule may be too slow to account for the observed reaction times (Lee and Fitzsimons, 1976). Electrical phenomena seem to be the fastest feasible mechanisms for sensory transduction and are likely to be involved in phototaxis and phobic responses of a number of organisms. This is especially true for multicellular organisms, such as filamentous blue-green algae, the transduction mechanism of which is described in Section 4.3.

The transport of information is not restricted to the outer membrane of a cell. In the phototactic zoospores of *Ulva lactuca,* microtubular junctions have been found connecting the flagellar base with the stigma region, which is supposed to be part of the apparatus for sensing the light direction in this organism (Melkonian, 1979, 1980; Robenek and Melkonian, 1981). Similar structures have also been found in other green algal species. In the case of *Euglena,* for example, the fibrils matrix, most likely of actomyosin, surrounding the canal has been suggested to play the role of an effector for the photomotile responses of this alga (Piccinni and Omodeo, 1975), but no effects of cytochalasins B and D, two well-known actomyosin blocking drugs, were observed on the step-up and step-down photophobic responses (Coppellotti *et al.,* 1979).

6. MOTOR RESPONSES

The final step in the sensory transduction chain of movement responses is the modulation of the motor activity, which leads to the observable behavioral response. In photokinesis, the speed at which the motor apparatus operates is a function of light intensity, whereas in phototaxis the direction of movement is constantly modulated by the direction of propagation of the light beam S. Since the mechanical aspects of the systems used by various microorganisms are very different, they are discussed separately.

6.1. Flagellated Bacteria

Flagellated bacteria are propelled by one or more helical semirigid filaments that either arise from one or two poles of the cell or are distributed over the entire cell surface. The filaments have long been believed to propagate helical waves, but it can be shown that each flagellum is driven by a rotatory motor (Fig. 10). This rotation cannot be detected with optical techniques, since the filaments have a diameter of 10 nm and rotate with a frequency of 40–60 Hz (Berg, 1975). Silverman and Simon (1974) have proved the rotation hypothesis by tethering a flagellum to a glass surface by antifilament antibodies. Under these conditions, the bacterium clearly rotates. A change in the direction of rotation reverses the direction of movement of the bacterium.

Rotational movement does not depend on ATP, but it is powered directly by a proton gradient generated by photosynthesis, respiration, or another energy-fixing mechanism (Sistrom, 1978). Protons are supposed to move through the interstice between the stator (S ring) and the rotor (M ring), along ion channels consisting of polar ligand groups, driving the rotor with the attached filament in a manner similar to the action of a turbine (Läuger, 1977).

Since it has been shown that the electrochemical gradient is under the influence of a light stimulus, a light-induced change in the proton gradient could have a very direct effect on the movement.

The action spectra determined for the step-up and the step-down photophobic responses of *Halobacterium halobium* (Hildebrand and Dencher, 1975; Dencher, 1978; Hildebrand, 1978) seem to suggest bacteriorhodopsin and a retinylidene protein plus a bacterioruberin (as an additive pigment) as photo-

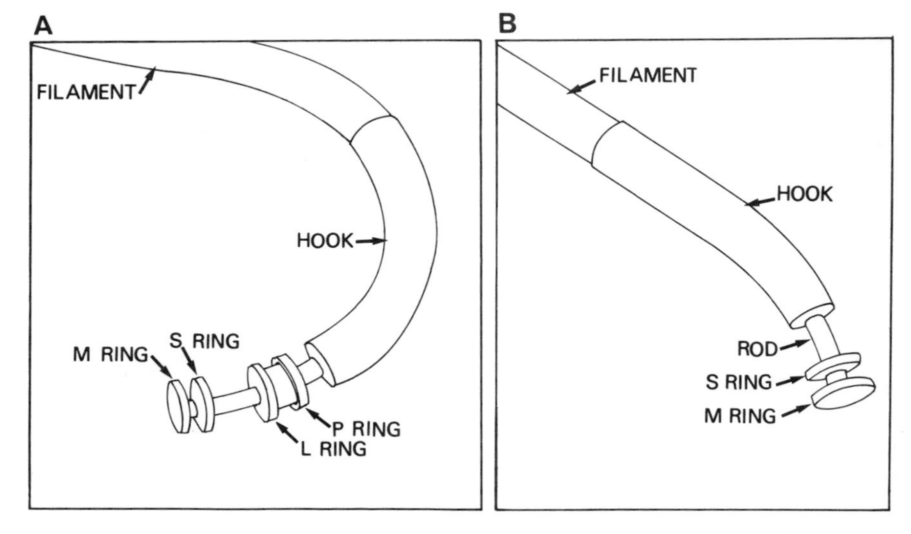

Figure 10. Outline of the structure of the flagellar rotatory motor in bacteria. (A) Gram-negative bacteria; (B) gram-positive bacteria. (Redrawn after Berg, 1975.)

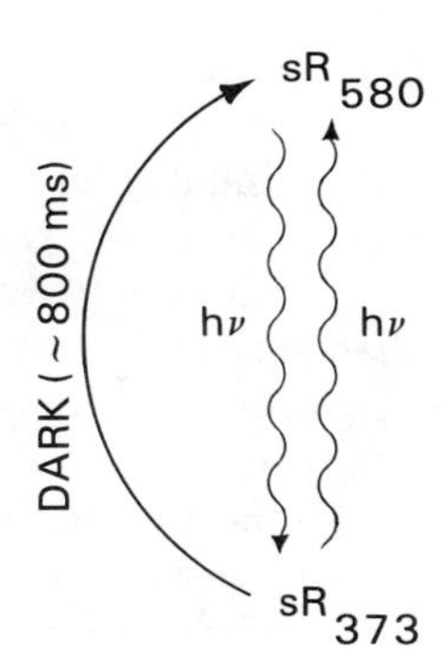

Figure 11. Photocycle of the rhodopsin-like phototactic pigment sR in *Halobacterium*. (Redrawn after Bogomolni and Spudich, 1982.)

receptor pigments, respectively, for the step-down (PS 565) and for the step-up responses (PS 370). Furthermore, inhibition of retinal synthesis by nicotine was found to block photomovement driven by PS 560 without inhibiting chemotaxis (Spudich and Stoeckenius, 1979). It was therefore reasonable to assume that the decrease in light intensity, interrupting the bacteriorhodopsin photocycle, caused a decrease in the proton translocation rate, which would have constituted an initial step in the sensory process, ultimately resulting in the step-down photophobic response (Hildebrand, 1978, 1980; Wagner, 1979). Recent studies on a mutant deficient in all previously reported rhodopsin-like pigments (Spudich and Spudich, 1982) and on artificially reconstituted photosystems (Schimz *et al.*, 1982) suggest that a retinal pigment different from both bacteriorhodopsin and halorhodopsin is responsible for the photosensory behavior of *Halobacterium*. On the basis of a series of spectroscopic measurements, Bogomolni and Spudich (1982) suggest a new rhodopsin-like pigment, sR_{580}, as the photosensory receptor for *Halobacterium*. At physiological light intensities, two photoactive forms of the pigment are present: sR_{580} (absorbing at 570 ÷ 590 nm) and S_{373} (absorbing at 373 nm). Illumination of sR_{580} generates the S_{373} species, which reconverts to sR_{580} either very slowly in the dark (800 msec) or upon photoexcitation (Fig. 11). Changes in the relative light intensities at 580 nm or 373 nm cause opposing shifts in the photostationary state, which would, in turn, be responsible for the photosensory responses of *Halobacterium*.

6.2. Gliding Movements

A number of species from different systematic groups move by means of a gliding mechanism. There seem to be differences, however, in the mechanical systems. Desmids and red algae are propelled by the extrusion of slime through cell wall pores (Mix, 1969; Kiermayer and Staehelin, 1972; Lin *et al.*, 1975). Simple extrusion of slime may account for forward and backward movement, but it cannot explain the fine-tuned changes in direction during oriented movements (Neuscheler, 1967*a,b*; Häder and Wenderoth, 1977). This requires a sophisticated activity of individual pores that cannot be explained by a uniform change in the membrane potential. Staining moving cells of the red alga *Porphyridium* with crystal violet showed similar traces (Nultsch and Schuchart, 1980) but, as in the case of desmids, this cannot account for the maneuvers that bring the cell back into the light field after they have crossed the light–dark border (Schuchart, 1980). Studies revealing the mechanism controlling the slime extrusion have not been carried out in either organism.

Prokaryotes seem to use a different mechanism of gliding movement. Although slime is produced by both blue-green algae and flexibacters (Humphrey *et al.*, 1979), it appears to provide only suitable conditions for temporary adhesion and to serve as a lubricant, without being a means of propulsion as was formerly believed. When flexibacters are allowed to move over a thin gold layer, they leave a pattern that might result from rhythmical contractions of the outer membrane. In blue-green algae Halfen and Castenholz (1971*a,b*) have described parallel microfilaments having a diameter of 5–8 nm, wrapped around the filaments in a helical fashion. Pitch and orientation coincide with the rotation of the organisms during forward movement. This indicates that a contraction of filaments might be involved in the mechanism, although cytochalasin B, which inhibits filament contraction in other systems, does not impair movement in blue-green algae (Halfen, 1973). Jarosch (1963) had proposed that the surface contraction waves were produced by the rotation of twisted pairs of protein fibrils. If the mechanism of movement is hypothetical, the coupling to the sensory transduction chain of photomovement is a totally open question.

6.3. Eukaryotic Cilia and Flagella

A number of excellent descriptions of structure and function of cilia and flagella have been published (e.g., Blum and Hines, 1979; Satir and Ojakian, 1979); we therefore limit our discussion to the possible mechanisms of triggering changes in the flagellar beat during a light-induced response. Cilia and flagella of eukaryotic organisms are identical organelles, although they may vary in length and in beating pattern.

A cylinder of nine doublet tubules is arranged around a central pair of single microtubules to which they are bound by radial spokes (Piperno *et al.*, 1981). Bending is brought about by an active sliding of some of the outer tubules along each other. The central pair determines the plane of beating, since the cilium moves perpendicular to the plane connecting the central tubules (Satir and Ojakian, 1979). In order to change the direction of beating, the plane through the central pair has to rotate, as has been demonstrated in ciliates (Omoto and Kung, 1980). This steering is under the control of ion-generated potential changes, making a modulation of movement by light plausible. Independent of a change in the direction of the power stroke, there is a wide variety of changes in the bending characteristics. These changes seem to depend on the ion concentration and are controlled by the inner Ca^{2+} concentration, but the site of reversible Ca^{2+} binding to the axoneme is unknown (Satir and Ojakian, 1979). It might be speculated that this role is played by calmodulin, the presence of which has been demonstrated in the flagellar apparatus of *Chlamydomonas* (Gitelman and Witman, 1980).

As we have seen in Section 4.2, Ca^{2+} controls the direction of flagellar beat in the flagellar apparatus of *Chlamydomonas* (Hyams and Borisy, 1978; Bessen *et al.*, 1980). Both calcium and magnesium are necessary to guarantee normal flagellar activity and are essential for phototaxis of this alga (Bean and Harris, 1979), but addition of a high Ca^{2+} concentration (10^{-2} M) causes a strong inhibition (Nultsch, 1979).

The dependence of phototaxis on a specific external cation concentration seems to be a general condition among flagellates and ciliates. Sagakuchi (1979) found a change from positive to negative phototaxis in *Volvox* upon addition of 30 mM K^+; in the case of *Stentor,* Colombetti *et al.* (1982*b*) found that extracellular K^+ ions above a critical concentration induce ciliary reversal in *Stentor* and suppress the step-up photophobic response, the K^+ threshold concentration depending on the extracellular Ca^{2+} concentration.

7. CONCLUDING REMARKS

Light-induced movement responses of freely motile microorganisms obviously do not follow one single scheme. The development of different strategies and molecular mechanisms for light-dependent behaviors is not limited to such diverse organisms as prokaryotes and eukaryotes; even more closely related organisms, such as green flagellates, seem to use different reaction mechanisms as well. Hidden behind the diversity of strategies and molecular mechanisms, however, there seem to be a number of common steps in the reaction chain used for light perception, signal amplification, and transmission, and probably for controlling the motor apparatus of the cell.

In addition to the absorption of light quanta, which follows the general rules of photophysics and photochemistry, we have pointed out electrical and ion-transport phenomena involving either internal membranes or the plasmalemma. The cell maintains a gated cation flux through voltage-dependent channels along a concentration gradient by means of energy-dependent membrane-bound ion pumps. This flux provides both a significant electrical amplification and a fast means of information transmission over considerable distances within a cell or between different cells in a multicellular organism. In addition, at least in ciliates and flagellates, the modulation of the locomotory organelle activity has been found to be strongly dependent on the extracellular concentration of cations, which can cause changes in the beating pattern and permit the reversal of movement. In many other cases, the way in which the motor apparatus is controlled still remains an open question; a concentration of research activity on this subject could provide rewarding results.

ACKNOWLEDGMENTS. We wish to express our gratitude to Professors E. Polacco, P. Omodeo, and P.-S. Song for their critical review of the manuscript. This work was made possible by a NATO travel grant.

REFERENCES

Barghigiani, C., Colombetti, G., Franchini, B., and Lenci, F., 1979, Photobehavior of *Euglena gracilis:* action spectrum for the step-down photophobic response of individual cells, *Photochem. Photobiol.* **29**:1015–1019.

Bean, B., and Harris, S., 1979, Selective inhibition of flagellar activity in *Chlamydomonas* by nickel, *J. Protozool.* **26**:235–240.

Benedetti, P. A., and Lenci, F., 1977, "In vivo" microspectrofluorometry of photoreceptor pigments in *Euglena gracilis, Photochem. Photobiol.* **26**:315–318.

Bensasson, R., 1977, Pigments involved in photomotion of microorganisms, in: *Research in Photobiology* (A. Castellani, ed.), pp. 85–94, Plenum Press, New York.

Bensasson, R., 1980, Aspects of photoreceptor function: carotenoids and rhodopsins, in: *Photoreception and Sensory Transduction in Aneural Organisms* (F. Lenci and G. Colombetti, eds.), pp. 211–234, Plenum Press, New York.

Berg, H. C., 1975, How bacteria swim, *Sci. Am.* **233**:36–44.

Berg, H. C., and Purcell, E. M., 1977, Physics of chemoreception, *Biophys. J.* **20**:193–219.

Bessen, M., Fay, R. B., and Witman, G. B., 1980, Calcium control of waveform in isolated flagellar axonemes of *Chlamydomonas, J. Cell Biol.* **86**:446–455.

Björn, G. S., 1978, Phycochrome d, a new photochromic pigment from the blue-green alga *Iolypothrix distorta, Physiol. Plant* **42**:321–323.

Björn, G. S., 1979, Action spectra for "in vivo" and "in vitro" conversions of phycochrome b, a reversibly photochromic pigment in a blue-green alga, and its separation from other pigments, *Physiol. Plant* **46**:281–286.

Björn, G. S., 1980, Phycochromes b and d: their occurrence in some phycoerythrocyanin-containing blue-green algae *(Cyanobacteria), Physiol. Plant* **48**:483–485.

Blum, J. J., and Hines, M., 1979, Biophysics of flagellar motility, *Q. Rev. Biophys.* **12**:103–180.

Blum, J. J., Hayes, A., Jamieson, G. A., Jr., and Vanaman, T. C., 1980, Calmodulin confers calcium sensitivity on ciliary dynein ATPase, *J. Cell Biol.* **87**:386–397.

Bogomolni, R. A., and Spudich, J. L., 1982, Identification of a third rhodopsin-like pigment in phototactic *Halobacterium halobium, Proc. Natl. Acad. Sci. USA* **79**:6250–6254.

Calero, F., Ullrich, W. R., and Aparicio, P. J., 1980, Regulation by monochromatic light of nitrate uptake in *Chlorella fusca,* in: *The Blue Light Syndrome* (H. Senger, ed.), pp. 411–421, Springer-Verlag, Berlin.

Carlile, M. J., 1980*a*, From prokaryote to eukaryote: gains and losses, in: *The Eukaryotic Microbial Cell* (G. W. Gooday, D. Lloyd, and A. P. J. Trinci, eds.), pp. 1–40, Cambridge University Press, London.

Carlile, M. J., 1980*b*, Positioning mechanisms—the role of motility, taxis and tropism in the life of microorganisms, in: *Contemporary Microbiological Ecology* (C. D. C. Ellwood, M. J. Lathan, J. N. Hedger, J. M. Lynch, and J. H. Slater, eds.), pp. 55–74, Academic Press, London.

Checcucci, A., 1976, Molecular sensory physiology of *Euglena, Naturwissenschaften* **63**:412–417.

Checcucci, A., Colombetti, G., Ferrara, R., and Lenci, F., 1976, Action spectra for photoaccumulation of green and colorless *Euglena:* evidence for identification of receptor pigments, *Photochem. Photobiol.* **23**:51–54.

Colombetti, G., 1984, Receptor pigments in light-induced behavior of microorganisms, in: *Photoreceptors* (A. Borsellino and L. Cervetto, eds.), pp. 3–28, Plenum, New York.

Colombetti, G., and Lenci, F., 1980*a*, Identification and spectroscopic characterization of photoreceptor pigments, in: *Photoreception and Sensory Transduction in Aneural Organisms* (F. Lenci and G. Colombetti, eds.), pp. 173–188, Plenum Press, New York.

Colombetti, G., and Lenci, F., 1980*b*, Photosensory transduction chains in eucaryotes, in: *Photoreception and Sensory Transduction in Aneural Organisms* (F. Lenci and G. Colombetti, eds.), pp. 341–354, Plenum Press, New York.

Colombetti, G., and Lenci, F., 1983, Photoreception and photomovements in microorganisms, in: *The Biology of Photoreceptors* (D. Cosens and D. Vince-Prue, eds.), pp. 393–416, Cambridge University Press, Cambridge.

Colombetti, G., Ghetti, F., Lenci, F., Polacco, E., Posudin, I., and Campani, E., 1981, Laser microspectrofluorometry of photopigments "in vivo," *Kvantovaia Elektron.* **8**:2680–2683.

Colombetti, G., Lenci, F., and Diehn, B., 1982*a*, Responses to photic chemical and mechanical stimuli, in: *The Biology of Euglena,* Vol. 3 (D. E. Buetow, ed.), pp. 169–195, Academic Press, New York.

Colombetti, G., Lenci, F., and Song, P.-S., 1982*b*, Effects of K^+ and Ca^{++} ions on motility and photosensory responses of *Stentor coeruleus, Photochem. Photobiol.* **36**:609–611.

Coppellotti, O., Piccinni, E., Colombetti, G., and Lenci, F., 1979, Responses of *Euglena gracilis* to Cytochalasins B and D, *Boll. Zool.* **46**:72–75.

Dencher, N. A., 1978, Light-induced behavioral reactions of *Halobacterium holobium:* evidence for two rhodopsins acting as photopigments, in: *Energetics and Structure of Halophilic Microorganisms* (S. R. Caplani and M. Grintburg, eds.), pp. 67–68, Elsevier, Amsterdam.

Diehn, B., 1969, Actions spectra of the phototactic responses in *Euglena, Biochim. Biophys. Acta* **177**:136–143.

Diehn, B., 1979, Photic responses and sensory transduction in motile protists, in: *Handbook of Sensory Physiology,* Vol. VII/6A (H. Autrum, ed.), pp. 23–68, Springer-Verlag, Berlin.

Diehn, B., 1980, Experimental determination and measurement of photoresponses, in: *Photoreception and Sensory Transduction in Aneural Organisms* (F. Lenci and G. Colombetti, eds.), pp. 107–126, Plenum Press, New York.

Diehn, B., Feinleib, M., Haupt, W., Hildebrand, E., Lenci, F., and Nultsch, W., 1977, Terminology of behavioral responses of motile microorganisms, *Photochem. Photobiol.* **26**:559–560.

Doughty, M. J., and Diehn, B., 1979, Photosensory transduction in the flagellated alga, *Euglena gracilis.* I. Action of divalent cations, Ca antagonists and Ca ionophore on motility and photobehavior, *Biochim. Biophys. Acta* **588**:148–168.

Doughty, M. J., and Diehn, B., 1980, Flavins as photoreceptor pigments for behavioral responses in motile microorganisms, especially in the flagellated alga, *Euglena* sp., in: *Structure and Bonding,* Vol. 41: *Molecular Structure and Sensory Physiology* (P. Hemmerich, ed.), pp. 45–70, Springer-Verlag, Berlin.

Doughty, M. J., Grieser, R., and Diehn, B., 1980, Photosensory transduction in the flagellated alga *Euglena gracilis.* II. Evidence that blue light affects alteration in Na–K permeability of the photoreceptor membrane, *Biochim. Biophys. Acta* **602**:10–23.

Eckert, R., 1977, Genes, channels, and membrane currents in *Paramecium, Nature (Lond.)* **268**:104–105.

Eckert, R., and Brehm, P., 1979, Ionic mechanisms of excitation in *Paramecium, Ann. Rev. Biophys. Bioeng.* **8**:353–383.

Feinleib, M. E., 1978, Photomovements of microorganisms, *Photochem. Photobiol.* **27**:849–854.

Feinleib, M. E., 1980, Photomotile responses in flagellates, in: *Photoreception and Sensory Transduction in Aneural Organisms* (F. Lenci and G. Colombetti, eds.), pp. 45–68, Plenum Press, New York.

Feinleib, M. E., and Curry, G. M., 1974, The nature of the photoreceptor in phototaxis, in: *Handbook of Sensory Physiology,* Vol. 1 (W. R. Lowenstein, ed.), pp. 366–395, Springer-Verlag, Berlin.

Fisher, P. R., Smith, E., and Williams, K. L., 1981, An extracellular chemical signal controlling phototactic behaviour by *Dictyostelium discoideum* slugs, *Cell* **23**:799–807.

Fisher, P. R., and Williams, K. L., 1981, Bidirectional phototaxis by *Dictyostelium discoideum slugs, FEMS Microbiol. Lett.* **12**:87–89.

Foster, K. W., and Smyth, R. D., 1980, Light antennas in phototactic algae, *Microbiol. Rev.* **44**:572–630.

Gerisch, G., 1978, Cell interactions by cyclic AMP in *Dictyostelium, Biol. Cell.* **32**:61–68.

Gitelman, S. E., and Witman, G. B., 1980, Purification of calmodulin from *Chlamydomonas:* calmodulin occurs in cell bodies and flagella, *J. Cell Biol.* **98**:764–770.

Häder, D.-P., 1973, Untersuchungen zur Photo-phobotaxis bei Phormidium uncinatum, doctoral thesis, Marburg, 1983.

Häder, D.-P., 1977, Influence of electric fields on photophobic reactions in blue-green algae, *Arch. Microbiol.* **114**:83–86.

Häder, D.-P., 1978*a*, Evidence of electrical potential changes in photophobically reacting blue-green algae, *Arch. Microbiol.* **118**:115–119.

Häder, D.-P., 1978*b*, Extracellular and intracellular determination of light-induced potential changes during photophobic reactions in blue-green algae, *Arch. Microbiol.* **119**:75–79.

Häder, D.-P., 1979*a*, Photomovement, in: *Encyclopedia of Plant Physiology,* new ser., vol. 7: *Physiology of Movements* (W. Haupt and M. Feinleib, eds.), pp. 268–309, Springer-Verlag, Berlin.

Häder, D.-P., 1979*b*, Effect of inhibitors and uncouplers on light-induced potential changes triggering photophobic responses, *Arch. Microbiol.* **120**:57–60.

Häder, D.-P., 1980, Photosensory transduction chains in procaryotes, in: *Photoreception and Sensory Transduction in Aneural Organism* (F. Lenci and G. Colombetti, eds.), pp. 355–372, Plenum Press, New York.

Häder, D.-P., 1981, Electrical and proton gradients in the sensory transduction of photophobic responses in blue-green alga, *Phormidium uncinatum, Arch. Microbiol.* **130**:63–86.

Häder, D.-P., and Burkart, U., 1982, Mathematical simulation of photophobic responses in blue-green algae, *Math. Biosci.* **58**:1–17.

Häder, D.-P., and Wenderoth, K., 1977, Role of three basic light reactions in photomovement of *Desmids, Planta* **137**:207–214.

Halfen, L. N., 1973, Gliding motility of *Oscillatoria:* ultrastructural and chemical characterization of the fibrillar layer, *J. Phycol.* **9**:248–253.

Halfen, L. N., and Castenholz, R. W., 1971*a*, Gliding motility in the blue-green alga *Oscillatoria Princeps, J. Phycol.* **7**:133–145.

Halfen, L. N., and Castenholz, R. W., 1971*b*, Energy expenditure for gliding motility in a blue-green alga, *J. Phycol.* **7**:258–260.

Halldal, P., 1980, Light and microbial activities, in: *Contemporary Microbial Ecology* (D. C. Ellwood, M. J. Latham, J. N. Hedger, J. M. Lynch, and J. H. Slater, eds.), pp. 1–13, Academic Press, London.

Hansma, H. G., 1979, Sodium uptake and membrane excitation in *Paramecium, J. Cell Biol.* **81**:374–381.

Harold, F. M., 1977, Ion currents and physiological functions in microorganisms, *Annu. Rev. Microbiol.* **31**:181–203.

Haupt, W., 1980, Localization and orientation of receptor pigments, in: *Photoreception and Sensory Transduction in Aneural Organisms* (F. Lenci and G. Colombetti, eds.), pp. 155–172, Plenum Press, New York.

Hemmerich, P., and Schmidt, W., 1980, Bluelight reception and flavin photochemistry, in: *Photoreception and Sensory Transduction in Aneural Organisms* (F. Lenci and G. Colombetti, eds.), pp. 271–284, Plenum Press, New York.

Henderson, R., 1977, The purple membrane from *Halobacterium halobium, Annu. Rev. Biophys. Bioeng.* **6**:87–109.

Hildebrand, E., 1978, Light-controlled behavior of bacteria, in: *Taxis and Behavior* (G. L. Hazelbauer, ed.), pp. 37–73, Chapman & Hall, London.

Hildebrand, E., 1980, Comparative discussion of photoreception in lower and higher organisms. Structural and functional aspects, in: *Photoreception and Sensory Transduction in Aneural Organisms* (F. Lenci and G. Colombetti, eds.), pp. 319–340, Plenum Press, New York.

Hildebrand, E., and Dencher, N., 1975, Two photosystems controlling behavioral responses of *Halobacterium halobium, Nature (Lond.)* **257**:46–48.

Hirschberg, R., and Hutchinson, W., 1980, Effect of chlorpromazine on phototactic behavior in *Chlamydomonas, Can. J. Microbiol.* **26**:265–267.

Hobbs, A. S., and Albers, R. W., 1980, The structure of proteins involved in active membrane transport, *Annu. Rev. Biophys. Bioeng.* **9**:259–291.

Hook, C., and Hildebrand, E., 1980, Excitability of *Paramecium* and the significance of negative surface changes, *J. Math. Biol.* **9**:347–360.

Hug, D. H., O'Donnell, P. S., and Hunter, J. K., 1980, Light activation of enzymes, *Photochem. Photobiol.* **32**:841–848.

Humphrey, B. A., Dickson, M. R., and Marshall, K. C., 1979, Physicochemical and "in situ" observation on the adhesion of gliding bacteria to surfaces, *Arch. Microbiol.* **120**:231–238.

Hyams, J. S., and Borisy, G. G., 1978, Isolated flagellar apparatus of *Chlamydomonas:* characterization of forward swimming and alternation of waveform and reversal of motion by calcium ions *in vitro, J. Cell Sci.* **33**:235–253.

Itano, T., Itano, R., and Penniston, J. T., 1980, Interactions of basic polypeptides and proteins with calmodulin, *Biochem. J.* **189**:455–459.

Jarosch, R., 1963, Gleitbewegung und torsion von oscillatorien, *Osterr. Bot. Z.* **11**:143–148.

Kiermayer, O., and Staehelin, A. L., 1972, Feinstruktur von zellwand und plasmamembran bei *Micrasterias denticulata* breb. nach gefrierätzung, *Protoplasma* **74**:227–237.

Knox, R. S., 1973, Transfer of excitation energy, in: *Primary Molecular Events in Photobiology* (A. Checcucci and R. A. Weale, eds.), pp. 45–77, Elsevier, Amsterdam.

Läuger, P., 1977, Ion transport and rotation of bacterial flagella, *Nature (Lond.)* **268**:360–362.

Lazarides, E., and Revel, J. P., 1979, The molecular basis of cell movement, *Sci. Am.* **240**:88–100.

Lee, A. G., and Fitzsimons, J. T. R., 1976, Motility in normal and filamentous forms of *Rhodospirillum rubrum, J. Gen. Microbiol.* **93**:346–354.

Lenci, F., 1982, Photomovements of microorganisms, in: *Trends in Photobiology* (C. Hélène, M. Charlier, Th. Montenay-Garestier, and G. Laustriat, eds.), pp. 421–435, Plenum Press, New York.

Lenci, F., and Colombetti, G., 1978, Photobehavior of microorganisms: a biophysical approach, *Annu. Rev. Biophys. Bioeng.* **7**:341–361.

Lin, H.-P., Sommerfeld, M. R., and Swafford, J. R., 1975, Light and electron microscope observation on motile cells of *Porphyridum purpureum (Rhodophyta), J. Phycol.* **11**:452–457.

Lipson, E. D., 1975, White noise analysis of *Phycomyces* light growth response system. III. Photomutants, *Biophys. J.* **15**:1033–1045.

Litvin, F. F., Sineshchekov, O. A., and Sineshchekov, V. A., 1978, Photoreceptor electric potential in the phototaxis of the alga *Haematococcus pluvialis, Nature (Lond.)* **271**:476–478.

Lundström, I., and Nylander, C., 1981, A possible model for receptors in excitable membranes, *J. Theor. Biol.* **88**:671–683.

Maeda, Y., 1977, Role of cyclic AMP in the polarized movement of the migrating pseudoplasmodium of *Dictyostelium discoideum, Dev. Growth Diff.* **19**:201–205.

Maeda, A., and Yoshizawa, T., 1982, Molecular transducing system in visual cells, *Photochem. Photobiol.* **35**:891–898.

Marbach, I., and Mayer, A. M., 1971, Effect of electric field on the phototactic response of *Chlamydomonas reinhardtii, Isr. J. Bot.* **20**:96–100.

Martonosi, A. M., 1980, Calcium pumps, *Fed. Proc.* **39**:2401–2402.

Melkonian, M., 1979, Structure and significance of cruciate flagellar root systems in green algae: zoospores of *Ulva lactuca* (Ulvaceae, Chlorophyceae), *Helg. Wiss. Meer.* **32**:425–435.

Melkonian, M., 1980, Flagellar roots, mating structure and gametic fusion in the green alga *Ulva lactuca* (Ulvales), *J. Cell Sci.* **46**:149–169.

Melkonian, M., and Robenek, H., 1980*a*, Eyespot membranes in newly released zoospores of the green alga *Chlorosarcinopsis gelatinosa* (Chlorosarcinales) and their fate during zoospore settlement, *Protoplasma* **104**:129–140.

Melkonian, M., and Robenek, H., 1980*b*, Eyespot membranes of *Chlamydomonas reinhardtii:* a freeze-fracture study, *J. Ultrastruct. Res.* **72**:90–102.

Mix, M., 1969, Zur feinsstruktur der zellwände in der gattung *Closterium* (Desmidiaceae) unter besonderer berücksichtigung des porensystems, *Arch. Mikrobiol.* **68**:306–325.

Neuscheler, W., 1967*a*, Bewegung und orientierung bei *Micrasterias denticulata* breb. im licht. I. Zur bewegung und orientierungsweise, *Z. Pflanzenphysiol.* **57**:46–59.

Neuscheler, W., 1967*b*, Bewegung und orientierung bei *Micrasterias denticulata* Breb. im licht. II. Photokinesis und phototaxis, *Z. Pflanzenphysiol.* **57**:151–172.

Ninnemann, H., 1980, Blue light photoreceptors, *Bioscience* **30**:166–170.

Ninnemann, H., and Klemm-Wolfgramm, E., 1980, Blue light controlled conidiation and absorbance change in *Neurospora* are mediated by nitrate reductase, in *The Blue Light Syndrome* (H. Senger, ed.), pp. 238–243, Springer-Verlag, New York.

Nultsch, W., 1962, Der einfluss des lichtes auf die bewegung der *Cyanophyceen.* III. Mitteilung: photophobotaxis bei *Phormidium uncinatum, Planta* **58**:647–663.

Nultsch, W., 1979, Effect of external factors on phototaxis of *Chlamydomonas* reinhardtii. III. Cations, *Arch. Microbiol.* **123**:93–99.

Nultsch, W., 1980, Photomotile responses in gliding organisms and bacteria, in *Photoreception and Sensory Transduction in Aneural Organisms* (F. Lenci and G. Colombetti, eds.), pp. 69–88, Plenum Press, New York.

Nultsch, W., and Häder, D.-P., 1979, Photomovement of motile microorganisms, *Photochem. Photobiol.* **29**:423–437.

Nultsch, W., and Häder, D.-P., 1980, Light perception and sensory transduction in photosynthetic prokaryotes, in: *Structure and Bonding,* Vol. 41: *Molecular Structure and Sensory Physiology* (P. Hemmerch, ed.), pp. 111–146, Springer-Verlag, New York.

Nultsch, W., and Hellmann, W., 1972, Untersuchugen zur photokinesis von *Anabaena variabilis, Arch. Mikrobiol.* **82**:76–90.
Nultsch, W., and Schuchart, H., 1980, Photomovement of the red alga *Porphyridium cruentum* (Ag) Naegeli. II. Phototaxis, *Arch. Microbiol.* **125**:181–188.
Omoto, C. K., and Kung, C., 1980, Rotation and twist of the centralpair microtubules in the cilia of *Paramecium, J. Cell Biol.* **87**:33–46.
Padan, E., and Schuldinger, S., 1978, Energy transduction in the photosynthetic membranes of the cyanobacterium (blue-green alga) *Plectomena boryanum, J. Biol. Chem.* **253**:3281–3286.
Paietta, J., and Sargent, M. L., 1982, Blue light responses in nitrate reductase mutants of *Neurospora crasse, Photochem. Photobiol.* **35**:853–855.
Pearlstein, R. M., 1982, Excitation migration and trapping in photosynthesis, *Photochem. Photobiol.* **35**:835–844.
Piccinni, E., and Coppellotti, O., 1979, Struttura dell'apparato fotomotore di *Ochromonas danica, XLVII Congr. Un. Zool. Ital.,* Abstr. pp. 175–176, Milano, Italy.
Piccinni, E., and Mammi, M., 1978, Motor apparatus of *Euglena gracilis:* ultrastructures of the basal portion of the flagellum and the paraflagellar body, *Boll. Zool.* **45**:405–414.
Piccinni, E., and Omodeo, P., 1975, Photoreceptors and phototactic programs in protista, *Boll. Zool.* **42**:57–79.
Piperno, G., Huang, B., Ramanis, Z., and Luck, D. J. L., 1981, Radial spokes of *Chlamydomonas flagella:* polypeptide composition and phosphorylation of stalk components, *J. Cell Biol.* **88**:73–79.
Poff, K. L., and Butler, W. L., 1975, Spectral characterization of the photoreducible b-type cytochrome of *Dictyostelium discoideum, Plant. Physiol.* **55**:427–429.
Poff, K. L., and Whitaker, B. D., 1979, Movement of slime molds, in: *Encyclopedia of Plant Physiology,* Vol. 7: *Physiology of Movements* (W. Haupt and M. E. Feinleib, eds.), pp. 355–382, Springer-Verlag, New York.
Ristori, T., Ascoli, C., Banchetti, R., Parrini, P., and Petracchi, D., 1981, Localization of photoreceptor and active membrane in the green alga *Haematococcus pluvialis, Proceedings of the Sixth International Congress on Protozoology, Warsaw,* p. 314.
Robenek, H., and Melkonian, M., 1981, Comparative ultrastructure of eyespot membranes in gametes and zoospores of the green alga *Ulva lactuca* (Ulvales), *J. Cell Sci.* **50**:149–164.
Sagakuchi, H., 1979, Effect of external ionic environment on phototaxis of *Volvox carteri, Plant Cell Physiol.* **20**:1643–1651.
Saimi, Y., and Kung, C., 1980, A Ca-induced Na-current in *Paramecium, J. Exp. Biol.* **88**:305–325.
Sarkar, H. K., and Song, P. S., 1982, Nature of phototransformation of phytochrome as probed by intrinsic tryptophane residues, *Biochemistry* **21**:1967–1972.
Satir, P., and Ojakian, G. K., 1979, Plant cilia, in: *Encyclopedia of Plant Physiology,* Vol. 7: *Physiology of Movements* (W. Haupt and M. E. Feinleib, eds.), pp. 224–249, Springer-Verlag, New York.
Schimz, A., Sperling, W., Hildebrand, E., and Köhler-Hahn, D., 1982, Bacteriorhodopsin and the sensory pigment of the photosystem 565 in *Halobacterium halobium, Photochem. Photobiol.* **36**:193–196.
Schmidt, J. A., and Eckert, R., 1976, Calcium couples flagellar reversal to photostimulation in *Chlamydomonas reinhardtii, Nature (Lond.)* **262**:713–715.
Schmidt, W., 1980, Physiological Bluelight reception, in: *Structure and Bonding,* Vol. 41: *Molecular Structure and Sensory Physiology* (P. Hemmerich, ed.), pp. 3–44, Springer-Verlag, Berlin.
Schuchart, H., 1980, Photomovement of the red alga *Porphyridium cruentum* (Ag) Naegli. III. Action spectrum of the photophobic response, *Arch. Microbiol.* **128**:105–112.

Senger, H., 1982, The effect of blue light on plants and microorganisms, *Photochem. Photobiol.* **35**:911–920.

Senger, H., and Briggs, W. R., 1981, The blue light receptor(s): primary reactions and subsequent metabolic changes, in: *Photochemical and Photobiological Reviews,* Vol. 6 (K. C. Smith, ed.), pp. 1–38, Plenum Press, New York.

Shusterman, C. L., Thiede, E. W., and Kung, C., 1978, K-resistant mutants and "adaption" in *Paramecium, Proc. Natl. Acad. Sci. USA* **75**:5645–5649.

Silverman, M., and Simon, M., 1974, Flagellar rotation and the mechanism of bacterial motility, *Nature (Lond.)* **249**:73–74.

Sistrom, W. R., 1978, Phototaxis and chemotaxis, in: *The Photosynthetic Bacteria* (R. K. Clayton and W. R. Sistrom, eds.), pp. 899–905, Plenum Press, New York.

Song, P.-S., 1980, Primary photophysical and photochemical reactions: theoretical background and general introduction, in: *Photoreception and Sensory Transduction in Aneural Organisms* (F. Lenci and G. Colombetti, eds.), pp. 189–210, Plenum Press, New York.

Song, P.-S., 1981, Photosensory transduction in *Stentor coeruleus* and related organisms, *Biochim. Biophys. Acta* **639**:1–29.

Song, P.-S., Chae, Q., and Gardner, J. D., 1979, Spectroscopic properties and chromophore conformations of the photomorphogenic receptor: phytochrome, *Biochim. Biophys. Acta* **576**:479–495.

Song, P.-S., and Walker, E. B., 1981, Molecular aspects of photoreceptors, in: *Biochemistry and Physiology of Protozoa,* Vol. 4 (M. Lewandowski and S. M. Hutner, eds.), pp. 199–233, Academic Press, New York.

Song, P.-S., Häder, D.-P., and Poff, K. L., 1980, Step-up photophobic response in the ciliate, *Stentor coeruleus, Arch. Microbiol.* **126**:181–186.

Spudich, E. N., and Spudich, J. L., 1982, Control of transmembrane ion fluxes to select halorhodopsin-deficient and other energy transduction mutants of *Halobacterium halobium, Proc. Natl. Acad. Sci. USA* **79**:4308–4312.

Spudich, J. L., and Stoeckenius, W., 1979, Photosensory and chemosensory behavior of *Halobacterium halobium, Photobiochem. Photobiophys.* **1**:43–53.

Stavis, R. L., 1974, The effect of azide on phototaxis in *Chlamydomonas reinhardtii, Proc. Natl. Acad. Sci. USA* **71**:1824–1827.

Stavis, R. L., and Hirschberg, R., 1973, Phototaxis in *Chlamydomonas reinhardtii, J. Cell Biol.* **59**:367–377.

Stroobant, P., Dame, J. B., and Scarborought, G. A., 1980, The *Neurospora* plasma membranes Ca pump, *Fed. Proc.* **39**:2437–2441.

Takagi, T., Nemoto, T., Konishi, K., Yazawa, M., and Yagi, K., 1980, The amino acid sequence of the calmodulin obtained from sea anemone *(Metridium senile)* muscle, *Biochem. Biophys. Res. Commun.* **96**:377–381.

Takeuchi, Y., Sugiyama, Y., Kaji, Y., Usukura, J., and Yamada, E., 1981, The white membranes of crystalline bacteriorhodopsin in *Halobacterium halobium* strain R1mW and its conversion into purple membranes by exogenous retinal, *Photochem. Photobiol.* **33**:587–592.

Throm, G., 1968, Untersuchungen zum reaktionsmechanismus von phototaxis und kinesis an *Rhodospirillum rubrum, Arch. Prot.* **101**:313–371.

Vanaman, T. C., 1980, Structure, function and evolution of calmodulin, in: *Calcium and Cell Function,* Vol. 1: *Calmodulin* (W. E. Chung, ed.), pp. 41–58, Academic Press, New York.

Vincenzi, F. F., and Larsen, F. L., 1980, The plasma membrane calcium pump: regulation by a soluble Ca binding protein, *Fed. Proc.* **39**:2427–2431.

Wagner, G., 1979, *Halobakterium:* Vordringen in biotische grenzbereiche, *Biol. Uns. Z.* **9**:171–179.

Walker, E. B., Lee, T. Y., and Song, P.-S., 1979, Spectroscopic characterization of the *Stentor* photoreceptor, *Biochim. Biophys. Acta* **587**:129–144.
Walne, P. L., and Arnott, H. J., 1967, The comparative ultrastructure and possible function of eyespots: *Euglena granulata* and *Chlamydomonas eugametos, Planta* **77**:325–353.
Wenderoth, K., and Häder, D.-P., 1979, Wavelength dependence of photomovement in desmids, *Planta* **145**:1–5.
Yamaki, T., and Hidaka, H., 1980, Ca-independent stimulation of cyclic GMP-dependent protein kinase by calmodulin, *Biochem. Biophys. Res. Commun.* **94**:727–733.

Chapter 7

Phototropism

Ulrich Pohl and Vincenzo E. A. Russo

1. INTRODUCTION

*Et comme les naturalistes remarquent que la fleur nommée héliotrope tourne sans cesse vers cet astre du jour. . . .**

Jean Molière, "Le malade imaginaire," 1673, act II, scene V

Not only is this the first citation we were able to find on *heliotropism* (renamed *phototropie,* or "phototropism," by Oltmanns, 1892), but even the players in the comedy reflect a characteristic that parallels the reaction to studies of this phenomenon. Namely, Argan *(le malade imaginaire)* is absolutely convinced of being sick, and the entourage is equally convinced of the opposite. Similarly, as we shall see, some investigators are absolutely convinced that one hypothesis is the right one, while others are equally convinced that another one is correct. Fortunately, sometimes there is common agreement, but not always.

Light can induce phototropism in a wide variety of higher and lower plants and in different tissues. Most of the effects are achieved with blue light, but in some cases red light is active. Figure 1 shows some of the plants considered.

Although the theme of this volume centers on the role of membranes in sensory transduction, we believe that a presentation of the basic phototropic phenomena is absolutely essential to an understanding of what little is known

*"And as naturalists observe that the flower called heliotrope (sunflower) bends always toward the sun. . . ."

Ulrich Pohl and Vincenzo E. A. Russo • Max-Planck Institut für Molekulare Genetik, D-1000 Berlin 33, Federal Republic of Germany.

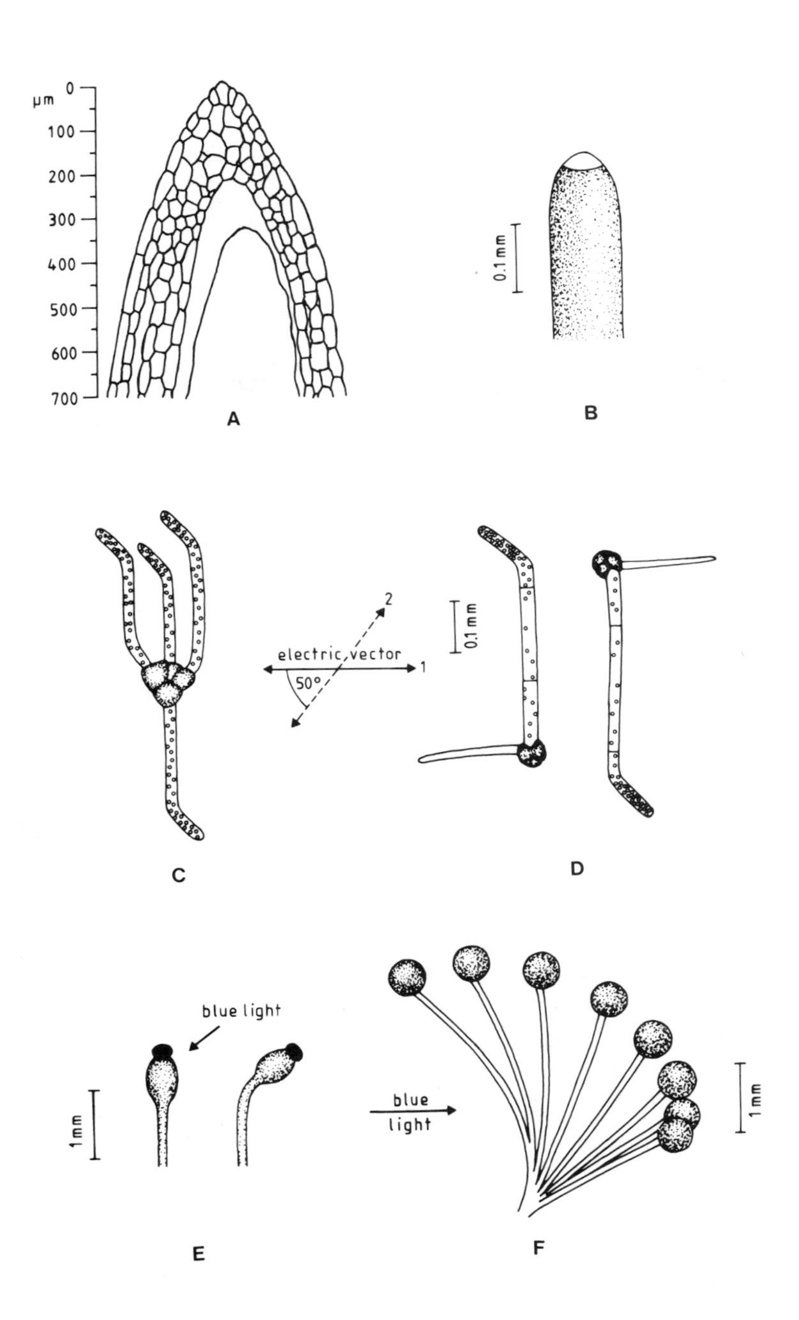
µm
0
100
200
300
400
500
600
700
A
0.1 mm
B
electric vector
50°
1
2
0.1 mm
C
D
blue light
1mm
blue
light
1mm
E
F

on the sensory transduction of phototropism. Toward that end, this chapter does not represent a complete compendium of all known phototropic reactions, but rather attempts to review as fully as possible the phototropic reactions in terms of selected examples. In higher plants, we concentrate mainly on the presentation and discussion of data regarding the phototropism of coleoptiles of *Avena sativa* L. and *Zea mays* L. and in some cases describe work in other plants or tissues. In lower plants, we present data on algae, liverworts, mosses, and ferns, but we concentrate mainly on fungi, especially *Phycomyces*.

Until 1935 the most comprehensive review was that of du Buy and Nuernbergk (1932, 1934, 1935). These workers gave a very good historical review from the year 1693, when Ray discussed the phototropic reaction of plants in his *Historia plantarum*. There have since been many reviews on the subject.

We have tried to be as critical as possible in this review. Any criticism is welcome that will contribute to a better understanding of the fascinating and still mysterious phenomenon of phototropism.

2. THE SENSORY TRANSDUCTION CHAIN

2.1. What Is It?

The series of events that begin with the stimulus perception (e.g., light) and end with the response (e.g., bending) are called the sensory transduction chain. The simplest chain is the linear one; an example for phototropism is shown in Fig. 2. A represents the photoreceptor, B is an unknown compound, the level of which we assume to be influenced by the photoreceptor (either increased or decreased), C is similarly influenced by B, and so on.

←

Figure 1. Phototropic/polarotropic species and organs. (A) Cross section through the tip of a coleoptile of *Avena sativa*. (Redrawn from Lange, 1927.) (B) Filamental tip of the alga *Vaucheria geminata* with hyaline cap (white). (Adapted from a photograph in Kataoka, 1975*a*.) (C) Germ tubes of a spore tetrad of the liverwort *Sphaerocarpus donnellii* 4 hr after induction of polarotropic bending. The direction of the electric vectors of preorienting (1) and stimulating (2) linearly polarized blue light are shown. (Redrawn from Steiner, 1969*c*.) (D) Two germinated spores of the fern *Dryopteris filix-mas* 8 hr after induction of polarotropic bending of the chloronemata. The rhizoids did not bend. The electric vectors between (C) and (D) relate to both, but for (D) linearly polarized red light was used for pretreatment and stimulation. (After a photograph in Etzold, 1965.) (E) Stage IV SPPH of the fungus *Pilobolus kleinii* before and about 2 hr after start of unilateral stimulation with blue light. (Adapted from a photomicrograph in Page, 1962.) (F) Stage IVb SPPH of *Phycomyces blakesleeanus* while bending toward a blue light source. Bending proceeds from the right. The single drawings represent the course of bending in 4-min intervals. Bending occurs at a constant distance from the sporangium, at the lower end of the subapical growing zone. (Adapted from a photograph in Dennison, 1979.)

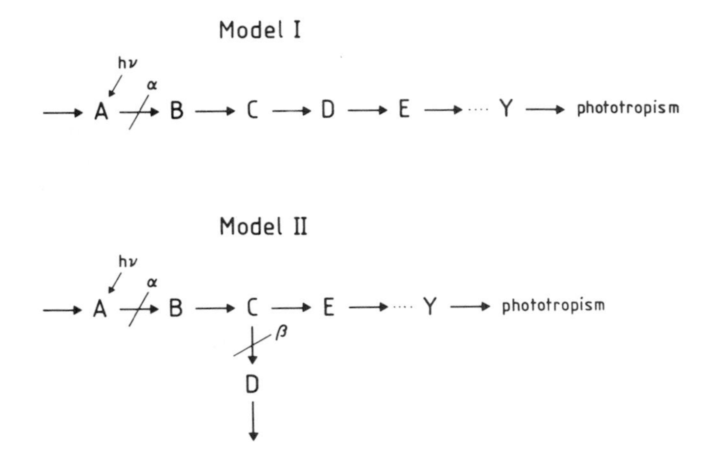

Figure 2. Hypothetical linear transduction chain for phototropism. *A* is by definition the photoreceptor. α and β are inhibitors of the photochange of the level of compound *D*. Only with the inhibitor β is it possible to distinguish between the two models ($h\nu$ = light).

Our knowledge of phototropism will be complete when we are able to give a name to each of these components. As soon as someone has the courage to propose a function for these components it will be possible to perform experiments designed to disprove the model. It may seem trivial to some, but we want to repeat that with experiments it is only possible to disprove a model, it is never possible to prove it. A model is true as long as it is not disproved. In model I of Fig. 2 we say, for example, that the change in the level of D is a *conditio sine qua non* for a response to occur; in other words, it is an essential link in the sensory transduction chain. In model II it is not.

For the photoreceptor A as well as for any other component of the model there are simple rules that must be followed by anyone interested in answering in a meaningful way the questions: Is A the photoreceptor? Is a change in D a *conditio sine qua non* for the response? Here we present what we consider some elementary rules. Following these rules is necessary but not sufficient to obtain the correct answer to the two previous questions.

2.2. Elementary Rules for Model Builders

2.2.1. Rules for Photoreceptor Analysis

1. The photoreceptor must have an absorption spectrum similar to the action spectrum of the response. In Section 5.2 we discuss why we cannot expect the two spectra to be identical.

2. The fluence–response curve must be a function of the concentration of the active photoreceptors. For any response we need a fluence–response curve that can be as complicated as in Fig. 3.

In general, we can say that

$$R = f(I)$$

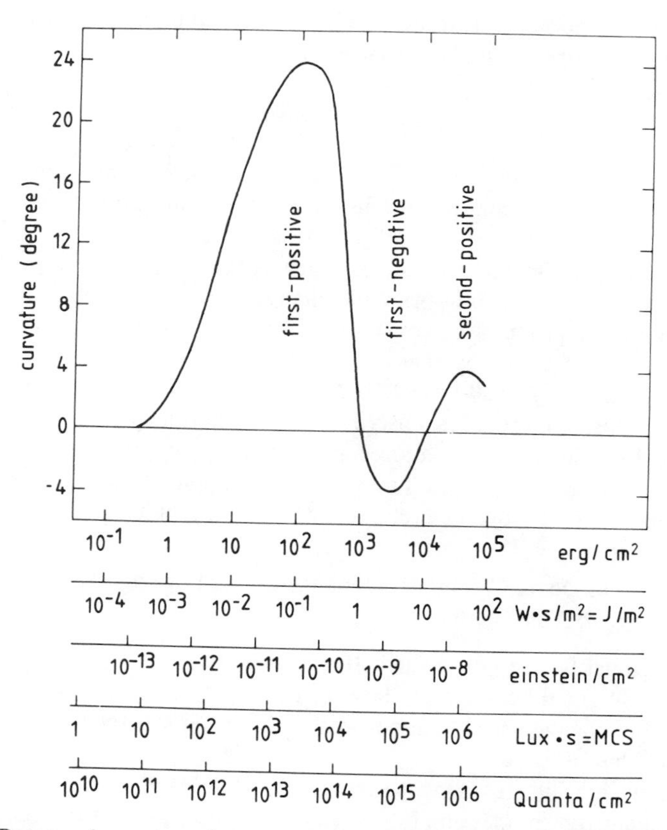

Figure 3. Degrees of curvature of *Avena* coleoptiles irradiated for 30 sec at different fluence rates. Monochromatic light at 458 nm was used; the curvature was measured 100 min after illumination. The fluence (in erg/cm^2) is given on the abscissa. We have added four scales with other fluence units to help the reader compare different works in which only one of these fluence units was used. To obtain the equivalence of Lux · sec, or meter candle · sec (mcs) with the other units is problematic, because investigators who used this unit were working with white light. We used the equivalence suggested by Bergman *et al.* (1969). Similar equivalences were suggested by du Buy and Nuernbergk (1929, 1934) and by Oppenoorth (1942). The equivalence of quanta/cm^2 and of einstein/cm^2 is valid only for blue light of 458 nm. (Redrawn from Blaauw and Blaauw-Jansen, 1970*b*.)

where R is the response, I is the fluence of light, and f is the complicated function that probably cannot be expressed in a simple mathematical formula. According to the Grotthuss-Draper law (enunciated by Grotthuss in 1818) that light can affect a system only if it is absorbed, we can revise the previous formula

$$R = f(N)$$

where N is the number of absorbed photons that induce a response. For low fluences we can write (e.g., Shropshire, 1972)

$$N = N_0 \delta d \phi I$$

where N_0 is the concentration of the photoreceptor, δ is the absorption cross section, d is the path length of the light, ϕ is the quantum efficiency, and I is the fluence.

As long as N is constant, the response is the same. N being the product of five parameters, we can in principle change some of these parameters without changing the product. For example, if we have fewer photoreceptors (e.g., 1% of the normal), either by chemical or genetic manipulation, we can increase the light fluence (e.g., 100 times), and we shall still obtain the same value for N and therefore the same response. $R = f(I)$ is normally plotted on a semilog paper; in the case of fewer photoreceptors, the fluence–response curve will be shifted toward higher fluences. The same will happen if we use a quencher that lowers the quantum efficiency (e.g., for riboflavin, iodide or xenon).

2.2.2. Rules to Decide Whether a Change of the Level of D Is a *Conditio Sine Qua Non* for Phototropism

1. D must be present in the cell or tissue.
2. Light should change the level of D.
3. This change should happen with a lag time shorter than the lag time of the response.
4. Any inhibitor of a photochange of the level of D should also inhibit phototropism (see inhibitors α and β of Fig. 2), and the concentration for half-maximal inhibition of phototropism should not be higher than that for photochange of D.
5. Unilateral application of D resulting in physiological concentrations in the tissue or cell should give a response.
6. The fluence–response curves for the response and for the change of the level of D should have similar thresholds.
7. Both action spectra should be similar.

Similar rules can be applied to decide whether some hormones (e.g., IAA) induce growth via a precise mechanism (e.g., proton secretion).

3. PHOTOTROPISM IN HIGHER PLANTS

3.1. Fluence–Response Curve

An *Avena* coleoptile unilaterally illuminated for a short time and with low fluence rate will bend toward the light after a lag time. The lag time is of the order of 25–100 min, depending on the total fluence (Arisz, 1915). With continuous illumination at high light fluence rate, the lag can be as short as 3.3 min (Pickard *et al.*, 1969). Once the bending reaction has started, it continues for several hours. It soon became quite clear that for *Avena* coleoptiles a short pulse of the right fluence rate was enough to induce a bending reaction (Blaauw, 1908). Blaauw further asked: Is a minimum time duration of the pulse necessary? The answer is that even if the pulse is as short as 1 msec there is still a response as long as the light fluence rate is high enough (Blaauw, 1908; Blaauw and Blaauw-Jansen, 1970*a,b*). As a matter of fact, it was found that for low fluence rates, the response is a function of the product *Ixt,* i.e., the number of photons that reach the coleoptile. This is called the *reciprocity law,* or the *Bunsen-Roscoe law*. This law is true for *Avena* and *Zea* for light pulses no longer than a few minutes and for light fluence rates that are not too high, i.e., for first-positive and first-negative responses as defined in Fig. 3 (Blaauw, 1908; Arisz, 1915; Briggs, 1964; Chon and Briggs, 1966; Zimmerman and Briggs, 1963; Blaauw and Blaauw-Jansen, 1970*b*). The next question is: What is the curvature of a coleoptile as a function of the light fluence like? Coleoptiles were illuminated with pulses of different duration and different fluence rate and scoring of curvature after ~ 2 hr. These studies were started by Arisz (1915) and continued by du Buy and Nuernbergk (1929), Briggs (1960), Zimmerman and Briggs (1963), Chon and Briggs (1966), Everett and Thimann (1968), Pickard *et al.* (1969), and Blaauw and Blaauw-Jansen (1970*a,b*).

The last two papers seem to be the most complete works on fluence–response curves of *Avena* coleoptile published until now. The picture that emerges is represented diagrammatically in Fig. 3. When the fluence rate is varied at a fixed exposure duration of 30 sec, there is a first-positive phototropic response with a threshold of ~ 1 erg/cm^2, then a first-negative response at ~ 10^3 erg/cm^2, and finally a second-positive at 10^4 erg/cm^2 (Fig. 3). If instead the light fluence rate is kept constant at $10^{2.9}$ $erg/cm^2 \cdot sec$ and the exposure duration is varied, it is then possible to see a third-positive response. It is strange that the fluence–response curve of *Avena* with ultraviolet (UV) light (280 nm) is different. The threshold is 20 erg/cm^2, after which there is a plateau at 3000–27,000 erg/cm^2 (Curry *et al.*, 1956). The negative response seems to be peculiar to *Avena* among the higher plants, but the two positive responses separated by a region of phototropic indifference are common to *Zea* coleoptiles (Briggs, 1960) and also to a series of other mono- and dicotyledonous plants (Steyer, 1965, 1967).

3.2. Phototropism and Light–Growth Response

The light–growth response in plants was discovered by Blaauw (1914*b*) for *Phycomyces* sporangiophores (SPPH) and by Vogt (1915) for *Avena* coleoptiles. After intensive studies 5 years later, Blaauw (1919) decided: *"Im Phototropismus selbst doch liegt kein Problem; denn der Phototropismus ist eine reine Wachstumserscheinung"* (Blaauw, 1919). More precisely, he stated: *"Die Reaktion des Wachstums auf Licht ist primär, der Phototropismus ist sekundär. Sie ist der notwendige Erfolg der Lichtwachstumsreaktion, wenn das Licht durch Ungleichseitigkeit eine ungleichseitige Lichtwachstumsreaktion bedingt"* (Blaauw, 1919).*

This very clear and daring statement that phototropism is only a differential light–growth response stimulated a lot of work in order to disprove the theory. The whole controversy can be studied historically in Cholodny (1933) and in du Buy and Nuernbergk (1935). The main points are given here. The most extensive work on the light–growth response of *Avena* coleoptiles is that of van Dillewijn (1927). He found that with pulses of different fluence, the growth is either partially inhibited, e.g., with 25 or 800 meter-candle-seconds (mcs), or it is inhibited and later promoted, e.g., with 10^6 mcs). Assuming a gradient of a factor 30 between the illuminated and the dark side of a unilaterally illuminated coleoptile, he could explain qualitatively the complicated curve of phototropism as a function of fluence (Fig. 3).

Went (1928) showed, however, a great discrepancy between the light–growth response and phototropism if one makes quantitative calculations for the first-positive response. Boysen Jensen (1928) showed with a simple experiment discussed in depth in Section 3.4.1 that the Blaauw theory is invalid for the second-positive response. Cholodny (1933) found conditions (coleoptiles submerged in water) where there was a normal phototropic response in the range of the first-positive response, but no growth inhibition (if any, a slight growth promotion).

All this indicates that the Blaauw theory *in sensu strictiori* is not acceptable. Phototropism is more than a simple differential growth response or, as

*"The growth is primary, phototropism is secondary. It is the necessary consequence of the light–growth reaction, while a gradient of light induces a gradient of growth."

Cholodny (1933) claimed, the different cells of a coleoptile do not react independently, but the coleoptile reacts as a whole. Relatively recent work of Gordon and Dobra (1972), using sensitive capacitance auxanometry, indicated that at very low fluences of blue light (1–50 erg/cm^2 at 480 nm), the *Avena* coleoptile elongation is enhanced, whereas at 100–300 erg/cm^2 it is inhibited. This is partially in contradiction with the work of van Dillewijn and incompatible with the Blaauw theory.

However, there may well be a correlation between light–growth response and phototropism. More evidence in support of this was brought forward by Elliott and Shen-Miller (1976), who found that the fluence–response curves in the first-positive range ($\leq 2 \times 10^{14}$ quanta/cm^2) and the action spectra for phototropism and photoinhibition of growth in *Avena* are very similar. But these data, in the range of 10^{11}–10^{13} quanta/cm^2, are in contradiction to the data of Gordon and Dobra (1972), which were obtained in the same institute but by a different method.

3.3. Light Perception

3.3.1. Action Spectra for *Avena* Phototropism

The first scientist to make phototropic experiments with light of different colors was Poggioli (1817). He observed that seedlings of *Raphanus* bent more with violet than with red light (6 hr illumination) using different colors obtained with a prism. In *Avena* the first action spectrum was made by Blaauw (1908). He was able to show that only the blue and the near-UV part of the visible spectrum is effective; he also used a prism.

Bachmann and Bergann (1930) and du Buy and Nuernbergk (1935) reviewed the early literature. Haig (1934) claimed different action spectra for the tip and the base responses (defined in Section 3.3.6). Haig measured as a response the reaction time of phototropism at an unspecified light fluence. Johnston (1934), using filters and a monochromator), investigated the range from 405 to 540 nm, using the null point method. He was the first to clearly find two peaks, at 440 and at 475 nm. It is not clear whether he worked in the first-positive or (more probably) in the second-positive range. Atkins (1936) measured the action spectra of *Avena sativa, Lepidium sativum,* and *Celosia cristata.*

Shropshire and Withrow (1958) have shown that the fluence–response curves of the first-positive response are linear with the log of the fluence, supporting the validity of the Weber-Fechner law (cf. Shropshire, 1979).* They

*The Weber-Fechner law states that the response is proportional to the logarithm of the stimulus.

are not parallel, however, at the different wavelengths tested (350–520 nm). Therefore, the action spectrum depends on the response measured (either the threshold or a given curvature: 5°, 10°, 20°). The peaks occur at about 370, 412, 442, and 472 nm, but the ratios between the peaks change depending on the response measured. Based on a mathematical analysis, they claimed, "the action spectrum obtained for the threshold response will correspond to the photoreceptor absorption." However, in a later paper on action spectroscopy, Shropshire (1972) did not repeat this claim.

Similar action spectra for the first-positive response were found by Thimann and Curry (1961), Shen-Miller *et al.* (1969), and Elliott and Shen-Miller (1976).

Everett and Thimann (1968) obtained the action spectrum for the second-positive response; it is similar but not identical to that of the first-positive response. It has only one peak in the blue region at 450 nm.

We do not know of any action spectrum for *Zea.* The action spectra should help decide which molecule acts as the photoreceptor of the sensory transduction chain. For *Avena* phototropism, the photoreceptor is most probably a molecule that absorbs blue light (see Section 5). For more information on action spectroscopy, see Shropshire (1972) and Hartmann (1977).

3.3.2. Effects of Red Light

Although some stimulating effect of red light on the growth of *Avena sativa* was already known (Vogt, 1915), no systematic work was started before the late 1950s. Curry (1969), citing his Ph.D. thesis of 1957 found that 1.2 mc (meter-candle) of red light given for 1 hr before the blue stimulus shifts the sensitivity of the first-positive response of *Avena* coleoptiles by a factor 5 (the threshold is $\sim$ 1 erg/cm^2 without red light and 5 erg/cm^2 with). Almost at the same time, Blaauw-Jansen (1958) carried out detailed studies on the influence of red and far-red light on growth and phototropism of *Avena* coleoptiles. She found that red light (given about 15 min) had no effect on the threshold of the first-positive response, but that it enhanced the total bending to blue light. The red light effect could not be reversed by far-red light (740 nm). Red light also inhibited the growth of the coleoptiles.

This was the beginning of a long scientific controversy. Zimmerman and Briggs (1963) did not only confirm the data of Curry, they even found that red light decreased the sensitivity to the first-negative response and shifted the second-positive response toward lower fluence; however, they did not find any enhancement in the total bending. Briggs (1963*a*) attributed the lack of threshold shift in Blaauw-Jansen's experiments to the short red light exposure (15 min) when compared with that (1 hr) of Curry and their own experiments. Blaauw and Blaauw-Jansen (1964) made threshold curves with different time exposures to red light and confirmed the enhancement of total bending of Blaauw-Jansen (1958). With a red light pretreatment of 2 min, they found a shift of the peak only if blue light was given after a dark interval of 1–2 hr, partially accepting the interpretation of Briggs (1963*a*). But a shift of the threshold of the first-positive response is not evident; furthermore, there is clearly no shift of the thresholds of the first-negative and second-positive responses. Blaauw and Blaauw-Jansen (1964) criticized the data of Zimmerman and Briggs (1963) saying that either their green light was impure or they had a green light effect. Chon and Briggs (1966) found a 100-fold shift in the first-positive response of *Zea mays* coleoptiles. Furthermore, they showed that this effect is reversed by far-red light, implying that phytochrome is involved. They criticized the results of Blaauw-Jansen (1958) because of the nonreversibility of the far-red light, explaining that her illumination by far-red light (40 min) was too long. They further stated that Blaauw and Blaauw-Jansen (1964) confirmed the shift for the first-positive response of *Avena* (which is only partially correct) and that they did not explore the first-negative curvature (which is invalid).

The matter becomes more complicated when we consider the results of Elliott and Shen-Miller (1976). They found with red light (660-nm) exposure 90 min before the blue that the pretreatment lowered the total bending in the first-positive response (in contrast with Blaauw and Blaauw-Jansen, 1964) and that it did not shift either the threshold or the peak (in contrast with Zimmerman and Briggs, 1963). Unfortunately, Elliott and Shen-Miller (1976) did not elaborate on this point, nor did they compare or discuss the data of the other two groups. Blaauw and Blaauw-Jansen (1970*b*) also found a red light effect on the third-positive response which results in an increase of the total bending.

It is clear that red light has some effect on phototropism. However, this effect varies from one laboratory to another.

3.3.3. Light Direction or Light Gradient?

Many years ago the question whether the coleoptile senses a light gradient or a light direction was discussed. Buder (1920) performed simple experiments indicating that the light gradient is sensed. He illuminated only one side of the *Avena* coleoptile from above and the coleoptile bent toward the illuminated side. Even more elegantly, he illuminated from the inside, with a "light pipeline," only one side of the tip. The coleoptile bent toward the illuminated side as if this side would have been illuminated from outside. Meyer (1969*a*) repeated the half-side experiments of Buder with light fluences of 60–190,000 erg/cm^2 and found that the Buder effect was valid for the first-positive response, which was extended up to 19,000 erg/cm^2, whereas the first-negative response disappeared.

3.3.4. Lens Effect or Absorption Gradient?

Blaauw (1914*b*) noted that the sporangiophore (SPPH) of *Phycomyces* behaves like a cylindrical lens (the sporangiophore clearly shows an intense band in the center when looked at against the light). In his work of 1919, he used this finding in order to explain according to his theory (Section 3.2) how both coleoptiles and *Phycomyces* SPPH are positively phototropic, even though coleoptiles have a negative light–growth response, while Phycomyces SPPH have a positive response. Buder (1918) suggested that the lens effect of *Phycomyces* SPPH does indeed have something to do with phototropism (Section 4.5.5.c). In his elegant experiment, Buder immersed the SPPH in paraffin oil (refractive index 1.47); in this case there is for optical reasons no concentration of light in the distal side of the SPPH, and the SPPH bends away from light (phototropic inversion). Cholodny (1933) showed that for the stumps of *Avena* coleoptiles, where an artificial lens effect was induced by filling the stumps with water, the lens effect plays no role (however, he gave no data to support his statement). So it seems that in the coleoptiles the light gradient is created by absorption and not by lens effect, if we take the work of Cholodny at face value. Strangely enough, there is a phototropic inversion in *Avena* coleoptiles immersed in paraffin oil (Ziegler, 1950). Specht (1960), a student of Buder, showed that this phototropic inversion is exhibited by many dicotyledonous seedlings if submerged in paraffin oil. She has shown, however, that this has nothing to do with the lens effect for two reasons: (1) the phototropic inversion happens even if the illumination is done in air immediately after 2 hr of immersion, and (2) if the infiltration of the oil into the seedlings is inhibited (by covering the seedlings with a gelatine film), the submerged seedlings bend toward the light and do not show any phototropic inversion. Humphry (1966) confirmed the phototropic inversion for *Avena* coleoptiles. All three experiments were done with continuous intense illumination. So they are probably valid for second-positive or third-positive responses. Meyer (1969*b*) found that the phototropic inversion is also true for first-positive and first-negative responses, but her results show that there is no lens effect because (1) the coleoptiles were irradicated in air after 2 hr of submersion in oil, and (2) illumination in oil just a few seconds after submersion gave normal phototropism. However, it was inverted only if the illumination occurred after a period of 1 min to 1 hr. Despite all we have said until now, Shropshire (1975) stated, "The lens effect also operates for higher plants although this point is sometimes overlooked in current reviews." In support of this statement, he cited the phototropic inversion in *Avena* coleoptiles as the work of Buder (1918); however, Buder did not work with *Avena,* but with *Phycomyces.* He quoted Meyer (1969*b*), stating that "[she] demonstrated that for immersion in paraffin oil the blue light stimulus had to be given during immersion to produce negative curvatures" (which is not correct, as explained). Finally, Shropshire presented a piece of experimental evidence of his own. He repeated with *Avena* similar experiments to those he had done with *Phycomyces* (Shropshire, 1962); in these experiments he created a diverging ray path with a cylindrical lens. He reported (1975) that the curvature was negative at a nonspecified light fluence that gave a positive curvature with a converging lens. If this single experiment reported by Shropshire (1975) can be confirmed, it might represent evidence for the fact that also in *Avena* the lens effect plays some role in phototropism.

3.3.5. Tip Sensitivity

Charles and Francis Darwin (1880) noted that "The cotyledons of *Avena,* like those of *Phalaris,* when growing in soft, damp, fine sand, leave an open crescentric furrow on the shaded side, after bending to a lateral light; and they become bowed beneath the surface at a depth to which, as we know, light cannot penetrate." Furthermore, they stated, "We do not know whether it is a general rule with seedling plants that the illumination of the upper part determines the curvature of the lower part. But as this occurred in the four species examined by us, belonging to such distinct families as the Gramineae, Cruciferae and Chenopodeae, it is probably of common occurrence" (C. Darwin and F. Darwin, 1880). After doing experiments with illumination of the base of the coleoptiles (the tip being covered with tin foil), they summarized, "All observers apparently believe that light acts directly on the part which bends, but we have seen with the above described seedlings that this is not the case. Their lower halves were brightly illuminated for hours, and yet did not bend in the least towards the light, though this is the part which under ordinary circumstances bends the most" (C. Darwin and F. Darwin, 1880). Finally, they concluded, "These results seem to imply the presence of some matter in the upper part which is acted on by light, and which transmits its effects to the lower part" (C. Darwin and F. Darwin). So it was already known that the tips are more sensitive than the bases. It was already postulated that there is a transmission of the light information from the tip (where light is sensed) to the base of the coleoptile (where the bending appears).

Sierp and Seybold (1926) were the first to determine quantitatively the sensitivity of the different parts of the *Avena* coleoptile by illuminating the whole coleoptile except for different lengths of the tip. Lange (1927) studied systematically the problem by illuminating different zones along the coleoptile (as small as 100 μm, starting from the very tip down to 1 mm) and determining the threshold. (Lange's definition of threshold is quite strange, i.e., the light fluence that gives 50% "clear" bending, meaning at least 15°. A similar definition was used by Sierp and Seybold, 1926, and by Koch, 1934). Finally, Koch (1934) did similar experiments, but with decapitated coleoptiles. It is surprising that three different laboratories with three different techniques obtained the same results: the threshold increases exponentially along the *Avena* coleoptile, from 5 mcs at the very tip to 500 mcs at 1 mm below the tip.

In contrast to these data are the works of Arisz (1915) and Reinders (1934), who found a threshold of 100–200 mcs for a coleoptile with 5 mm of its tip covered with tin foil. Furthermore, both Brauner (1922) and Reinders (1934) found that illuminating a stump with subsequent immediate addition of the tip gives a phototropic response (at 200 mcs) similar to that of intact coleoptiles, opposing the data of Sierp and Seybold (1926). Meyer (1969*a*) made a rather crude fluence–response curve from 60–19,000 erg/cm^2 of blue light (300–550 nm) or *Avena* coleoptiles. Either the tip (the first 350 μm) or the next region (350–700 μm) was illuminated. From these experiments it is difficult to decide whether the threshold is slightly shifted, but it seems that the first-positive and the first-negative peaks are not shifted and that the total bending is less at any light fluence. As discussed in Section 2.2.1, this is an indication that the difference in sensitivity along the coleoptile may not be attributable to a different concentration of the photoreceptor, but to a gradient of concentration of some other component of the sensory transduction chain.

3.3.6. Tip and Base Responses

In their review, du Buy and Nuernbergk (1934) stated that in the first-positive response the point of bending is just below the tip, while in the second-positive and third-positive responses it is nearer the base. This was generally accepted until recently (e.g., Thimann and Curry, 1960; Curry, 1969); however, Blaauw and Blaauw-Jansen (1970*a*) have shown that a tip response occurs only at high fluence as long as no red light is given before the blue-light irradiation.

3.4. Stimulus Transmission

3.4.1. Theories

As early as 1880, the Darwins had realized that light can be sensed in the tip of a coleoptile but that the effect (bending) is far away (in the base), requiring that the stimulus be transmitted (Section 3.3.5). A central point of investigation in phototropism over the last 104 years has been and remains: How is the light information transduced in biochemical information, and what is transmitted and how?

Boysen Jensen (1910) stimulated the research in this field by his discovery that decapitated *Avena* coleoptiles also react to light (probably in a second-positive response) if illuminated after the tip had been put on the stump again. So the organic continuity of the tissue is not necessary for stimulus transmission. He found also that if only one-half the coleoptile is cut horizontally and a thin piece of mica is put in this wound, phototropism occurs only if the unilateral illumination is from the side of the cut containing the mica (Fig. 4A). Boysen Jensen concluded that transmission of the stimulus occurs on the dark side.

Paál (1914, 1919) confirmed that the stimulus can be transmitted over a total cut of the coleoptile, but he could not repeat Boysen Jensen's experiment on the partial cut illustrated in Fig. 4A. In the belief that light acts on the illuminated side, either destroying the growth substance or inhibiting its transport, Paál (1919) wrote

> *Man kann nun die Annahme machen, daß eben dieser Korrelationsträger, der unter normalen Umständen fortwährend und allseitig gleichmäßig aus der Spitze nach unten wandert, bei Belichtung der Spitze in seiner Enstehung gestört oder photochemisch zersetzt oder in seiner Wanderung etwa durch eine Änderung des Plasmas gehemmt wird und zwar an der besser beleuchteten Seite in stärkerem Maße.**

Boysen Jensen and Nielsen (1926) and Boysen Jensen (1928) performed an elegant experiment in order to prove the Paál theory incorrect, at least for the second-positive response. They split the tip of *Avena* coleoptile longitudinally for 2–3 mm and inserted a thin plate of platinum or glass; they then illuminated either parallel to the platinum or glass plate (showing bending) or perpendicularly (showing no bending, Fig. 4B). The light fluence was in the second-positive range. The results show that there must be a physical continuity between the dark and the illuminated side of the tip in order for phototropic bending to occur. The Boysen Jensen theory states that light enhances the level of the growth substance on the illuminated side and this will migrate to the dark side, thereby creating a bend.

Ramaer (1926) found that splitting the tip has no effect on phototropism in the first-positive (800 mcs) and in the first-negative ranges (80,000 mcs). That is, at these two light fluences there is normal bending even when light is given perpendicularly to the platinum plate. Ramaer's paper does not present enough data in order to convince us of the validity of his results, but it represents an important experiment that must be repeated. For *Zea* coleoptiles the platinum plate experiments were done by Briggs (1960) using glass instead of platinum; in the first-positive range there was no bending if light was perpendicular to the plate, while in the second-positive range there was less (50%), but there was bending—the very opposite of the results achieved with *Avena.*

*"One can make the assumption that the growth substance, which under normal conditions is transported equally from the tip to the bottom, by illumination of the tip will be either inhibited in its production or photochemically destroyed or inhibited in its transport through a change in the plasma and this effect is more pronounced in the illuminated site than in the dark site."

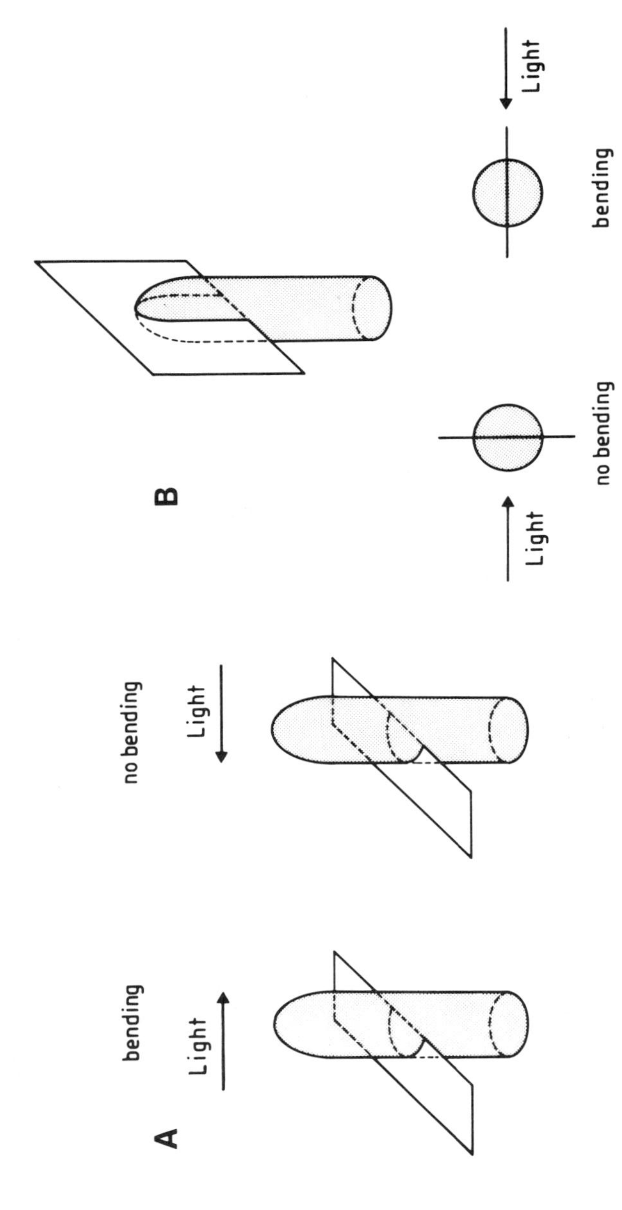

Figure 4. (A) Diagram illustrating the effect on phototropism of a partial cut of an *Avena* coleoptile (with insertion of a piece of mica). (According to Boysen Jensen, 1910.) (B) Diagram illustrating the effect on phototropism of splitting the tip of an *Avena* coleoptile longitudinally for 2–3 mm (with insertion of a thin plate of platinum or glass). (According to Boysen Jensen and Nielsen, 1926; Boysen Jensen, 1928.)

If Ramaer's results are true, they challenge the Cholodny–Went theory for the first-positive and the first-negative response.

On purely theoretical grounds, Cholodny (1927) suggested that the total amount of the growth substance remains constant but that a stimulus (either gravity in geotropism or light in phototropism) induces a lateral transport so that the dark side has more growth substance than the illuminated side. He wrote that when the manuscript was ready he became aware of Went's work. In a beautiful work of 116 pages, which is essentially his Ph.D. thesis, Went (1928) demonstrated several findings:

1. The tip continuously produces a growth substance.
2. The molecular weight of this substance is 350–400 (measured by diffusion).
3. The transport of this substance is polar in coleoptile stumps, from the tip to the base.
4. Bilateral illumination with light of 10^3 mcs diminishes the transport of this substance.
5. Unilateral illumination (first-positive range) transports more substance in the dark than in the illuminated side of the coleoptile.

The amount of growth substance was measured according to the bending test of *Avena* (developed by Went). A block of agar with a specific amount of diffused growth substance is put on one side of a coleoptile stump, the curvature of which is measured after a given period. The bending is linear within a given concentration range of the diffusate. He suggested the same model as Cholodny, hence reference to the Cholodny–Went theory. Went quoted Ramaer's (1926) results without saying that they were wrong and without explaining how Ramaer's results could fit the Cholodny–Went theory. The predictions of the three theories on the distribution of the growth substance in the coleoptile are graphically represented in Fig. 5:

1. Boysen Jensen: Made only for the second-positive range
2. Cholodny–Went: Suggested by Went only for the first-positive range
3. Paál: Fluence unspecified

3.4.2. Chemical Nature of the Growth Substance

A growth substance made in the tip and transported into the base was postulated by Paál (1914, 1919). Söding (1923, 1925) demonstrated its existence in *Avena* coleoptile tips. Went (1928) introduced the quantitative measurement of this substance with the *Avena* bending test explained in Section 3.4.1. The word *auxin* (from the Greek *auxano,* "to increase") was suggested instead of *growth substance* by Kögl and Haagen-Smit (1931); they also sug-

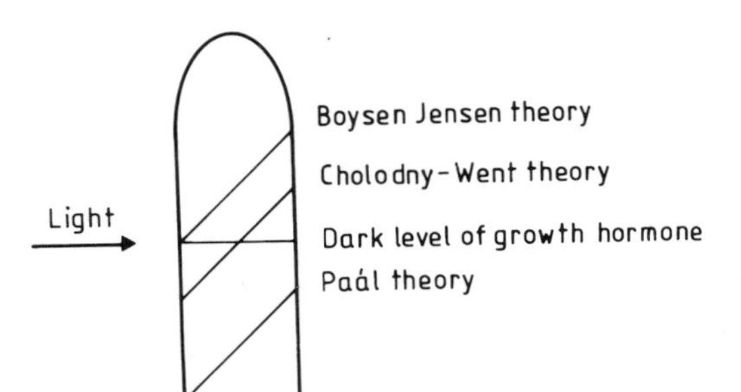

Figure 5. Diagram illustrating the predictions of the three theories on the growth hormone concentration in the coleoptile after illumination and before phototropism.

gested the word *phytohormone* for the growth substance that they had isolated from, e.g., urine, *Zea*, and yeast. It turned out to be two auxins: auxin a and auxin b (see structure in Kögl, 1937). In 1933 investigators in Kögl's laboratory observed that the so-called heteroauxin IAA (3-indolylacetic acid) also has a growth-promoting activity.

The discussion on which is the real growth hormone continued for about 20 years before it became clear that the auxin of *Avena* coleoptiles is IAA. Diffusible auxin was found to be IAA by Wildman and Bonner (1948) with a colorimetric method and by Shiboaka and Yamaki (1959) with chromatographic methods. Greenwood *et al.* (1972) found the same in *Zea* coleoptiles using mass spectrometry. With mass spectrometry of acetone extracts from seedlings (coleoptiles, primary leaves, mesocotyls) of *Avena* and *Zea*, Bandurski and Schulze (1974) found IAA and ester IAA. In a later work (Bandurski and Schulze, 1977), they also found peptidyl IAA in a series of plants and organs including *Avena* and *Zea*. The concentration of IAA is $\sim 10^{-7}$ *M* for both plants, while that of the ester IAA is much higher in *Zea* than in *Avena* seedlings. It is ironic that even though Kögl and co-workers invented the word *auxin* and stimulated the research in this field, they discovered auxins a and b, which do not exist in nature, according to Bearder (1980). For more information on natural auxins, see MacMillan (1980), and for natural and synthetic auxins and antiauxins, see Gordon (1954), Bentley (1958) and Jönsson (1961).

3.4.3. Action of Light on Auxin and IAA Transport

In this section we use the word *auxin* to indicate the natural growth substance and IAA for the synthetic one. Of the three theories presented in Fig. 5, that of Cholodny-Went became the most popular. Asana (1938) showed that the diffusible auxin behaves according to the theory in the first-positive response. Wilden (1939) showed the same for the first- and second-positive and first-negative responses. Oppenoorth (1939, 1942) measured separately the amount of endogenous auxin in the tissue (ether extraction) in the dark and in the illuminated halves of the *Avena* coleoptile (split at different times after the stimulus). For the first-positive response (330 erg/cm^2 with blue light at 436 nm), he found that about 35% disappeared from both sides after only 5 min; after 1 or 2 hr there is more on the dark than on the illuminated side. For the first-negative response (3000 erg/cm^2 with the same blue light) he found an increase on both sides (already after 5 min), but there was more of an increase on the illuminated than on the dark side. He still subscribed to the Cholodny–Went theory, at least for the first-positive response, but he could not prove it because of the destruction of the auxin. Went (1942) claimed that there are two types of auxins in *Avena* coleoptiles: the bound auxin, which is responsible for the growth of the stumps, and the free-moving auxin, which is responsible for the phototropic response of the stumps. If the work of Went is correct, the work of Oppenoorth is irrelevant. Briggs *et al.* (1957) and Briggs (1963*b*) studied the diffusion of auxin from the dark and the illuminated side of phototropically induced *Zea* coleoptiles. They did not find destruction either in the first-positive or second-positive responses and the results for both responses are in agreement with the theory.

After World War II the technology to obtain radioactive compounds was developed. Consequently the transport of externally applied radioactive IAA was widely studied. It was shown that the transport of radioactive IAA is polar in coleoptiles of *Avena* (Goldsmith and Thimann, 1962; Goldsmith, 1966) and *Zea* (Hertel and Leopold, 1963) with a velocity of 12–15 mm/hr in *Zea* (Hertel and Leopold, 1962; Goldsmith, 1967) and 15 mm/hr in *Avena* (Thornton and Thimann, 1967). For more information on polar transport of IAA see Goldsmith (1977).

A controversy that has lasted for more than 25 years arose over the question: Does light influence the transport of radioactive IAA?

Bünning *et al.* (1956) and Gordon and Eib (1964) could not find any effect in *Avena*. Pickard and Thimann (1964) found a lateral transport in *Zea* coleoptiles for the first- and second-positive response and in *Zea* stumps for the second-positive response. Shen-Miller and Gordon (1966) found a lateral transport of endogenous auxin but not of radioactive IAA in *Zea* coleoptiles (first-positive response). Naqvi and Gordon (1967) found that bilateral white light (1000 mcs) diminished the amount of diffusible auxin from the tip and also the transport of IAA in *Zea* stumps. Similarly, Thornton and Thimann (1967), in work on *Avena,* showed a transient inhibition of IAA transport (3×10^5 erg/cm^2 of blue light for 15 min in the second- or third-positive range). Gordon and Shen-Miller (1968) suggested that the effect of unilateral illumination with blue light in the first-positive response is a differential inhibition in the rate at which the hormone (IAA) is transported basipetally. This is none other than the Paál theory, as discussed in Section 3.4.1. Shen-Miller *et al.* (1969) presented data that are not in contradiction with this model: The fluence–response curve for IAA transport in *Avena* coleoptiles (maximum inhibition 13%) and the action spectrum are very similar (but not identical) to the fluence–response curve and action spectrum of phototropism.

The most complete work on the lateral and longitudinal transport of IAA seems to be that of Gardner *et al.* (1974) in Wilkins's laboratory. As Wilkins (1971) pointed out, symmetrical application of IAA cannot establish whether or not lateral transport takes place. Therefore, Gardner *et al.* (1974) applied IAA asymmetrically with a glass micropipette within 200 μm of the coleoptile apex and illuminated unilaterally with blue light (440 nm).

For our purposes, the main results can be summarized as follows:

1. In *Zea* seedlings they found no effect of blue light on the *longitudinal* transport for the first-positive and the second-positive response. With and without red light pretreatment (not inhibiting phototropism significantly) the *lateral* transport was always evident for the second-positive response, but for the first-positive response it was not seen after a red light pretreatment. For excised coleoptile apices the situation is different, and this is not explained.
2. In *Avena* seedlings pretreated with red light (which permitted normal phototropism), there was no lateral transport either in the first- or in the second-positive response (the same was true for the second-positive response without red light). However, there was inhibition of longitudinal transport in both responses. For unknown reasons, they have not done this last experiment with red light.

It seems from these data that for *Avena* the Paál theory reproposed by Gordon and Shen-Miller could be valid. In *Zea* coleoptiles there is no light effect either on *longitudinal* or on *lateral* transport during the first-positive response.

The most interesting and disturbing result of Gardner *et al.* (1974) is the finding that in some cases it was possible to have normal phototropism without a blue light effect on IAA transport. In these cases it seems that the differential IAA transport is not a *conditio sine qua non* for phototropism. But what about the other cases?

It may be interesting to note that Bruinsma *et al.* (1975) found no change of the amount of endogenous auxin in the illuminated and dark side of *Helianthus annuus* during phototropism. These workers concluded that IAA has nothing to do with phototropism in this plant. Phillips (1972) suggested that in *Helianthus annuus* a gibberellin is the hormone involved in photo- and geotropism because there is a lateral transport of gibberellin during phototropism. On the other hand Bruinsma *et al.* (1980) suggested a gradient of inhibitor (with xanthoxin as a possible candidate) to be involved in *Helianthus* phototropism.

This raises the question: Is IAA necessary for phototropism? Von Guttenberg (1959) answered the question for *Avena* in the second-positive response. He illuminated the stumps with 780 Lux (white light) for 3 hr, which were then covered with blocks of agar, some containing IAA at 500 μg/liter (3×10^{-6} *M*), and then put in the dark. The stumps with IAA bent much more than the controls without IAA (28° to 8°). The amazing thing is that there is an effect even when the blocks are put on the stumps 8 hr after the end of the illumination. This shows that IAA is useful for phototropism and that light does not have the function to destroy IAA but to create somehow a physiological gradient (at least in these particular conditions).

Pohl (1960) confirmed the results of von Guttenberg with blue light in the second-positive range; however, he stated (without giving any data) that in the first-positive (75 erg/cm^2) and the first-negative (4500 erg/cm^2) ranges no bending occurred on stumps that were covered with agar blocks containing 200 μg/liter of IAA ($\sim 10^{-6}$ *M*). Ball (1962) repeated von Guttenberg's experiments but immersed the stumps in IAA solution (5×10^{-6}) and found inhibition of phototropism. These three works indicate that IAA is necessary for phototropism of *Avena* coleoptiles in the second-positive response, but nothing can be said for the other responses or for *Zea* phototropism.

It seems that the role of IAA in the phototropism of *Avena* and *Zea* coleoptiles is still an open question.

3.4.4. Effect of Light on IAA Metabolism

Galston (1949) was the first to show that light could destroy IAA *in vitro* if riboflavin was present. That finding was very important for speculations on the nature of the photoreceptor (as we will see in Section 5.2), and at the same time provided adherents of the Paál theory with a molecular basis for the destruction of auxin. Zenk (1968) was the first to examine IAA destruction *in vivo*. He used radioactive 1-naphthalene acetic acid (1-NAA), a synthetic compound with auxin activity, instead of IAA, because IAA is too unstable in light. He found that 1-NAA is photolytically decarboxylated in *Avena* by blue light, but it was too little to explain phototropism. Zenk suggested that the triplet state of flavin photooxidizes an unknown compound necessary for growth. At about the same time Hager and Schmidt (1968*a*) found that the transport of radioactive IAA through *Zea* stumps (in the dark) can be inhibited by a light *pretreatment* of 1 hr at 13,000 Lux (second- or third-positive range), but only if the stumps were provided with cold IAA during the illumination. They were also able to find the compound 3-methyloxindole (3-M), made from radioactive IAA in the illuminated crude extracts. 3-M inhibited both growth and IAA transport (Hager and Schmidt, 1968*b*). The conclusion was that light would create 3-M by photooxidation of IAA and 3-M would inhibit the transport of IAA. In a subsequent paper by Hager's group Menschick *et al.* (1977) did not find any effects of light on IAA decarboxylation *in vivo* (second- to third-positive range) in *Avena* coleoptiles and they stated that the content of 3-M is the same in the dark as in the light.

At the same time Bandursky *et al.* (1977) reported that in *Zea* seedlings blue light ($4 \cdot 10^5$ erg/cm^2) decreased the amount of IAA by 10 μg/kg fresh weight while at the same time the amount of ester IAA is increased by 9 μg/kg fresh weight, both measured after 90 min. They suggested that this change could be a link in the sensory transduction chain of *Zea* phototropism. It is interesting that ester IAA and peptidyl IAA were found in many species and tissues including *Avena* (Bandursky and Schulze, 1977).

3.4.5. Light- and IAA-Induced Electrical Potentials

Brauner (1927) observed that shoots of different plants (including *Helianthus*) develop transverse electrical potentials if put in the horizontal position; the top side is more electronegative than the bottom side by 5–9 mV. Brauner and Bünning (1930) showed that a coleoptile of *Avena* will bend in an electrical field of 640 V/cm toward the positive pole. Cholodny (1927) and Went (1932) suggested that both geotropism and phototropism are due to a transverse electrical polarization of the coleoptile which will create an electrophoretic movement of auxin toward the positive side. (It was not known at that time that auxin is IAA, an acid, negatively charged.)

Clark (1935) was the first to measure a longitudinal potential change by illumination with 600,000 mcs. In a subsequent paper, Clark (1937) stated that with his method he could not see any correlation between potential differences and polar auxin transport. Schrank (1946) was the first to see a transverse electrical polarity upon unilateral illumination with white light (160 mc continuous) in *Avena* coleoptiles. The lag for the potential rise was 30 min and bending started at 47 min. The light intensity was in the second- or third-positive range and the dark side was electropositive with respect to the illuminated side, in agreement with the Cholodny–Went hypothesis. Backus and Schrank (1952) repeated the same type of experiment for the first-positive (200 mcs) and the first-negative response (31,200 mcs). In the first case, the dark side was electropositive, in the second it was electronegative. So the Cholodny–Went electrophoretic hypothesis seemed confirmed for all three responses. Furthermore, an electrical current (10 μA for 10 min) induced bending of *Avena* coleoptiles toward the positive pole (Schrank, 1948). It was possible to induce tropism of coleoptile stumps covered with an agar block containing IAA (0.5–1 mg/liter) by applying a current of 20 μA for 2 min (Webster and Schrank, 1953). Interestingly, IAA could be added 120–150 min after the 2 min of current; there was still a tropism, indicating that the current somehow polarizes the coleoptiles, and this polarity stays for a long time. The observation that coleoptiles filled with Shive's solution did not respond to light, while a nonelectrolyte solution of glycerol did not inhibit phototropism was considered further evidence (Wilks and Lund, 1947; Schrank, 1950, 1953). The interpretation was that the Shive's electrolyte solution shunts the surface electropotential differences. But this last observation became less significant after the experiments of Bridges and Wilkins (1971), who showed that nonelectrolyte solutions of mannitol or carbowax 1540 also inhibit phototropism and geotropism.

The electrophoretic hypothesis is weakened by the experiments of Johnsson (1965). He showed that in *Zea* coleoptiles white light induced a transverse electrical potential (40 mV positive on the shadow side) with a lag of 30–40 min. There were no potential differences on stumps illuminated with white light in the second-positive range; but they were present if the stumps were covered with agar containing IAA (5×10^{-5} *M*). Unfortunately, it is not possible to evaluate the statistical significance of these results. If they are true, the best interpretation would be that light in the second-positive range induces electrical potentials in coleoptile stumps only in the presence of IAA. So for the second-positive range, the sensory transduction chain cannot be considered operative, i.e., unilateral illumination, transverse potential, and electrophoretic transverse movement of IAA. Probably, the chain events are as follows: light creates a polarity, IAA is laterally transported, and on the side where there is IAA, a positive potential exists with respect to the other side.

The last part of this chain was shown to be true in *Zea* by unilateral application of IAA; the side with IAA became more electropositive than the other side (Grahm, 1964; Morath and Hertel, 1978). Note that it is not the presence of IAA per se that creates the potential difference, because IAA is negative while the side receiving IAA becomes electropositive. Presumably there must be some interaction of IAA with the cell membrane. Johnsson's experiments have not yet been carried out on *Avena.*

Recently it was found that in *Zea* coleoptiles blue light induces a membrane potential hyperpolarization on the dark side after 5–10 min. The response is quicker (1–2 min) if the blue light is given after red light. The red light (probably via phytochrome) depolarizes the membrane potential very quickly, in less than 1 min (Racusen and Galston, 1980).

3.4.6. Short-Term Effects of IAA on Coleoptiles

Bottelier (1934) has shown that blue light induces changes in the velocity of protoplasmic streaming in epidermal cells of *Avena* coleoptiles. Light fluences in the first-positive range transiently inhibit the streaming, while light fluences in the first-negative range accelerate it. The lag was about 4 min and the region observed was somewhere in the first 15 mm from the tip. This was the first effect of light on some cellular parameter that correlated, as far as the fluence–response curve is concerned, with the phototropism of *Avena* coleoptiles.

Thimann and Sweeney (1938) found, in the same plant and tissue, that IAA at low concentration (3×10^{-7} *M*) accelerates the streaming, while at high concentration it inhibits it (the null point being at $1\text{–}2 \times 10^{-6}$ *M*). The observed region was at ~ 1 cm below the tip of the coleoptile. The lag in both cases was less than 2 min and the effect was transient, disappearing after 30 min. In a subsequent paper, Sweeney and Thimann (1938) studied the effect of oxygen and sugar on this process and suggested that (1) the transient nature of this effect is attributable to temporary exhaustion of carbohydrates from the tissue, and (2) auxin accelerates an oxygen-consuming process that controls the rate of protoplasmic streaming, which controls growth.

That IAA induces growth in coleoptiles was known since the early 1930s, but Ray and Ruesink (1962) were the first to do short-term kinetic studies with *Avena* stumps; they found that IAA (2×10^{-5} *M*) induces elongation with a lag of 13 min at 23°C, while at 13°C the lag was 28 min. The IAA induction was inhibited both by cyanide as well as by anaerobiosis. Evans and Ray (1969) found a lag of 10 min in *Avena* and of 10–15 min in *Zea*, at the same concentration of IAA. Cycloheximide inhibits the elongation with a lag of 5 min. Cycloheximide is a potent inhibitor of protein synthesis in eukaryotes, but it may be not very specific in plants (MacDonald and Ellis, 1969; Ellis and MacDonald, 1979; Timberlake *et al.*, 1972; Timberlake and Griffin, 1974). Hager *et al.* (1971), working with *Avena*, reported a lag of 10 min in stimulating elongation with 10^{-5} *M* IAA at 30°C. They also found that in anaerobiosis there is no stimulation by IAA unless ITP or GTP are present. Their interpretation of these and other data are discussed later in this section. Philipson *et al.* (1973*a*,*b*) reported a lag of 10–11 min with 10^{-5} *M* IAA at pH 7 and 25°C. Tietze-Haß and Dörffling (1977), using a sensitive apparatus, could show that the elongation of *Avena* stumps after 120 min in buffer was 0.36 ± 0.02 μm/min · segment (the segments were 5 mm long). Addition of IAA after these 120 min, at a given concentration ($10^{-9}\text{–}10^{-3}$ *M*), induced an initial inhibition of elongation that increased with increasing IAA concentrations (50% at 10^{-9} *M* up to complete inhibition at 10^{-3} *M*). The lag was shorter than 2 min. After 10 min, at all concentrations above 10^{-8} *M* IAA, there was an acceleration the maximum of which was as large as 5 μm/min · segment (almost 14 times faster than the control) at 10^{-6} *M* and 30 min. The results of Evans and Ray (1969) in *Zea* were confirmed by Hertel *et al.* (1969), who also showed the specificity of IAA both in transport and in elongation (using different molecules that display auxin or nonauxin activity). Rayle *et al.* (1970*a*) found that the early effect in *Zea* is dependent on IAA concentration. At 2×10^{-7} *M* IAA 4-mm-long segments are accelerated in their elongation after 2 min (results reproducible only 50% of the time according to the authors)

At 5×10^{-5} *M* IAA, there is a 50% inhibition after less than 2 min, and an acceleration after 15 min.

Ullrich (1978) measured bending instead of growth in intact coleoptiles of *Zea* applying IAA on one side with a small paper strip. He observed a bending toward the side of application in less than 3 min and then away from the IAA side after about 30 min; he also checked the specificity using auxins and antiauxins.

To conclude this overview on the effects of IAA on coleoptile elongation, we can say that there is a rapid but slight inhibition in elongation followed by a large increase after 10–15 min.

How does IAA induce elongation? There are two hypotheses:

1. The gene-expression hypothesis (see the reviews of Trewavas, 1968; Key, 1969; Stoddart and Venis, 1980).
2. The wall-acidification hypothesis independently proposed by Hager *et al.* (1971) and Cleland (1971).

The first hypothesis states that IAA induces some new proteins necessary for elongation. The main drawback of this hypothesis is that the shortest lag of an IAA-induced change in the pattern of protein synthesis is at least 3 hr (Zurfluh and Guilfoyle, 1980), while the IAA effect on elongation is much sooner. This might be explained by insufficiently sensitive measurement of IAA-induced changes in protein synthesis.

The wall-acidification hypothesis states that IAA induces an acidification of the wall (proton secretion) and the low pH induces elongation. Hager *et al.* (1971) were even more specific and suggested that IAA activates an anisotropic ATPase or proton pump.

That low pH can accelerate the elongation of *Helianthus* roots and hypocotyls and the *Avena* coleoptiles has been known for some time (Strugger, 1932; Bonner, 1934, respectively).

Rayle and Cleland (1970) have shown that a change of pH from 7 to 3 accelerates the elongation of *Avena* coleoptiles after only 1 min, but the pH does not work as a trigger; a change back to pH 7 will immediately restore the original elongation rate. This induced elongation can also happen *in vitro* with frozen-thawed *Avena* coleoptile segments in which the turgor is replaced by a constant applied force (Rayle *et al.*, 1970*b*). Rayle and Cleland (1972) have evidence that protons hydrolyse the cell wall of *Avena* coleoptiles *in vitro* in a nonenzymatic way. The acid-induced growth is also present in anaerobiosis (Hager *et al.*, 1971). With peeled *Avena* coleoptiles (without cuticle and epidermis), the IAA-induced growth can be inhibited by 10 m*M* buffer concentration (Durand and Rayle, 1973). With normal coleoptiles, this is not true, which was regarded as an indication that cuticle and epidermis are a barrier for H^+ ions.

The next step in order to test the wall-acidification hypothesis was to see whether IAA induces H^+ secretion. That was indeed found for peeled *Avena* coleoptiles by Clelend (1973) and by Rayle (1973); both investigators reported a lag of 20 min. Rayle also checked the specificity of this effect showing that the antiauxin TIBA (2,3,5-triiodobenzoic acid), which is known to inhibit IAA transport (Hertel and Leopold, 1963) and IAA stimulation of elongation (his own paper), inhibits H^+ secretion. Both Cleland and Rayle found that cycloheximide inhibits this secretion. Similar experiments were done with *Zea* coleoptile segments (Jacobs and Ray, 1976). Cycloheximide blocks the H^+ secretion in *Avena* after 5 min (Cleland, 1975). IAA (10^{-5} *M*) can cause an hyperpolarization of the membrane potential of *Avena* coleoptiles of 25 mV after a lag of 7–8 min (Cleland *et al.*, 1977). Finally, Bates and Cleland (1979) showed that for peeled *Avena* coleoptiles three different protein synthesis inhibitors can inhibit IAA-induced H^+ secretion with 5–8 min lag and IAA-induced growth after 8–10 min lag (the three inhibitors have no effect on respiration). This last work seems to exclude the possibility that inhibition is caused by an unknown side effect, hence one has to assume that protein synthesis is necessary both for the IAA-induced H^+ secretion as well as for the IAA-induced elongation. As Bates and Cleland pointed out, it is still not clear whether IAA induces the production of some protein(s) or whether the protein synthesizing machinery has to function well in order to have such IAA effects. One year later, Bates and Cleland (1980) reported data that are not in disagreement with the model of Ray (1977), described in Section 3.4.7.

We are not aware of any experiments in which a dose–response curve of IAA-induced H^+ secretion was done and compared with the same curve for IAA-induced elongation. The only data on this subject are the unpublished data from Hertel's laboratory quoted by Hertel (1981). He claimed that at 10^{-6} *M* the auxin analogue 1-NAA does not induce proton secretion, while there is good induction of elongation in *Zea* coleoptiles.

The IAA-stimulated ATPase hypothesis of Hager *et al.* (1971) was challenged experimentally by Cross *et al.* (1978). However, their data (which show partially separated peaks of 1-NAA binding sites and of ATPase) are of limited significance in view of the fact that they used a microsomal fraction that probably contains different types of membranes instead of plasmamembrane alone.

3.4.7. Subcellular Sites of IAA Action

How is IAA polarly transported in the coleoptile? What is its molecular interaction with subcellular components of the coleoptile? Thomson *et al.* (1973) provided data that indicate that in *Zea* coleoptiles the site of action of IAA in elongation is different from the site of transport. Hertel *et al.* (1972), in a pioneer work, showed that there is binding of IAA and of α-NAA 1-naphthalene acetic acid (1-NAA), which has auxin properties in elongation, to particulate cell fractions of *Zea* coleoptiles. The isomer 2-NAA, 2-naphthalene acetic acid, (β-NAA), without auxin properties in elongation, did not bind. At that time, it was suggested that the binding, specific for auxin molecules, was at the plasmamembrane. It was already clear that even if this binding site had something to do with the physiological effects of IAA, it would not be the only binding site that would explain all the auxin and antiauxin effects. For example, two synthetic antiauxins for transport *N*-1-naphthylphthalamic acid (NPA) and TIBA (Thomson *et al.*, 1973), did not compete with the IAA binding. NPA was already known to bind to plasmamembranes (Lembi *et al.*, 1971). Batt *et al.* (1976) and Batt and Venis (1976) found two binding sites for 1-NAA in *Zea* coleoptiles, which could be distinguished kinetically and by sucrose gradient fractionation. Site 1 should be associated with the Golgi membranes and/or endoplasmic reticulum (ER), while site 2 should be associated with the plasmamembrane. Ray *et al.* (1977*a*) improved techniques in *Zea* after they found a supernatant factor (SF) that inhibited the binding of 1-NAA. The SF was identified as benzoxazolinones by Venis and Watson (1978), who also suggested that the SF cannot have a general role in auxin physiology because of its limited distribution in nature. Ray (1977) found that most of the 1-NAA binding is on the rough endoplasmic reticulum (RER) and suggested that the physiological effect of auxin is to bind to the RER, which should induce an increase of H^+ uptake in the ER cisternal space. The ER, probably via the Golgi apparatus, should fuse with the plasmamembrane and excrete the protons, which induce growth. Ray also suggested that for unknown reasons any protein synthesis inhibitor should stop the secretory outward flow of products from the ER space. So the Ray model could explain both the relatively long latency (10–20 min) in IAA-induced H^+ secretion and elongation and the effect of protein synthesis inhibitors on these two IAA effects. Ray *et al.* (1977*b*) checked the specificity of 1-NAA binding and found that out of 27 synthetic compounds, which have either auxin or antiauxin properties in coleoptile elongation, at least two-thirds of these compete with 1-NAA binding with a similar K_D. Dohrmann *et al.* (1978) found also a second binding site of 1-NAA (site II), which they suggested to be on the tonoplast. Jacobs and Hertel (1972) andd Hertel (1979, 1981) claimed to have observed a weak binding site also on the plasmamembrane (site III). The auxin receptor(s) could be solubilized and could be shown to be proteins (Venis, 1977; Cross *et al.*, 1978; Cross and Briggs, 1978, 1979).

To complicate the picture, Murphy (1980) also found binding of 1-NAA to proteins in a membrane-free preparation from *Zea* coleoptiles. It is too early to say whether these proteins are new binding sites or sites "involuntarily solubilized" from membranes.

The work of Weigl (1969*a*,*b*) is also interesting. He found a specific binding of IAA *in vitro* to pure lecithin (a major constituent of membranes) with a stoichiometric relation of 0.8 moles of IAA to 1 mole of lecithin. Veen (1974) confirmed the data of Weigl, but he stated that the binding of IAA to lecithin is not related to the primary action of the hormone in plant growth because biological inactive substances (e.g., β-NAA) also bind to lecithin. He also confirmed the stoichiometric binding of IAA and found that α-NAA has different binding properties than those of IAA, there were at least 8 α-NAA molecules per lecithin molecule. The fact that α-NAA binds in a way different from IAA indicates that the physiological effect could also be different. For example, the membrane could be more permeable with IAA than with α-NAA, and the latter could act on a different site. There is no strong physiological evidence that IAA and α-NAA act on the same site, *at least in elongation.* A recent review on molecular and subcellular aspects of hormone action (including auxin binding) is that of Stoddard and Venis (1980). They presented evidence that in other organisms there is also specific binding of IAA.

To conclude, it seems to us that the binding approach is a very exciting and rewarding means of understanding the molecular basis of IAA effects on coleoptiles, but it is still in its initial phase. All the work presented here was done with *Zea;* information on *Avena* coleoptiles would be useful.

4. PHOTOTROPISM IN LOWER PLANTS

4.1. Types of Phototropic Reactions

In contrast with higher plants, in which only a multicellular type of phototropism occurs, lower plants exhibit both multi- and unicellular types of phototropism (Page, 1968). In the latter type, perception of the light stimulus and bending response occurs within the same cell.

All bending mechanisms in tropistic responses of plants fall into two categories (Fig. 6), first expounded by Jaffe and Etzold (1965; but cf. Stadler, 1952; Green *et al.*, 1970):

1. *Bowing* proceeds by differential growth on both sides of the organ or cell. If the diameter does not change considerably, the faster-growing side will become the convex side, so that the axis of the organ or cell is simply bent away from this side. In higher plants only this type of bending occurs.
2. *Bulging* means that the prospective concave side will achieve a more rapid growth either by a shift of the growth center from the original tip or by formation of a new growth center excentrically to the original axis. The latter type was proposed to occur in all organs or cells with

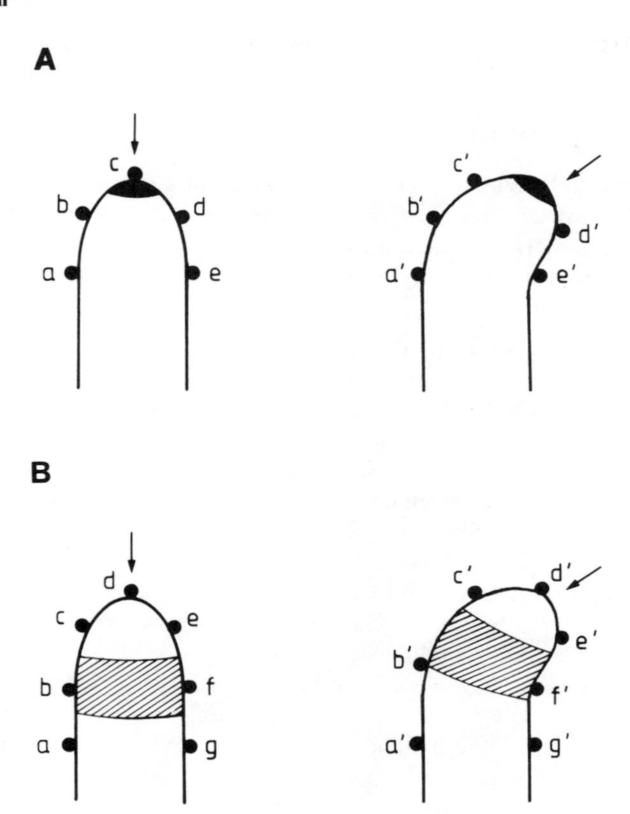

Figure 6. Comparison of growth and bending types. The black spots a–g and a′–g′ represent markers before and after bending has occurred, respectively. The arrows indicate the direction from which the stimulus (e.g., light) is given in a positive response. On the left side we show the characteristics of straight and on the right side those of bending growth. Examples for both modes of bending are given in Sections 4.2–4.5. (A) Extreme tip growth (the growing apex is marked black), which often, but possibly not always, implies a bulging type of bending. Note that the distance between c′ and d′ greatly exceeds the distance between c and d before bending. The apical growing zone has been shifted to the proximal side. (B) Growth in a subapical zone (hatched area), which always implies a bowing type of bending. Note that the distances between the distal markers a′, b′, and c′ exceed those between the proximal markers e′, f′, and g′ (differential growth).

tip growth (Haupt, 1965; but cf. Page and Curry, 1966; see also our comments in the following sections). For a mathematical treatment of the bending modes, see Hejnowicz and Sievers (1971). Phototropism in lower plants can be caused by both bending mechanisms. The occurrence of bulging makes the estimation of bending rates nearly impossible (Steiner, 1969*a,c*) and often hinders the determination of fluence–response curves (Nebel, 1968).

4.2. Algae

4.2.1. General Survey

Positive and negative phototropism occurs in algae. In *Fucus* and *Mougeotia,* polarotropism has been detected.* In most cases investigated a unicellular type of phototropic reaction is involved. It is to this type of reaction that our consideration is restricted.

4.2.2. Growth and Bending Mechanisms

4.2.2.a. Growth Mechanism and Light–Growth Response. Filamentous algae grow mostly by tip (terminal) and rarely by intercalary (subapical) growth. The apical cells of red-light-grown filaments of *Mougeotia* are assumed to grow at the extreme tip, but the intercalary cells show intercalary growth, as the whole filament grown under blue light (Neuscheler-Wirth, 1970; Haupt, 1977). The polarotropic reactions point to a stimulation of growth by light (Neuscheler-Wirth, 1970; Haupt, 1977).

Tip growth was proved for filaments of *Vaucheria geminata* (Vauch.) De Condolle by two methods (Kataoka, 1975*a*; Fig. 1B). There was no cellulose staining by Calcofluor in walls of growing tips, indicating that this is the region of wall synthesis. Experiments with marker particles showed maximum growth at the extreme tip, at the hyaline cap (Kataoka, 1980). No growth occurred behind the hemispherical apex (Kataoka, 1975*a*). Under split-field illumination, the tip always bends to the side irradiated by blue light (Kataoka, 1975*b*). Formerly such behavior was regarded as indicative of a negative light–growth response (Dassek, 1939; Gettkandt, 1954); however, this is only true for a differential growth mechanism (cf. Kataoka, 1975*b*, 1980). Exposure of the whole tip to different regimes of red and blue light with equal fluence rates invariably caused a higher growth rate under blue light (Kataoka, 1975*b*, i.e., a positive blue-light–growth response).

The rhizoids of *Boergesenia forbesii* (Harvey) Feldmann the growing zone is restricted to the apical 20 μm, as shown by Ishizawa and Wada (1979*a*) in experiments with marker particles. Nearly uniform irradiation with phototropically inactive red light promoted growth, but blue light inhibited it. This growth inhibition is even more evident under unilateral irradiation from the front of the tip (Ishizawa and Wada, 1979*a*).

*In polarotropism the organ or cell grows in a plane perpendicular to the direction of irradiation; its growth direction is not oriented by direction of light, but by its plane of polarization (Jaffe, 1958; Ezold, 1961; Jaffe and Etzold, 1965; Haupt, 1977; Dennison, 1979).

4.2.2.b. Bending Mechanism. In many cases, algal phototropism proceeds by the bulging mechanism. Neuscheler-Wirth (1970) proposed this for red-light photo- and polarotropism of apical *Maougeotia* cells. In unpolarized red light only the intercalary cells of *Mougeotia* filaments bend by bowing. However, in unpolarized or polarized blue light all cells bend by bowing. Therefore, only *Mougeotia* and possibly the fungus *Pilobolus* display both kinds of bending mechanisms (Page and Curry, 1966; Neuscheler-Wirth, 1970).

Experiments using markers on the growing tip of *Vaucheria* filaments showed a bulging mechanism under split-field spot or spiral stripe illumination: The growth center shifts to the illuminated flank (Kataoka, 1975*b*, 1980). The positive blue-light–growth response is most probably directly involved in the positive phototropism by bulging. This is in contrast with the interpretation of Briggs and Blatt (1980). Moreover, *Dryopteris* filaments exhibit exactly such growth and bending responses under red-light irradiation (Etzold, 1965).

Bulging, but by the probable formation of a new growth center on the shaded side, was demonstrated in experiments with marker particles for the negative phototropism of *Boergesenia* rhizoids coinciding with the observed growth inhibition by blue light (Ishizawa and Wada, 1979*a*).

4.2.3. Kinetics of Bending

In algal phototropism, almost no kinetic data on bending are available, but observations done hours to days after start of illumination point to a rather slow process.

Vaucheria phototropism is an exception: After a lag time of 1–2 min the hyaline cap shifts rapidly (3–4°/min) under continuous split-field irradiation. This continues more slowly about 1°/min) for 10–20 min after turning off the light (Kataoka, 1975*a*).

The negative phototropism in *Boergesenia* rhizoids is strongly affected by temperature (Ishizawa and Wada, 1979*a*). At 26°C the lag period is about 1 hr, bending proceeds under continuous irradiation (467 nm, 74 mW/m^2) with about 10°/hr; on turning off the light, a spontaneous reverse bending occurs. The same conditions at 32°C result in no apparent lag time, a bending speed of about 35°/hr and a decrease or stop of bending after turning off the light. At 26°C a 10-fold increase of the fluence rate diminishes the lag time and accelerates the bending speed considerably.

4.2.4. Fluence–Response Behavior

For the positive phototropism of *Vaucheria* filaments, Kataoka (1975*a*) established a fluence rate–response curve only at 450 nm with his split-field illumination technique.* This method is an improvement of the half-side illumination technique of Buder (1920). Hence Kataoka is not measuring phototropism in the strict sense; but, because unilateral illumination evokes a true positive phototropic response (Kataoka, 1975*b*), the bending in the split field may be considered phototropism in a broad sense (Briggs and Blatt, 1980; Kataoka, 1980). It follows the Bunsen-Roscoe law for threshold bending ($30 J/m^2$) and the Weber-Fechner law below 26° (Kataoka, 1975*a*, 1980). Use of higher fluence rates revealed a complex curve with a second positive and a negative curvature (Kataoka, 1977*a*; cf. Section 3.1 for *Avena* coleoptiles and Section 4.4.4 for *Dryopteris chloronemata*). Stimulation by intermittent blue light revealed a summation of the light pulses, i.e., phototropic response to continuous light equals that to intermittent light of the same fluence (Nathansohn and Pringsheim, 1908) in certain time intervals (Kataoka, 1979), but within a narrow range of pulse durations and dark intervals, the dark period is not perceived.

Validity of the Weber-Fechner law for curvatures between 0° and 40° and the Bunsen-Roscoe law for a considerable range of fluence rates and irradiation durations was shown in the negative phototropism of *Boergesenia* rhizoids at 467 nm and 32°C (Ishizawa and Wada, 1979*b*). The slopes of the fluence–response curves change with wavelength similar to liverwort and fern polarotropism curves (cf. Sections 4.3.4 and 4.4.4) and similar to the *Avena* coleoptile curves (cf. Section 3.3.1; but see Curry and Gruen, 1961).

Table I shows the types of bending and the thresholds of the filaments and rhizoids discussed.

4.2.5. Photoreception

4.2.5.a. Spectral Sensitivity and Action Spectra. Different phototropic and polarotropic responses to red and blue light were shown in filamental cells of *Mougeotia* (Neuscheler-Wirth, 1970). In unpolarized unilateral blue light the apolar cells are negatively phototropic, whereas in red light the apical polar cells are positively phototropic, while the intercalary cells are negatively phototropic. In polarized red light, growth is promoted perpendicular to the electrical vector. However, in polarized white light (high content of blue) it is parallel to the vector.

In *Vaucheria sessilis* positive phototropism of filaments was caused by white, monochromatic blue and green, but not red light (Weber, 1958).

A typical blue-light action spectrum for positive phototropism of *V. gem-*

*This is not a true fluence–response curve, because at some fluences the measurement took place immediately after the completion of irradiation, permitting insufficient time for the completion of bending.

Table I
Bending Types and Phototropic Thresholds in Algae

Species and organ	Type of bending	Threshold (J/m²) at λ (nm)	Reference
Filaments of *Mougeotia*	Red light	—	Neuscheler-Wirth (1970)
	Bulging (apical cells)		
	Bowing (intercalary cells)		
	Blue light	—	
	Bowing		
Filaments of *Vaucheria geminata*	Bulging	30 at 450	Kataoka (1975*a*, 1977*a*)
Rhizoids of *Boergesenia forbesii*	Bulging	30 at 467	Ishizawa and Wada (1979*b*)

inata (Kataoka, 1975*a*, 1980) has been determined by a null-point balancing method in the split-field arrangement (for a description of the null-point balancing method, see Shropshire, 1972). A similar action spectrum was found for the related phenomenon of lateral branching (Kataoka, 1975*b*) and, in *V. sessilis* for chloroplast aggregation (Haupt, 1965; Briggs and Blatt, 1980). For the phototropic action spectrum, the fluence–rate responses have only been determined for one wavelength; furthermore, only one reference wavelength and one reference fluence rate have been used (cf. Shropshire, 1972; Hartmann, 1977).

For the typical blue-light action spectrum of negative phototropism of *Boergesenia* rhizoids (Ishizawa and Wada, 1979*b*), only a fixed response was used, though the slopes of the fluence–response curves change with wavelength. Hence selection of other curvatures should change the spectrum (cf. Steiner, 1969*b*,*d*). Red light promoted rhizoidal growth and altered the phototropic sensitivity. The inhibition of straight rhizoidal growth (Ishizawa and Wada, 1979*a*) and the induction of polarity of newly formed cells from artificially produced protoplasmic fragments (Ishizawa *et al.*, 1979) showed similar spectral sensitivities.

Table II shows the types of phototropism/polarotropism and the spectral sensitivities in four algal species.

4.2.5.b. Type and Localization of Photoreceptor. The involvement of a blue-light-type photoreceptor in algal phototropism has already been shown by the rough spectral data of Hurd (1920) *in Fucus*. The outer cytoplasmic layer was assumed to be the sensitive site in phototropism of the coencytic algae *Dryopsis* and *Derbesia* as early as 1887 by Noll.

In blue-light phototropism and polarotropism of apolar cells of *Mougeotia*

Table II
Types of Phototropism/Polarotropism and Spectral Sensitivities in Algae

Species and organ	Type of phototropism/ polarotropism	Spectral sensitivities, action maxima at λ (nm)	Reference
Filaments of *Mougeotia*	Negative	Blue	Neuscheler-Wirth (1970)
	Positive (apical cells) Negative (intercalary cells) Polarotropism: Perpendicular to electrical red vector (apical cells)	Red	
	Parallel to electric vector	White (with high content of blue light)	
Filaments of *Vaucheria sessilis*	Positive	Blue, green	Weber (1958)
Filaments of *Vaucheria geminata*	Positive	415, 450, 480	Kataoka (1975*a*)
Rhizoids of *Boergesenia forbesii*	Negative	380, 430sh[a], 443, 470sh	Ishizawa and Wada (1979*b*)

[a]sh, Shoulder.

the unknown photoreceptor is assumed to be dichroic and oriented in a surface-parallel way. A similar conclusion was drawn for the photoreceptor in red-light photo- and polarotropism (Neuscheler-Wirth, 1970). These speculations were based upon the hypotheses set up for the similar red-light photo- and polarotropism of the chloronemata of *Funaria* (Jaffe and Etzold, 1965) and of *Dryopteris* (Etzold, 1965). However, in *Mougeotia* only the effect of red, but not of far-red light could be tested (Neuscheler-Wirth, 1970). This similarity may be more pronounced: Phytochrome could be the photoreceptor in all three cases. In *Mougeotia* it controls the low-fluence chloroplast movement: its involvement, probable localization close to or associated with the plasmalemma, and change of orientation between the P_r (red-light-absorbing) and P_{fr} (far-red-light-absorbing) forms are discussed by Haupt (1965, 1968, 1972, 1977, 1980), Haupt *et al.* (1969), Hartmann and Haupt (1977), and Shropshire (1979).

Action spectra indicated blue-light photoreceptors in phototropism of filaments of *V. geminata* (Kataoka, 1975*a*,*b*, 1980) and of rhizoids of *B. forbesii* (Ishizawa and Wada, 1979*b*). The same, or the same kind, of photoreceptor is thought to be involved in the aggregation of chloroplasts in *Vaucheria* (Haupt, 1965; Briggs and Blatt, 1980). In *V. geminata* the phototropic sensitivity is localized in the outermost cytoplasmic layer of the apical half of the hemispherical tip (Kataoka, 1975*a*) cf. photoreceptor localization of chloroplast movement (Fischer-Arnold, 1963). This points to a probable important role of the apical plasmalemma (Kataoka, 1980). On the basis of these results and assuming a dichroic photoreceptor, an influence of differently polarized blue light might be expected (cf. Bentrup, 1968), but this has not been detected (Kataoka, 1975*a*,*b*). Briggs and Blatt (1980) speculated that this failure could have been due to the use of split-field instead of unilateral illumination.

Concerning the chemical nature (flavinoid or carotenoid type) of the "unique" blue-light photoreceptor, which is also found in algae, refer to Section 5.

4.2.5.c. Optics and Formation of Gradient of Light Intensity. In *Mougeotia,* Neuscheler-Wirth (1970) proposed that no lens effect would work and that screening, mainly by the chloroplast, created the light-intensity gradient for red- and blue-light phototropism in unilateral light. Alternatively, Haupt (1977) proposed a lens effect for the red-light phototropism of apical cells. The absorption gradient should be attributable to the principle of the dead space (i.e., the incident beams bypass the sensitive apical part of the cell) as well as to the principle of dichroic orientation (i.e., involvement of oriented dichroic photoreceptor molecules) (cf. Jaffe, 1958; Etzold, 1965). The absorption gradient for blue- and red-light polarotropism results probably from the latter principle with both kinds of photoreceptor molecules periclinally oriented (Neuscheler-Wirth, 1970; in partial analogy to Etzold, 1965; cf. Haupt, 1977).

In media of different refractive indices, filaments of *V. geminata* curve toward unilateral blue light, although the hyaline cap should act as a lens and the refractive index of the tip should allow an inversion of phototropism (Kataoka, 1975*b*). But the apical half of the dome—the only sensitive part for the bending response—is apparently bypassed by the beams (principle of the dead space), so that the side facing the light always receives more light. In contrast to Dennison (1979), we believe the situation seems clear.

A lens effect in negative phototropism of the rhizoids of *Boergesenia* was excluded (Ishizawa and Wada, 1979*a*): The rhizoid is hardly transparent and has only a small hyaline cap (Kataoka, 1980). Thus, the intensity gradient seems to be established by screening (absorption, reflection and/or scattering).

4.2.6. Transduction of Stimulus

In the examples given of algal phototropism, there is no need for a spatial transmission of the stimulus, because perception and reaction occur in the same region (e.g., Kataoka, 1977*a*). Without doubt, auxin occurs in algae and affects growth in some cases. Therefore, it could be involved in the phototropic transduction process (Banbury, 1959). But up to now this problem seems to be unsolved.

Electro- and phototropism in *Fucus* rhizoids are correlated (Bentrup, 1968). The transduction of the light stimulus was proposed to proceed via changes in the potential difference at the plasmalemma. The occurrence of light-stimulated K^+ fluxes in the red alga *Rhodymenia palmata* (L.) Grev. (MacRobbie and Dainty, 1958) and of blue-light-induced local electrical current effluxes in *V. sessilis,* supposedly an early photosensory transduction step in chloroplast aggregation (Blatt and Weisenseel, 1980), possibly support this view.

In *Mougeotia* phototropism phytochrome could be involved as it is in the low fluence rate chloroplast movement (Neuscheler-Wirth, 1970). If this were true, the hypotheses on that sensory transduction chain might be transferable to the red-light phototropism: The change in the dichroic orientation of the phytochrome molecules on interconversion from P_r to P_{fr} may bring about a physical change in the membrane, thereby influencing permeability and bioelectrical potentials (Haupt, 1968, 1972, 1980).

The exclusion of a red-light (phytochrome) influence on the phototropic transduction process in *Vaucheria* (Kataoka, 1975*a*) may be premature. In a very similar case, *Sphaerocarpus* polarotropism (cf. Sections 4.3.4 and 4.3.6), Steiner (1969*c*,*d*, 1970) presented evidence for a phytochrome effect. Moreover, phototropically ineffective red light affects phototropic sensitivity in higher plants (cf. Section 3.3.2.) and in *Boergesenia.*

Exogenously supplied IAA and zeatin affected neither growth nor phototropism, but cAMP, which is present in the cells at about 3×10^{-6} *M*, seems to regulate phototropic sensitivity or amplification mechanisms in *Vaucheria* (Kataoka, 1977*b*). Supposedly increased intracellular levels of cAMP lowered sensitivity and vice versa, whereas growth was hardly influenced. So the change of growth direction (steering process) is separate from growth itself. These, although somewhat speculative (Briggs and Blatt, 1980), results suggest that here, as in *Phycomyces* (Cohen, 1974*a*), cAMP could be a part of the sensory transduction chain. Pulsating irradiations revealed two different dark reactions in the phototropic transduction chain of *Vaucheria* (Kataoka, 1979).

In the negative phototropism of *Boergesenia* rhizoids, the steering process and the growth itself were separated by use of different temperatures for irradiation (18°C) and the bending response (25°C; Ishizawa and Wada, 1979*a*). The steering process of unknown chemical or physical nature is less sensitive to temperature than growth, so that at 18°C only photoreception and steering processes take place. At 18°C the effect could be preserved by prolonged irradiation (called *stored bending*) and decayed with a half-life of 1.5 hr. Ishizawa and Wada (1979*b*) and Ishizawa *et al.* (1979) stated without giving data that red light, possibly via phytochrome, altered the sensitivity in phototropism and polarity induction, respectively, although it is phototropically inactive (Ishizawa and Wada, 1979*a*).

4.3. Liverworts and Mosses

4.3.1. General Survey

According to the reviews of du Buy and Nuernbergk (1932) and Banbury (1959), negative and positive phototropism and plagiophototropism occur in the Bryophytes (liverworts and mosses). Polarotropism is seen according to Haupt (1965), Thomas (1968), Briggs and Rice (1972), Dennison (1979), and Hertel (1980). Both the multi- and the unicellular types of phototropic reaction are observed in the Bryophytes, but the following discussion is restricted to the latter type.

4.3.2. Growth and Bending Mechanisms

4.3.2.a. Growth Mechanism and Light–Growth Response. As early as 1889, Haberlandt detected the exclusive growth at the 10–13 μm long rhizoid apices of the liverworts *Marchantia polymorpha* and *Lunularia vulgaris.* Subsequent workers and reviewers in the field agreed with this view (e.g., du Buy and Nuernbergk, 1932; Dassek, 1939; Banbury, 1959).

In analogy to other cryptogam germlings, the germ tubes of *Sphaerocarpus donnellii* Aust. were assumed to posses an apical growth center (Steiner, 1969*c*; Fig. 1C).

In the chloronema of *Physcomitrium turbinatum* (Michx.) Britt. only red and moderately UV, but not blue light promoted growth (Nebel, 1968). At 730 nm there was no growth promotion; an influence of two red-light-absorbing systems (possibly the photosynthetic and a photomorphogenetic one) was supposed.

4.3.2.b. Bending Mechanism. Generally bulging mechanisms have been assumed to evoke bending in moss and fern chloronemata by Jaffe and Etzold (1965) and by Haupt (1965). Specifically this was also assumed for *Sphaerocarpus* germ tubes (Steiner, 1969*c*) and chloronemata of *Funaria* (Jaffe and Etzold, 1965) and *Physcomitrium* (Nebel, 1968). On the other hand, this mechanism has only been proved unequivocally for chloronemata of the ferns *Dryopteris* (Etzold, 1965) and *Pteridium* (Davis, 1975) and for *Pilobolus* stage I sporangiophores (Page, 1962, but cf. Page and Curry, 1966). Probably the assumptions are correct but, because a bowing mechanism can not be totally excluded (see, e.g., Banbury, 1959), further investigations on this topic are desirable.

4.3.3. Kinetics of Bending

No kinetics for the bending rate can be given in *Sphaerocarpus* polarotropism because of the bulging mechanism involved (Steiner, 1969*c*); the same should hold true for *Physcomitrium* phototropism (Nebel, 1968); but the times for completed bending (4 and 1.75 hr after start of irradiation, respectively) indicate that the lag time of the growth reorientation is relatively short and reorientation itself relatively fast.

4.3.4. Fluence–Response Behavior

The fluence–response curves for polarotropism of the germ tubes of the liverwort *Sphaerocarpus* were of similar shape irrespectively of whether the fluence rate at 460 nm was held constant and the duration of illumination varied, or *vice versa* (Steiner, 1967*a*). For certain ranges of both relationships, the Weber-Fechner law, but not the Bunsen-Roscoe law, is valid. There is no red light effect on the fluence–response curves at 392–460 nm (Steiner, 1969*c*,*d*). The fluence–response curves exhibited steeper slopes in the blue than in UV and blue-green ranges and blue-light threshold values of $\sim 10^{-2}$ J/m^2 (Steiner, 1969*c*). A suspected influence of phytochrome on the fluence–response behavior in the UV range has been suggested (Steiner, 1969*d*, 1970).

Table III shows the types of bending and the thresholds of the two objects discussed.

4.3.5. Photoreception

4.3.5.a. Spectral Sensitivity and Action Spectra. Only wavelengths below 500 nm were effective in negative phototropism of some liverwort rhizoids (Dassek, 1939). Red light evoked polarotropism in chloronemata of *F. hygrometrica* (Bünning and Etzold, 1958; Jaffe and Etzold, 1965).

The polarotropism in germ tubes of *S. donnellii* is caused only by blue-light and UV irradiation (Steiner, 1967*a*,*b*, 1969*c*,*d*, 1970). The shape of the action spectrum differs according to the chosen response level because the slopes of the fluence–response curves change with wavelength (Steiner, 1969*d*, 1970).

Table III
Bending Types and Phototropic Thresholds in Bryophytes

Species and organ	Type of bending	Threshold (J/m^2) at λ (nm)	Reference
Germ tubes of *Sphaerocarpus donnellii*	(Bulging)[a]	10^{-3} at 442 nm	Steiner (1969*c*,*d*)
Chloronemata of *Physcomitrium turbinatum*	(Bulging)[a]	—	Nebel (1968)

[a]Parentheses, assumed.

Table IV
Types of Phototropism/Polarotropism and Spectral Sensitivities in Bryophytes

Species and organ	Types of phototropism/polarotropism	Spectral sensitivities, action maxima at λ (nm)	Reference
Germ tubes of *Sphaerocarpus*	Polarotropism, perpendicular to the electrical vector	392, 442, 460sh[a]	Steiner (1969*c*)
Chloronemata of *Funaria hygrometrics*	Polarotropism, perpendicular to the electrical vector	Red	Bünning and Etzold (1958); Jaffe and Etzold (1965)
Chloronemata of *Physcomitrium*	Positive	(365)[b], 680, 730[c]	Nebel (1968)

[a]sh, shoulder.
[b]Parentheses, Minimal action.
[c]Null-point method with standard irradiation at 665 nm.

A totally different action spectrum was determined for the positive phototropism of chloronemata of the moss *Physcomitrium* with a null-point balancing method (Nebel, 1968). With the standard light at 665 nm there was a high activity in the red and far-red, no activity in the blue, and a very moderate one in the UV range. The action spectrum for photogrowth, though mainly with activity in the red, differed markedly. A different shape for the phototropic action spectrum above 680 nm was obtained, if the standard irradiation was at 730 nm: no bending took place with test irradiations above 700 nm because there was no growth. Interpretation of the action spectra was complicated by the screening by chlorophyll and by the assumed importance of cycling of the suggested photoreceptor, phytochrome (Nebel, 1968, 1969).

Table IV shows the types of phototropism/polarotropism and the spectral sensitivities of three liverworts and mosses discussed here.

4.3.5.b. Type and Localization of Photoreceptor. Only a blue-light-type photoreceptor is directly involved in polarotropism *Sphaerocarpus* germ tubes (Steiner, 1967*a*,*b*, 1969*c*,*d*). Its identification is difficult because of the fluence–response behavior at different wavelengths pointing to an influence of phytochrome (Steiner, 1969*d*, 1970), but from the action spectrum at medium response levels, Steiner (1969*d*) speculated that a flavin may be a candidate. In analogy to other polarotropic systems dichroic-oriented photoreceptor molecules in the outer cytoplasm and close to the cell wall were postulated (Steiner, 1969*d*).

Jaffe and Etzold (1965) speculated that phytochrome could be the photoreceptor for positive phototropism of *Funaria* chloronemata. Nebel (1968, 1969) detected only a red-light-type photoreceptor, probably phytochrome (based on action spectrum and synergism of red and far-red light for proposed cycling, but cf. Dennison, 1979), for positive phototropism of *Physcomitrium* chloronemata. Results of irradiations with polarized light were interpreted as evidence for dichroic, disc-shaped photoreceptor molecules oriented parallel and close to the cell surface. He found no evidence for a different orientation of red- and far-red absorbing molecules as in *Mougeotia* (Haupt, 1968; Haupt *et al.*, 1969) and *Dryopteris* (Etzold, 1965), but cf. Shropshire (1979). More investigations are needed to settle the nature and orientation of this photoreceptor.

4.3.5.c. Optics and Formation of Gradient of Light Intensity. Analogous to *Mougeotia* and fern polarotropism (Etzold, 1965; Neuscheler-Wirth, 1970; Haupt, 1977), an absorption gradient mechanism can explain polarotropism in *Sphaerocarpus* best (Steiner, 1969*d*).

Neither paraffin nor the screening by chlorophyll (red-light-type photoreceptor) reversed or lowered the positive phototropism in *Physcomitrium* chloronemata (Nebel, 1968, 1969). Further investigations with polarized light (Nebel, 1969) showed that most absorption takes place at the proximal flank of the apex.

4.3.6. Transduction of Stimulus

Little is known on stimulus transduction in photo/polarotropism of *Bryophytes.* In *Sphaerocarpus* polarotropism sensitivity did not change after dark periods (Steiner, 1969*c*) and a phytochrome interaction (probably on the transduction) was suggested (Steiner, 1969*d*, 1970). Simultaneous excitation and consequent cycling of P_r and P_{fr} was assumed to be important in the phototropic transduction in *Physcomitrium* (Nebel, 1968, 1969).

4.4. Ferns

4.4.1. General Survey

Fern gametophytes and sporophytes exhibit phototropic responses: negative and positive phototropism, plagiophototropism, and polarotropism. The sporophytes' behavior was thought to be almost identical to that of Spermatophytes (Banbury, 1959). In ferns both the multi- and the unicellular type of phototropism are observed. Here a detailed view is given only for the *Dryopteris* gametophyte (unicellular type; Fig. 1d).

4.4.2. Growth and Bending Mechanisms

4.4.2.a. Growth Mechanism and Light–Growth Response. Rhizoids of protonemata of *Pteridium aquilinum* (L.) Kuhn exhibit a very extreme tip growth within the first apical 1 μm (Takahashi, 1961). Red light inhibited and blue light promoted rhizoid growth as compared with dark controls in *Dryopteris filix-mas* (L.) Schott; apical growth in the first 20 μm of the chloronema tip was suggested on the basis of bending measurements (Mohr, 1956*b*).

Tip growth in *Pteridium* chloronemata has been reported twice: Takahashi (1961) showed by micromanipulator marking that in multicellular chloronemata only the apical parts of the first three cells grow. Most growth occurred in the first 10 μm of the tip cell. But the growth conditions differed from those used for photo- and polarotropic experiments. Davis's results (1975) under red-light-bending conditions coincided. Red light promoted and blue light inhibited chloronemata growth in *Dryopteris* (Mohr, 1956*b*). A bending toward the illuminated flank of *Dryopteris* chlorenemata indicated a positive light–growth response on partial blue- and red-light illumination, and a bulging mechanism was proved independently (Etzold, 1965). By adoption of this hypothesis (i.e., the place of highest red-light absorption is the place of most vigorous growth), Hartmann *et al.* (1965) explained branching in chloronemata of *Dryopteris* and *Struthiopteris filicastrum* All. The obvious contradiction between Etzold's positive blue-light–growth response (1965) and Mohr's blue-light–growth inhibition (1956*b*; cf. Banbury, 1959) may be due to the much higher fluences Mohr used (Steiner, 1969*a*).

For further information on the light influence on fern growth, see Miller (1968) and Wada *et al.* (1978).

4.4.2.b. Bending Mechanism. Among photo- and polarotropic fern organs, which react according to the unicellular type, the mechanism of bending has only been investigated in chloronemata and has been found to be exclusively of the bulging type (in contradiction to Banbury, 1959).

Etzold (1965) proved for the first time the existence of bulging in fern phototropism, by observing rice grain marks on *Dryopteris* chloronemata. Preliminary electron microscopic observations revealed no concomitant changes in subcellular structures on bulging (Falk and Steiner, 1968). In *Struthiopteris* and *Dryopteris* chloronemata, polarized red light shifted the apical growth point to the flanks of the apex and also induced new growth centers on subapical filament flanks (Hartmann *et al.*, 1965). This was a confirmation of the bulging type reorientation of apical growth and offered a common explanation for both effects (Hartmann *et al.*, 1965). By the use of markings, phototropic bulging was proved in *Pteridium* chloronemata (Davis, 1975).

4.4.3. Kinetics of Bending

Hardly any data are available on the kinetics of photo- and polarotropic bending in fern. The bulging mechanism prevents the estimation of a bending rate (Steiner, 1969*a*; cf. Section 4.3.3). Thus, bending is in practice measurable only after subsequent straight growth. The final bending angles of *Dryoteris* chloronemata have been measured between 8 hr and 2 days (Mohn, 1956*b*; Etzold, 1965; Hartmann *et al.*, 1965; Steiner, 1969*a*,*b*, 1970) after change of irradiation conditions. Only Davis (1975) reported initial kinetics in bending of *Pteridium* chloronemata and found visible bulging 3 hr after a 90° shift of the incident unilateral light.

4.4.4. Fluence–Response Behavior

The only investigations of fluence–response relationships in ferns are those of Etzold (1965) and Steiner (1967*a*, 1969*a*,*b*, 1970), working on polarotropism in *Dryopteris* chloronemata. Both investigators found the Weber-Fechner law valid in lengthy irradiations of 8 hr (Etzold, 1965) and 4 hr (Steiner 1969*a*). The Bunsen-Roscoe law was invalid under similar conditions (Steiner, 1969*a*). The slopes of the fluence–response curves varied with wavelength (Etzold, 1965; Steiner, 1969*a*,*b*). Both workers determined threshold values of blue and red light. Although their data differ (Table V), both workers found that blue is 100 times more effective than red light.

A fluence–response curve with two rising parts separated by a minimum was indicated for 15-min irradiation with different fluence rates of blue light (Etzold, 1965; his Fig. 15 and discussion). This fluence–response behavior, now known from algal (cf. Section 4.2.4) and higher plant (cf. Section 3.1) phototropism and from polarity induction (e.g., Haupt, 1957, 1958; Haupt and Bentrup, 1961; Bentrup, 1963), was first published for short blue and red irradiations in *Dryopteris* polarotropism by Steiner (1967*a*). Moreover, variation in irradiation time and fluence rate gave rise to different shapes for the fluence–

Table V
Bending Types and Polarotropic Thresholds in Ferns

Species and organ	Type of bending	Threshold (J/m^2) at λ (nm)	Reference
Chloronemata of *Dryopteris*	Bulging	7×10^{-2} at 640[a]	Etzold (1965); Hartmann *et al.* (1965); Steiner (1969*a*,*b*)
		4 at 665[a]	
		3×10^{-2} at 376[b]	
		4×10^{-3} at 442, 483[b]	
		0.3 at 668[b]	
Chloronemata of *Struthiopteris*	Bulging	—	Hartmann *et al.* (1965)
Chloronemata of *Pteridium*	Bulging	—	Davis (1975)

[a]Data of Etzold (1965).
[b]Data of Steiner (1969*b*).

response curves: from ascending curves without maxima and minima to curves with two rising and two descending parts (Steiner, 1969*a*,*b*).

Different temperatures and pretreatments with darkness or red light affected the fluence–response behavior (Etzold, 1965; Steiner, 1969*a*). Red light given simultaneously with blue light revealed a phytochrome interaction in the blue range (Steiner, 1970; cf. the mirror-image behavior of *Sphaerocarpus* in the same work).

The types of bending and the thresholds of the chlorenemata of three species are given in Table V.

4.4.5. Photoreception

4.4.5.a. Spectral Sensitivity and Action Spectra. Dryopteris chloronemata exhibited phototropism in blue and red light (in 90% of cases it was positive and in 10%, negative; Mohr, 1956*b*; cf. Hartmann *et al.*, 1965), whereas the rhizoids did not react (Mohr, 1956*b*; Bünning and Etzold, 1958; Etzold, 1965) or were negatively phototropic in red light (Hartmann *et al.*, 1965). Red-light-positive phototropism also occurred in chloronemata of both *Struthiopteris* (Hartmann *et al.*, 1965) and *Pteridium* (Davis, 1975).

Polarotropism (chloronema growth perpendicular to the electrical vector) was described for *Struthiopteris* and *Dryopteris* in red light (Bünning and Etzold, 1958; Hartmann *et al.*, 1965) and for *Dryopteris* in red and blue light (Etzold, 1965; Steiner, 1967*a*,*b*, 1969*a*,*b*). On the other hand, *Dryopteris* rhizoids grew parallel to polarized red light (Hartmann *et al.*, 1965).

Action spectra exist only for polarotropism of *Dryopteris* chloronemata (Etzold, 1965; Steiner, 1967*b*, 1969*b*). Etzold's rough spectrum revealed maximum activity of blue and red light (but the latter 100 times less). The more detailed action spectra of Steiner (1967*b*, 1969*b*) generally coincide, but here a near UV peak was separated from two blue peaks, indicating a combination of a blue-light and a phytochrome action spectrum. Because the slopes of the temporally lengthly fluence–response curves varied with wavelength, the shapes of the spectra depended on the chosen response level (Etzold, 1965; Steiner 1969*a*,*b*). Some influence of phytochrome action also in the blue region was assumed (Steiner, 1970).

The types of phototropism/polarotropism and the spectral sensitivities of the species discussed here are shown in Table VI.

4.4.5.b. Type and Localization of Photoreceptor. All information on photoreceptors for fern photo- and polarotropism stems from investigations in *Dryopteris* chloronemata, except for one study on *Struthiopteris* (Hartmann *et al.*, 1965), in which red- and blue-light-type photoreceptors are involved.

The spectral sensitivities (Mohr, 1956*b*; Hartmann *et al.*, 1965), the action spectra (Etzold, 1965; Steiner, 1967*b*, 1969*b*), and the red/far-red

Table VI
Types of Phototropism/Polarotropism and Spectral Sensitivities in Ferns

Species and organ	Types of phototropism/ polarotropism	Spectral sensitivities, action maxima at λ (nm)	Reference
Chloronemata of *Pteridium*	Positive	Red	Davis (1975)
Chloronemata of *Struthiopteris*	Positive; polarotropism, perpendicular to the electrical vector	Red	Hartmann *et al.* (1965)
Rhizoids of *Dryopteris*	Negative; polarotropism, parallel to the electrical vector	Red	Hartmann *et al.* (1965)
Chloronemata of *Dryopteris*	Positive (10% negative)	Red, blue	Mohr (1956*b*)
		Red	Bünning and Etzold (1958); Hartmann *et al.* (1965)
	Polarotropism, perpendicular to the electric vector	Red, blue	Etzold (1965); Steiner (1967*a,b*, 1969*a,b*)
		460, 665	Etzold (1965)
		376, 442/451,[a] 483, 668	Steiner (1969*b*)

[a]Maximum shifted with increasing response level from 451 to 442 nm.

interrelationships (Etzold, 1965; Steiner, 1969*a*, 1970) established phytochrome as the red-light photoreceptor. In *Dryopteris* phytochrome is involved in the promotion of germination (Mohr, 1956*a*) and in morphogenesis (Mohr, 1956*b*; Etzold, 1965) and influenced polarotropism in the blue range (Steiner, 1970).

In photo- and polarotropism the additional and more effective participation of a blue-light photoreceptor is evident (Mohr, 1956*b*; Etzold, 1965; Steiner, 1967*a,b*, 1969*a,b*, 1970). Steiner's speculations (1967*b*, 1969*a*) that this blue-light photoreceptor could possibly be phytochrome in a high-irradiance reaction (HIR, cf. Hartmann and Haupt, 1977) were probably wrong. The action spectrum and the different efficacies of red and blue light disagree with this concept (Hartmann and Haupt, 1977; Haupt, 1977; Dennison, 1979). Probably a blue-light photoreceptor (flavinoid type), as already proposed by Mohr, is active (Mohr, 1956*b*; Etzold, 1965; Steiner, 1969*b*, 1970), but no unequivocal decision can be made because of the complex, phytochrome influenced fluence–response behavior (Steiner, 1969*a,b*, 1970). On the other hand, Hartmann (1977), based on Etzold's and Steiner's results, proposed after theoretical considerations a photochromic blue-light photoreceptor system (cf. Section 4.5.5.b concerning a similar proposal for *Phycomyces* phototropism). Etzold's speculations (1965) on different photoreceptors are not conclusive because of different thresholds in morphogenesis and polarotropism. Various blue-light effects in *Phycomyces* exhibit very different thresholds, although on the basis of genetic analysis, the same photoreceptor was suggested for all effects (Russo *et al.,* 1980). So this behavior may be caused by different sensitivities of separate transduction steps or by different localizations of the photoreceptor, Etzold's second proposal.

The blue- and red-light photoreceptor molecules were assumed to be dichroic and located close to or within the cell wall (Etzold, 1965; Hartmann *et al.,* 1965; Steiner, 1969*b*): the plasmalemma is a candidate as the site for phytochrome (Briggs and Rice, 1972). This localization seems to exclude carotenoids as blue-light photoreceptors, because in ferns they are exclusively found in plastids (Steiner, 1969*b*). Phytochrome does not seem to be restricted to the chloronemal tip, because branching and bending were induced under the same red-light conditions, but an action spectrum is lacking (Hartmann *et al.,* 1965). Irradiations with various combinations of unpolarized and differently polarized red and far-red light revealed that the P_r molecules are oriented parallel to the cell surface (Etzold, 1965; cf. Hartmann *et al.,* 1965; Steiner, 1969*b*) and the P_{fr} molecules perpendicular to it (Etzold, 1965): Etzold proposed a 90° rotation of the main absorption axis on interconversion of P_r to P_{fr} (cf. *Mougeotia* chloroplast movement: Haupt, 1968; Haupt *et al.,* 1969; and Section 4.2.5.b).

4.4.5.c. Optics and Formation of Gradient of Light Intensity. The light gradient for phototropism in unpolarized blue and red light is caused in *Dryopteris* chloronemata by a weak screening effect, a lens effect (bypassed subapical zone, probably only in air), and possibly by the orientation of the dichroic photoreceptor molecules. The last effect is (probably of particular importance in water Etzold, 1965; Haupt, 1977). All these effects create a gradient with more absorption on the proximal side of the tip than on the distal (Etzold, 1965; Hartmann *et al.,* 1965; Haupt, 1977; cf. proposal of Banbury, 1959).

The absorption gradient in polarotropism of *Dryopteris* is brought about by dichroic, surface-parallel blue and red-light-absorbing (P_r) photoreceptor molecules. So absorption in filamental wall regions parallel to the electric vector is maximum (Etzold, 1965; Hartmann *et al.,* 1965; Haupt, 1977): On a 50° rotation of the plane of the electrical vector, maximum absorption is at the flank of the tip parallel to the vector (Etzold, 1965); on a 90° rotation, both flanks of the tip and filament absorb maximally, as indicated by occasional subapical branching (Hartmann *et al.,* 1965).

Thus there are considerable analogies with the corresponding mechanisms in photo- and polarotropism of some algae and mosses (see Sections 4.2.5.c and 4.3.5.c).

4.4.6. Transduction of Stimulus

Mohr (1956*b*; cf. Banbury, 1959) speculated on auxin involvement in *Dryopteris* phototropism. Red and blue light were assumed either to destroy auxin or to inhibit its production. Mohr attempted to explain the aberrant negative phototropism (cf. Section 4.4.5.a) by assuming supraoptimal auxin concentrations in some chloronemata.

Etzold (1965) found that a red-light pretreatment reduced the sensitivity in red- and blue-light polarotropism. He assumed this to be caused by saturation with intermediate products formed subsequent to P_{fr}. In 1970 Steiner confirmed his previous assumption (1969*b*) that phytochrome interacts with blue-light polarotropism (possibly via the transduction) as similarly proposed for *Sphaerocarpus* (1969*d*, 1970).

There is no information on transduction in ferns at biochemical and biophysical levels. But an influence of phytochrome on membrane properties (Fondeville *et al.*, 1967; Briggs and Rice, 1972; Hartmann and Haupt, 1977) could play some role. The rotation of the molecule axis on P_f/P_{fr} could cause a change in membrane permeability of chloroplast movement in *Mougeotia* (Haupt, 1968, 1972, 1980).

4.5. Fungi

4.5.1. General Survey

Fungi exhibit positive and rarely negative phototropism. Only in *Botrytis* and *Penicillium glaucum* is polarotropism known. However, polarized light affects phototropism in other fungi. The reproductive structures are there, which are mostly phototropic, but sometimes vegetative and germ hyphae are phototropic as well. Multicellular and unicellular, mostly coenocytic, structures can be phototropically active. Attention will be paid here only to the unicellular type with some emphasis on *Pilobolus* and mostly on *Phycomyces,* the most intensively investigated fungal object in phototropism, also because it is amenable to genetic analysis (cf. Bergman *et al.,* 1973; Ootaki *et al.,* 1974, 1977; Cerdá-Olmedo, 1977; Russo, 1980; Russo and Galland, 1980). For the species problem in the early works (*Phycomyces nitens* Kunze and *P. blakesleeanus* Burgeff) and the life cycle of *P. blakesleeanus,* see Bergman *et al.* (1969). Starting from 1956, stage IVb SPPH of the *wt* (−) strain NRRL 1555 of *P. blakesleeanus* and its mutant derivatives were predominantly used.

4.5.2. Growth and Bending Mechanisms

4.5.2.a. Growth Mechanism and Light–Growth Response. In wall growth of fungal hyphae and of yeast cells, there is good evidence for membrane involvement and models on the subcellular and molecular mechanisms by which it proceeds. So membrane vesicles (from ER/Golgi apparatus) are assumed to transport chitinolytic and chitin-synthesizing enzymes and wall precursors to the plasmamembrane and to extrude them after fusion with it into the cell wall space. For more information on fungal growth and cell wall synthesis, see Castle (1958), Moor (1967), Page (1968), Trinci and Banbury (1969), Bartnicki-Garcia (1973), Smith (1975), and Bartnicki-Garcia *et al.* (1978).

In most hyphae, apical growth takes place at the very tip, a fact mentioned earlier by Nägeli (1846) and de Bary (1884), both cited by Reinhardt (1892). Reinhardt described tip growth in *Peziza* hyphae and had supposed that new wall particles are laid down between the existing wall material mainly at the extreme tip. Tip growth was proved then by Stameroff (1897) on vegetative hyphae of *Mucor mucedo* (L.) Fresenius and *Saprolegnia.* Sometimes a subapical growth mechanism is evident. Examples are stage IV or IVa/b SPPH of *Pilobolus* and *Phycomyces.*

Light can, but does not have to, influence the longitudinal growth rate of hyphae. The first description of an effect stems from Stameroff (1897), who found a growth inhibition by light in *Mucor* SPPH. Banbury (1959) stated that in the Zygomycetes *Entomophthora* spp. and *Basidiobolus ranarum* there was no phototropism of SPPH, although there was a positive light–growth response. He assumed a balance between focusing and screening cause this.

Jolivette (1914) found in a *Pilobolus* sp. that stage I SPPH showed tip growth and stage IV SPPH subapical growth (below the subsporangial swelling; cf. Foster, 1977; also for stages). Page (1962) suggested for stage I SPPH of *Pilobolus kleinii* van Tieghem and *P. crystallinus* Tode an extreme tip growth on the basis of phototropic bending studies and the occasional occurrence of branching at the tip (cf. Haupt, 1965; Page, 1965; Section 4.5.2.b). Page and Curry (1966) found in stage I SPPH of *P. kleinii* indicated that in most conditions the growing zone was not so restricted (cf. Carlile, 1970). Thus, the situation is not clear.

Jacob (1959) claimed the occurrence of spiral growth in stage I SPPH of *P. crystallinus* as did Schneider (1943) in unpublished results cited by Jacob (1959), since there was an aiming error in phototropism under weak illumination. Pringsheim and Czurda (1927) assumed a light inhibition of growth in stage I and IV SPPH of *P. crystallinus.* Van der Wey (1929) proposed a positive light–growth response for stage IV SPPH of *Philobolus* (presumably *P. kleinii*) and rejected the assumption of Pringsheim and Czurda as did du Buy and Nuernbergk (1935). Paul (1950, cited by Jacob, 1959) found by half-side illumination a positive light–growth response in stage I SPPH of *P. umbonatus* Buller. Jacob (1964) reported further unpublished results of Paul (1950) on positive light–growth responses in stage IV SPPH of *P. kleinii, P. sphaerosporus* Palla, and *P. crystallinus* and on negative light–growth responses in stage I SPPH of the first two species (cf. also Bünning, 1956; Banbury, 1959). Using half-side illumination, he also found a negative and a positive light–growth response in stage I and IV SPPH of *P. longipes* van Tieghem, respectively. So a reversal of the light–growth responses during SPPH development seemed to occur in three species. Page and Curry (1966) found a positive light–growth response with small grazing beams of light in stage I SPPH of *P. kleinii*. So there are still doubts as to the nature of the light–growth response in stage I SPPH of *P. kleinii*.

Carnoy was the first to describe the development of *Phycomyces* SPPH (1870, cited in Errera, 1884). Errera divided the SPPH development into four stages of which the first (stage I) and the fourth (stage IV) showed elongation growth (growth rates 1–2 mm/hr and 3 mm/hr, respectively, according to Bergman *et al.,* 1969). A subdivision of stage IV into stages IVa and b was introduced Castle (1942) on the basis of different spiral growth. These stages became established (see, e.g., Bergman *et al.,* 1969; Johnson and Gamow, 1972; Ortega and Gamow, 1974; and Foster, 1977). However, in a preliminary paper, Koske (1977) found a counterclockwise rotation (as assumed for stage IVa) only in three uncertain cases among 200 examined SPPH from seven *Phycomyces* isolates. He concluded that "stage IVa was never a regular developmental stage in the maturation of a sporangium." So the situation concerning a stage IVa, incorporated in recent theories on spiral growth, has become doubtful and should be carefully controlled.

Possibly the first to note a torsion of a part of a *Phycomyces* SPPH was Burgeff (1914, cited in Oort, 1931). Oort named the phenomenon *spiral growth,* which he investigated in stage IV SPPH, and found that elongation and rotation were always coupled. He assumed counterclockwise rotation (viewed from above) to be abnormal. Intensive investigations on spiral growth were then carried out by Castle, concerning several factors: steepness and direction of the spiral and origin of spiral growth (1934*b*, 1936*b*, 1953); a mechanical model (1936*a*); spiral growth in stage I 1936*b*; not detected; 1942 and 1958, slow clockwise rotation at about 0.5°/min, also found by Koske, 1977); direct coupling of twist and stretch at a constant ratio throughout the growing zone and their distribution in the growing zone (highest values about 0.3 mm below the sporangium, Castle, 1937*a*); orientation of micelles, composed of chitin fibrils (cf. Oort and Roelofsen, 1932), in primary and secondary walls of stages I and IV (Castle, 1937*b*, 1938); detection of stage IVa with counterclockwise rotation (total amount 50°–100°, Castle, 1942). Preston (1948; cf. Castle, 1953) then presented a theoretical quantitative model for spiral growth including reversal between stages IVa and IVb. Cohen and Delbrück (1958) found that in the growing zone of stage IVb SPPH stretch and twist maxima did not coincide and that in the lower part twist was negligible compared to stretch (but see below). So they refuted Castle's (1937*a*) assumption on a constant ratio between twist and stretch throughout the growing zone. The maximum rotational speed occurs at the sporangium (12°–18°/min) and decreases to zero at the lower end of the growing zone (Delbrück and Reichardt, 1956; Bergman *et al.*, 1969). Dennison (1959*b*) explained tropic oscillations around the phototropic equilibrium direction by spiral growth with improved methods, Ortega *et al.* (1974) confirmed Castle's results (1937*a*) on a roughly constant ratio between twist and stretch in the growing zone except for the lowest part where rotation, but hardly any elongation occurred. The authors explained the discrepancy to the results of Cohen and Delbrück (1958) by use of their own improved method. Ortega *et al.* (1974) and Ortega and Gamow (1974) developed a novel theory on spiral growth in stages I and IVb: It should be attributable to longitudinal reorientation of transversely laid down fibrils during pressure-induced extension. Ortega and Gamow (1974) extended this model also to stage IVa by introduction of a transient interfibril slippage that should explain the counterclockwise spiraling and its reversal between stages IVa and IVb. Gamow and Böttger (1979) found some more indications for this model. But the last assumption of the model may be unnecessary if the above-mentioned preliminary results of Koske (1977) could be verified.

Errera (1884) stated that the growing zone of stage I SPPH should be confined to the tip and determined its length as varying between 0.1 and 1.1 mm. Castle (1942) found apical growth in the first 1 mm of stage I SPPH by the use of markers in the first one mm and extended this observation (Castle, 1958): Growth should take place at the extreme tip at maximum rate and should decrease to zero at 1–2 mm below the apex. Recently, Trinci and Halford (1975) found, by determination of the region of phototropic bending, variations in length of the growing zones of 0.05–0.95 mm. They detected a good correlation between the length of the growing zone and the length of the tapered apical region, and proposed a possible relationship between the length of the growing zone and the growth rate. Castle (1942) found that the growth after stage III is resumed at a point 0.6 mm below the sporangium and that this short zone extends to the final length of stage IVb of 1.5–2 mm. Errera (1884) cited the work of Carnoy (1870) for the location of the growing zone in stage IV SPPH just below the sporangium. Errera determined the length of the growing zone to be 0.2–2 mm. Other workers presented similar data, also for the region, the extremes being 0.1–3 mm below the sporangium (Castle, 1937*a*, 1959; Cohen and Delbrück, 1958; Ortega *et al.*, 1974).

Stage IVb SPPH of *Phycomyces* elongate in dark or light up to 1 W/m^2 at the same speed, ~ 50 $\mu m/min$ (e.g., Castle, 1932*b*; Bergman *et al.*, 1969; Russo, 1980).

Blaauw (1914*a*,*b*) was the first to describe transient accelerations of *Phycomyces* SPPH growth (stage IVb) when equilateral white light pulses (pulse-up) were given; he named this the *primary photogrowth reaction.* He considered it the basis of phototropic bending (Blaauw, 1914*a*,*b*, 1915, 1919), measured lag times, durations, and magnitudes of accelerations. The parameters changed with the fluence used and that the growth acceleration was followed by a deceleration (Blaauw, 1914*b*, 1919). In 1915 Blaauw named the behavior detected in *Phycomyces* positive *Photowachstumsreaktion* (positive light–growth response) in contrast to the negative light–growth response, which he found in *Helianthus* germlings. Tollenaar (1923) described then a "dark–growth response" (transient less intensive deceleration of growth) when illuminated SPPH were darkened (step-down). Oort (1931) found that light affected not only elongation, but rotation rate as well. Wiechulla (1932) found coincidence in spectral sensitivities for light–growth response and phototropism. His fluence–response results agreed with Blaauw's results. A thorough analysis of the light–growth and dark–growth responses was done by Delbrück and Reichardt (1956). With a pulse-up, there was first an increase and then a decrease in growth rate before the baseline level was reached again (the opposite happened when a pulse-down was given). With a step-up (or step-down), there was only a transient increase (decrease) in growth rate. With pulse programs there were no net gains or net losses in growth.

Buder (1920) showed for stage I SPPH (half-side illumination technique) that the illuminated side grew more than the shaded side (regarded as proof for a positive light–growth response). This method was later often used to determine light–growth responses, but it is only conclusive when growth in an apical or subapical zone occurs. The results of Buder were confirmed by Banbury (1952*a*), who stimulated with small light beams grazing the edges of the growing zones of stage I and IVb SPPH. A second confirmation came from experiments by Meistrich *et al.* (1970) by the use of microbeam laser irradiation (cf. Sections 4.5.2.b and 4.5.5.b).

Cohen and Delbrück described that with blue light (1958) and UV light spots (1959) there was neither a twist nor a stretch response where stretch in continuous growth was maximum and that the sensitivity for light–growth response and phototropism coincided in the region from 0.65 to ∽ 2 mm below the sporangium. Castle (1959), using another method, rejected these observations. He found that the whole growing zone showed the same kind of growth distribution after a pulse-up as in the steady state.

Castle (1930*a*,*b*) found that the reaction time (time between onset of stimulus and start of response) increased below a critical exposure duration and that it was composed of a series of components. For the dark–growth response, he found (Castle, 1932*b*), that the reaction time decreased in a regular way with increasing fluence rates of the adapting illumination (see also Castle and Honeyman, 1934). On the other hand, Delbrück and Reichardt (1956) found no dependency of the lag time on the stimulus size, but a constant lag of 2.5 min (confirmed by Foster and Lipson, 1973).

Castle (1931*a*) discovered that the light–growth response can still take place at fluence rates where phototropic indifference occurs. The Bunsen-Roscoe law holds for short stimulations (a few seconds, confirmed by Delbrück and Reichardt, 1956, for up to 1 min). Delbrück and Reichardt found furthermore that the light–growth responses are graded. Foster and Lipson (1973) studied the light–growth response with an automated tracking system. The variation of the response magnitude in a wide range of levels of adaptation paralleled that of the phototropic response. Some phototropic *mad* mutants also showed changed light–growth responses (for the genetic analysis of *Phycomyces,* see Section 4.5.4). At high adapting fluence rates a step-up produced a long lasting increase in growth rate that seemed not to return to the normal level (see also Bergman *et al.,* 1969). Russo (1980) and Russo and Galland (1980) confirmed this for *wt* and a *mad*A mutant.

Curry and Gruen (1957) discovered that a pulse of UV light also provoked a transient positive light–growth response. These workers (1959) and Delbrück and Shropshire (1960) found an action spectrum for the light–growth response similar to that for phototropism. Delbrück *et al.* (1976) calculated an action maximum for the light–growth response at about 595 nm and regarded this as evidence for a flavin type photoreceptor (cf. Sections 4.5.5.a and 4.5.5.b). Studies on the effectivities of stimulations by polarized light in dependence on the plane of the electrical vector were performed by Shropshire (1959; cf. theoretical considerations by Jaffe, 1960) and more extensively by Jesaitis (1974). (For more information, see Section 4.5.5.b.)

The term "adaptation" was brought into the investigations on light–growth responses by Castle (1929) and concerned the transient nature of the response. Delbrück and Reichardt (1956) introduced the concept of a "level of adaptation." This level was postulated to be a biochemical or biophysical status of the organism. Its magnitude should be somehow dependent on the fluence rate used during the pretreatment. After a long pretreatment at constant fluence rate, the adaptation level was assumed to reach the equilibrium corresponding to this fluence rate. After a change of the fluence rate, the growth output should depend on a subjective fluence rate. But this model fails at low and high fluence rates (cf. fluence rate behavior) and big stimuli (cf. Foster and Lipson, 1973; Russo and Galland, 1980).

Castle (1961*a*) doubted the Blaauw theory on phototropism in the strict sense when he found that phototropism occurred without a change of the light fluence rate (on change from equilateral to unilateral illumination). He suggested a preliminary model postulating a central pool of a substance involved in growth for both responses. The elaborated kinetic model (Castle, 1966*b*) consisted of two sequential reactions. These were thought to be connected by two parallel reaction pathways, one of which should be regulated by light. On changes of the fluence rates the character of the light dependent reaction should be altered leading to a readjustment of the levels of two intermediate substances. But Ortega and Gamow (1970) found that after a saturating equilateral light stimulus, where the SPPH does not respond to an additional light stimulus, it is able to react to a symmetrical avoidance stimulus. So the prediction of the Castle model, that an increased growth rate depletes a substrate pool in a way that no further growth response by any stimulus can be evoked, was refuted. Therefore, the avoidance stimulus must act at some point past or parallel to the adaptation mechanism in the light–growth response. Foster and Lipson (1973) proposed that functional photoreceptors should be growth inhibitors that can be bleached at high fluence rates (unfunctional state). Under these conditions, their growth rate control is canceled. Lipson (1975*a,b,c*) used white gaussian noise stimulation for the analysis of the light–growth response system with the "tracking machine." He studied responses in the normal fluence rate range (at about 10^{-2} W/m^2), in extended fluence rate ranges (10^{-8}–10 W/m^2) (Lipson, 1975*b*), and in several phototropically defective *mad* mutant strains. His model (1975*b*) concerns the molecular nature of the level of adaptation. This should be solely a function of the fraction of the active photoreceptors. Dennison and Foster (1977) and Medina and Cerdá-Olmedo (1977*a*) proposed similar models for phototropism based on differential light–growth responses (see Section 4.5.5.b). Recently, Russo (1980) proposed a new qualitative model that hypothesizes two sequential types of adaptation. The first adaptation site (adaptation I) should be responsible for the adjustment of the light sensitivity and the second adaptation site (adaptation II) for the adjustment (adaptation) of the growth output.

But what happens at the output site on the molecular level? Ortega and Gamow (1974) speculated that the light stimulus should increase the plasticity of the cell wall in the growing zone, possibly by the activation of a cell wall loosening enzyme. By tensile tests on stage IVb SPPH Ortega *et al.* (1975) discovered indeed such increases in extensibility of the growing zone. The increase in extensibility started 2 min after stimulation and continued for about 15 min (Ortega and Gamow, 1976). These kinetics are comparable to those of the light–growth response and indicate a possible connection between both processes. A saturating light stimulus decreased the ratio between stretch and twist in strain-hardened spph (Instron technique; Gamow and Böttger, 1980). They regarded this as evidence for the formerly suggested model of spiral growth especially for what concerned fibril slippage.

Hardly any information is available on the biochemistry of SPPH growth (see also Section 4.5.6). An approach directed to the biochemical output of cell wall (and elongation) growth was undertaken by Jan (1974), who discovered a chitin synthetase in *Phycomyces* mycelium and SPPH of all stages. The mycelial enzyme seemed to be plasmamembrane bound according to cell fractionation and marker enzyme studies. Autoradiography showed higher chitin synthetase activity in the growing zone than in nongrowing parts of SPPH, pointing to the possible importance of this enzyme in growth. The particulate nature of chitin synthetase in *Phycomyces* was confirmed by van Laere and Carlier (1977). The occurrence and properties of chitinase, a chitinolytic enzyme, in stage IVb SPPH of *Phycomyces* was described by R. J. Cohen (1974*b*). He discussed the possible role of this enzyme in the growth system: It should transiently loosen the chitinous cell wall. A delicate balance between cell wall lysis and synthesis had already been claimed in the above mentioned more general works on fungal wall growth, but there is only restricted knowledge on the biochemical level.

4.5.2.b. Bending Mechanism. In fungal phototropism, bending occurs mainly by bowing, although an extreme apical growth was proved in some instances (cf. Page, 1962, 1968), and for some time bulging was favored as the bending mechanism (cf., e.g., Haupt, 1965; Carlile, 1970). Stadler (1952) was the first who considered bending of fungal hyphae not to be caused exclusively by bowing, but by a "bulging forth" of protoplasm at the weakest point of the tip. The growth rate difference between proximal and distal sides in phototropism may be the result of an asymmetrical supply of cell wall precursors, probably contained in membrane vesicles (Trinci and Banbury, 1969; Bartnicki-Garcia, 1973), the so-called chitosomes (Bartnicki-Garcia *et al.,* 1978) to the growing zone, created by the light gradient. But how this differential preference is established at the molecular level is unknown. Light could stimulate somehow chitin synthetase and/or chitinase activities at the plasmamembrane and so cause the difference in growth rate.

In *Pilobolus* Jolivette (1914) had already found that the phototropic bending is confined to the growing tip of stage I SPPH and to a growing zone below the subsporangial swelling of stage IV SPPH (bowing at least in stage IV). Although bending proceeded mainly by bowing stage I SPPH of *P. kleinii* and *P. crystallinus* showed bulging in some instances (Page, 1962, 1968; Page and Curry, 1966). Pringsheim and Czurda (1927) assumed that light should inhibit growth in stage I SPPH of *Pilobolus* (probably *P. crystallinus*). Inhibition on the proximal side of the strongly absorbing tip should produce the positive phototropism. In stage IV SPPH a hypothetical lens action should cause more illumination on the proximal side below the swelling and should inhibit growth there (bowing in both cases). Van der Wey (1929) doubted these results at least for stage IV SPPH of *P. kleinii*. He suggested that a positive light–growth response and more illumination (by a calculated lens action; see Section 4.5.5.c) on the distal side below the swelling should cause positive phototropism. Jacob (1959) cited unpublished results of Paul (1950) and Schneider (1953) on positive light–growth response and positive phototropism by lens action in stage I SPPH of *P. crystallinus* and negative light–growth response and positive phototropism by screening mechanism in stage I

SPPH of *P. kleinii, P. sphaerosporus,* and *P. longipes.* He found for *P. umbonatus* the same characteristics as described for *P. crystallinus* by the former investigators. Positive light–growth responses in stage IV SPPH of all *Pilobolus* spp. investigated by them should cause positive phototropism by lens action (Jacob, 1964; Paul, 1950, cited by Jacob). Page and Curry (1966), working with stage I SPPH of *P. kleinii,* assumed a positive light–growth response to cause positive phototropism via lens effect. This is in strong contrast to the results of Jacob (1959, 1964) (see Sections 4.5.2.a and 4.5.5.c). Bünning (1938*b*) suggested that in stage IV SPPH of *P. kleinii* light should increase the cell wall extensibility on the distal side below the swelling.

Since the early investigations (Errera, 1884; Oltmanns, 1897) it has generally been accepted that phototropic bending in stage I and IVb spph of *Phycomyces* occurs only via a bowing mechanism. For stage I SPPH, Errera (1884) found the maximum of phototropic bending in the growing zone, 175–200 μm from the tip. Oltmanns (1897), working with stage IVb SPPH, discovered that the curvature in phototropism started close to the sporangium and then migrated downwards. Cohen and Delbrück (1959) confirmed this observation by illumination of small test areas of the growing zone by small light beams.

Blaauw (1914*a,b,* 1919) believed that in almost all cases the phototropic curvature had to be preceded by a positive light–growth response. The use of light pulses favored this assumption. He made kinetic measurements of both light–growth response and positive phototropism and tried to connect both responses by proposal of a lens action (cf. Sections 4.5.2.a and 4.5.5.c). Blaauw, (1914*b*) stated: *"daß der ganze Phototropismus von Phycomyces nichts anderes bedeutet als die Resultate der ungleichen Photowachstumsreaktion der ungleich belichteten Vorder- und Rückseite der Zelle."** He did not become aware of the paradox: The light–growth response is transient, while phototropism is not (see, e.g., Castle, 1961*a;* Carlile, 1965; Haupt, 1965; Dennison and Foster, 1977; Foster, 1977; Haupt, 1977; Hertel, 1980; Russo, 1980; Russo and Galland, 1980). On the other hand, several workers confirmed that there is more growth where there is more light (cf. Section 4.5.2.a). Close connections along the sensory transduction chain from input to output are obvious between light–growth response and phototropism, although Castle (1961*a*) stated the contrary. This evidence stems from similar action spectra, from similar fluence rate–response curves, from similar lag times, and from genetic analysis.

One model to overcome this dilemma was presented by Castle (1961*a,b,* 1966*b*). Based on phototropic inversions caused by step-ups during bending (1961*b*) and occurrence of phototropic bending without change of fluence rates

*" . . . that the phototropism of *Phycomyces* is nothing but a differential light-induced growth due to a light gradient in the cell.

on switching from equilateral to unilateral illumination (1961*a*), Castle rejected the Blaauw theory in the strict sense. His model of limited supply (cf. Section 4.5.2.a) claims in its asymmetrical version for phototropism an unequal competitive distribution of cell wall precursors from a common pool. The observation of equally increased and decreased growth rates on distal and proximal sides, respectively, supported this claim (Castle, 1961*a;* Shropshire, 1971). The elaborate model (1966*b*) explained furthermore qualitatively all the known phototropic inversions (Reichardt and Varjú, 1958; Castle, 1961*b;* Dennison, 1965). Some doubts as to the accuracy of this model were expressed by Russo and Galland (1980).

The Dennison–Foster model (1977; cf. quantitative version by Medina and Cerdá-Olmedo, 1977*a*) assumed that the spatial light gradient under unilateral illumination should be converted to a temporary gradient by rotation of the photoreceptors (see also Section 4.5.5.b).

The most recent model in this context is that of Russo (1980; cf. Section 4.5.2.a). It assumed that the ratio between an activator and an inhibitor of an enzyme involved in growth is important. In constant bending the ratio of activator to inhibitor is thought to exhibit a transversal gradient with a maximum on the distal side. The different kinetics of adaptation sites I and II (II being faster) can qualitatively explain all observed inversion and adaptation phenomena. Therefore, this working hypothesis offers a good basis for future genetic and physiological investigations.

4.5.3. Kinetics of Bending

Kinetic data on phototropic bending in fungi except *Phycomyces* are rarely found in the literature. From the results of Jolivette (1914), we deduced phototropic lag times of ~ 30–45 min and bending rates of ~ 1°/min for stage I SPPH of a *Pilobolus* spp. Stage I SPPH of *Pilobolus kleinii* and *P. crystallinus* were reported by Page (1962) to stop growth after start of unilateral illumination and to resume it after a lag time of ~ 1 hr in the new direction (bulging). For stage IV SPPH of *P. kleinii,* Page (1962) found a lag time of ~ 10 min, followed by a rapid increase in bending rate to a constant value of 1.1°–4°/min. After overshooting beyond the direction of incident light, the bending slowed down and reversed direction. After ~ 1 hr the new growth direction was established. In maturing SPPH, presumably beginning stage IV, this process lasted longer and the overshooting phenomenon assumed the shape of a damped oscillation. By spot illumination the bowing (perpendicularly to the light direction: A kind of half-side illumination) was clearly visible after 30 min in stage I SPPH of *P. kleinii* (Page and Curry, 1966).

Kinetic measurements in phototropism of *Phycomyces* SPPH have been carried out only in stage IVb. The following discussion is therefore restricted to these SPPH. Blaauw (1914*a,b*) found lag times in positive phototropism of *Phycomyces* SPPH in the range of 5–9 min. Castle (1930*a,b*) determined increasing lag times of 2.6–3.9 min for positive phototropism. Reichardt and Varjú (1958) reported a lag time of 5–7 min when the fluence rates were halved on switching from equilateral to unilateral illumination. The first worker, who stimulated with the same continuously given fluence rate as was used for the pretreatment, was Castle (1961*a*). Thus he excluded any light–growth response in the strict sense. He found a true phototropic lag time of ~ 6 min. For phototropic responses with a step-down of the light, fluence rate concomitant with the start of the unilateral illumination the lag time increases as a function of the ratio between the equilateral and unilateral fluence rates in a biphasic way (Russo, 1980; Russo and Galland, 1980; see there also for behavior of *mad* mutants). In principle the same behavior was observed in experiments with phototropic stimulation in conjunction with a step-up of the fluence rate (Russo, 1980; Russo and Galland, 1980).

Reichardt and Varjú (1958) found (under the above-mentioned conditions) that the bending rate after the lag became rapidly constant (~ 7°/min). It hardly varied with fluence rates of 10^{-4}–0.4 W/m^2. Lower and higher fluence rates caused lower bending rates. Dennison (1965) discovered that phototropism can last for many hours when the right geometric relationship between SPPH and incident light is held on. The bending rates observed varied in different SPPH of 1.8°–4.5°/min. He found no correlation between bending and growth rates. But Shropshire (1971) succeeded in finding some correlation between ratio of average growth rate to radius of the SPPH and the bending rate, as was suggested before in a model of Bergman *et al.* (1969). Reichardt (1961) reported on high bending rates of 20°/min under unilateral stimulation with UV light of 280 nm.

Cohen and Delbrück (1959) for visible light and Delbrück and Varjú (1961) for UV light based their conclusions on the sensitivity distribution in the growing zone on measurements of lag times and bending rates. For the kinetics in phototropic inversions due to changes in fluence rates during bending we refer to the works of Reichardt and Varjú (1958), Castle (1961*b*, 1966*b*), and Dennison (1965) and to the reviews of Castle (1966*a*), Bergman *et al.* (1969), Russo (1980), and Russo and Galland (1980).

Generally accepted are lag times of 6 ± 2 min and bending rates of 2°–6°/min (in visible light), when the unilateral stimulation is made without a change of the light–fluence rate and in a range of 1–10^{-5} W/m^2 (Russo, 1980; Russo and Galland, 1980).

4.5.4. Fluence–Response Behavior

Among fungi, the phototropic fluence–response behavior has been investigated only for *Pilobolus* (in white light) and *Phycomyces*. Under continuous unilateral white light (12 hr), Jacob (1959) determined sigmoid (if semilogarithmically plotted) fluence rate–response curves for stage I SPPH of different *Pilobolus* species. The fluence rate thresholds differed by a factor 10^3 between the most and the least sensitive species. Using the same methods, Jacob (1964) found that in stage I SPPH of *P. kleinii,* when grown on medium containing diphenylamine, the phototropic fluence rate–response curve was shifted in one strain by about a factor 15, while a second was not affected (cf. Section 4.5.5.c). *Pilobolus* stage IV SPPH are reported to be 10^5 times less sensitive than stage I SPPH (Banbury, 1959).

The first phototropic fluence–response data of dark-adapted stage IVb SPPH of *Phycomyces* are given by Blaauw (1908) for white light pulses. The Bunsen-Roscoe law was valid in a considerable range of combinations of fluence and illumination durations but he considered the Weber-Fechner law invalid (but cf. Massart, 1888). Blaauw's thresholds (1908, 1914*b*) are listed in Table VII. His data (1908), indicating indifference and furthermore nega-

Table VII
Bending Types and Phototropic Thresholds in Fungi

Species and organ	Type of bending	Threshold (J/m^2) at λ (nm)	Reference
Stage I SPPH of *Pilobolus kleinii* and *P. crystallinus*	Bowing and bulging	—	Page (1962, 1968); Page and Curry (1966)
Stage IV SPPH of *Pilobolus* spp.	Bowing	—	e.g., Jolivette (1914); Jacob (1964)
Stage I SPPH of *Pilobolus longipes*	Bowing (?)	3×10^{-7} with white light[a,b]	
Stage I SPPH of *P. sphaerosporus*	Bowing (?)	3×10^{-4} with white light[a,b]	Jacob (1959)
Stage I SPPH of *Phycomyces blakesleeanus*	Bowing	10^{-6} with white light[a,b]	
Stage I SPPH of *Phycomyces* sp.	Bowing	5×10^{-3} with white light[a,b,c]	Blaauw (1908)
		4×10^{-5} with white light[a,d]	Blaauw (1914*b*)
Stage IVb SPPH of *P. blakesleeanus*	Bowing	7×10^{-2} at 280	Curry and Gruen (1957)
Stage IVb SPPH of *wt* of *P. blakesleeanus*	Bowing	10^{-5} with blue light[e]	Bergman *et al.* (1969); Russo (1980)

[a] J/m^2 of equivalent blue light.
[b] Maximum threshold value (see text) for just visible reaction.
[c] Same as b, but for 50% reaction.
[d] Stimulation with a pulse of light.
[e] A lag time of 10 min was used for calculation. (?) questionable.

tive curvatures, at higher fluence rates are incorrect, because infrared light was not excluded causing growth inhibition (Castle, 1932*a*). Castle (1931*a*) found indifference to unilateral white light starting at ~ 10^3 mcs ($4 \cdot 10^{-2}$ J/m^2) in a short pulse for bending of stage IVb SPPH of *Phycomyces*. The mean white-light phototropic threshold in stage I SPPH of four *Phyiomyces* strains was determined as described above for *Pilobolus* (Jacob, 1959) (Table VII).

Varjú *et al.* (1961) stated that stage I SPPH are about 100 times more sensitive to unilateral blue light than are those in stage IVb. But for negative phototropism in continuous unilateral UV light (280 nm), dark-adapted stage I SPPH seem to be less sensitive than those in stage IVb (Curry and Gruen, 1957).

Since 1958 two main methods have been employed for phototropic investigations on stage IVb SPPH of *Phycomyces:* Under continuous unilateral irradiation either initial phototropic bending rates or photogeotropic equilibrium angles (Medina and Cerdá-Olmedo, 1977*b*) after long-term irradiations (5–8 hr) have been measured (cf. Lipson and Terasaka, 1981). The fluence rate–response curves obtained by both methods agree reasonably well. Bending rate estimations for dark-grown and subsequently blue-light preadapted wild-type (*wt*) SPPH gave fluence rate thresholds at 3×10^{-9} W/m^2, about the same rate of bending between 10^{-4} and 0.4 W/m^2, and indifference beyond 2.3 W/m^2 (Reichardt and Varjú, 1958). Bergman *et al.* (1969) reported a drop in the bending rate at 0.1 W/m^2 and no bending beyond 9 W/m^2. For light-grown *wt* SPPH, Russo (1980) found threshold bending at 10^{-8} W/m^2, between 10^{-6} and 6 W/m^2, only a difference of a factor 2 in the bending rate, and for *wt* and the phototropic defective *mad*A and *mad*B mutants (cf. Section 4.5.6), no bending at 10 W/m^2. *Mad*A mutants exhibit the same maximum bending rate as the *wt,* but a threshold of 10^{-4} W/m^2, whereas *mad*B mutants show a diminished bending rate and a threshold of ~ 3×10^{-3} W/m^2 (Russo, 1980).

At 280, 380, and 480 nm, Varjú *et al.* (1961) determined photogeotropic fluence rate–response curves for dark-adapted *wt* (−) SPPH. They found about the same thresholds of 10^{-9} W/m^2, a range at which the angles depended on the fluence rates and saturations for reaching the maximum, i.e., the photogeotropic equilibrium, angles (~ 60–75° from the vertical) at ~ 5×10^{-6} W/m^2. Besides the indicated fluence rates, this is the general shape of the corresponding curves for *wt* and the *mad* mutants of class 1. Similar data for blue light with maximum deviations of a factor 10 for *wt* photogeotropism were reported by Bergman *et al.* (1973), Ootaki *et al.* (1974, 1977), and Presti *et al.* (1977). This threshold is similar to that for the light–growth response (Foster and Lipson, 1973).

Bergman *et al.* (1973) and Ootaki *et al.* (1974) investigated phototropic defective *mad* mutants and found two classes (cf. Section 4.5.6). Class 1 mutants ("night-blind" genes *mad*A, *mad*B, and *mad*C) exhibited up to 5×10^5 times higher thresholds than did *wt* and nearly the same saturation as found for *wt* (Lipson and Terasaka, 1981). Class 2 mutants ("stiff" genes *mad*D, *mad*E, and later detected *mad*F and *mad*G, Eslava *et al.*, 1976; Ootaki *et al.*, 1977) showed differently diminished slopes of the fluence rate–response curves (some not reaching the *wt* equilibrium angle under standard procedure, Lipson and Terasaka, 1981), but have about the same threshold as *wt*.

Lipson and Terasaka (1981) constructed all possible double mutants and one triple *mad* mutant. For unknown reasons, their class 2 mutant results do not coincide with the results of other authors. The thresholds were raised in comparison with the corresponding single mutants in double mutants within class 1 or equal to the thresholds of the corresponding class 1 mutants in class ½ double mutants. Double mutants within class 2 exhibited thresholds as the less sensitive parent or higher.

Löser and Schäfer (1980) investigated photogeotropism with the opposed beams inclined 20° down from horizontal. At high fluence rates at 605 nm, bending occurred (cf. Delbrück *et al.*, 1976), but there were oscillations at the saturation level below that at 450 nm. Simultaneous irradiation with light of these two wavelengths showed an inhibitory influence of 605-nm light. These results were interpreted as hints for the occurrence of two photoreceptors in *Phycomyces* in some analogy to hypotheses in *Dryopteris* polarotropism (Steiner, 1969*a,b,* 1970).

Table VII shows the types of bending and the phototropic thresholds of some *Pilobolus* and *Phycomyces* species.

4.5.5. Photoreception

4.5.5.a. Spectral Sensitivity and Action Spectra. In positive phototropism of conidiophores of *Entomophthora coronata* (Constantin) Kevorkian *(Conidiobolus villosus)* a divergent, although very crude, action spectrum resulted: Wavelengths of 400–650 nm were effective with maxima at 400 and 490–650 nm (Page and Brungard, 1961). The sensitivity was ~ 25 to 50 times greater at 405 than at 630 nm (Page, 1965, 1968).

The spectral sensitivity curve for stage I SPPH of a *Pilobolus* sp. by Parr (1918) rose from red to violet without maxima. It suffered obviously from experimental errors (Castle, 1931*b;* Bünning, 1937; Banbury, 1959) and has now been disregarded. The action spectrum (null-point balancing method) for positive phototropism of stage IV SPPH of *Pilobolus kleinii* showed maximum activity in the blue (peaks at 450 and 480 nm), but measurable efficiency at least up to 650 nm (Bünning, 1937). For stage I SPPH of the same species (Bünning, 1938*b*), measured between 440 and 520 nm, the phototropic efficiency peaked only at about 450 nm. But the red-light sensitivity was rejected by Flint (1942), who found no trace of sensitivity to unilateral 680-nm light supposedly working with stage IV SPPH of *P. longipes*. Jacob (1959) investigated stage I SPPH of six *Pilobolus* spp. for red spectral limits of positive phototropism and cited some corresponding unpublished works. The sensitivity vanished between 570 nm *(P. sphaerosporus)* and 660 nm *(P. longipes),* which did not contradict Flint's results. Jacob's results are doubtful, because stage I SPPH of *Phycomyces* reacted up to 625 nm, in strong contrast at least to stage IVb SPPH (Curry and Gruen, 1959; Delbrück and Shropshire, 1960; Page, 1968). Page (1962) found for stage I and IV SPPH of different *Pilobolus* spp. by rough spectrum projection no evidence for a phototropic reaction at wavelengths greater than 580 nm. This was confirmed by an action spectrum for stage I SPPH of *P. kleinii* with the more precise null-point balancing method (Page and Curry, 1966). They got a blue-light action spectrum with an upper limit at 530 nm. Negative phototropism in the UV range started at 300 nm and was pronounced at 280 nm. An almost typical blue-light action spectrum (but with three peaks at 400–500 nm) for trophocyste induction in *P. kleinii* was measured by Page (1950) between 380 and 530 nm. Up to now it is not totally clear whether the differences in spectral limits for phototropism of *Pilobolus* SPPH are attributable to different species and strains and/or mainly to different methods (cf. Banbury, 1959; Page and Curry, 1966; Page, 1968). But nowadays a cutoff in the range of 500–540 nm is assumed probable for most blue-light phototropic responses in fungi, including *Pilobolus* (Page, 1968; Carlile, 1970; Presti and Delbrück, 1978).

Initially there was also some confusion concerning the spectral limits and maxima in phototropism of stage IVb SPPH of *Phycomyces*. So the data of Blaauw (1908) on fluence variation for 50% reaction with maximum sensitivity around 495 nm and some activity even at 615 nm were rejected concordantly by Castle (1931*a*), Buder (1932), and Bünning (1938*b*). For experimental reasons (e.g., restricted spectral range, broad spectral bands, possible stray light) they found only one maximum in the blue at 420–430 nm and a very moderate sensitivity at 580 nm using null-point balancing (Castle, 1931*a* Bünning, 1938*b*) or threshold fluence–response methods. With a cutoff at 625 nm for stage I SPPH, Jacob (1959) cast some doubt on the upper spectral limit (but cf. Page, 1968). But the most reliable phototropic action spectra (by null-point balancing methods) in the normal fluence rate, the range for stage IVb SPPH by Curry and Gruen (1959) and Delbrück and Shropshire (1960) exhibited a clear cutoff at ~ 500 nm. (For action maxima see Table VIII.) Negative phototropism in the UV range was detected by Curry and Gruen (1957), also for stage I SPPH, starting at ~ 300–310 nm (indifference range) and being strong at 280 nm. It is caused by gallic acid screening (Dennison, 1959*a*).

The action spectra for phototropism and positive light–growth response by Delbrück and Shropshire (1960) coincided almost perfectly, but in the UV there was no negative light–growth response and the positive peak at 280 nm was smaller than the negative peak for phototropism. Rough action spectra with grossly consistent features were found for the photogeotropic equilibrium

Table VIII
Types of Phototropism and Spectral Sensitivities in Fungi

Species and organ	Type of phototropism	Spectral sensitivities, action maxima at λ (nm)	Reference
Conidiophores of *Entomophthora coronata*	Positive	(650–400)[a] 650–490, 400	Page and Brungard (1961), Page (1965, 1968)
Stage IV SPPH of *Pilobolus kleinii*	Positive	480, 450 (<650)[a]	Bünning (1937)
Stage I SPPH of *P. kleinii*	Positive	450	Bünning (1938*b*)
		480, 450, 420sh,[b] 370 (<530)[a]	Page and Curry (1966)
	Negative	280 (<300)[a]	
Stage I SPPH of *P. sphaerosporus*	Positive	(<570)[a]	Jacob (1959)
Stage I SPPH of *P. longipes*	Positive	(<660)[a]	
Stage I SPPH of *P. kleinii* and *P. crystallinus*	Positive	(<580)[a]	Page (1962)
Stage IVb SPPH of *Phycomyces* sp.	Positive	495 (<615)[a]	Blaauw (1908)
		430/420 (<580)[a]	Castle (1931*a*); Buder (1932); Bünning (1938*b*)
Stage I SPPH of *P. blakesleeanus*	Positive	(<625)[a]	Jacob (1959)
	Negative	280 (<310/300)[a]	Curry and Gruen (1957, 1959); Delbrück and Shropshire (1960)
Stage IVB SPPH of *P. Blakesleeanus*	Positive	470, 445, 430sh,[b] 370 (<500)[a]	Curry and Gruen (1959)
		485, 455, 430sh,[b] 385 (<500)[a]	Delbrück and Shropshire (1960)
		595	Delbrück *et al.* (1970)
		621–573	Löser and Schäfer (1980)

[a]( ), spectral region or upper wave length limits of action.
[b]sh, Shoulder.

by a threshold fluence-rate method (Varjú *et al.*, 1961) and for phototropism and light–growth response in the high fluence rate range (Bergman *et al.*, 1969). Experiments with very intensive laser light enabled Delbrück *et al.* (1976) to calculate an additional minute action peak at 595 nm (~ 10^9 times smaller than in blue) from calculated phototropic gaussians and experimental light–growth response data: The authors claimed a direct excitation of the lowest triplet state of riboflavin. In general agreement with this, spectral sensitivity changes were observed (Otto *et al.*, 1981) in riboflavin auxotrophs doped with roseoflavin (exhibiting a different absorption spectrum than that of riboflavin: Kasai *et al.*, 1975; Song *et al.*, 1980*b*). Löser and Schäfer (1980) reported furthermore on bending in strong light at 573–621 nm. But they speculated on a photochromic system. If this is verified, the blue-light action spectrum in *Phycomyces* phototropism would indeed extend to the red light range.

Table VIII shows the types of phototropism as well as the spectral sensitivities of some fungal species discussed here.

4.5.5.b. Type and Localization of Photoreceptor. Gettkandt (1954) claimed that carotenoids cannot be the photoreceptor chromophores in germ hyphae of some parasitic fungi, because no carotenoids were microscopically detectable in the apices, which were solely phototropically sensitive. Although this statement was premature, because the photoreceptor concentration may well be far below microscopic and even spectroscopic detectability (cf. Banbury, 1959; Delbrück and Shropshire, 1960; Thimann and Curry, 1960; Page, 1968; Bergman *et al.*, 1969), she correctly rejected the opinion that blue-light sensitivity is always localized in regions with visibly high carotenoid contents (e.g., Bünning, 1937, 1938*b*).

Because of the coincidence of the obviously not blue-light-type action spectrum with the absorption spectrum of an acetone extract, characteristic of porphyrins, the photoreceptor chromophore in *Entomophthora coronata* was assumed to be some porphyrin (Page and Brungard, 1961; Page, 1965, 1968); cf. the assumption of a hemoprotein photoreceptor in *Dictyostelium* phototaxis (Poff and Butler, 1974*a*).

Zopf (1892) managed the first identification of yellow pigments in fungi as carotenoids in *Pilobolus kleinii*. The pigment localization in the stimulus-perceiving SPPH region, the properties of pigment crystals, and the better (stage IV) or worst (stage I) coincidence of the phototropic action spectra with *in vivo* and *in vitro* absorption spectra led Bünning (1937, 1938*a,b*) to postulate carotenoids (especially β-carotene) as photoreceptor chromophores in *P. kleinii* (cf. also Flint, 1942). The first indication for a flavin type photoreceptor in *P. kleinii* by Page (1956) came from studies on trophocyste photoinduction. An action spectrum was not decisive for either carotenoids or flavins. Inhibitor studies with lyxoflavin (flavins) and diphenylamine (DPA), or colored carotenoids, showed only inhibition by lyxoflavin that could be decreased by longer light exposures and cancelled by additional supply with riboflavin. Jacob (1959) excluded carotenoids as photoreceptor chromophores in *Pilobolus* stage I SPPH and regarded them as shading pigments, because he detected in some species phototropism up to red-light ranges (but cf. Section 4.5.5.a). But he did not reject the (also yellow) riboflavin and speculated that the differing red-light spectral limits may suggest that either more than one substance works as photoreceptor or additional pigments absorbing at longer wavelengths activate the photoreceptor proper, but cf. Carlile's proposal (1965), on the possible occurrence of a red-light-sensitive flavoprotein photoreceptor. In 1964 Jacob regarded results of DPA studies with stage I SPPH of *P. kleinii* as evidence against a carotenoid photoreceptor. Also, Page and Curry (1966; cf. Page, 1968) rejected a carotenoid and favored a flavin as photoreceptor chromophore after DPA tests without influence on the phototropic behavior of stage I and IV SPPH of *P. kleinii.* But the DPA-based arguments of both works alone would not have been conclusive enough, because DPA does not totally inhibit the synthesis of colored carotenoids (Goodwin, 1952) and small residual amounts may suffice as photoreceptor chromophore. Page and Curry (1966) found three independent evidences for a restriction of the light sensitivity to the terminal 50 μm of the stage I SPPH. Banbury (1959) cited unpublished results of Schneider on phototropic sensitivity in the orange carotene-rich ring below the subsporangial swelling in stage IV SPPH as hypothesized by Pringsheim and Czurda (1927) and van der Wey (1929).

When the yellow pigments in *Phycomyces* spph were identified as carotenoids (Castle, 1935; Schopfer, 1935; Bünning, 1938*a*) they were regarded as being the photoreceptor chromophore because of the coincidence of the absorption spectra with the phototropic action spectra (Castle, 1935; Bünning, 1938*a*). Difficulties (action spectra with only one maximum) were discussed on the basis of additional assumptions (Bünning, 1938*b*). The idea of a carotenoid photoreceptor chromophore was reproposed by Curry and Gruen (1959) based on their improved phototropic action spectrum with three maxima at 400–500 nm, thought to be indicative for carotenoids.

Zankel *et al.* (1967) tried to rule out β-carotene as primary photoreceptor chromophore on the basis of *in vivo* absorption spectra of the sensitive zone of stage IVb SPPH. The *in vivo* red shift of the β-carotene maxima (compared with spectra in hexane) destroyed the coincidence with the action spectra of Delbrück and Shropshire (1960). Meissner and Delbrück (1968) ruled out retinol as a possible photoreceptor chromophore by biochemical analysis of albino mutants and of *wt* after DPA addition. *In vivo* spectra of albino mutants showed that the photoreceptor has to be very dilute, but the minimum concentration for β-carotene was still high enough (here 10^{-7} *M* regarded as minimum photoreceptor concentration; cf. Varjú *et al.*, 1961; Bergman *et al.*, 1969) not to rule it out conclusively as photoreceptor chromophore (see also Bergman *et al.*, 1973). Microspectrophotometric methods were used by Wolken (1969) for stage IVb SPPH of *wt* and albino mutants. In the albino mutant, the theoretically minimum photoreceptor concentration should have been detected, but since β-carotene was not, it was excluded as the primary photoreceptor chromophore. Presti *et al.* (1977) ruled out *cis*-β-carotene as photoreceptor chromophore in *Phycomyces* (not detectable). Furthermore, mutants blocked in all six steps of carotenoid synthesis (i.e., with less than 4×10^{-5} times the β-carotene amount of *wt* so that the theoretically lowest limit of photoreceptor concentration was not reached) behaved phototropically like *wt,* hence β-carotene was excluded as the photoreceptor chromophore (but cf. de Fabo, 1980). Also the observation of different main absorption axes of the photoreceptor was contrary to a carotenoid (Jesaitis, 1974), but agreed with a flavin-type photoreceptor. Before and after this conclusive work, the results of polarized light experiments (Castle, 1934*a;* Shropshire, 1959; Haupt and Buchwald, 1967; Hertel, 1980) pointed—although from some of the workers falsely interpreted as due to refraction effects—to dichroic photoreceptors as theoretically claimed by Jaffe (1960).

The first to propose riboflavin as photoreceptor in *Phycomyces* was Reinert (1952). He observed that the carotenoid-decreased fungus (on lactate medium) was phototropically more sensitive than the one containing carotenoid. Thus, he considered carotenoids as interfering with phototropism. Based on the very preliminary spectral sensitivity data for phototropism of Curry and Gruen (1957), Carlile (1957; cf. Banbury and Carlile, 1958) speculated that riboflavin could be the chromophore in a flavoprotein photoreceptor both in the visible and in the UV range, because these data resembled the absorption spectrum of riboflavin. Thus, he rejected the proposal by Curry and Gruen (1957) on an unknown special photoreceptor in the UV range (see also Shropshire, 1979). This proposal has been renewed by Schmidt (1980), who proposed IAA for this purpose. Carlile (1962) tried to support his hypothesis on the flavin/flavoprotein nature of the photoreceptor by results of experiments with flavoprotein inhibitors, which affected mycelial growth. Also, Delbrück and Shropshire (1960) favored flavins as possible candidates for the photoreceptor chromophore because of the action peaks at 385 and 280 nm. At least for the 280-nm peak, this may have been premature, because this peak could be due to a protein moiety of either a caroteno- or a flavoprotein. For the same reasons, a flavin was also favored by Bergman *et al.* (1969), but they did not deny the possible existence of more than one photoreceptor chromophore. Wolken (1969) and Ootaki and Wolken (1973) reported on crystals in the growing zone of stage IVb SPPH with absorption spectra resembling riboflavin. Furthermore, Wolken (1969) showed that *wt* SPPH exhibited *in vivo* absorption spectra in the lower part of the growing zone characteristic for flavins. Albino mutant SPPH spectra were comparable to lumichrome or reduced flavins. Berns and Vaughn (1970) speculated on a flavoprotein as photoreceptor candidate, because they were able to induce changes in absorbance at 345 and 460 nm and in fluorescence emission at 450 and 540 nm by irradiance of *wt* and albino mutant mycelia, SPPH, and extracts from both, a *mad*C mutant did not show this. For unknown reasons these changes could not be repeated by other workers (Poff and Butler, 1974*b*). Lipson (1975*b*) determined data on inactivation and regeneration kinetics of the photoreceptor chromophore in the light–growth response. He considered the closeness of the partial absorption coefficient, observed for its inactivation, to the absorption coefficient of oxidized riboflavin as support for the hypothesis of a flavin photoreceptor chromophore. Also, the calculated action peak at 595 nm (Delbrück *et al.*, 1976; cf. Section 4.5.5.a) was regarded as strong evidence for riboflavin (not some kind of flavin) as photoreceptor chromophore in *Phycomyces* (but cf. criticism by de Fabo, 1980).

Hertel (1980) stated that the amount of tightly membrane-bound, nonmitochondrial flavoproteins was lower than the theoretically needed lowest photoreceptor concentration. Therefore, he suggested (not speculating on a mitochondrial localization) a new type of flavin photoreceptor with reversibly bound flavin, based on riboflavin-binding data from *Zea* and *Cucurbita* preparations (Hertel *et al.*, 1980), which were interpreted as evidence for such a receptor. In connection with the hypothesis of Löser and Schäfer (1980), Hertel claimed a photochromic flavin system of, e.g., oxidized versus half or fully reduced flavin with a short-lived active state (cf. Hartmann, 1977).

The latest strong support for a flavin photoreceptor in *Phycomyces* phototropism came from experiments of Otto *et al.* (1981) with a riboflavin auxotrophic mutant (cf. recent riboflavin auxotrophic mutant work in *Neurospora crassa* by Paietta and Sargent, 1981). The riboflavin analogue roseoflavin is taken up by the mycelium, translocated into the SPPH, and effectively phosphorylated. They postulated that riboflavin in the photoreceptor was substituted by roseoflavin to about 80% (lowering its effectiveness to only 0.1% of normal), producing an increase of threshold and the now higher effectiveness of 529-nm light (where roseoflavin absorbs 180 times stronger than riboflavin). They interpreted these results as proof of riboflavin as the normal photoreceptor chromophore. Although some doubts remain, e.g., on substitution in other flavoproteins possibly involved in phototropism and on the statement that riboflavin and not possibly some derivative is the photoreceptor chromophore, the higher effectiveness at 529 nm in the roseoflavin-doped mutant seems to be conclusive for the replacement of some other flavin in the normal photoreceptor.

Löser and Schäfer (1980) suggested that phototropism in *Phycomyces* is not mediated by a single photoreceptor type. Based on a theory of Hartmann (1977) they proposed a photochromic system of at least two photoreceptors with possibly overlapping absorption bands and different quantum efficiencies at different wavelengths. But caution is necessary because of some doubts that a light quality, which in mono- and dichromic irradiation at higher fluence rates evokes positive phototropism, should inhibit phototropism when applied dichromically at fluence rates which are monochromically ineffective.

That the question of the nature of the "unique" blue-light photoreceptor

(Does it exist at all? cf. Senger and Briggs, 1981) is still open, was eloquently treated by de Fabo (1980). Concerning *Phycomyces,* we point to two of his arguments. The work by Presti *et al.* (1977) on carotene-less mutants was considered less than foolproof, because the possible linkage to a protein and/or binding to membranes could have prevented the detection of low carotene amounts. Furthermore, the results of Delbrück *et al.* (1976) interpreted as evidence for triplet excitation of riboflavin were obtained only by calculation using some unproved assumptions.

Other blue-light phenomena in *Phycomyces* that are supposedly attributable to the same photoreceptor involved in phototropism (and light–growth response), are light-induced SPPH formation (Bergman, 1972; Bergman *et al.,* 1973; Thornton, 1973; Galland and Russo, 1979), light-induced sporangium formation (Russo *et al.,* 1980), and light-induced absorbance changes (LIAC) (Poff and Butler, 1974*b;* Lipson and Presti, 1977).

Lipson and Presti (1977) argued that the very low quantum efficiency of the LIAC made its involvement in phototropism or light–growth response improbable. In blue-light-induced β-carotene synthesis (LICS), Jayaram *et al.* (1979) speculated that riboflavin and β-carotene could possibly both act as photoreceptor chromophores and that β-carotene could possibly act in the low-fluence range (de Fabo *et al.*, 1976; Shropshire, 1980). But only 1 year later, Jayaram *et al.* (1980) favored one common flavin-type photoreceptor chromophore. Curiously enough, just 1 year later, Whitaker and Shropshire (1981) favored a general carotenoid photoreceptor for *Phycomyces* LICS. Thus, in *Phycomyces* LICS the nature of the photoreceptor chromophore is obscure, and in phototropism and light–growth response it is also somewhat ambiguous, although a flavin-type photoreceptor seems more probable.

The photoreceptors in *Phycomyces* were previously thought to be restricted to the growing zones of stage I and IVb SPPH. But after the detection of blue-light responses in the mycelium (cf. Russo and Galland, 1980), it became evident that they are distributed throughout the whole organism, if one assumes that indeed the different responses were mediated by the same photoreceptor.

Blaauw (1919) stated that the light sensitivity for phototropism and light–growth response was restricted to the growing zone in stage IVb SPPH of *Phycomyces* (cf. Section 4.5.2.2). Generally, this was supported by all subsequent investigations (e.g., Delbrück and Reichardt, 1956; Castle, 1959; R. Cohen and Delbrück, 1959). But there was some uncertainty as to the exact distribution of sensitivity, adaptation, and growth. R. Cohen and Delbrück (1959) stated that under continuous blue-light irradiation, the sensitive zone started 0.5 mm below the sporangium and did not include the zone which stretches most without stimulation. Castle (1959), however, claimed that all the growing zone should perceive light and should react locally and considered the differences to be attributable to dissimilar experimental conditions. Delbrück and Varjú (1961) excluded this possibility. With narrow lines of continuous UV light, they found the sensitive and adapting zone to start closer to the sporangium than the responding zone. In contrast, Dennison and Foster (1977) assumed the most effective photoreceptors to be located in the upper part of the growing zone. In stage I SPPH Page (1968) found the first 200 μm of the tip to be light sensitive for bending in a plane perpendicular to small grazing light beams.

A longitudinal movement and possible changes in the qualities of the photoreceptors during growth were first proposed by R. Cohen and Delbrück (1958, 1959). They suggested a movement of the stimulus receiving structures (the photoreceptor molecules) relative to the cell wall. Because of a movement of adaptation relative to the wall, Varjú *et al.* (1961) suggested a movement of photoreceptors, which would distribute them widely. Delbrück and Varjú (1961) furthermore found that the sensitive and adapting zones did not move relative to the sporangium, so indicating that they moved relatively both to the growth system and the cell wall.

From UV adaptation experiments with stage IVb SPPH rotated for 90° around the long axis between an adapting and a stimulating light, Dennison and Bozof (1973) suggested a local autonomy and a rotation of the photoreceptors (cf. also Dennison, 1965). Dennison and Foster (1977; cf. Foster, 1977; Medina and Cerdá-Olmedo, 1977*a*; Dennison, 1979) derived a "rotation of photoreceptor hypothesis" from data on rotation experiments. This hypothesis should overcome the paradox of the adapting and therefore temporary light–growth response and the nonadapting and hence in principle continuous phototropism (cf. Dennison, 1965). A rotation of the photoreceptors in the growing zone through the focused bright light band on the rear side should covert the spatial gradient into a strong temporary one. By taking into account the lag time between stimulus perception and the start of bending, it should also explain the observed phototropic "aiming error" (Varjú *et al.*, 1961; Dennison, 1965). But besides the fact that some doubts on the analysis of the experimental data exist (cf. their Fig. 5; dealing with: How to fit here a line with confidence; and why only after 20 min?) there are other discrepancies. The minimum responses in UV (0°/min) and blue light (10°/min) did not occur at the same external rotation speeds, as if UV and visible light would be effective differently in the different longitudinal regions of the growing zone. The strong negative phototropism to a diverging beam (Shropshire, 1962), where no bright light band and hence no strong intensity gradient exist, cannot be explained, as Dennison pointed out in 1965. For some further serious objections, see the review by Hertel (1980). The attractive but not convincing hypothesis of Dennison and Foster (1977) should be tested.

Castle (1933*b*) suggested in connection with his *path-length theory* (cf. Section 4.5.5.c) that the photoreceptors should not lie at the wall, but should lie uniformly distributed in the cytoplasm of the growing zone of stage IVb SPPH, which was assumed not to possess a vacuole (but cf. R. Cohen and Delbrück, 1959; Castle, 1965; Bergman *et al.*, 1969). Delbrück and Shropshire (1960), assumed that the photoreceptors could not be localized at the wall, but at some distance inside in the cytoplasm or at the tonoplast (cf. Zankel *et al.*, 1967; Bergman *et al.*, 1969). The latter investigators and Zalokar (1969) found no coincidence of the absorption spectra in centrifuged layers of single SPPH and the action spectra. So they speculated that the photoreceptor either was not concentrated in a particular organelle, or had a very low absorption, or was attached to the cell wall. These speculations interfered with the observation of birefringent crystals in the growing zone located near the vacuole with riboflavin-like absorption spectra (Wolken, 1969; Ootaki and Wolken, 1973). Meistrich *et al.* (1970) found by horizontal tracing with laser microbeams the phototropic sensitivity (but the response occurred perpendicularly to the incident light as in the half-side illumination method of Buder, 1920) spread all over the cytoplasmic part of the diameter, indicating that the photoreceptor molecules could not be restricted to either tonoplast or plasmalemma.

All experiments with polarized light were performed on stage IVb SPPH. Castle (1934*a*) detected that for phototropic balance the fluence rate of vertically polarized light (i.e., electric vector parallel to the long axis of the SPPH) had to be 10–15% higher than that of horizontally polarized light (i.e., electrical vector perpendicular to the long axis). He speculated on different reflection losses to explain this, as did Shropshire (1959), when he obtained similar results for light–growth response and phototropism. These "reflection explanations" were rejected by Jaffe (1960), (see also Jaffe, 1958) in a purely theoretical publication. He analyzed their results and showed that different reflection losses could not explain the observed different effectivities. Thus he proposed anisotropic absorption by uniaxial (but cf. Jesaitis, 1974) oriented dichroic photoreceptors linked to the cell wall (?). The theory was calculated only for tally transparent and totally opaque cells and interpolated linearly for unknown real intermediate conditions. Jesaitis (1974) found this interpolation questionable. Etzold (1961) could not detect true polarotropism (cf. Section 4.2.1), which could either be due to the subterminal growing zone (Etzold, 1961) or could argue against a perfectly uniform photoreceptor orientation (cf. Hertel, 1980). This was supported by Haupt and Buchwald (1967), who supposed that the photoreceptors were probably oriented partially horizontally, and the rest at random. They excluded a spiral orientation with an inclination of more than 5–10°. Jaffe's hypothesis was tested by Jesaitis (1974) for the light–growth response. He confirmed the different effectivities of horizontally and vertically polarized light and showed that *wt* and mutants with more or less carotenoids than *wt* behaved in the same way. He concluded that the effectiveness of differently polarized light should be a consequence of dichroic-oriented photoreceptor chromophores (action dichroism). For blue light, the main absorption axis should be nearly horizontal (equatorial and tangential) and for 280 nm inclined at 45°. This general belief was recently shaken by Hertel's short report on Wulff's results (Hertel, 1980). In strict contradiction, he inferred a radial orientation of the photoreceptor chromophores, because of higher adaptation, i.e., more light absorption took place on the flanks and not on the proximal and distal sides, when unpolarized and horizontally polarized, but not when vertically polarized light was applied. Further work is needed.

The results of the polarized light experiments indicate that the photoreceptor is probably more or less rigidly bound to a cell structure (Delbrück *et al.*, 1976) which is not in quick and random motion in the protoplasmic stream (cf. Bergman *et al.*, 1969). Otherwise the indicated relatively stable orientation cannot be maintained. The best guess for such a structure is some kind of membraneous component. So nearly all more recent reviews, which have dealt more or less with *Phycomyces* phototropism, considered a membrane localization as most probable (e.g., Dennison, 1979; Shropshire, 1979; Haupt, 1980; Hertel, 1980; Song, 1980*a,b*; Senger and Briggs, 1981). Also, Lipson (1975*a*) speculated extensively on a membrane association of the photoreceptor chromophore.

The very first hint for a localization of the plasmalemma, as seen from today's viewpoint, came from the phototropic inversion experiments of Buder (1918). He believed the fluence rates on the proximal and distal walk, of stage IVb SPPH were important for the direction of bending (cf. Buder, 1946, cited in Bergman *et al.*, 1969). The plasmalemma as photoreceptor site was recently proposed by Hertel (1980), who cited a model (H. Sierp and R. Hertel, in press) that postulates a multimeric macromolecule integrated into the plasmalemma and a freely diffusing ligand, that should bind tightly and in a cooperative way to form the functional photoreceptor.

On the other hand, the tonoplast was a candidate for photoreceptor localization (Delbrück and Shropshire, 1960; Zankel *et al.*, 1967).

Another possibility would be the mitochondria, which may somehow be oriented in the growing zone (Foster, 1977). A localization on mitochondrial membranes could explain both the relatively uniform distribution between cell wall and vacuole (Meistrich *et al.*, 1970) and the membrane association. Moreover, probably enough flavoproteins were then available as photoreceptor candidates in contrast to the situation with other membranes (Hertel, 1980). But then the stimulus had somehow to be transmitted to the plasmalemma, where the growth output takes place (Haupt, 1977).

Up to now, the subcellular localization of the photoreceptor responsible for phototropism on the organelle or membrane level is totally obscure.

4.5.5.c. Optics and Formation of Gradient of Light Intensity. Lens effects are often assumed or proved to play rolls in diverse fungal phototropisms.

Doubts concerning the lens effects in themselves (cf. Section 4.3.5.c) may in some cases be ruled out by experiments such as Shropshire's (1962) in *Phycomyces*.

The phototropically active stages I and IV of all studied *Pilobolus* spp. exhibit positive phototropism in visible light. But the mechanisms in stage I SPPH of different species differ markedly and may involve positive or negative light–growth responses, bulging or bowing mechanisms, and lens or screening actions. In stage IV SPPH, a lens action is generally accepted (cf. Page, 1962; Jacob, 1964; Foster, 1977), but Page (1968) put more emphasis on the screening by the carotenoid ring below the subsporangial swelling. He proposed that the main function of the swelling could be the formation of a jet for propulsion of the sporangium on shooting (Page, 1964).

Pringsheim and Czurda (1927) assumed a screening mechanism by yellow pigments (carotenoids) in stage I SPPH of *Pilobolus* (probably *P. crystallinus*) and hypothesized some kind of lens action by the subsporangial swelling in stage IV SPPH.* By taking into account the measured or assumed (contents and wall, respectively) refractive indices of swellings of stage IV SPPH of probably *P. kleinii*, van der Wey (1929) rejected their path of rays assumed for stage IV SPPH of *P. crystallinus*, but also suggested a lens effect (the contradictions of which may be explained by morphological differences between the species). Jacob (1959) postulated a lens action in stage I SPPH of *P. umbonatus* because of inversion of positive phototropism under liquid paraffin. He reported unpublished results of Paul and Schmidt on a screening effect by carotenoids in stage I SPPH of *P. kleinii, P. sphaerosporus*, and *P. longipes*. Experiments with DPA addition (inhibition of synthesis of colored carotenoids) confirmed these results for stage I SPPH of *P. kleinii* and *P. longipes* (Jacob, 1964). But whereas the residual screening by the opaque cytoplasm was regarded as sufficient to prevent a lens action in *P. kleinii*, in *P. longipes* the screening obviously failed. The now dominating lens effect created negative phototropism that could be inverted under paraffin (inversion of inversion). He claimed that screening by carotenoids was necessary for normal phototropism in *P. longipes*, but not in *P. kleinii*. In stage IV SPPH of both species DPA addition did not influence positive phototropism, indicating that the carotenoids are possibly located in a region where they do not interfere with the lens action (Dennison, 1979). Page and Curry (1966), in contradiction to Jacob; proposed a lens action in stage I SPPH of *P. kleinii*. They found negative phototropism in UV, which was assumed to indicate an internal screen of unknown chemical nature. The failure of phototropic inversion in mineral oil (refractive index,

*But for stage I of this species, contradicting unpublished results on a lens action of Paul (1950) were reported by E. Bünning (1956), G. H. Banbury (1959), and F. Jacob (1959).

1.47), as was found by Jacob (1964), was not explained by screening, but by the very high refractive index of the SPPH tip (1.57). Their first result indicates a lens action, but a separate UV photoreceptor, although improbable, cannot be excluded. Up to now clear and unsolved contradictions persist on the kind of light–growth response (and bending mechanism; cf. Page and Curry, 1966; Carlile, 1970) between the results of Paul and Jacob and those of Page and Curry (cf. Sections 4.5.2.a and 4.5.2.b). These must be solved, before one can decide for lens actions or for screening effects in phototropism of stage I SPPH of *Pilobolus* spp. (cf. Carlile, 1965; Dennison, 1979; Hertel, 1980).

Buder (1920) proved by his ingenious half-side illumination technique that in *Phycomyces* phototropism (stage I) the light gradient, rather than the light direction, determines the bending direction. This gradient is produced by the counteraction of a lens effect with an attenuation by absorption, reflection, and scattering (cf., e.g., Bergman *et al.*, 1969; Foster, 1977).

Blaauw (1914*a,b*) was the first to claim that the hyaline growing zone of stage IVb SPPH of *Phycomyces* acts as a cylinder lens with a focal line just behind the SPPH (later found by Castle, 1933*a*). In air the light should be concentrated in a bright band at the distal side. So Blaauw tried to make positive light–growth response and positive phototropism coincident. By immersion in liquid paraffin Buder (1918) inverted the phototropic bending (confirmed by Banbury, 1952*a*) and regarded this as conclusive proof for the lens action. In paraffin (refractive index, 1.47) the lens was changed from convex (concentrating) to concave (diverging). Oehlkers (1927) doubted the importance of a lens action and proposed a total reflection on the distal side. These doubts were ruled out by Castle (1933*b*). Further doubts were evoked by Ziegler (1950) by detection of inversion by paraffin also in *Avena* coleoptiles. This inversion was not regarded as due to a lens effect, but to altered surface properties by the dielectricity of the oil. Shropshire (1962) proved that in *Phycomyces* a lens action is the cause of inversion, and hence of phototropism, in visible light. He produced a diverging beam of light by putting a small glass cylinder lens in the incident beam near the SPPH. He obtained in air a strong phototropic inversion in blue light, because the SPPH could not focus the diverging rays within the SPPH.

Castle (1933*a*) determined the overall refractive index of the growing zone as 1.38. He assumed that in the growing zone no vacuole was present, but its occurrence in this region (Castle and Honeyman, 1934; Bergman *et al.*, 1969) should not change too much. Reichardt and Varjú (1958) calculated the intensity distribution of visible light caused by lens action and showed that only one-fifth of the distal side is illuminated. Theoretical polar logarithmic plots were presented by Dennison and Foster (1977) for the light intensity distribution on the inner wall surfaces when unilateral visible or UV light is given perpendicularly to the longitudinal axis. Banbury (1952*a*) calculated that a

10% difference in light fluence rates applied antagonistically, because of lens action led to a 0.5% difference in total absorbed energy between both halves of the SPPH. So the discriminatory capacity should be very sensitive. Castle (1965) elaborated, on the basis of lens effect data, that for two light beams 180° opposed to each other the resulting bending speed is given by the difference-to-sum-ratio of the two light fluence rates.

All above cited works are restricted to light incident perpendicularly to the longitudinal SPPH axis. Varjú *et al.* (1961) considered cases of obliquely incident light beams. They explained the photogeotropic equilibrium angle in visible light of ~ 30° above horizontal by the interplay between weak negative geotropism and rapidly decreasing phototropic stimulus at increasing bending angles. This decrease should be produced by a rapid change in the light distribution on the distal side brought about by the obliquely irradiated cylinder lens. That the final bend in UV and under paraffin is horizontal was consequently explained by a lack of lens action in these cases. Dennison (1965) found that the bending speed was constant at angles of 90–45° between incident light and SPPH axis and decreased at angles below 45°, reaching zero at 14°. He suggested this was due to the optical properties of the obliquely irradiated cylinder lens and proposed an amplification model. On the dynamics of the lens action during bending, see also Foster (1977).

Blaauw (1919) considered light attenuation not as important for the different light action in the SPPH halves. Oehlkers (1927), on the contrary, thought that much light should be absorbed. Castle (1933*b*) hypothesized that the focusing advantage should fail in cases of great absorption or scattering. For the inversion of phototropism in UV Curry and Gruen (1957) proposed no need to a lens effect. That this inversion was due to the failure of the focusing was proposed by Carlile (1957). Banbury and Carlile (1958) explained so the lack of adaptation on the distal side found by Curry and Gruen (1957). The effective screening in UV, causing inversion below ~ 320–310 nm, is due to high concentrations of gallic acid quite uniformly distributed in stage IVb SPPH (Dennison, 1959*a*). This concentration should give about 1% transmission of light at wavelengths below 290 nm, when a path length of about 100 μm was assumed. A gallic acid localization in the vacuole (Delbrück and Shropshire, 1960) or in the vacuole and autophagic vesicles (Bergman *et al.*, 1969) should be of no importance for the effectiveness of attenuation. So no need for Foster's (1977) assumption of UV screening by large amounts of carotenoids is given. Foster (1977) cited his own thesis (1972) to draw attention to the suggestion that, as attenuation factor in visible light, scattering should be more important than absorption and that it has the dual effect of diminishing the intensity inside the focused light band and raising it outside.

The small absorption around the inversion point (from positive to negative phototropism) between 320 and 310 nm points to a small focusing advantage (Delbrück and Shropshire, 1960). Buder (1920) found phototropic indifference of stage I SPPH in water (i.e., balance between focusing advantage and screening), Stifler (1961) found phototropic inversion in stage IVb SPPH in water, although its refractive index (1.33) is slightly lower than that of the stage IVb SPPH (1.38). He explained this by an internal absorption of 0.1 (Delbrück and Shropshire, 1960; in the visible) overcompensating the slight residual lens effect. This interpretation was supported by Shropshire (1962). He determined with miscible fluorochemicals of different refractive indices the balance point between focusing and attenuation in blue light at a refractive index of 1.295 and calculated a focusing advantage of about 14% to balance attenuation. The critical refractive indices (for no bending to occur) hardly differed between wavelengths (495 and 510 nm) where great differences in attenuation were present (Zankel *et al.*, 1967). They tried to explain this by the assumption of a photoreceptor localization at the tonoplast (cf. Section 4.5.5.b).

One important paradox is evident: There is a focusing advantage although the distal half of the SPPH receives at all wavelengths at least a little bit less light than the proximal half because of screening. Blaauw (1919) proposed that the strong illumination of the middle part of the distal wall was most important and should overcompensate the loss by screening (a precursor to the *lever arm theory*). Oehlkers (1927) doubted the prominent role of the lens focusing. He thought that the distal half received more light than the proximal because in the distal half the light paths should be longer and because by total reflection there should be more "absorption points" on the distal wall (*path length theory*). Castle (1933*b*) modified this *path length theory*. By lens action and equally distributed photoreceptors in the cytoplasm, more light should be absorbed on a longer path (factor 1.26) in the distal half than in the proximal. He supported therefore Buder's (1918) interpretation of the inversion of action of a concave lens that provokes shorter paths in the distal half. Buder (1947; cited in Bergman *et al.*, 1969; Foster, 1977) suggested the *lever arm theory*: This hypothesis inferred (photoreceptor localization close to the wall and lens action assumed) that the light absorbed in the distal half is more effective, because there the average distance from the midline is greater than in the proximal half.

Both theories give identical results for the focusing advantage if attenuation is ignored (Bergman *et al.*, 1969), but both cannot tie together the transient light–growth response and the (in principle) continuous phototropism. This was tried in the model by Dennison and Foster (1977; cf. Section 4.5.5.b, also for criticism): The sharp focused light band in the distal half was assumed to evoke, via the rotating photoreceptors, a much greater light–growth response than the unfocused light in the proximal half.

4.5.6. Transduction of Stimulus

Reports on stimulus transmission and transduction are very rare in fungi other than *Phycomyces*. Early general speculations on fungal growth regulation (including differential growth regulation in phototropism) can be found in the works of Bünning (1938*b*), Jeffreys and Greulach (1956); Banbury (1959), Carlile (1965, 1970), and Page (1968).

For *Phycomyces* it is assumed that the sensory transduction chains for the light–growth response and phototropism have many components in common (see, e.g., Bergman *et al*., 1969; Dennison and Foster, 1977; Medina and Cerdá-Olmedo, 1977*a*; Russo, 1980; Russo and Galland, 1980).

Banbury (1952*b*) tested IAA for induction of tropism in *Phycomyces*. The results were negative. But griseofulvin, applied to one side of the SPPH at a concentration of 100 μg/ml, produced a local increase in the rate of extension of the wall, with consequent curvature. It is not known whether griseofulvin is produced by *Phycomyces*. Goodell (1971) found that the elongation and the branching of mature SPPH were controlled by the sporangium. He claimed that the controlling substance is produced by the spores.

Phycomyces is the only organism in which a genetic dissection of the sensory transduction chain of phototropism was undertaken. Bergman *et al.* (1973) isolated mutants abnormal in phototropism. These mutants were divided into three classes: class 1.1, class 1.2, and class 2. The mutants in class 1.1 were also blind for the photoinduction of β-carotene synthesis (LICS) and for the photoinitiation of SPPH. The mutants in class 1.2 were specific for phototropism. The mutants in class 2 were abnormal also in autochemotropism and geotropism. These workers suggested a model for the sensory transduction chain, consisting of a photoreceptor, early and late transducers, a logarithmic transducer, and a growth-controlling element. Ootaki *et al.* (1974, 1977) have complemented these mutants and found that they belong to seven genes: in class 1.1 there are the mutants in the genes *mad*A, and *mad*B; in class 1.2 the mutants in gene *mad*C; and in class 2 the mutants in the genes *mad*D, E, F, G. Bergman *et al.* (1973) suggested that the gene products of *mad*A and *mad*B are needed at the beginning of the sensory transduction chain, while the gene products of *mad*D, E, F, G are needed at the end of the chain. Eslava *et al.* (1976) and Lipson *et al.* (1980) found that all seven genes are unlinked. Russo *et al.* (1980) suggested that the same blue light photoreceptor is responsible for all blue light responses in *Phycomyces*. For more information on the genetic analysis, see Cerda-Olmedo (1977), Lipson (1980), Russo (1980), Russo and Galland (1980).

Biochemical studies on the sensory transduction chain of phototropism in *Phycomyces* have also been done. Jan (1974) partially characterized the chitin synthetase in *Phycomyces* (cf. Section 4.5.2.a). In addition, he reported preliminary results on a blue light effect *in vitro*: 30% more chitin synthesis than was found in the control. This type of experiment has not been continued, however.

R. J. Cohen (1974*a*) discovered that light induced a drop of 50% in the level of cAMP of SPPH within 1 min and cAMP added to a submersed SPPH induced a transient growth inhibition. He therefore suggested that the change in the level of cAMP is a link in the sensory transduction chain of light–growth response (and of phototropism). R. J. Cohen (1978) found that the level of cAMP in one *mad*D mutant was not regulated by light, i.e., the light–growth response of *mad*D is much smaller than that of *wt* (Foster and Lipson, 1973). This result did not contradict the previous hypothesis. Unfortunately, no other mutants have been checked until recently. R. J. Cohen and Atkinson (1978) found an *in vitro* photoactivation of *Phycomyces* cAMP phosphodiesterase (50%) by blue light (6 $\mu W/cm^2$, continuous illumination for 20 min). This enzyme was further characterized by R. J. Cohen (1979). R. J. Cohen *et al.* (1980) found an *in vitro* photoactivation (50%) of *Phycomyces* adenylate cyclase by blue light (1 $\mu W/cm^2$). In both cases the *in vitro* activation is at the limit of significance. It is nevertheless a very interesting approach. It might prove rewarding to compare the behavior of the different mutants.

A different biochemical approach was undertaken by Fischer and Thomson (1979). These workers partially characterized *Phycomyces* serine proteinase and their inhibitors in *wt*. They stated, without giving data, that preliminary results indicated different properties of the enzymes in mutants of class 2. These proteinases could be of interest because they are probably membrane bound and might activate the chitin synthase.

5. QUEST FOR THE PHOTORECEPTOR RESPONSIBLE FOR PHOTOTROPISM

5.1. Criteria for Identifying the Photoreceptor Suggested by Physiological Studies

Phototropism is initiated in different plants primarily by UV blue light, but not exclusively so, as was erroneously stated by Galston (1974). In fact, red light can induce phototropism of filaments of *Mougeotia* (Section 4.2), of chloronemata of *Funaria* and *Physcomitrium* (Section 4.3), of rhizoids of *Dryopteris*, and of chloronemata of *Pteridium, Struthiopteris*, and *Dryopteris* (Section 4.4). At least in the last case, the photoreceptor is most probably phytochrome. This is based not only on the action spectrum of polarotropism, but also on the far-red reversibility typical of the phytochrome effects.

For blue light phototropism the only criterion is the action spectrum. The action spectra for blue light phototropism are very similar, an indication that perhaps the same chromophore, if not the same photoreceptor, is involved in the phototropic response. Some investigators stated either explicitly or implicitly that the tip sensitivity effect in *Avena* and *Zea* (Section 3.3.5) suggest that instead of an even distributor of photoreceptor in the coleoptile, there must be a gradient of photoreceptor concentration (Curry, 1969; Senger and Briggs, 1981). However, the tip sensitivity can be attributed to a concentration gradient of any component of the sensory transduction chain and not necessarily to a photoreceptor gradient. Furthermore, if it were caused by such a gradient, the fluence–response curve of coleoptiles partially illuminated at different distances from the tip would be shifted toward higher fluences, as discussed in Section 2.2.1. The only data we have on this subject are those of Meyer (1969*a*), which indicate no shift, but rather a partial inhibition of the response at any fluence (Section 3.3.5). These latter data run counter to the claim that the tip sensitivity is evidence for a gradient of photoreceptor concentration.

In those plants in which polarotropism and action dichroism occur, it is safe to say that the photoreceptor is oriented. But is this evidence that the photoreceptor is on the plasmamembrane? Can this be used as a criterion for photoreceptor isolation? We shall now concentrate on the blue light photoreceptor of *Zea* and *Avena*, because most of the work on the isolation and identification of the photoreceptor was carried out with these organisms. Some works and the many speculations on lower plants have already been discussed in Section 4.

5.2. The Action Spectrum Approach

This approach is taken by those who believe that an action spectrum should be similar to, but not necessarily identical with, the absorption spectrum of the photoreceptor. One difficulty in this approach is that it predicts that in the ideal case the fluence–response curves at any wavelength should be parallel to each other (Shropshire, 1972). In reality, this is never the case as we have shown. Why is there no parallelism? Some investigators suggest that other pigments are screening in some region of the spectrum. But what then is the parameter that should be used in the construction of the action spectrum?

Wun *et al.* (1977) found that the absorption spectrum of the purified photoreactivating enzyme (photolyase) of *Escherichia coli* is not similar at all to the action spectrum of photoreactivation (maximum at 380 nm), unless the enzyme is bound to UV-treated DNA. In this case, the chromophore would be the DNA–enzyme complex.

Historically, the first inquiries into the nature of the blue light photoreceptor were made by Castle (1935), who suggested α-carotene for *Phycomyces*, and by Bünning (1938*a,b*), who suggested β-carotene for *Avena*. In *Avena* coleoptiles, carotenoids were discovered by Wald and du Buy (1936).

Galston (1949) was the first to suggest that riboflavin might be the chromophore of the blue light photoreceptor. Flavins and carotenoids have similar but not identical absorption spectra. However, it is difficult on the basis of action spectra alone to decide whether one or the other is the chromophore (on this problem, see Schmidt, 1980).

Even if the chromophore is a flavin, certain questions remain: How can we decide which flavoprotein out of the many present in any cell is the photoreceptor? Must the photoreceptor be a flavoprotein, or could it be simply a flavin? Any cell contains riboflavin, FAD, and FMN, but new flavins are continually being discovered. In *Avena* Zenk (1967) found a new flavin, called FX, which he suggested to be a riboflavin ester. In microorganisms several other flavins have been found (see the reviews of Hemmerich, 1976; Müller, 1983). The absorption spectra of flavoproteins and of these new flavins are so different from that of riboflavin that an action spectrum in a given organism is difficult to interpret as long as no accurate analysis of flavins is made in the organism under consideration.

Gressel (1979) suggested the term *cryptochrome* for the UV blue light photoreceptor. An action spectrum is certainly useful for distinguishing between a cryptochrome and a cytochrome as photoreceptor.

5.3. The Inhibitor Approach

The philosophy behind the inhibitor approach is to use specific inhibitors of carotenoid synthesis or specific flavin quenchers.

Bara and Galston (1968) were able to obtain conditions, using inhibitors of carotenoid synthesis, in which the carotenoids of *Avena* were diminished to 20% of the control. The fluence–response curve was like that of the control. These workers concluded that "riboflavin seems at present to be the likeliest candidate for the role of photoreceptor." Similar results were obtained in *Zea* by Vierstra and Poff (1981*b*). Schmidt *et al.* (1977), using specific quenchers of flavins–(azide, phenylacetic acid, and iodide), suggested that in *Zea* the chromophore of the photoreceptor is a flavin and that the flavin triplet state is the active species. Vierstra *et al.* (1981) criticized the latter work saying that their conclusions were based essentially on the quenching effect of KI and that at the concentration used by Schmidt *et al.* (1977) it is not a specific quencher for the triplet state of flavin. Vierstra *et al.* (1981) used xenon as a specific quencher of the flavin triplet state. They found no xenon inhibition of phototropism in *Zea* coleoptiles at 2 $\mu W/cm^2$ of continuous blue light and at a xenon concentration that inhibits the triplet state in an *in vitro* photooxidation of NADH. They concluded that the flavin singlet state is more likely to be involved in the primary photoprocess of phototropism in *Zea*. Vierstra and Poff (1981*a*) suggested the use of phenylacetic acid as a photoaffinity label of the photoreceptor. However, they obtained a fluence–response curve of *Zea*, which is completely different from the one they reported on a few months later (Vierstra and Poff, 1981*b*).

We do not consider the statements of Schmidt *et al.* (1977) and Vierstra and Poff (1981*a*) acceptable as long as they do not show a fluence–response curve for any given quencher they use. According to our discussion in Section 2.2.1, any quencher of the photoreceptor should shift the fluence–response curve. The only data on this subject are those of Meyer (1969*a*), who used KI at 10^{-4} *M* in *Avena* phototropism in the first-positive and in the first-negative range. From the crude fluence–response curve, no shift is apparent but there is a general inhibition of phototropism at the three fluences used.

The interpretation by Vierstra *et al.* (1981) of their own xenon experiments is also unacceptable. Xenon could partially quench the photoreceptor, but Vierstra *et al.* could not have seen this hypothetical quenching because they measured at a fixed fluence rate of 2 $\mu W/cm^2$ for 3 hr. This fluence should give just a saturation (according to the fluence–response curve of Vierstra and Poff, 1981*a*), or it is more than 100 times higher than the fluence needed to reach saturation (according to Vierstra and Poff, 1981*b*). Also in this case only a fluence–response curve could permit any conclusions.

5.4. The Genetic Approach

The genetic approach has been used up to the present only to exclude β-carotene as a chromophore. Galston (1950) stated that a mutant strain of *Zea* that had less than 0.01 the normal concentration of carotenoids is normally phototropic at a fluence of 600 mcs. The work of Presti *et al.* (1977), which excluded carotenoids as photoreceptors for *Phycomyces* phototropism, has already been discussed in Section 4.5.5.b.

5.5. The LIAC Approach

A light-induced absorbance change (LIAC) was first found by Lewis *et al.* (1961) in *Euglena.* These workers showed that cytochrome 552 could be photooxidized *in vitro* by a flavoprotein of *Euglena.*

Poff and Butler (1974*b*) obtained opposite results with *Phycomyces*, and also with *Dictyostelium discoideum*, for which no blue light effects are known. The LIAC was measured *in vivo* and an endogenous b-type cytochrome was photoreduced. Lipson and Presti (1977) confirmed the data for *Phycomyces* and found a LIAC as well in HeLa cells (a human cervical carcinoma in which no blue light effects are either known or expected). The quantum yield for LIAC was 0.015, while it is believed to be 1 for the light–growth response

(Lipson, 1975*b*). Lipson and Presti therefore questioned the physiological relevance of this LIAC.

A b-type cytochrome was found in a plasmamembrane fraction of *Zea* (Jesaitis *et al.*, 1977). Also in *Zea* coleoptile extracts a LIAC was found (Brain *et al.*, 1977). In coleoptiles of *Triticum aestivum*, a LIAC was found *in vivo* with an action spectrum, which suggested a cytochrome as photoreceptor (Widell and Björn, 1976; it is interesting that iodide inhibited 50% of this LIAC, challenging the specificity of iodide as a quencher of only flavin triplet). Schneider and Bogorad (1978) obtained similar results for *Zea in vivo*. Britz *et al.* (1979) and Widell *et al.* (1980) showed with *Zea* extracts that a photoreduction of the b-type cytochrome can be obtained with red light in the presence of methylene blue. The LIAC effect in *Zea* extracts was better characterized and the flavin–cytochrome complex was partially purified (Goldsmith *et al.*, 1980; Widell, 1980; Leong and Briggs, 1981). Lipson and Presti (1980) calculated that for this LIAC of *Zea* extracts the quantum yield could be 1.

Is the flavoprotein in the plasma membrane, which is responsible for the LIAC of the b-type cytochrome, also the blue light photoreceptor for phototropism? There is no evidence either for or against. It is interesting that the photoreceptor for the LIAC of *Zea* and *Triticum* coleoptiles *in vivo* seems to be a hemoprotein.

5.6. The Physicochemical Approach

Schmidt and Butler (1976) suggested that a flavin-mediated photoreduction in an artificial system, causing production of superoxide and hydrogen peroxide, could be "a model for the blue light photoreceptor pigment in living systems."

On the basis of physicochemical arguments and comparative photobiochemistry, Song (1980*a,b*) and Song *et al.* (1980*a*), suggested several criteria for the photoreceptor: (1) it should be a flavin covalently bound to an apoprotein, (2) the flavoprotein should be membrane bound, and (3) it should be fairly fluorescent, with a relatively short fluorescence lifetime. The photochemical role of flavin in triggering the phototropic event should be a photoreduction. Song and co-workers reported preliminary results on the discovery of such a flavoprotein.

ACKNOWLEDGMENTS. We would like to thank the Deutsche Forschungsgemeinschaft for generous financial support during the preparation of this contribution, which is dedicated to Professor Max Delbrück (1906–1981). We also appreciate the technical help of Christiane Ernsting, Marion Kamke, and Dagmar Manglitz and the devoted secretarial work of Ms. R. Fischer. Finally, we would like to extend our gratitude to our families for their understanding during the preparation of this manuscript.

REFERENCES

Abbot, M. T. J., and Grove, J. F., 1959, Uptake and translocation of organic compounds by fungi. I. Microspectrophotometry in the study of translocation, *Exp. Cell Res.* **17**:95–104.

Arisz, W. H., 1915, Untersuchungen über den Phototropismus, *Recl. Trav. Bot. Neerl.* **12**:44–216.

Asana, R. D., 1938, On the relation between the distribution of auxin in the tip of the *Avena* coleoptile and the first negative phototropic curvature, *Ann. Bot.* **2**:955–975.

Atkins, G. A., 1936, The effect of pigment on phototropic response: a comparative study of reactions to monochromatic light, *Ann. Bot.* **50**:197–218.

Bachmann, F., and Bergann, F., 1930, Über die Wertigkeit von Strahlen verschiedener Wellenlänge für die phototropische Reizung von *Avena sativa, Planta* **10**:744–755.

Backus, G. E., and Schrank, A. R., 1952, Electrical and curvature responses of the *Avena* coleoptile to unilateral illumination, *Plant Physiol.* **27**:251–262.

Ball, N. G., 1962, The effects of externally applied 3-indolylacetic acid on phototropic induction and response in the coleoptile of *Avena, J. Exp. Bot.* **13**:45–60.

Banbury, G. H., 1952*a*, Physiological studies in the Mucorales. Part I. The phototropism of sporangiophores of *Phycomyces blakesleeanus, J. Exp. Bot.* **3**:77–85.

Banbury, G. H., 1952*b*, Physiological studies in the Mucorales. Part II. Some observations on growth regulation in the sporangiophore of *Phycomyces, J. Exp. Bot.* **3**:86–94.

Banbury, G. H., 1959, Phototropism of lower plants, in: *Encyclopedia of Plant Physiology,* Vol. 17 Part 1 (W. Ruhland, ed.), pp. 530–578, Springer-Verlag, Berlin.

Banbury, G. H., and Carlile, M. J., 1958, Phototropism of *Phycomyces* sporangiophores, *Nature (Lond.)* **181**:358–359.

Bandurski, R. S., and Schulze, A., 1974, Concentrations of indole-3-acetic acid and its esters in *Avena* and *Zea, Plant Physiol.* **54**:257–262.

Bandurski, R. S., and Schulze, A., 1977, Concentration of indole-3-acetic acid and its derivatives in plant, *Plant Physiol.* **60**:211–213.

Bandurski, R. S., Schulze, A., and Cohen, J. D., 1977, Photoregulation of the ratio of ester to free indole-3-acetic acid, *Biochem. Biophys. Res. Commun.* **79**:1219–1223.

Bara, M., and Galston, A. W., 1968, Experimental modification of pigment content and phototropic sensitivity in excised *Avena* coleoptiles, *Physiol. Plant.* **21**:109–118.

Bartnicki-Garcia, S., 1973, Fundamental aspects of hyphal morphogenesis, in: *Microbial Differentiation, Twenty-Third Symposium of the Society for General Microbiology,* (J. M. Ashworth and J. E. Smith, eds.), pp. 245–267, Cambridge University Press, Cambridge.

Bartnicki-Garcia, S., Bracker, C. E., Reyes, E., and Ruiz-Herrera, J., 1978, Isolation of chitosomes from taxonomically diverse fungi and synthesis of chitin microfibrils *in vitro, Exp. Mycol.* **2**:173–192.

Bates, G. W., and Cleland, R. E., 1979, Protein synthesis and auxin-induced growth: inhibitor studies, *Planta* **145**:437–442.

Bates, G. W., and Cleland, R. E., 1980, Protein patterns in the oat coleoptile as influenced by auxin and by protein turnover, *Planta* **148**:429–436.

Batt, S., and Venis, M. A., 1976, Separation and localization of two classes of auxin binding sites in corn coleoptile membranes, *Planta* **130**:15–21.

Batt, S., Wilkins, M. B., and Venis, M. A., 1976, Auxin binding to corn coleoptile membranes: kinetics and specificity, *Planta* **130**:7–13.

Bearder, J. R., 1980, Plant hormones and other growth substances—Their background, structures and occurrence, in: *Encyclopedia of Plant Physiology,* new ser., Vol. 9: *Hormonal Regulation of Development,* I (J. MacMillan, ed.), pp. 11–112, Springer-Verlag, Berlin.

Bentley, J. A., 1958, The naturally-occurring auxins and inhibitors, *Annu. Rev. Plant Physiol.* **9**:47–80.

Bentrup, F. W. (Meyer zu), 1963, Vergleichende Untersuchungen zur Polaritätsinduktion durch das Licht an der *Equisetum*-Spore und der *Fucus*-Zygote, *Planta* **59**:472–491.

Bentrup, F. W., 1968, Die Morphogenese pflanzlicher Zellen im elektrischen Feld, *Z. Pflanzenphysiol.* **59**:309–339.

Bergman, K., 1972, Blue-light control of sporangiophore initiation in *Phycomyces, Planta* **107**:53–67.

Bergman, K., Burke, P. V., Cerdá-Olmedo, E., David, C. N., Delbrück, M., Foster, K. W., Goodell, E. W., Heisenberg, M., Meissner, G., Zalokar, M., Dennison, D. S., and Shropshire, W., Jr., 1969, *Phycomyces, Bacteriol. Rev.* **33**:99–157.

Bergman, K., Eslava, A. P., and Cerdá-Olmedo, E., 1973, Mutants of *Phycomyces* with abnormal phototropism, *Mol. Gen. Genet.* **123**:1–16.

Berns, D. S., and Vaughn, J. R., 1970, Studies on the photopigment system in *Phycomyces, Biochem. Biophys. Res. Commun.* **39**:1094–1103.

Blaauw, A. H., 1908, Die Perzeption des Lichtes, *Recl. Trav. Bot. Neerl.* **5**:209–372.

Blaauw, A. H., 1914*a*, The primary photo-growth reaction and the cause of positive phototropism in *Phycomyces nitens, Proc. Kon. Akad. Wet. Amst.* **16**:774–786.

Blaauw, A. H., 1914*b*, Licht und Wachstum. I, *Z. Bot.* **6**:641–703.

Blaauw, A. H., 1915, Licht und Wachstum. II, *Z. Bot.* **7**:465–532.

Blaauw, A. H., 1919, Licht und Wachstum. III, (Die Erklärung des Phototropismus), *Meded. Landbouwhogesch. Wageningen* **15**:89–204.

Blaauw, O. H., and Blaauw-Jansen, G., 1964, The influence of red light on the phototropism of *Avena* coleoptile, *Acta Bot. Neerl.* **13**:541–552.

Blaauw, O. H., and Blaauw-Jansen, G., 1970*a*, Third positive (C-type) phototropism in the *Avena* coleoptile, *Acta Bot. Neerl.* **19**:764–776.

Blaauw, O. H., and Blaauw-Jansen, G., 1970*b*, The phototropic responses of *Avena* coleoptiles, *Acta Bot. Neerl.* **19**:755–763.

Blaauw-Jansen, G., 1958, The influence of red and far red light on growth and phototropism of the *Avena* seedling, *Acta Bot. Neerl.* **8**:1–39.

Blatt, M. R., and Weisenseel, M., 1980, Blue light stimulates a local electrical current efflux in the alga *Vaucheria sessilis, Carnegie Inst. Wash. Year Book* **1979**:123–125.

Bonner, J., 1934, The relation of hydrogen ions to the growth rate of the *Avena* coleoptile, *Protoplasma* **21**:406–423.

Bottelier, H. P., 1934, Über den Einfluss äusserer Faktoren auf die Protoplasmaströmung in der *Avena*-Koleoptile, *Recl. Trav. Bot. Neerl.* **31**:474–582.

Boysen Jensen, P., 1910, Über die Leitung des phototropischen Reizes in *Avena* Keimpflanzen, *Ber. Dtsch. Bot. Ges.* **28**:118–120.

Boysen Jensen, P., 1928, Die phototropische Induktion in der Spitze der *Avena*coleotile, *Planta* **5**:464–477.

Boysen Jensen, P., and Nielsen, N., 1926, Studien über die hormonalen Beziehungen zwischen Spitze und Basis der *Avena*coleoptile, *Planta* **1**:321–331.

Brain, R. D., Freeberg, J. A., Weiss, C. V., and Briggs, W. R., 1977, Blue light-induced absorbance changes in membrane fractions from corn and *Neurospora, Plant Physiol.* **59**:948–952.

Brauner, L., 1922, Lichtkrümmung und Wachstumsreaktion, *Z. Bot.* **14**:497–547.

Brauner, L., 1927, Untersuchungen über das geoelektrische Phänomen, *Jahrb. Wiss. Bot.* **66**:381–428.

Brauner, L., and Bünning, E., 1930, Geoelektrischer Effekt und Elektrotropismus, *Ber. Dtsch. Bot. Ges.* **48**:470–476.

Bridges, I. G., and Wilkins, M. B., 1971, Effects of electrolyte and nonelectrolyte solutions on the tropic responses of *Avena* coleoptiles, *J. Exp. Bot.* **22**:208–212.

Briggs, W. R., 1960, Light dosage and phototropic responses of corn and oat coleoptiles, *Plant Physiol.* **35**:951–962.

Briggs, W. R., 1963*a*, Red light, auxin relationships, and the phototropic responses of corn and oat coleoptiles, *Am. J. Bot.* **50**:196–207.

Briggs, W. R., 1963*b*, Mediation of phototropic responses of corn coleoptiles by lateral transport of auxin, *Plant Physiol.* **38**:237–247.

Briggs, W. R., 1964, Phototropism in higher plants, in: *Photophysiology*, Vol. 1 (A. C. Giese, ed.), pp. 223–271, Academic Press, New York.

Briggs, W. R., and Blatt, M. R., 1980, Blue light responses in the siphonaceous alga *Vaucheria*, in: *The Blue Light Syndrome* (H. Senger, ed.), pp. 261–268, Springer-Verlag, Berlin.

Briggs, W. R., and Rice, H. V., 1972, Phytochrome: Chemical and physical properties and mechanism of action, *Annu. Rev. Plant Physiol.* **23**:293–334.

Briggs, W. R., Tocher, R. D., and Wilson, J. F., 1957, Phototropic auxin redistribution in corn coleoptiles, *Science* **126**:210–212.

Britz, S. J., Schrott, E., Widell, S., and Briggs, W. R., 1979, Red-light-induced reduction of a particle-associated b-type cytochrome from corn in the presence of methylene blue, *Photochem. Photobiol.* **29**:359–365.

Bruinsma, J., Karssen, C. M., Benschop, M., and Van Dort, J. B., 1975, Hormonal regulation of phototropism in the light-grown sunflower seedling, *Helianthus annuus* L.: Immobility of endogenous indoleacetic acid and inhibition of hypocotyl growth by illuminated cotyledons, *J. Exp. Bot.* **26**:411–418.

Bruinsma, J., Franssen, J. M., and Knegt, E., 1980, Phototropism as a phenomenon of inhibition, in: *Plant Growth Substances* 1979, (F. Skoog, ed), pp. 444–450, Springer-Verlag, New York.

Buder, J., 1918, Die Inversion des Phototropismus bei *Phycomyces, Ber. Dtsch. Bot. Ges.* **36**:104–105.

Buder, J., 1920, Neue phototropische Fundamentalversuche, *Ber. Dtsch. Bot. Ges.* **38**:10–19.

Buder, J., 1932, Über die phototropische Empfindlichkeit von *Phycomyces* für verschiedene Spektralgebiete, *Beitr. Biol. Pflanz.* **19**:420–435.

Bünning, E., 1937, Phototropismus und Carotinoide. I. Phototropische Wirksamkeit von Strahlen verschiedener Wellenlänge und Strahlungsabsorption im Pigment bei *Pilobolus, Planta* **26**:719–736.

Bünning, E., 1938*a*, Phototropismus und Carotinoide. II. Das Carotin der Reizaufnahmezone von *Pilobolus, Phycomyces* und *Avena, Planta* **27**:148–158.

Bünning, E., 1938*b*, Phototropismus und Carotinoide. III. Weitere Untersuchungen an Pilzen und höheren Pflanzen, *Planta* **27**:583–610.

Bünning, E., 1956, Bewegungen, *Fortschr. Bot.* **18**:347–364.

Bünning, E., and Etzold, H., 1958, Über die Wirkung von polarisiertem Licht auf keimende Sporen von Pilzen, Moosen und Farnen, *Ber. Dtsch. Bot. Ges.* **71**:304–306.

Bünning, E., Reisener, H. J., Weygand, F., Simon, H., and Klebe, J. F., 1956, Versuche mit radioaktiver Indolylessigsäure zur Prüfung der sogenannten Ablenkung des Wuchshormonstromes durch Licht, *Z. Naturforsch.* **11b**:363–364.

Carlile, M. J., 1957, Phototropism of*Phycomyces* sporangiophores, *Nature (Lond.)* **180**:202.

Carlile, M. J., 1962, Evidence for a flavoprotein photoreceptor in *Phycomyces, J. Gen. Microbiol.* **28**:161–167.

Carlile, M. J., 1965, The photobiology of fungi, *Annu. Rev. Plant Physiol.* **16**:175–202.

Carlile, M. J., 1970, The photoresponses of fungi, in: *Photobiology of Microorganisms* (P. Halldal, ed), pp. 309–344, Wiley–Interscience, New York.

Castle, E. S., 1929, Dark adaptation and the light-growth responses of *Phycomyces, J. Gen. Physiol.* **12**:391–400.

Castle, E. S., 1930*a*, The light-sensitive system as the basis of the photic responses of *Phycomyces, Proc. Natl. Acad. Sci. USA* **16**:1–6.

Castle, E. S., 1930*b*, Phototropism and the light-sensitive system of *Phycomyces, J. Gen. Physiol.* **13**:421–435.

Castle, E. S., 1931*a*, Phototropic "indifference" and the light-sensitive system of *Phycomyces, Bot. Gaz.* **91**:206–212.

Castle, E. S., 1931*b*, The phototropic sensitivity of *Phycomyces* as related to wave-length, *J. Gen. Physiol.* **14**:701–711.

Castle, E. S., 1932*a*, On "reversal" of phototropism in *Phycomyces, J. Gen. Physiol.* **15**:487–489.

Castle, E. S., 1932*b*, Dark Adaptation and the dark growth response of *Phycomyces, J. Gen. Physiol.* **16**:75–88.

Castle, E. S., 1933*a*, The refractive indices of whole cells, *J. Gen. Physiol.* **17**:41–47.

Castle, E. S., 1933*b*, The physical basis of the positive phototropism of *Phycomyces, J. Gen. Physiol.* **17**:49–62.

Castle, E. S., 1934*a*, The phototropic effect of polarized light, *J. Gen. Physiol.* **17**:751–762.

Castle, E. S., 1934*b*, The spiral growth of single cells, *Science* **80**:362–363.

Castle, E. S., 1935, Photic excitation and phototropism in single plant cells, *Cold Spring Harbor Symp. Quant. Biol.* **3**:224–229.

Castle, E. S., 1936*a*, A model imitating the origin of spiral wall structure in certain plant cells, *Proc. Natl. Acad. Sci. USA* **22**:336–340.

Castle, E. S., 1936*b*, the origin of spiral growth in *Phycomyces, J. Cell. Comp. Physiol.* **8**:493–502.

Castle, E. S., 1937*a*, The distribution of velocities of elongation and of twist in the growth zone of *Phycomyces* in relation to spiral growth, *J. Cell. Comp. Physiol.* **9**:477–489.

Castle, E. S., 1937*b*, Membrane tension and orientation of structure in the plant cell wall, *J. Cell. Comp. Physiol.* **10**:113–121.

Castle, E. S., 1938, Orientation of structure in the cell wall of *Phycomyces, Protoplasma* **31**:331–345.

Castle, E. S., 1942, Spiral growth and reversal of spiraling in *Phycomyces,* and their bearing on primary wall structure, *Ann. J. Bot.* **29**:664–672.

Castle, E. S., 1953, Problems of oriented growth and structure in *Phycomyces, Q. Rev. Biol.* **28**:364–372.

Castle, E. S., 1958, The topography of tip growth in a plant cell, *J. Gen. Physiol.* **41**:913–926.

Castle, E. S., 1959, Growth distribution in the light-growth responses of *Phycomyces, J. Gen. Physiol.* **42**:697–702.

Castle, E. S., 1961*a*, Phototropism, adaptation, and the light–growth response of *Phycomyces, J. Gen. Physiol.* **45**:39–46.

Castle, E. S., 1961*b*, Phototropic inversion in *Phycomyces, Science* **133**:1424–1425.

Castle, E. S., 1965, Differential growth and phototropic bending in *Phycomyces, J. Gen. Physiol.* **48**:409–423.

Castle, E. S., 1966*a*, Light responses of *Phycomyces*, *Science* **154**:1416–1420.

Castle, E. S., 1966*b*, A kinetic model for adaptation and the light responses of *Phycomyces*, *J. Gen. Physiol.* **49**:925–935.

Castle, E. S., and Honeyman, A. H. M., 1934, The light growth response and the growth system of *Phycomyces*, *J. Gen. Physiol.* **18**:385–397.

Cerdá-Olmedo, E., 1977, Behavioral genetics of *Phycomyces*, *Annu. Rev. Microbiol.* **31**:535–547.

Cholodny, N., 1927, Wuchshormone und Tropismen bei den Pflanzen, *Biol. Zentralbl.* **47**:604–626.

Cholodny, N., 1933, Beiträge zur Kritik der Blaauwschen Theorie des Phototropismus, *Planta* **20**:543–576.

Chon, H. P., and Briggs, W. R., 1966, Effect of red light on the phototropic sensitivity of corn coleoptiles, *Plant Physiol.* **41**:1715–1724.

Clark, W. G., 1935, Note on the effect of light on the bioelectric potentials in the *Avena* coleoptile, *Proc. Natl. Acad. Sci. USA* **21**:681–684.

Clark, W. G., 1937, Polar transport of auxin and electrical polarity in coleoptile of *Avena*, *Plant Physiol.* **12**:737–754.

Cleland, R., 1971, Cell wall extension, *Annu. Rev. Plant Physiol.* **22**:197–222.

Cleland, R., 1973, Auxin-induced hydrogen ion excretion from *Avena* coleoptile, *Proc. Natl. Acad. Sci. USA* **70**:3092–3093.

Cleland, R. E., 1975, Auxin-induced hydrogen ion excretion: correlation with growth, and control by external pH and water stress, *Planta* **127**:233–242.

Cleland, R. E., Prins, H. B. A., Harper, J. R., and Higinbotham, N., 1977, Rapid hormone-induced hyperpolarization of the oat coleoptile transmembrane potential, *Plant Physiol.* **59**:395–397.

Cohen, R., and Delbrück, M., 1958, Distribution of stretch and twist along the growing zone of the sporangiophore of *Phycomyces* and the distribution of response to a periodic illumination program, *J. Cell. Comp. Physiol.* **52**:361–388.

Cohen, R., and Delbrück, M., 1959, Photoreactions in *Phycomyces:* growth and tropic responses to the stimulation of narrow test areas, *J. Gen. Physiol.* **42**:677–695.

Cohen, R. J., 1974*a*, Cyclic AMP levels in *Phycomyces* during a response to light, *Nature (Lond.)* **251**:144–146.

Cohen, R. J., 1974*b*, Some properties of chitinase from *Phycomyces blakesleeanus*, *Life Sci.* **15**:289–300.

Cohen, R. J., 1978, Aberrant cyclic nucleotide regulation in a behavioral mutant of *Phycomyces blakesleeanus*, *Plant Sci. Lett.* **13**:315–319.

Cohen, R. J., 1979, Adenosine 3′,5′-cyclic monophosphate phosphodiesterase from *Phycomyces blakesleeanus*, *Phytochemistry* **18**:943–948.

Cohen, R. J., and Atkinson, M. M., 1978, Activation of *Phycomyces* adenosine 3′,5′ monophosphate phosphodiesterase by blue light, *Biochem. Biophys. Res. Commun.* **83**:616–621.

Cohen, R. J., Ness, J. L., and Whiddon, S. M., 1980, Adenylate cyclase from *Phycomyces* sporangiophore, *Phytochemistry* **19**:1913–1918.

Cross, J. W., and Briggs, W. R., 1978, Properties of a solubilized microsomal auxin-binding protein from coleoptiles and primary leaves of *Zea mays*, *Plant Physiol.* **62**:152–157.

Cross, J. W., and Briggs, W. R., 1979, Solubilized auxin-binding protein, *Planta* **146**:263–270.

Cross, J. W., Briggs, W. R., Dohrmann, U. C., and Ray, P. M., 1978, Auxin receptor of maize coleoptile membranes do not have ATPase activity, *Plant Physiol.* **61**:581–584.

Curry, G. M., 1969, Phototropism, in: *The Physiology of Plant Growth and Development* (M. B. Wilkins, ed.), pp. 243–273, McGraw-Hill, London.

Curry, G. M., and Gruen, H. E., 1957, Negative phototropism of *Phycomyces* in the ultra-violet, *Nature (Lond.)* **179**:1028–1029.

Curry, G. M., and Gruen, H. E., 1959, Action spectra for the positive and negative phototropism of *Phycomyces* sporangiophores, *Proc. Natl. Acad. Sci. USA* **45**:797–804.

Curry, G. M., and Gruen, H. E., 1961, Dose response relationships at different wave lengths in phototropism of *Avena,* in: *Progress in Photobiology* (B. C. Christensen and B. Buchmann, eds.), pp. 155–157, Elsevier, Amsterdam.

Curry, G. M., Thimann, K. V., and Ray, P. M., 1956, The base curvature response of *Avena* seedlings to the ultraviolet, *Physiol. Plant.* **9**:429–440.

Darwin, C., and Darwin, F., 1880, *The Power of Movement in Plants,* John Murray, London.

Dassek, M., 1939, Der Phototropismus der Lebermoosrhizoide, *Beitr. Biol. Pflanz.* **26**:125–200.

Davis, B. D., 1975, Bending growth in fern gametrophyte protonema, *Plant Cell Physiol.* **16**:537–541.

de Fabo, E., 1980, On the nature of the blue light photoreceptor: still an open question, in: *The Blue Light Syndrome* (H. Senger, ed.), pp. 187–197, Springer-Verlag, Berlin.

de Fabo, E. C., Harding, R. W., and Shropshire, W., Jr., 1976, Action spectrum between 260 and 800 nanometers for the photoinduction of carotenoid biosynthesis in *Neurospora crassa, Plant Physiol.* **57**:440–445.

Delbrück, M., and Reichardt, W., 1956, System analysis for the light growth reactions of *Phycomyces,* in: *Cellular Mechanisms in Differentiation and Growth* (D. Rudnick, ed.), pp. 3–44, Princeton University Press, Princeton, N.J.

Delbrück, M., and Shropshire, W., Jr., 1960, Action and transmission spectra of *Phycomyces, Plant Physiol.* **35**:194–204.

Delbrück, M., and Varjú, D., 1961, Photoreactions in *Phycomyces:* responses to the stimulation of narrow test areas with ultra-violet light, *J. Gen. Physiol.* **44**:1177–1188.

Delbrück, M., Katzir, A., and Presti, D., 1976, Responses of *Phycomyces* indicating optical excitation of the lowest triplet state of riboflavin, *Proc. Natl. Acad. Sci. USA* **73**:1969–1973.

Dennison, D. S., 1959*a,* Gallic acid in *Phycomyces* sporangiophores, *Nature (Lond.)* **184**:2036.

Dennison, D. S., 1959*b,* Phototropic equilibrium in *Phycomyces, Science* **129**:775–777.

Dennison, D. S., 1965, Steady-state phototropism in *Phycomyces, J. Gen. Physiol.* **48**:393–408.

Dennison, D. S., 1979, Phototropism, in: *Encyclopedia of Plant Physiology,* new ser., vol. 7 (W. Haupt and M. E. Feinleib, eds.), pp. 506–566, Springer-Verlag, Berlin.

Dennison, D. S., and Bozof, R. P., 1973, Phototropism and local adaptation in *Phycomyces* sporangiophores, *J. Gen. Physiol.* **62**:157–168.

Dennison, D. S., and Foster, K. W., 1977, Intracellular rotation and the phototropic response of *Phycomyces, Biophys. J.* **18**:103–123.

Dohrmann, U., Hertel, R., and Kowalik, H., 1978, Properties of auxin binding sites in different subcellular fractions from maize coleoptiles, *Planta* **140**:97–106.

du Buy, H. G., and Nuernbergk, E., 1929, Weitere Untersuchungen über den Einfluss des Lichtes auf das Wachstum von Koleoptile und Mesokotyl bei *Avena sativa* II, *Proc. Kon. Akad. Wet. Amst.* **32**:808–817.

du Buy, H. G., and Nuernbergk, E., 1932, Phototropismus und Wachstum der Pflanzen, *Ergeb. Biol.* **9**:358–544.

du Buy, H. G., and Nuernbergk, E., 1934, Phototropismus und Wachstum der Pflanzen. Zweiter Teil, *Ergeb. Biol.* **10**:207–322.

du Buy, H. G., and Nuernbergk, E. L., 1935, Phototropismus und Wachstum der Pflanzen. Dritter Teil, *Ergeb. Biol.* **12**:325–543.

Durand, H., and Rayle, D. L., 1973, Physiological evidence for auxin-induced hydrogen-ion secretion and the epidermal paradox, *Planta* **114**:185–193.

Elliot, W. M., and Shen-Miller, J., 1976, Similarity in dose responses, action spectra and red light responses between phototropism and photoinhibition of growth, *Photochem. Photobiol.* **23**:195–199.

Ellis, R. J., and MacDonald, I. R., 1970, Specificity of cycloheximide in higher plant systems, *Plant Physiol.* **46**:227–232.

Errera, L., 1884, Die grosse Wachsthumsperiode bei den Fruchtträgern von *Phycomyces, Bot. Zeitung.* **42**:497–503, 513–522, 529–537, 545–552, 561–566.

Eslava, A. P., Alvarez, M. I., Lipson, E. D., Presti, D., and Kong, K., 1976, Recombination between mutants of *Phycomyces* with abnormal phototropism, *Mol. Gen. Genet.* **147**:235–241.

Etzold, H., 1961, Die Wirkungen des linear polarisierten Lichtes auf Pilze und ihre Beziehungen zu den tropistischen Wirkungen des einsteitigen Lichtes, *Exp. Cell Res.* **25**:229–245.

Etzold, H., 1965, Der Polarotropismus und Phototropismus der Chloronemen von *Dryopteris filix mas* (L.) Schott, *Planta* **64**:254–280.

Evans, M. L., and Ray, R. M., 1969, Timing of the auxin response in coleoptiles and its implications regarding auxin action, *J. Gen. Physiol.* **53**:1–20.

Everett, M., and Thimann, K. V., 1968, Second positive phototropism in the *Avena* coleoptile, *Plant Physiol.* **43**:1786–1792.

Falk, H., and Steiner, A. M., 1968, Phytochrome-mediated polarotropism: an electron microscopical study, *Naturwissenschaften* **55**:500.

Fischer, E.-P., and Thomson, K. S., 1979, Serine proteinase and their inhibitors in *Phycomyces blakesleeanus, J. Biol. Chem.* **254**:50–56.

Fischer-Arnold, G., 1963, Untersuchungen über die Protoplastenbewegung bei *Vaucheria sessilis, Protoplasma* **56**:495–520.

Flint, L. H., 1942, Note on phototropism in *Pilobolus, Am. J. Bot.* **29**:672–674.

Fondeville, J. C., Schneider, M. J., Barthwick, H. A., and Hendricks, S. B., 1967. Photocontrol of *Mimosa pudica* L. leaf movement, *Planta* **75**:228–238.

Foster, K. W., 1977, Phototropism of coprophilous Zygomycetes, *Annu. Rev. Biophys. Bioeng.* **6**:419–443.

Foster, K. W., and Lipson, E. D., 1973, The light growth response of *Phycomyces, J. Gen. Physiol.* **62**:590–617.

Galland, P., and Russo, V. E. A., 1979, Photoinhibition of sporangiophores in *Phycomyces* mutants deficient in phototropism and in mutants lacking β-carotene, *Photochem. Photobiol.* **29**:1009–1014.

Galston, A. W., 1949, Riboflavin-sensitized photooxidation of indoleacetic acid and related compounds, *Proc. Natl. Acad. Sci. USA* **35**:10–17.

Galston, A. W., 1950, Riboflavin, light, and the growth of plants, *Science* **111**:619–624.

Galston, A. W., 1974, Plant photobiology in the last half-century, *Plant Physiol.* **54**:427–436.

Gamow, R. I., and Böttger, B., 1979, *Phycomyces:* Modification of spiral growth after mechanical conditioning of the cell wall, *Science* **203**:268–270.

Gamow, R. I., and Böttger, B., 1980, *Phycomyces.* Modification of light-induced spiral growth after mechanical conditioning of the cell wall, *Plant Physiol.* **66**:525–527.

Gardner, G., Shaw, S., and Wilkins, M. B., 1974, IAA transport during the phototropic responses of intact *Zea* and *Avena* coleoptiles, *Planta* **121**:237–251.

Gettkandt, G., 1954, Zur Kenntnis des Phototropismus der Keimmyzelien einiger parasitischer Pilze, *Wiss. Z. Martin Luther Univ. Halle-Wittenberg Math–Naturwiss. Reihe* **3**:691–709.

Goldsmith, M. H. M., 1966, Movement of indoleacetic acid in coleoptiles of *Avena sativa* L. II. Suspension of polarity by total inhibition of the basipetal transport, *Plant Physiol.* **41**:15–27.

Goldsmith, M. H. M., 1967, Movement of pulses of labeled auxin in corn coleoptiles, *Plant Physiol.* **42**:258–263.

Goldsmith, M. H. M., 1977, The polar transport of auxin, *Annu. Rev. Plant Physiol.* **28**:439–478.

Goldsmith, M. H. M., and Thimann, K. V., 1962, Some characteristics of movement of indoleacetic acid in coleoptiles of *Avena*—I. Uptake, destruction, immobilization, and distribution of IAA during basipetal translocation, *Plant Physiol.* **37**:492–505.

Goldsmith, M. H. M., Caubergs, R. J., and Briggs, W. R., 1980, Light-inducible cytochrome reduction in membrane preparations from corn coleoptiles, I. Stabilization and spectral characterization of the reaction, *Plant Physiol.* **66**:1067–1073.

Goodell, E. W., 1971, "Apical dominance" in the sporangiophore of the fungus *Phycomyces, Planta* **98**:63–75.

Goodwin, T. W., 1952, Studies in carotenogenesis. 3. Identification of the minor polyene components of the fungus *Phycomyces blakesleeanus* and a study of their synthesis under various cultural conditions, *Biochem. J.* **50**:550–558.

Gordon, S. A., 1954, Occurrence, formation, and inactivation of auxins, *Annu. Rev. Plant Physiol.* **5**:341–378.

Gordon, S. A., and Dobra, W. A., 1972, Elongation responses of the oat shoot to blue light, as measured by capacitance auxanometry, *Plant Physiol.* **50**:738–742.

Gordon, S. A., and Eib, M., 1964, Hormonal relation in the phototropic responses, II. translocation of C^{14}-labeled indoleacetic acid in irradiated *Avena* coleoptile segments, *Argonne Natl. Lab. Annu. Rep.* **6971**:176–181.

Gordon, S. A., and Shen-Miller, J., 1968, Auxin relations in phototropism of the coleoptile: a reexamination, in: *Biochemistry and Physiology of Plant Growth Substances, Proceedings of the Sixth International Conference on Plant Growth Substances, July 24–29, 1976* (F. Wightman and G. Setterfield, eds.), pp. 1097–1108, Runge, Ottawa.

Grahm, L., 1964, Measurements of geoelectric and auxin-induced potentials in coleoptiles with a refined vibrating electrode technique, *Physiol. Plant.* **17**:231–261.

Green, P. B., Erickson, R. O., and Richmond, P. A., 1970, On the physical basis of wall morphogenesis, *Ann. NY Acad. Sci.* **175**:712–731.

Greenwood, M. S., Shaw, S., Hillman, J. R., Ritchie, A., and Wilkins, M. B., 1972, Identification of auxin from *Zea* coleoptile tips by mass spectrometry, *Planta* **108**:179–183.

Gressel, J., 1979, Blue light photoreception, *Photochem. Photobiol.* **30**:749–754.

Haberlandt, G., 1889, Über das Längenwachsthum und den Geotropismus der Rhizoiden von *Marchantia* und *Lunularia, Oesterr. Bot. Z.* **39**:93–98.

Hager, A., and Schmidt, R., 1968*a,* Auxintransport und Phototropismus, I. Die lichtbedingte Bildung eines Hemmstoffes für den Transport von Wuchsstoffen in Koleoptilen, *Planta* **83**:347–371.

Hager, A., and Schmidt, R., 1968*b,* Auxintransport und Phototropismus. II. Der Hemmechanismus des aus IES gebildeten Photooxidationsproduktes 3-Methylen-oxindol beim Transport von Wuchsstoffen, *Planta* **83**:372–386.

Hager, A., Menzel, H., and Krauss, A., 1971, Versuch und Hypothese zur Primärwirkung des Auxins beim Streckungswachstum, *Planta* **100**:47–75.

Haig, C., 1934, The spectral sensibility of *Avena, Proc. Natl. Acad. Sci. USA* **20**:476–479.

Hartmann, K. M., 1977, Aktionsspektrometrie, in: *Biophysik* (W. Hoppe, W. Lohmann, H. Markl, and H. Ziegler, eds.), pp. 197–222, Springer-Verlag, Berlin.

Hartmann, K. M., and Haupt, W., 1977, Photomorphogenese, in: *Biophysik* (W. Hoppe, W. Lohmann, H. Markl, and H. Ziegler, eds.), pp. 449–468, Springer-Verlag, Berlin.

Hartmann, K. M., Menzel, H., and Mohr, H., 1965, Ein Beitrag zur Theorie der polarotropischen und phototropischen Krümmung, *Planta* **64**:363–375.

Haupt, W., 1957, Die Induktion der Polarität bei der Spore von *Equisetum, Planta* **49**:61–90.

Haupt, W., 1958, Über den Primärvorgang bei der polarisierenden Wirkung des Lichtes auf keimende *Equisetum*-Sporen, *Planta* **51**:74–83.

Haupt, W., 1965, Perception of environmental stimuli orienting growth and movement in lower plants, *Annu. Rev. Plant Physiol.* **16**:267–290.

Haupt, W., 1968, Die Orientierung der Phytochrom-Moleküle in der *Mougeotia*zelle: Ein neues Modell zur Deutung der experimentellen Befunde, *Z. Pflanzenphysiol.* **58**:331–346.

Haupt, W., 1972, Localization of phytochrome within the cell, in: *Phytochrome* (K. Mitrakos and W. Shropshire, Jr., eds.), pp. 553–569, Academic Press, New York.

Haupt, W., 1977, *Bewegungsphysiologie der Pflanzen,* Georg Thieme Verlag, Stuttgart.

Haupt, W., 1980, Localization and orientation of photoreceptor pigments, in: *Photoreception and Sensory Transduction in Aneural Organisms* (F. Lenci and G. Colombetti, eds.), pp. 155–172, Plenum Press, New York.

Haupt, W., and Bentrup, F.-W., (Meyer zu), 1961, Versuch zur Polaritätsinduktion durch Licht bei *Equisetum*sporen und *Fucus*-Zygoten, *Naturwissenschaften* **48**:723.

Haupt, W., and Buchwald, M., 1967, Die Orientierung der Photorezeptor-Moleküle im Sporangienträger von *Phycomyces, Z. Pflanzenphysiol.* **56**:20–26.

Haupt, W., Mörtel, G., and Winkelnkemper, I., 1969, Demonstration of different dichroic orientation of phytochrome P_R and P_{FR}, *Planta* **88**:183–186.

Hejnowicz, Z., and Sievers, A., 1971, Mathematical model of geotropically bending *Chara* rhizoids, *Z. Pflanzenphysiol.* **66**:34–48.

Hemmerich, P., 1976, The present status of flavin and flavocoenzyme chemistry, in: *Fortschritte der Chemie organischer Naturstoffe,* Vol. 33 (W. Herz, H. Grisebach, and G. W. Kirby, eds.), pp. 451–527, Springer-Verlag, Vienna.

Hertel, R., 1979, Auxin binding sites: subcellular fractionation and specific binding assays, in: *Plant Organelles, methodological Surveys (B) Biochemistry,* Vol. 9 (E. Reid, ed.), pp. 173–183, Ellis Horwood Ltd. Chichester, West Sussex, England.

Hertel, R., 1980, Phototropism of lower plants, in: *Photoreception and Transduction in Aneural Organisms* (F. Lenci and G. Colombetti, eds.), pp. 89–105, Plenum Press, New York.

Hertel, R., 1981, Zur Auxinproblematik: Primäre Wirkung, Transport und *in-vitro*-Bindung, *Biochem. Physiol. Pflanz.* **176**:495–506.

Hertel, R., and Leopold, A. C., 1962, Auxintransport und Schwerkraft, *Naturwissenschaften* **49**:377–378.

Hertel, R., and Leopold, A. C., 1963, Versuche zur Analyse des Auxintransportes in der Koleoptile von *Zea mays* L., *Planta* **59**:535–562.

Hertel, R., Evans, M. L., Leopold, A. C., and Sell, H. M., 1969, The specificity of the auxin transport system, *Planta* **85**:238–249.

Hertel, R., Thomson, K. S., and Russo, V. E. A., 1972, *In-vitro* auxin binding to particulate cell fractions from corn coleoptiles, *Planta* **107**:325–340.

Hertel, R., Jesaitis, A. J., Dohrmann, U., and Briggs, W. R., 1980, *In vitro* binding of riboflavin to subcellular particles from maize coleoptiles and *Cucurbita* hypocotyls, *Planta* **147**:312–319.

Humphry, V. R., 1966, The effects of paraffin oil on phototropic and geotropic responses in *Avena* coleoptiles, *Ann. Bot.* **30**:39–45.

Hurd, A. M., 1920, Effect of unilateral monochromatic light and group orientation on the polarity of germinating *Fucus* spores, *Bot. Gaz.* **70**:25–50.

Ishizawa, K., and Wada, S., 1979*a,* Growth and phototropic bending in *Boergesenia* rhizoid, *Plant Cell Physiol.* **20**:973–982.

Ishizawa, K., and Wada, S., 1979*b,* Action spectrum of negative phototropism in *Boergesenia forbesii, Plant Cell Physiol.* **20**:983–987.

Ishizawa, K., Enomoto, S., and Wada, S., 1979, Germination and photoinduction of polarity in the spherical cells regenerated from protoplasma fragments of *Boergesenia forbesii, Bot. Mag. Tokyo* **92**:173–186.

Jacob, F., 1959, Vergleichende Studien über die phototropische Empfindlichkeit junger Sporangienträger der Gattung *Pilobolus, Arch. Protistenkd.* **103**:531–572.

Jacob, F., 1964, Über die Funktion eines Karotin-Lichtschirmes bei dem Phototropismus von Sporangienträgern chromosporer *Pilobolus*-Arten, *Flora (Jena)* **155**:209–222.

Jacobs, M., and Hertel, R., 1978, Auxin binding to subcellular fractions from *Cucurbita* hypocotyls: *in vitro* evidence for an auxin transport carrier, *Planta* **142**:1–10.

Jacobs, M., and Ray, P. M., 1976, Rapid auxin-induced decrease in free space pH and its relationship to auxin-induced growth in maize and pea, *Plant Physiol.* **58**:203–209.

Jaffe, L. F., 1958, Tropistic responses of zygotes of the Fucaceae to polarized light, *Exp. Cell Res.* **15**:282–299.

Jaffe, L. F., 1960, The effect of polarized light on the growth of a transparent cell. A theoretical analysis, *J. Gen. Physiol.* **43**:897–911.

Jaffe, L., and Etzold, H., 1965, Tropic responses of *Funaria* spores to red light, *Biophys. J.* **5**:715–742.

Jan, Y. N., 1974, Properties and cellular localization of chitin synthetase in *Phycomyces blakesleeanus, J. Biol. Chem.* **249**:1973–1979.

Jayaram, M., Presti, D., and Delbrück, M., 1979, Light-induced carotene synthesis in *Phycomyces, Exp. Mycol.* **3**:42–52.

Jayaram, M., Leutwiler, L., and Delbrück, M., 1980, Light-induced carotene synthesis in mutants of *Phycomyces* with abnormal phototropism, *Photochem. Photobiol.* **32**:241–245.

Jeffreys, D. B., and Greulach, V. A., 1956, The nature of tropism of *Coprinus sterquilinus, J. Elisha Mitchell Sci. Soc.* **72**:153–158.

Jesaitis, A. J., 1974, Linear dichroism and orientation of the *Phycomyces* photopigment, *J. Gen. Physiol.* **63**:1–21.

Jesaitis, A. J., Heners, P. R., Hertel, R., and Briggs, W. R., 1977, Characterization of a membrane fraction containing a b-type cytochrome, *Plant Physiol.* **59**:941–947.

Johnson, D. L., and Gamow, R. I., 1972, *Phycomyces:* growth responses of the sporangium, *Plant Physiol.* **49**:898–903.

Johnsson, A., 1965, Photoinduced lateral potentials in *Zea mays, Physiol. Plant.* **18**:574–576.

Johnston, E. S., 1934, Phototropic sensitivity in relation to wave length, *Smithsonian Misc. Collect.* **92**:1–17.

Jolivette, H. D. M., 1914, Studies on the reactions of *Pilobolus* to light stimuli, *Bot. Gaz.* **57**:89–121.

Jönsson, A., 1961, Chemical structure and growth activity of auxins and antiauxins, *Encycl. Plant Physiol.* **14**:959–1006.

Kasai, S., Miura, R., and Matsui, K., 1975, Chemical structure and some properties of roseoflavin, *Bull. Chem. Soc. Jpn.* **48**:2877–2880.

Kataoka, H., 1975*a*, Phototropism in *Vaucheria geminata.* I. The action spectrum, *Plant Cell Physiol.* **16**:427–437.

Kataoka, H., 1975*b*, Phototropism in *Vaucheria geminata.* II. The mechanism of bending and branching, *Plant Cell Physiol.* **16**:439–448.

Kataoka, H., 1977*a*, Second positive and negative phototropism in *Vaucheria geminata, Plant Cell Physiol.* **18**:473–476.

Kataoka, H., 1977*b*, Phototropic sensitivity in *Vaucheria geminata* regulated by 3′,5′-cyclic AMP, *Plant Cell Physiol.* **18**:431–440.

Kataoka, H., 1979, Phototropic responses of *Vaucheria geminata* to intermittent blue light stimuli, *Plant Physiol.* **63**:1107–1110.

Kataoka, H., 1980, Phototropism: determination of an action spectrum in a tip-growing cell, in: *Handbook of Phycological Methods,* Vol. 3, pp. 205–218, (E. Gautt, ed.) Cambridge University Press, Cambridge.

Key, J. L., 1969, Hormones and nucleic acid metabolism, *Annu. Rev. Plant Physiol.* **20**:449–474.

Koch, K., 1934, Untersuchungen über den Quer- und Längstransport des Wuchsstoffes in Pflanzenorganen, *Planta* **22**:190–220.

Kögl, F., 1937, Wirkstoffprinzip und Pflanzenwachstum, *Naturwissenschaften* **29**:456–470.

Kögl, F., and Haagen-Smit, A. J., 1931, Über die Chemie des Wuchsstoffs, *Proc. Kon. Akad. Wet. Amst.* **34**:1411–1416.

Koske, R. E., 1977, Spiral growth of *Phycomyces:* some new observations in the northern and southern hemispheres, *Mycologia* **69**:189–193.

Lange, S., 1927, Die Verteilung der Lichtempfindlichkeit in der Spitze der Haferkoleoptile, *Jahrb. Wiss. Bot.* **67**:1–51.

Lembi, C. A., Morré, D. J., Thomson, K. S., and Hertel, R., 1971, N-1-Naphthylphthalamic-acid-binding activity of a plasma membrane-rich fraction from maize coleoptiles, *Planta* **99**:37–45.

Leong, T.-Y., and Briggs, W. R., 1981, Partial purification and characterization of a blue light-sensitive cytochrome–flavin complex from corn membranes, *Plant Physiol.* **67**:1042–1046.

Lewis, S. C., Schiff, J. A., and Epstein, H. T., 1961, Photooxidation of cytochromes by a flavoprotein from *Euglena, Biochem. Biophys. Res. Commun.* **5**:221–225.

Lipson, E. D., 1975*a,* White noise analysis of *Phycomyces* light growth response system. I. Normal intensity range, *Biophys. J.* **15**:989–1011.

Lipson, E. D., 1975*b,* White noise analysis of *Phycomyces* light growth response system. II. Extended intensity ranges, *Biophys. J.* **15**:1013–1031.

Lipson, E. D., 1975*c,* White noise analysis of *Phycomyces* light growth response system. III. Photomutants, *Biophys. J.* **15**:1033–1045.

Lipson, E. D., 1980, Sensory transduction in *Phycomyces* photoresponses, in: *The Blue Light Syndrome* (H. Senger, ed.), pp. 110–118, Springer-Verlag, New York.

Lipson, E. D., and Presti, D., 1977, Light-induced absorbance changes in *Phycomyces* photomutants, *Photochem. Photobiol.* **25**:203–208.

Lipson, E. D., and Presti, D., 1980, Graphical estimation of cross sections from fluence-response data, *Photochem. Photobiol.* **32**:383–391.

Lipson, E. D., and Terasaka, D. T., 1981, Photogeotropism in *Phycomyces* double mutants, *Exp. Mycol.* **5**:101–111.

Lipson, E. D., Terasaka, D. T., and Silverstein, P. S., 1980, Double mutants of *Phycomyces* with abnormal phototropism, *Mol. Gen. Genet.* **179**:155–162.

Löser, G., and Schäfer, E., 1980, Phototropism in *Phycomyces:* a photochromic sensor pigment?, in: *The Blue Light Syndrome* (H. Senger, ed.), pp. 244–250, Springer-Verlag, New York.

MacDonald, I. R., and Ellis, R. J., 1969, Does cycloheximide inhibit protein synthesis specifically in plant tissues?, *Nature (Lond.)* **222**:791–792.

MacMillan, J. (ed.), 1980, *Encyclopedia of Plant Physiology,* new ser., Vol. 9, Springer-Verlag, New York.

MacRobbie, E. A. C., and Dainty, J., 1958, Sodium and potassium distribution and transport in the seaweed *Rhodymenia palmata* (L.). Grev., *Physiol. Plant.* **11**:782–801.

Massart, J., 1888, Recherches sur les organismes inférieurs.—1. La loi de Weber vérifiée pour l'héliotropisme du champignon, *Bull. Acad. R. Sci. Belg.* **16**:590–601.

Medina, J. R., and Cerdá-Olmedo, E., 1977*a,* A quantitative model of *Phycomyces* phototropism, *J. Theor. Biol.* **69**:709–719.

Medina, J. R., and Cerdá-Olmedo, E., 1977*b,* Allelic interaction in the photogeotropism of *Phycomyces, Exp. Mycol.* **1**:286–292.

Meissner, G., and Delbrück, M., 1968, Carotenes and retinal in *Phycomyces* mutants, *Plant Physiol.* **43**:1279–1283.
Meistrich, M. L., Fork, R. L., and Matricon, J., 1970, Phototropism in *Phycomyces* as investigated by focused laser radiation, *Science* **169**:370–371.
Menschick, R., Hild, V., and Hager, A., 1977, Decarboxylierung von Indolylessigsäure im Zusammenhang mit dem Phototropismus in *Avena*-Koleoptilen, *Planta* **133**:223–228.
Meyer, A. N., 1969*a*, Versuche zur 1. positiven und zur negativen phototropischen Krümmung der *Avena*-Koleoptile: I. Lichtperception und Absorptionsgradient, *Z. Pflanzenphysiol.* **60**:418–433.
Meyer, A. M., 1969*b*, Versuche zur 1. positiven und zur negativen phototropischen Krümmung der *Avena*-Koleoptile: II. Die Inversion durch Paraffinöl, *Z. Pflanzenphysiol.* **61**:129–134.
Miller, J. H., 1968, Fern gametophytes as experimental material, *Bot. Rev.* **34**:361–440.
Mohr, H., 1956*a*, Die Beeinflussung der Keimung von Farnsporen durch Licht und andere Faktoren, *Planta* **46**:534–551.
Mohr, H., 1956*b*, Die Abhängigkeit des Protonemawachstums und der Protonemapolarität bei Farnen vom Licht, *Planta* **47**:127–158.
Moor, H., 1967, Endoplasmic reticulum as the initiator of bud formation in yeast *(S. cerevisiae)*, *Arch. Mikrobiol.* **57**:135–146.
Morath, M., and Hertel, R., 1978, lateral electrical potential following asymmetric auxin application to maize coleoptiles, *Planta* **140**:31–35.
Müller, F., 1983, The flavin redox system and its biological function, in: *Topics in Current Chemistry,* Vol. 108, pp. 71–107, Springer-Verlag, New York.
Murphy, G. J. P., 1980, Napthaleneacetic acid binding by membrane-free preparations of cytosol from the maize coleoptile, *Plant Sci. Lett.* **19**:157–168.
Naqvi, S. M., and Gordon, S. A., 1967, Auxin transport in *Zea mays* coleoptiles. II. Influence of light on the transport of indoleacetic acid-2-C^{14}, *Plant Physiol.* **42**:138–143.
Nathansohn, A., and Pringsheim, E., 1908, Über die Summation intermittierender Lichtreize, *Jahrb. Wiss. Bot.* **45**:137–190.
Nebel, B. J., 1968, Action spectra for photogrowth and phototropism in protonemata of the moss *Physcomitrium turbinatum, Planta* **81**:287–302.
Nebel, B. J., 1969, Responses of mossprotomenata to red and far red polarized light: evidence for discshaped phytochrome photoreceptors, *Planta* **87**:170–179.
Neuscheler-Wirth, H., 1970, Photomorphogenese und Phototropismus bei *Mougeotia, Z. Pflanzenphysiol.* **63**:238–260.
Noll, F., 1887, Über Membranwachsthum und einige physiologische Erscheinungen bei Siphoneen, *Bot. Z.* **45**:473–482.
Oehlkers, F., 1927, Phototropische Untersuchungen an *Phycomyces nitens, Z. Bot.* **19**:1–44.
Oltmanns, F., 1892, Über die photometrischen Bewegungen der Pflanzen, *Flora (Jena)* **75**:183–266.
Oltmanns, F., 1897, Über positiven und negativen Heliotropismus, *Flora (Jena)* **83**:1–32.
Oort, A. J. P., 1931, The spiral growth of *Phycomyces, Proc. Kon. Akad. Wet. Amst.* **34**:564–575.
Oort, A. J. P., and Roelofsen, P. A., 1932, Spiralwachstum, Wandbau und Plasmaströmung bei *Phycomyces, Proc. Akad. Wet. Amst.* **35**:898–908.
Ootaki, T., and Wolken, J. J., 1973, Octahedral crystals in *Phycomyces*. II, *J. Cell Biol.* **57**:278–288.
Ootaki, T., Fischer, E. P., and Lockhart, P., 1974, Complementation between mutants of *Phycomyces* with abnormal phototropism, *Mol. Gen. Genet.* **131**:233–246.
Ootaki, T., Kinno, T., Yoshida, K., and Eslava, A. P., 1977, Complementation between *Phycomyces* mutants of mating type (+) with abnormal phototropism, *Mol. Gen. Genet.* **152**:245–251.

Oppenoorth, W. F. F., 1939, Photo-inactivation of auxin in the coleoptile of *Avena* and its bearing on phototropism, *Proc. Kon. Akad, Wet. Amst.* **42**:902–915.

Oppenoorth, W. F. F., Jr., 1942, On the role of auxin in phototropism and light growth reaction of *Avena* coleoptiles, *Recl. Trav. Bot. Neerl.* **38**:287–372.

Ortega, J. K. E., and Gamow, R. I., 1970, *Phycomyces:* habituation of the light growth response, *Science* **168**:1374–1375.

Ortega, J. K. E., and Gamow, R. I., 1974, The problem of handedness reversal during the spiral growth of *Phycomyces, J. Theor. Biol.* **47**:317–332.

Ortega, J. K. E., and Gamow, R. I., 1976, An increase in mechanical extensibility during the period of light-stimulated growth, *Plant Physiol.* **57**:456–457.

Ortega, J. K. E., Harris, J. F., and Gamow, R. I., 1974, The analysis of spiral growth in *Phycomyces* using a novel optical method, *Plant Physiol.* **53**:485–490.

Ortega, J. K. E., Gamow, R. I., and Ahlquist, C. N., 1975, *Phycomyces:* a change in mechanical properties after a light stimulus, *Plant Physiol.* **55**:333–337.

Otto, M. K., Jayaram, M., Hamilton, R. M., and Delbrück, M., 1981, Replacement of riboflavin by an analogue in the blue-light photoreceptor of *Phycomyces, Proc. Natl. Acad. Sci. USA* **78**:266–269.

Paál, A., 1914, Über phototropische Reizleitungen, *Ber. Dtsch. Bot. Ges.* **32**:499–502.

Paál, A., 1919, Über phototropische Reizleitung, *Jahrb. Wiss. Bot.* **58**:406–458.

Page, R. M., 1956, Studies on the development of asexual reproductive structures in *Pilobolus, Mycologia* **48**:206–224.

Page, R. M., 1962, Light and the asexual reproduction of *Pilobolus, Science* **138**:1238–1245.

Page, R. M., 1964, Sporangium discharge in *Pilobolus:* a photographic study, *Science* **146**:925–927.

Page, R. M., 1965, The physical environment for fungal growth, in: *The Fungi,* Vol. 1 (G. C. Ainsworth and A. S. Sussman, eds.), pp. 559–574, Academic Press, London.

Page, R. M., 1968, Phototropism in fungi, in: *Photophysiology,* Vol. 3 (A. C. Giese, ed.), pp. 65–90, Academic Press, New York.

Page, R. M., and Brungard, J., 1961, Phototropism in *Conidiobolus,* some preliminary observations, *Science* **134**:733–734.

Page, R. M., and Curry, G. M., 1966, Studies on the phototropism of young sporangiophores of *Pilobolus kleinii, Photochem. Photobiol.* **5**:31–40.

Paietta, J., and Sargent, M. L., 1981, Photoreception in *Neurospora crassa:* correlation of reduced light sensitivity with flavin deficiency, *Proc. Natl. Acad. Sci, USA* **78**:5573–5577.

Parr, R., 1918, The response of *Pilobolus* to light, *Ann. Bot.* **32**:177–205.

Philipson, J. J., Hillmann, J. R., and Wilkins, M. B., 1973*a,* Studies on the actions of abscisic acid on IAA-induced rapid growth of *Avena* coleoptile segments, *Planta* **114**:87–93.

Philipson, J. J., Hillmann, J. R., and Wilkins, M. B., 1973*b,* The effects of temperature and IAA concentration on the latent period for IAA-induced rapid growth of *Avena* coleoptile segments, *Planta* **114**:323–329.

Phillips, I. D. J., 1972, Diffusible gibberellins and phototropism in *Helianthus annuus, Planta* **106**:363–367.

Pickard, B. G., and Thimann, K. V., 1964, Transport and distribution of auxin during tropistic responses. II. The lateral migration of auxin in phototropism of coleoptiles, *Plant Physiol.* **39**:341–350.

Pickard, B. G., Dutson, K., Harrison, W., and Donegan, E., 1969, Second positive phototropic response patterns of the oat coleoptile, *Planta* **88**:1–33.

Poff, K. L., and Butler, W. L., 1974*a,* Spectral characteristics of the photoreceptor pigment of phototaxis in *Dictyostelium discoideum, Photochem. Photobiol.* **20**:241–244.

Poff, K. L., and Butler, W. L., 1974*b,* Absorbance changes induced by blue light in *Phycomyces blakesleeanus* and *Dictyostelium discoideum, Nature (Lond.)* **248**:799–801.

Poggioli, S., 1817, Della influenza che ha il raggio magnetico sulla Vegetazione delle piante, *Opusc. Scient. Fasc.* **I**:9–23.

Pohl, R., 1960, Beiträge zum Phototropismus der *Avena*-Koleoptile I, *Phyton. Rev. Int. Bot. Ext.* **15**:145–157.

Presti, D., and Delbrück, M., 1978, Photoreceptors for biosynthesis, energy storage and vision, *Plant Cell Environ.* **1**:81–100.

Presti, D., Hsu, W.-J., and Delbrück, M., 1977, Phototropism in *Phycomyces* mutants lacking β-carotene, *Photochem. Photobiol.* **26**:403–405.

Preston, R. D., 1948, Spiral growth and spiral structure. I. Spiral growth in sporangiophores of *Phycomyces, Biochem. Biophys. Acta* **2**:155–166.

Pringsheim, E. G., and Czurda, V., 1927, Phototropische und ballistische Probleme bei *Pilobolus, Jahrb. Wiss. Bot.* **66**:863–901.

Racusen, R. H., and Galston, A. W., 1980, Phytochrome modifies blue-light-induced electrical changes in corn coleoptiles, *Plant Physiol.* **66**:534–535.

Ramaer, H., 1926, Phototropical curvature of seedlings of *Avena* which appear when reaction of the distal side is excluded, *Proc. Kon. Akad. Wet. Amst.* **29**:1118–1121.

Ray, P. M., 1977, Auxin-binding sites of maize coleoptiles are localized on membranes of the endoplasmic reticulum, *Plant Physiol.* **59**:594–599.

Ray, P. M., and Ruesink, A. W., 1962, Kinetic experiments on the nature of the growth mechanism in oat coleoptile cells, *Dev. Biol.* **4**:377–397.

Ray, P. M., Dohrmann, U., and Hertel, R., 1977*a*, Characterization of naphthaleneacetic acid binding to receptor sites on cellular membranes of maize coleoptile tissue, *Plant Physiol.* **59**:357–364.

Ray, P. M., Dohrmann, U., and Hertel, R., 1977*b*, Specificity of auxin-binding sites on maize coleoptile membranes as possible receptor sites for auxin action, *Plant Physiol.* **60**:585–591.

Rayle, D. L., 1973, Auxin-induced hydrogen-ion secretion in *Avena* coleoptiles and its implications, *Planta* **114**:63–73.

Rayle, D. L., and Cleland, R., 1970, Enhancement of wall loosening and elongation by acid solutions, *Plant Physiol.* **46**:250–253.

Rayle, D. L., and Cleland, R., 1972, The *in-vitro* acid-growth responses: relation to *in-vivo*-growth responses and auxin action, *Planta* **104**:282–296.

Rayle, D. L., Evans, M. L., and Hertel, R., 1970*a*, Action of auxin on cell elongation, *Proc. Natl. Acad. Sci. USA* **65**:184–191.

Rayle, D. L., Haughton, P. M., and Cleland, R., 1970*b*, An *in-vitro* system that simulates plant cell extension growth, *Proc. Natl. Acad. Sci. USA* **67**:1814–1817.

Reichardt, W., 1961, Die Lichtreaktionen von *Phycomyces, Kybernetik* **1**:6–21.

Reichardt, W., and Varjú, D., 1958, Eine Inversionsphase der phototropischen Reaktion (Experimente an dem Pilz *Phycomyces blakesleeanus), Z. Physik. Chem.* **15**:297–320.

Reinders, D. E., 1934, The sensibility for light of the base of normal and decapitated coleoptiles of *Avena, Proc. Kon. Akad. Wet. Amst.* **37**:308–314.

Reinert, J., 1952, Über die Bedeutung von Carotin und Riboflavin für die Lichtreizaufnahme bei Pflanzen, *Naturwissenschaften* **39**:47–48.

Reinhardt, M. O., 1892, Das Wachsthum der Pilzhyphen. Ein Beitrag zur Kenntnis des Flächenwachstums vegetabilischer Zellmembranen, *Jahrb. Wiss. Bot.* **23**:479–566.

Russo, V. E. A., 1980, Sensory transduction in phototropism: genetic and physiological analysis in *Phycomyces,* in: *Photoreception and Sensory Transduction in Aneural Organisms* (F. Lenci and G. Colombetti, eds.), pp. 373–395, Plenum Press, New York.

Russo, V. E. A., and Galland, P., 1980, Sensory physiology of *Phycomyces blakesleeanus,* in: *Structure and Bonding,* Vol. 41 (P. Hemmerich, ed.), pp. 71–110, Springer-Verlag, New York.

Russo, V. E. A., Galland, P., Toselli, M., and Volpi, L., 1980, Blue light induced differentiation in *Phycomyces blakesleeanus*, in *The Blue Light Syndrome* (H. Senger, ed.), pp. 563–569, Springer-Verlag, New York.

Schmidt, W., 1980, Physiological blue-light reception, in: *Structure and Bonding*, Vol. 41 (P. Hemmerich, ed.), pp. 1–44, Springer-Verlag, New York.

Schmidt, W., and Butler, W. L., 1976, Flavin-mediated photoreactions in artificial systems: a possible model for the blue-light photoreceptor pigment in living systems, *Photochem. Photobiol.* **24**:71–75.

Schmidt, W., Hart, J., Filner, P., and Poff, K. L., 1977, Specific inhibition of phototropism in corn seedlings, *Plant Physiol.* **60**:736–738.

Schneider, H. A. W., and Bogorad, L., 1978, Light-induced, dark-reversible absorbance changes in roots, other organs, and cell-free preparations, *Plant Physiol.* **62**:577–581.

Schopfer, W.-H., 1935, Etude et identification d'un carotinoide de champignon, *C. R. Soc. Biol. (Paris)* **118**:3–5.

Schrank, A. R., 1946, Note on the effect of unilateral illumination on the transverse electrical polarity in the *Avena* coleoptile, *Plant Physiol.* **21**:362–365.

Schrank, A. R., 1948, Electrical and curvature responses of the *Avena* coleoptile to transversely applied direct current, *Plant Physiol.* **23**:188–200.

Schrank, A. R., 1950, Inhibition of curvature responses by shunting the inherent electrical field, *Plant Physiol.* **25**:583–593.

Schrank, A. R., 1953, Effect of inorganic ions and their conductances on geotropic curvature of the *Avena* coleoptile, *Plant Physiol.* **28**:99–104.

Senger, H., and Briggs, W. R., 1981, The blue light receptor(s): primary reactions and subsequent metabolic changes, in: *Photochemical and Photobiological Reviews*, Vol. 6 (K. C. Smith, ed.), pp. 1–38, Plenum Press, New York.

Shen-Miller, J., and Gordon, S. A., 1966, Hormonal relations in the phototropic response: III. The Movement of C^{14}-labeled and endogenous indoleacetic acid in phototropically stimulated *Zea* coleoptiles, *Plant Physiol.* **41**:59–65.

Shen-Miller, J., Cooper, P., and Gordon, S. A., 1969, Phototropism and photoinhibition of basipolar transport of auxin in oat coleoptiles, *Plant Physiol.* **44**:491–496.

Shibaoka, H., and Yamaki, T., 1959, A sensitized *Avena* curvature test and identification of the diffusible auxin in *Avena* coleoptile, *Bot. Mag. Tokyo* **72**:152–158.

Shropshire, W., Jr., 1959, Growth responses of *Phycomyces* to polarized light stimuli, *Science* **130**:336.

Shropshire, W., Jr., 1962, The lens effect and phototropism of *Phycomyces*, *J. Gen. Physiol.* **45**:949–958.

Shropshire, W., Jr., 1971, Phototropic bending rate in *Phycomyces* as a function of average growth rate and cell radius, in: *First European Biophysics Congress, Sept.* 14–17, 1971 (E. Broda, A. Locker, and H. Springer-Lederer, eds.), pp. 111–114, Verlag der Wiener Medizinischen Akademie, Vienna.

Shropshire, W., Jr., 1972, Action spectroscopy, in *Phytochrome* (K. Mitrakos and W. Shropshire, Jr., eds.), pp. 161–181, Academic Press, New York.

Shropshire, W., Jr., 1975, Phototropism, in: *Progress in Photobiology, Proceedings of the Sixth International Congress on Photobiology, Deutsch Gesellschaft für Lichtforschung e.V., Frankfurt, 1974* (G. O. Schenk, ed.), pp. 1–6, Springer-Verlag, Berlin.

Shropshire, W., Jr., 1979, Stimulus perception, in: *Encyclopedia of Plant Physiology*, new ser., Vol. 7 (W. Haupt and M. E. Feinleib, eds.), pp. 10–41, Springer-Verlag, New York.

Shropshire, W., Jr., 1980, Carotenoids as primary photoreceptors in blue-light responses, in: *The Blue Light Syndrome* (H. Senger, ed.), pp. 172–186, Springer-Verlag, New York.

Shropshire, W., Jr., and Withrow, R. B., 1958, Action spectrum of phototropic tip-curvature of *Avena*, *Plant Physiol.* **33**:360–365.

Sierp, H., and Seybold, A., 1926, Untersuchungen über die Lichtempfindlichkeit der Spitze und des Stumpfes in der Koleoptile von *Avena sativa, Jahrb. Wiss. Bot.* **65**:592–610.

Smith, J. E., 1975, The structure and development of filamentous fungi, in: *The Filamentous Fungi,* Vol. 1 (J. E. Smith and D. R. Berry, eds.), pp. 1–15, Edward Arnold, London.

Söding, H., 1923, Werden von der Spitze der Haferkoleoptile Wuchshormone gebildet?, *Ber. Dtsch. Bot. Ges.* **41**:396–400.

Söding, H., 1925, Zur Kenntnis der Wuchshormone in der Haferkoleoptile, *Jahrb. Wiss. Bot.* **64**:587–603.

Song, P. S., 1980*a,* Spectroscopic and photochemical characterization of flavoproteins and carotenoproteins as blue light photoreceptors, in: *The Blue Light Syndrome* (H. Senger, ed.), pp. 157–171, Springer-Verlag, New York.

Song, P. S., 1980*b,* Primary photophysical and photochemical reactions: theoretical background and general introduction, in: *Photoreception and Sensory Transduction in Aneural Organisms* (F. Lenci and G. Colombetti, eds.), pp. 189–210, Plenum Press, New York.

Song, P. S., Fugate, R. D., and Briggs, W. R., 1980*a,* Flavin as a photoreceptor for phototropic transduction: fluorescence studies of model and corn coleoptile systems, in: *Flavins and Flavoproteins* (K. Yagi and T. Yamano, eds.), pp. 443–453, University Park Press, Baltimore.

Song, P. S., Walker, E. B., Vierstra, R. D., and Poff, K. L., 1980*b,* Roseoflavin as a blue light receptor analog: spectroscopic characterization, *Photochem. Photobiol.* **32**:393–398.

Specht, J., 1960, Die phototropische Inversion dikotyler Keimlinge in ölartigen Medien, *Flora (Jena)* **149**:106–161.

Stadler, D. R., 1952, Chemotropism in *Rhizopus nigricans:* the staling reaction, *J. Cell Comp. Physiol.* **39**:449–474.

Stameroff, K., 1897, Zur Frage über den Einfluss des Lichtes auf das Wachstum der Pflanzen, *Flora (Jena)* **83**:135–150.

Steiner, A. M., 1967*a,* Dose-response curves for polarotropism in germlings of a fern and a liverwort, *Naturwissenschaften* **54**:497.

Steiner, A. M., 1967*b,* Action spectra for polarotropism in germlings of a fern and a liverwort, *Naturwissenschaften* **54**:497–498.

Steiner, A. M., 1969*a,* Dose response behaviour for polarotropism of the chloronema of the fern *Dryopteris filixmas* (L.) Schott, *Photochem. Photobiol.* **9**:493–506.

Steiner, A. M., 1969*b,* Action spectrum for polarotropism in the chloronema of the fern *Dryopteris filix-mas* (L.) Schott, *Photochem. Photobiol.* **9**:507–513.

Steiner, A. M., 1969*c,* Dose response behaviour for polarotropism of the germ tube of the liverwort *Sphaerocarpos donnellii* Aust., *Planta* **86**:334–342.

Steiner, A. M., 1969*d,* Action spectrum for polarotropism of the germ tube of the liverwort *Sphaerocarpos donnellii* Aust., *Planta* **86**:343–352.

Steiner, A. M., 1970, Red light interaction with blue and ultraviolet light in polarotropism of germlings of a fern and a liverwort, *Photochem. Photobiol.* **12**:169–174.

Steyer, B., 1965, Der Phototropismus dicotyler Keimpflanzen, *Wiss. Z. Univ. Rostock Math.–Naturwiss, Reihe* **14**:493–502.

Steyer, B., 1967, Die Dosiswirkungsrelationen bei geotroper und phototroper Reizung: Vergleich von Mono- mit Dicotyledonen, *Planta* **77**:277–286.

Stifler, R. D., 1961, Growth of sporangiophores of *Phycomyces* immersed in water, *Science* **133**:1022.

Stoddart, J. L., and Venis, M. A., 1980, Molecular and subcellular aspects of hormone action, in: *Encyclopedia of Plant Physiology,* new ser., Volume 9 (J. MacMillan, ed.), pp. 445–510, Springer-Verlag, New York.

Strugger, S., 1932, Die Beeinflussung des Wachstums und des Geotropismus durch Wasserstoffionen, *Ber. Dtsch. Bot. Ges.* **50**:77–92.

Sweeney, B. M., and Thimann, K. V., 1938, The effect of auxins on protoplasmic streaming II, *J. Gen. Physiol.* **22**:439–461.

Takahashi, C., 1961, The growth of protonema cells and rhizoids in bracken, *Cytologia (Tokyo)* **26**:62–66.

Thimann, K. V., and Curry, G. M., 1960, Phototropism and Phototaxis, in: *Comparative Biochemistry,* Vol. 1 (M. Florkin, and H. S. Mason, eds.), pp. 243–309, Academic Press, New York.

Thimann, K. V., and Curry, G. M., 1961, Phototropism, in: *Light and Life* (W. D. McElroy and B. Glass, eds.), pp. 646–672, Johns Hopkins University Press, Baltimore.

Thimann, K. V., and Sweeney, B. M., 1938, The effect of auxins upon protoplasmic streaming, *J. Gen. Physiol.* **21**:123–135.

Thomas, J. B. (ed.), 1968, *Einführung in die Photobiologie,* Georg Thieme Verlag, Stuttgart.

Thomson, K.-S., Hertel, R., Müller, S., and Tavares, J. E., 1973, 1-N-Naphthylphthalamic acid and 2,3,5-triiodobenzoic acid. *In vitro* binding to particulate cell fractions and action on auxin transport in corn coleoptiles, *Planta* **109**:337–352.

Thornton, R. M., 1973, New photoresponses of *Phycomyces, Plant Physiol.* **51**:570–576.

Thornton, R. M., and Thimann, K. V., 1967, Transient effects of light on auxin transport in the *Avena* coleoptile, *Plant Physiol.* **42**:247–257.

Tietze-Haß, E., and Dörffling, K., 1977, Initial phases of indoleacetic acid induced growth in coleoptile segments of *Avena sativa* L., *Planta* **135**:149–154.

Timberlake, W. E., and Griffin, D. H., 1974, Differential effects of analogs of cycloheximide on protein and RNA synthesis in *Achlya, Biochim. Biophys. Acta* **349**:39–46.

Timberlake, W. E., McDowell, L., and Griffin, D. H., 1972, Cycloheximide inhibition of the DNA-dependent RNA-polymerase I of *Achlya bisexualis, Biochim. Biophys. Res. Commun.* **46**:942–947.

Tollenaar, D., 1923, Dark growth-responses, *Proc. Kon. Akad. Wet. Amst.* **26**:378–389.

Trewavas, A., 1968, Relationship between plant growth hormones and nucleic acid metabolism, *Prog. Phytochem.* **1**:113–160.

Trinci, A. P. J., and Banbury, G. H., 1969, Phototropism and light-growth responses of the tall conidiophores of *Aspergillus giganteus, J. Gen. Microbiol.* **54**:427–438.

Trinci, A. P. J., and Halford, E. A., 1975, The extension zone of stage I sporangiophores of *Phycomyces blakesleeanus, New Phytol.* **74**:81–83.

Ullrich, C.-H., 1978, Continuous measurement of initial curvature of maize coleoptiles induced by lateral auxin application, *Planta* **140**:201–211.

van der Wey, H. G., 1929, Über die phototropische Reaktion von *Pilobolus, Proc. Kon. Akad. Wet. Amst.* **32**:65–77.

van Dillewijn, C., 1927, Die Lichtwachstumsreaktionen von *Avena, Recl. Trav. Bot. Neerl.* **24**:307–581.

van Laere, A. J., and Carlier, A. R., 1977, Chitin synthetase in *Phycomyces blakesleeanus* Burgeff., *Arch. Int. Physiol. Biochim.* **85**:1025–1026.

Varjú, D., Edgar, L., and Delbrück, M., 1961, Interplay between the reactions to light and to gravity in *Phycomyces, J. Gen. Physiol.* **45**:47–58.

Veen, H., 1974, Specificity of phospholipid binding to indoleacetic acid and other auxins, *Z. Naturforsch.* **29c**:39–41.

Venis, M. A., 1977, Solubilisation and partial purification of auxin binding sites of corn membranes, *Nature (London)* **266**:268–269.

Venis, M. A., and Watson, P. J., 1978, Naturally occurring modifiers of auxin-receptor interaction in corn: identification as benzoxazolinones, *Planta* **142**:103–107.

Vierstra, R. D., and Poff, K. L., 1981*a,* Mechanism of specific inhibition of phototropism by phenylacetic acid in corn seedlings, *Plant Physiol.* **67**:1011–1015.

Vierstra, R. D., and Poff, K. L., 1981*b*, Role of carotenoids in the phototropic response of corn seedlings, *Plant Physiol.* **68**:798–801.

Vierstra, R. D., Poff, K. L., Walker, E. B., and Song, P. S., 1981, Effect of xenon on the excited states of phototropic receptor flavin in corn seedlings, *Plant Physiol.* **67**:996–998.

Vogt, E., 1915, Über den Einfluss des Lichtes auf das Wachstum der Koleoptile von *Avena sativa, Z. Bot.* **7**:193–270.

von Guttenberg, H., 1959, Perzeption des phototropen Reizes, *Planta* **53**:412–433.

Wada, M., Kadota, A., and Furuya, M., 1978, Apical growth of protonemata in *Adiantum capillus-veneris.* II. Action spectra for the induction of apical swelling and the intracellular photoreceptive site, *Bot. Mag. Tokyo* **91**:113–120.

Wald, G., and du Buy, H. G., 1936, Pigments of the oat coleoptile, *Science* **84**:247.

Weber, W., 1958, Zur Polarität von *Vaucheria, Z. Bot.* **46**:161–198.

Webster, W. W., Jr., and Schrank, A. R., 1953, Electrical induction of lateral transport of 3-indoleacetic acid in the *Avena* coleoptile, *Arch. Biochem. Biophys.* **47**:107–118.

Weigl, J., 1969*a*, Einbau von Auxin in gequollene Lecithin-Lamellen, *Z. Naturforsch.* **24b**:365–366.

Weigl, J., 1969*b*, Spezifität der Wechselwirkung swischen Wuchsstoffen und Lecithin, *Z. Naturforsch.* **24b**:367–368.

Went, F. W., 1928, Wuchsstoff und Wachstum, *Recl. Trav. Bot. Neerl.* **25**:1–116.

Went, F. W., 1932, Eine botanische Polaritätstheorie, *Jahrb. Wiss. Bot.* **76**:528–557.

Went, F. W., 1942, Growth, auxin, and tropisms in decapitated *Avena* coleoptiles, *Plant Physiol.* **17**:236–249.

Whitaker, B. D., and Shropshire, W., Jr., 1981, Spectral Sensitivity in the blue and near ultraviolet for light-induction of carotene synthesis in *Phycomyces* mycelia, *Exp. Mycol.* **5**:243–252.

Widell, S., 1980, The effect of detergent treatment on methylene blue sensitized cytochrome b photoreduction in fractions from corn coleoptiles, *Physiol. Plant.* **48**:353–360.

Widell, S., and Björn, L. O., 1976, Light-induced absorption changes in etiolated coleoptiles, *Physiol. Plant.* **36**:305–309.

Widell, S., Britz, S. J., and Briggs, W. R., 1980, Characterization of the red light induced reduction of a particle associated b-type cytochrome from corn in the presence of methylene blue, *Photochem. Photobiol.* **32**:669–677.

Wiechulla, O., 1932, Beiträge zur Kenntnis der Lichtwachstumsreaktion von *Phycomyces, Beitr. Biol. Pflanz.* **19**:371–419.

Wilden, M., 1939, Zur Analyse der positiven und negativen phototropischen Krümmungen, *Planta* **30**:286–288.

Wildman, S. G., and Bonner, J., 1948, Observation on the chemical nature and formation of auxin in the *Avena* coleoptile, *Am. J. Bot.* **35**:740–746.

Wilkins, M. B., 1971, Hormone movements in geotropism, in: *Gravity and the Organism* (S. A. Gordon and M. J. Cohen, eds.), pp. 107–124, University of Chicago Press, Chicago.

Wilks, S. S., and Lund, E. J., 1947, The electric correlation field and its variation in the coleoptile of *Avena sativa,* in: *Bioelectric Fields and Growth* (E. J. Lund *et al,* eds.), pp. 24–74, University of Texas Press, Austin.

Wolken, J. J., 1969, Microspectrophotometry and the photoreceptor of *Phycomyces* I, *J. Cell Biol.* **43**:354–360.

Wun, K. L., Gih, A., and Sutherland, J. C., 1977, Photoreactivating enzyme from *E. coli:* appearance of new absorption on binding to ultraviolet irradiated DNA, *Biochemistry* **16**:921–924.

Zalokar, M., 1969, Intracellular centrifugal separation of organelles in *Phycomyces, J. Cell Biol.* **41**:494–509.

Zankel, K. L., Burke, P. V., and Delbrück, M., 1967, Absorption and screening in *Phycomyces, J. Gen. Physiol.* **50**:1893–1906.

Zenk, M. H., 1967, Untersuchungen zum Phototropismus der *Avena*-Koleoptile: II. Pigmente, *Z. Pflanzenphysiol.* **56**:122–140.

Zenk, M. H., 1968, The action of light on the metabolism of auxin in relation to phototropism, in: *Biochemistry and Physiology of Plant Growth Substances, Proceedings of the Sixth International Conference on Plant Growth Substances, July 24–29,* 1967 (F. Wight and G. Setterfield, eds.), pp. 1109–1128, Runge, Ottawa.

Ziegler, H., 1950, Inversion phototropischer Reaktionen, *Planta* **38**:474–498.

Zimmerman, B. K., and Briggs, W. R., 1963, Phototropic dosage-response curves for oat coleoptiles, *Plant Physiol.* **38**:248–253.

Zopf, W., 1892, Zur Kenntnis der Färbungsursachen niederer Organismen (Zweite Mittheilung), *Beitr. Physiol. Morphol. Nied. Organ.* **2**:3–12.

Zurfluh, L. L., and Guilfoyle, T. J., 1980, Auxin-induced changes in the patterns of protein synthesis in soybean hypocotyl, *Proc. Natl. Acad. Sci. USA* **77**:357–361.

SELECTED READINGS

Briggs, W. R., 1963, The phototropic responses of higher plants, *Annu. Rev. Plant Physiol.* **14**:311–352.

Briggs, W. R., 1976, The nature of the blue light photoreceptor in higher plants and fungi, in: *Light and Development* (H. Smith, ed.), pp. 7–18, Butterworths, London.

Briggs, W. R., 1980, A blue light photoreceptor system in higher plants and fungi, in: *Photoreceptors and Plant Development* (J. DeGreef, ed.), pp. 17–28, Antwerp University Press, Antwerp.

Carlile, M. J., 1975, Taxes and tropisms: diversity, biological significance and evolution, in: *Primitive Sensory and Communication Systems: The Taxes and Tropisms of Micro-Organisms and Cells* (M. J. Carlile, ed.), pp. 1–28, Academic Press, New York.

Carlile, M. J., 1980, Sensory transduction in aneural organisms, in: *Photoreception and Sensory Transduction in Aneural Organisms* (F. Lenci and G. Colombetti, eds.), pp. 1–22, Plenum Press, New York.

Crane, F. L., Goldenberg, H., Morré, J. D., and Löw, H., 1979, Dehydrogenases of the plasma membrane, *Subcell. Biochem.* **6**:345–399.

Curry, G. M., and Thimann, K. V., 1961, Phototropism: the nature of the photoreceptor in higher and lower plants, in: *Progress in Photobiology* (B. C. Christensen and B. Buchmann, eds.), pp. 127–134, Elsevier, New York.

Delbrück, M., 1962, Der Lichtsinn von *Phycomyces, Ber. Dtsch. Bot. Ges.* **75**:411–430.

Galston, A. W., 1950, Phototropism. II., *Bot. Rev.* **16**:361–378.

Galston, A. W., 1959, Phototropism of stems, roots and coleoptiles, in: *Encyclopedia of Plant Physiology,* Vol. 17, Part 1 (W. Ruhland, ed.), pp. 492–529, Springer-Verlag, New York.

Goldsmith, M. H. M., 1969, Transport of plant growth regulators, in: *The Physiology of Plant Growth and Development* (M. B. Wilkins, ed.), pp. 125–162, McGraw-Hill, New York.

Green, P. B., 1969, Cell morphogenesis, *Annu. Rev. Plant Physiol.* **20**:365–394.

Hager, A., 1971, Das differentielle Wachstum bei photo- und geotropischen Krümmungen, *Ber. Dtsch. Bot. Ges.* **84**:331–350.

Hemmerich, P., and Schmidt, W., 1980, Blue light reception and flavin photochemistry, in: *Photoreception and Sensory Transduction in Aneural Organisms* (F. Lenci and G. Colombetti, eds.), pp. 271–283, Plenum Press, New York.

Jaffe, L. F., 1958, Morphogenesis in lower plants, *Annu. Rev. Plant Physiol.* **9**:359–384.
Löw, H., and Crane, F. L., 1978, Redox function in plasma membranes, *Biochim. Biophys. Acta* **515**:141–161.
Malhotra, S. K., 1978, Molecular structure of biological membranes: functional characterization, *Subcell. Biochem.* **5**:221–259.
Marmé, D., and Schäfer, E., 1972, On the localization and orientation of phytochrome molecules in corn coleoptiles (*Zea mays* L.), *Z. Planzenphysiol.* **67**:192–194.
Massey, V., 1980, Possible photoregulation by flavoproteins, in: *Photoreception and Sensory Transduction in Aneural Organisms* (F. Lenci and G. Colombetti, eds.), pp. 253–268, Plenum Press, New York.
Quail, P. H., 1979, Plant cell fractionation, *Annu. Rev. Plant Physiol.* **30**:425–484.
Reinert, J., 1959, Phototropism and phototaxis, *Annu. Rev. Plant Physiol.* **10**:441–458.
Russo, V. E. A., and Pohl, U., 1980, *Phycomyces blakesleeanus*. Ein Modell für die Sinnesphysiologie aneuraler Organismen, *Naturwissenschaften* **67**:296–300.
Schneider, H. A. W., 1980, Visible and spectrophotometrically detectable blue light responses of maize roots, in: *The Blue Light Syndrome* (H. Senger, ed.), pp. 614–621, Springer-Verlag, New York.
Shropshire, W., Jr., 1963, Photoresponses of the fungus, *Phycomyces, Physiol. Rev.* **43**:38–67.
Shropshire, W., Jr., 1975, Unicellular-plant transducers, in: *Interdisciplinary Aspects of General Systems Theory, Proceedings of the Third Annual Meeting of the Middle Atlantic Regional Division, Sept. 21, 1974, The Society for General Systems Research,* pp. 50–57.
Song, P.-S., Moore, T. A., and Sun, M., 1972, Excited states of some plant pigments, in: *The Chemistry of Plant Pigments* (C. O. Chichester, ed.), pp. 33–74, Academic Press, London.
Tan, K. K., 1978, Light-induced fungal development, in: *The Filamentous Fungi,* Vol. 3 (J. E. Smith and D. R. Berry, eds.), pp. 334–357, Edward Arnold, London.
van Overbeek, J., 1966, Plant hormones and regulators, *Science* **152**:721–731.
Virgin, H. I., 1964, Some effects of light on chloroplasts and plant protoplasm, in: *Photophysiology,* Vol. 1 (A. C. Giese, ed.), pp. 273–303, Academic Press, New York.

Chapter 8

Chloroplast Movement

Wolfgang Haupt and Gottfried Wagner

1. INTRODUCTION

In textbooks, figures depicting a cell give the naive reader the impression that the cell is a static system in terms of the pattern of distribution of its organelles and contents. In the living organism, however, cells are highly dynamic, and the patterns change continuously; the only difference between different types of cells and different organisms is the number of these changes and the ease with which they can be observed. As long as only colorless cell structures are involved (as in animal cells), their movement is not so obvious. However, the plastids of plant cells, which because of their color can be observed in the microscope even at low magnification and under low-intensity illumination, can provide an excellent demonstration of the transiency of the pattern of cell structures as seen at a given moment. It is therefore not at all surprising that movement of chloroplasts is probably the most thoroughly investigated type of intracellular movement (cf. Senn, 1908; Haupt, 1959*a*, 1982; Zurzycki, 1962*b*; Britz, 1979; Seitz, 1979*b*).

The most conspicuous type of chloroplast movement in the cell is rotational streaming or cyclosis, occurring mainly in cells that have a large central vacuole (Kamiya, 1962; Seitz, 1979*b*). Here, nearly the whole cytoplasm is in continuous motion, with roughly one-half the cell exhibiting this movement in one direction, and the other portion in the opposite direction. It is true, there

Wolfgang Haupt • Institut für Botanik und Pharmazeutische Biologie, University of Erlangen-Nürnberg, D-8520 Erlangen, Federal Republic of Germany. **Gottfried Wagner** • Botanisches Institut I, D-6300 Giessen, Federal Republic of Germany.

are exceptional cases, in which chloroplasts do not participate in the overall rotational streaming. But this is restricted, among the well-investigated plant systems, to the characean algae, *viz. Nitella, Chara* (Kamiya, 1962, 1981). Otherwise, chloroplasts are moving with the cytoplasm, at least to some degree, as explained in Section 2.1, making it easy to observe and to measure parameters of cyclosis (e.g., its speed). However, since in these cases the chloroplasts are usually more or less evenly distributed over the whole surface of the cell, cyclosis does not change the overall pattern of chloroplast arrangement, even though the single chloroplast is moving. Accordingly, internal or external factors that affect cyclosis are without influence on the chloroplast pattern at a given moment.

The orientation movement of chloroplasts is in strong contrast to cyclosis. In this case, there exist two or several well-defined quasistationary patterns of chloroplast distribution that are formed depending on the conditions (Senn, 1908; Mayer, 1971; Haupt, 1982) (see Fig. 1). For a given pattern, the orientation is controlled by internal or external factors, e.g., cell anatomy, relationship to neighboring cells, and light direction. On the other hand, internal or external factors (e.g., developmental stage, circadian rhythm, light intensity, temperature, or a combination of those factors) also determine which of the possible patterns are formed. Besides the fundamental difference in the spatial patterns of cyclosis versus reorientation, there is also an obvious difference in the time course: Whereas in cyclosis chloroplasts move fast and continuously over a long period without a predetermined stop, in orientation responses they move more slowly and only until the new quasistationary pattern is reached, usually remaining in this pattern until a change of conditions calls for a rearrangement, i.e., a transient movement to the new pattern.

In spite of these important differences, there seem to be causal relationships between the two types of movement. This will become evident, if we analyze the factors that control the movements. The most thoroughly investigated factor is light; it can control cyclosis and can orient the chlorplast distribution. As a basis for the following sections, a short survey about these effects is given.

One of the classical examples for light effects on cyclosis is the freshwater plant *Vallisneria spiralis*. In darkness, only cytoplasm with small and colorless inclusions (e.g., mitochondria, microsomes) is streaming, but chloroplasts are resting. Upon illumination, streaming is accelerated, and chloroplasts begin to participate in the movement. This complex response to a light signal is called *photodinesis*. Depending on the measured parameter, photodinesis can be defined as an acceleration of existing streaming, as an increase of the portion of cytoplasm that is streaming, or as an induction of movement of hitherto resting organelles. Besides *Vallisneria* and the similarly behaving *Elodea*, there are only few plant systems in which photodinesis has been investigated at least superficially (Haupt, 1959*b*; Seitz, 1979*b*).

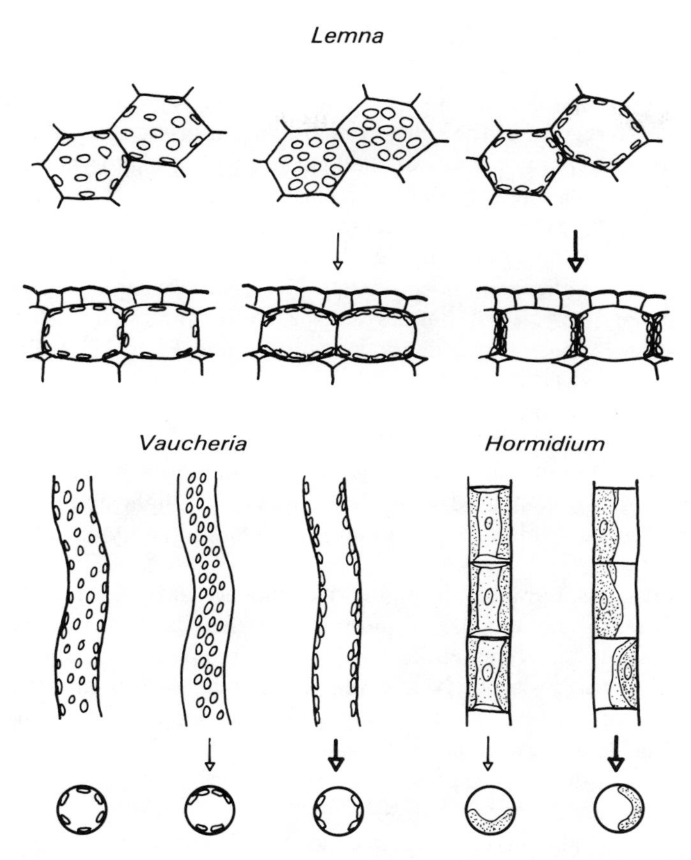

Figure 1. Chloroplast arrangement in *Lemna, Vaucheria,* and *Hormidium,* in darkness, low-intensity light (light arrow), and in high-intensity light (heavy arrow), respectively. Cells or tissue shown in surface view and in cross section. (Redrawn from Haupt, 1968b.)

In contrast, reorientation of chloroplasts by light has been studied in a wide variety of systems throughout the plant kingdom (Senn, 1908; Lechowski, 1972; Inoue and Shibata, 1973; Nultsch and Pfau, 1979). As a typical case, Fig. 1A shows the chloroplast distribution in frond cells of the duckweed *Lemna trisulca.* In unidirectional light the chloroplasts gather at the periclinal walls if the light intensity is moderate, but they move to the anticlinal walls in high-intensity light. The cells of many plants of a wide range of taxonomic groups behave very similarly (Fig. 1B); frequently there is a third possible pattern, viz. random distribution at all cell walls in darkness (Fig. 1A,B) These two (or three) types of chloroplast arrangement, viz. low-intensity, high-intensity (and dark) arrangement can be distinguished even in those cells in which, as a result of different anatomy, the patterns are modified (Fig. 1C, *Hormidium*).

As a general rule, in low- or moderate-intensity light, chloroplasts are found in those regions of the cell in which light absorption is highest; accordingly, optimal use can be made of the light energy for photosynthesis. On the other hand, in high-intensity light, the least absorbing parts of the cytoplasm are occupied by the cloroplasts, suggesting that here they are protected against damage by strong light. It should be added, however, that the interpretation of the biological significance for optimizing photosynthesis and protecting chloroplasts, respectively, is still not beyond speculation, and doubt is cast on this interpretation by recent investigations (Nultsch and Pfau, 1979; Britz, 1979; Nultsch *et al.*, 1981).

This ecological uncertainty in no way hampers the physiological analysis. The latter has to consider light as the signal which by its direction orients the possible chloroplast distribution, and which by its intensity determines the type of response. Thus, compared with photodinesis, with light acting as a scalar factor only, in chloroplast reorientation the vectorial quality of light is important too.

Nevertheless, there is a formal correspondence in both types of response. In every case, the light signal in question must first be perceived by the cell; the result of this signal perception serves to initiate processes called transduction *(sensu stricto)*, ultimately affecting the mechanism of movement or its energy-providing system in such a way as to result in a visible or measurable response (for details, see Lenci and Colombetti, 1980). This chain of events—perception–transduction–response—is also called the transduction chain, the term now used *sensu lato*.

Full understanding of a transduction chain requires knowledge of the localization of its steps as it occurs in the cell. If we realize how sensitive and precise the systems are, we have to assume a high degree of order in the components involved in the transduction chain. This in turn suggests that important steps are associated with well-defined cell structures; biomembranes are prospective candidates for those structures.

The main goal of this chapter is therefore to investigate to what degree biomembranes are involved in light-controlled cyclosis and in light-oriented chloroplast rearrangement. It will turn out that membranes can be important for all primary steps of the transduction chain: photoperception, transduction *(sensu stricto)*, energetics, and mechanics.

2. PERCEPTION OF THE LIGHT SIGNAL

If membranes are involved in the perception process, a close spatial relationship would be expected to exist between the photoreceptor pigment and membranous structures. The pigment in question might be bound to a membrane or may even be part of a membrane. It is not possible, however, to look for such associations without knowledge of the pigment's chemical nature.

Several issues therefore have to be tackled for photodinesis as well as for chloroplast orientation: (1) Action spectroscopy and other methods should give information about the nature of the photoreceptor pigment; (2) localized responses upon partial irradiation of cells should show its intracellular localization or distribution; and (3) specific responses to polarized light (action dichroism) should demonstrate the molecular orientation of photoreceptor pigments, if such an orientation exists. Along this line, evidence will be presented that photoreceptor pigments are specifically associated with biomembranes, although nothing can be said yet about the nature of this association, i.e., whether a pigment is part of a membrane, bound to it, or localized in a cytoplasmic fraction that always keeps close association with a membrane.

2.1. Photoreceptor Pigments in Photodinesis

Although it is not a typical photodinesis, we will begin with a light effect in the coenocytic alga *Vaucheria*. Here, the peripheral cytoplasm together with its organelles continuously moves in darkness or in red light (which is not "seen" by this organism). This movement, which can easily be followed by observing the chloroplasts, proceeds approximately in longitudinal directions, but these directions reverse, independently of each other, in different tracks after random time intervals on the order of minutes (Fischer-Arnold, 1963).

If part of the cell is illuminated with a white or blue microbeam, movement ceases in this region. This might be taken as a negative photodinesis. Upon closer inspection, it can be seen that the movement does not stop immediately, but after a lag period ranging from seconds to minutes, depending on the fluence rate of light. The chloroplasts that cross the illuminated region during the lag period are not influenced in their movement. But those that enter the field after the lag period stop immediately (Fig. 2), even if no other chloroplast has been illuminated before (Fischer-Arnold, 1963). This clearly demonstrates that the photoreceptor pigment is neither in the chloroplasts nor in the moving portion of the cytoplasm. Instead, a stationary "cortical layer" of cytoplasm, whatever this means, must be the site of the photoreceptor pigment. Hence, there is at least a close spatial relationship between this pigment and the cell membrane (the plasmalemma). This conclusion is supported by an action dichroism of the light effect (Zurzycki and Lelatko, 1969).

The action spectrum of this light effect reveals two maxima at about 370 and 450 nm (Fig. 3). This spectrum closely resembles the absorption spectrum of riboflavin, hence a flavin compound is assumed to be the photoreceptor pigment (Fischer-Arnold, 1963).

In the discussion of transduction processes (Section 4) we will learn that

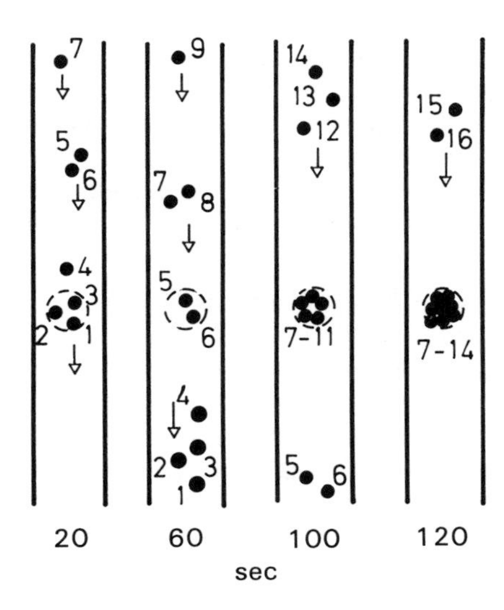

Figure 2. Movement of some chloroplasts in *Vaucheria* during microbeam irradiation with blue light (encircled). Individual chloroplasts are numbered; arrows indicate overall streaming of cytoplasm. Four consecutive positions are shown, with the time after beginning of irradiation given below (in seconds). (After Fischer-Arnold, 1963.)

in *Vaucheria*, light initiates fast reactions that are clearly related to the cell membrane. This once more supports the assumption of a close association between photoreceptor pigment and cell membrane. More detailed information has been obtained in a different system, which exhibits a classical photodinesis, in *Vallisneria* leaves.

In the mesophyll cells of dark-adapted *Vallisneria* leaves, rotational streaming of cytoplasm can be observed, but the chloroplasts usually do not participate in this rotation. Upon illumination, streaming is enhanced within a few minutes, and the chloroplasts begin to move as well. This effect is also found, although transiently, in darkness after a short light pulse (Seitz, 1967*a*).

The action spectrum of this photodinesis, as given in Fig. 4 (△--△) cf. also Fig. 5 (△—△), is similar to that of *Vaucheria* and has been interpreted as representing a flavin compound as well. Extrapolating from many blue-light responses in lower and higher plants, the photoreceptor molecule belongs to the flavoproteins, but no individual compound can yet be favored out of the many flavoproteins in cells (Schmidt, 1980; Seitz, 1980).

Additional evidence in favor of a flavin comes from inhibitor experiments. Potassium iodide (KI), known as a quencher of the triplet-excited state of flavins, strongly reduces or even eliminates the light effect without damaging the cells. However, extension of these experiments to the action spectrum revealed an interesting complication: The KI effect depends on the wavelength, and there are spectral regions in which KI is relatively ineffective. This is taken as evidence for the involvement of a second photoreceptor pigment (Seitz, 1967*b*). Its effect should be reflected by an action spectrum obtained under KI treatment, and the maxima point to the photosynthetic pigments in the blue region (Fig. 4, ●—●). Interestingly, there is always a small photodinetic effect of red light, and this, too, is uninfluenced by KI. Finally, treatment with the inhibitor of photosynthesis, 3-(3′,4′-dichlorophenyl)-1,1-dimethylurea (DCMU) abolishes the effects of these spectral regions that are attributed to the photosynthetic pigments (Fig. 4, ○—○) (Seitz, 1967*b*).

Taken altogether, photodinetic effects seem to be mediated by two different photoreceptor pigments independent of each other. As will be shown next, they are located in different compartments of the cell.

Using linearly polarized light, an action dichroism has been found in the blue region: Light is more effective if its **E** vector is oriented parallel to the long axis of the cell than if oriented perpendicularly. This points strongly to an orientation of the photoreceptor molecules, which is stable in spite of the rotational movement of the cytoplasm. Thus, as in *Vaucheria,* the cortical cytoplasm appears to be the site of photoreceptor molecules, and in fact this may mean the cell membrane (Seitz, 1967*a*). On the other hand, no action dichro-

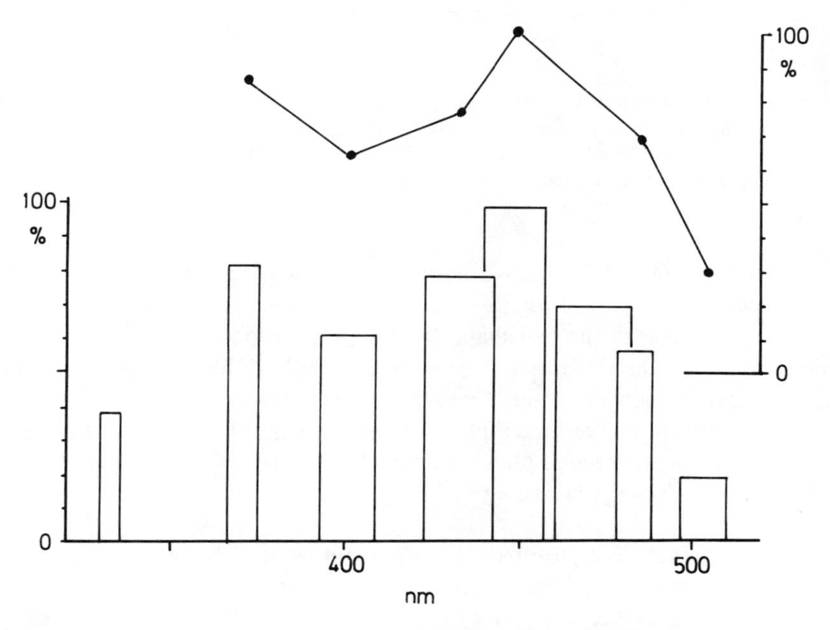

Figure 3. Action spectrum for local chloroplast accumulation in *Vaucheria* as induced by a microbeam (below). The width of the columns denotes the half-bandwidth of the interference filters. For comparison, the curve shows the action spectrum of high-intensity chloroplast orientation (above). Relative quantum effectiveness is given on the ordinates. (Redrawn from Fischer-Arnold, 1963.)

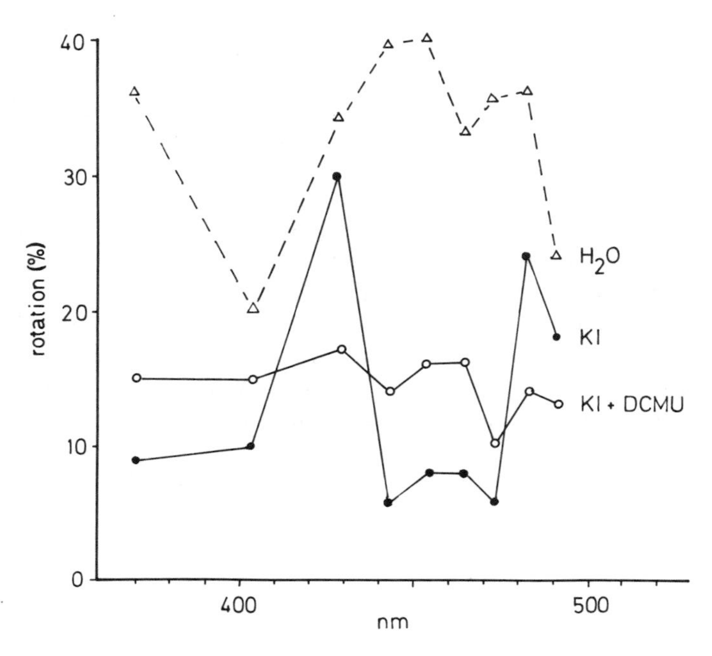

Figure 4. Photodinesis in *Vallisneria spiralis* as dependent on the wavelength and on inhibitors at equal quantum-flux density. The leaves were incubated in (△--△) water, (●—●) 0.2 *M* potassium iodide, and (○—○) additional 10^{-7} *M* DCMU. The ordinate denotes the percentage of cells with chloroplast rotation. (Redrawn from Seitz, 1967*b*.)

ism could be detected for the red-light effect, consistent with the second photoreceptor being localized in the chloroplasts, without any preferential orientation with respect to the cell shape (Seitz, 1972). Since strong association of chlorophyll with the thylakoid membrane is a well-established fact, and since the very early effects of the light reaction in photosynthesis are bound to that membrane, we may take for granted the involvement of membrane structures for the action of the second photoreceptor. Similar results have been obtained in *Elodea canadensis,* concerning action spectrum and action dichroism (Seitz, 1964).

There is another light effect in *Vallisneria,* which obviously comes about via the same photoreceptor systems: a change of the strength of anchoring of the chloroplasts to the cortical cytoplasm. This effect can be measured by exposing the tissue to centrifugal forces, and thus the mobility of the chloroplasts can be determined. The action spectrum is identical to that of photodinesis (Fig. 5, ○—○, △—△); evidence is also found for two different photoreceptors and the action dichroism can be demonstrated as well (Fig. 6; Seitz,

1967*a*, 1971). There is good evidence that these two responses not only share the photoreceptor systems, but that they are also causally connected in the transduction chain. More detailed information is given in Section 4.1.

2.2. Photoreceptor Pigments in Chloroplast Reorientation

In addition to photodinesis and the influence of light on chloroplast mobility, *Vallisneria* fortunately also exhibits reorientation of chloroplasts to light. The action spectrum of this response (Fig. 5, ●-●) is nearly identical to the other two action spectra (Seitz, 1967*a*). Moreover, chloroplast rearrangement can be reversibly inhibited by KI, and this inhibition depends on the wavelength as in photodinesis (Seitz, 1972). Finally, a conspicuous action dichroism has been found, with the chloroplasts accumulating at the cell walls oriented parallel to the **E** vector in moderate light, but at those oriented perpendicular in high-intensity light (Fig. 7). Since in the former case chloroplasts gather in the strongest absorbing regions, but in the least absorbing regions in the latter case, the transition moments of the photoreceptor molecules have to be oriented parallel to the cell surface (Seitz, 1967*a*).

There are a great number of action spectra in a variety of plants belonging to different taxonomic groups. They are all very similar to each other, and their shape is consistent with flavin compounds as the molecular base of photoper-

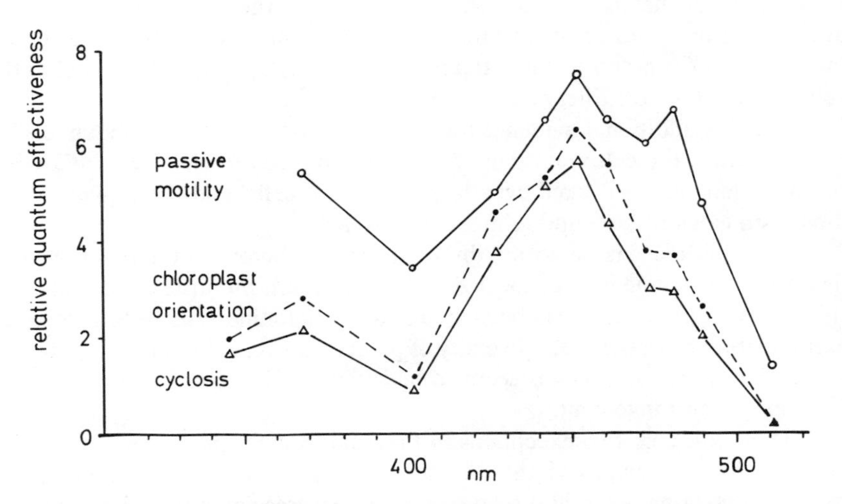

Figure 5. Action spectra of light responses in *Vallisneria spiralis*. (○—○) Mobility (passive motility) of chloroplasts as measured by displacing them with centrifugation; (●-●) chloroplast orientation to high-intensity light; (△—△) photodinesis as the light effect inducing cyclosis of chloroplasts. (After Seitz, 1967*a*.)

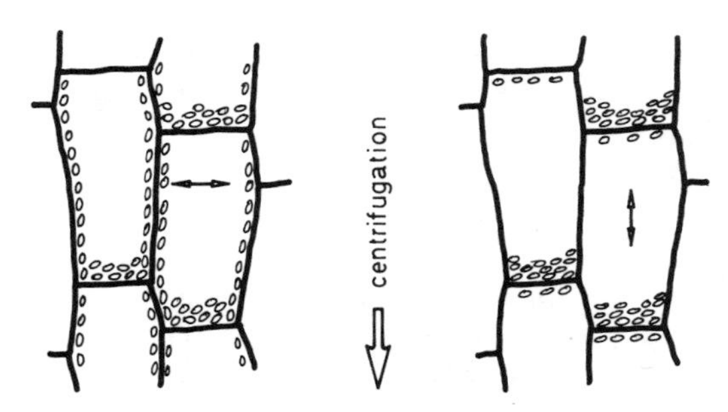

Figure 6. Action dichroism of light-induced chloroplast mobility in *Vallisneria spiralis* (cells shown in surface view). Irradiation 30 min with high-intensity blue light, linearly polarized, with the **E** vector as shown by the double-headed arrows. Thereafter, centrifugal force applied in the indicated direction. (Redrawn from Seitz, 1979*b*.)

ception (Zurzycki, 1967*a*; Lechowski, 1972; Inoue and Shibata, 1973). Moreover, the two types of rearrangement—low-intensity and high-intensity response—share the same action spectrum in all the systems studied so far (Zurzycki, 1962*a*; Haupt and Schönfeld, 1962; Fischer-Arnold, 1963; Mayer, 1964; cf. Fig. 3). Finally, the action dichroism as shown in *Vallisneria* seems to be a common feature in chloroplast rearrangement (Fig. 8), and a few exceptions can be explained on the basis of ultrastructural pecularities (Zurzycki, 1969; Zurzycki and Lełatko, 1969; Pfau *et al.*, 1979).

In the moss *Funaria,* the action dichroism has been found to depend on the wavelength; it disappears below 400 nm (Fig. 9; Zurzycki, 1967*b*). If indeed a flavin is the photoreceptor pigment, this effect could be predicted, because in the flavin molecule the orientation of the transition moment at 365 nm is tilted with respect to that at 450 nm; thus a surface-parallel orientation at 450 nm is consistent with an orientation at 365 nm not parallel to the cell surface (Fig. 10; Zurzycki, 1972).

There is additional evidence for a strong association of the photoreceptor pigments with the cell membrane or with its immediate vicinity. In *Funaria,* reorientation can be evoked even in plasmolyzed cells, and the typical action dichroism can still be found (Zurzycki, 1968).

It seems reasonable to attribute the main photoreceptor function to a flavin and to assume it to be localized at (or even in), but at least very close to the cell membrane, as it has been assumed in photodinesis as well. Moreover, extrapolating from the great diversity of plants with very similar action spectra and action dichroism, flavins seem to be the general photoreceptor system in chloroplast rearrangement.

There are only two exceptions to this rule to date, the green algae *Mougeotia* and *Mesotaenium.* In these algae, phytochrome is the photoreceptor pigment for detecting the light direction; it is closely associated with the cell membrane as well (Section 2.3). But even here, the ubiquitous blue-light receptor seems to operate in addition, although with a slightly different function. This becomes evident in comparing the two types of movement in *Mougeotia.*

The cell architecture of *Mougeotia* is unique insofar as there is only one flat chloroplast, which divides the cell into two half-cylinders. This chloroplast, merging with its edges into the parietal cytoplasm, can rotate in the cell (Fig. 11) so as to orient its face to low-intensity light or its edge to high-intensity light (Senn, 1908). For the low-intensity movement light absorption in phytochrome is sufficient, hence the action spectrum peaks in the red (Section 2.3). The high-intensity movement, however, is restricted to the blue region of the spectrum. As Schönbohm (1966) discovered, this response requires light absorption simultaneously in two photoreceptor pigments. Light direction again is perceived by phytochrome, but the transduction chain leads to the high-intensity response only if blue light is also absorbed in a yellow pigment. Interestingly, this latter effect is independent of the light direction, hence is called a tonic effect (Fig. 12).

Action spectra for this tonic effect point to a flavin as the photoreceptor pigment, and a pronounced action dichroism clearly reveals its association with stable cell structures, i.e., very probably with the cell membrane (Schönbohm, 1971*a*,*b*). The nature of the tonic effect, which switches the system from low-intensity response to high-intensity response, is still unknown.

A very similar interaction of phytochrome photoconversion and absorption in a yellow pigment has been found in *Mesotaenium* (Gärtner, 1970). In addition, a tonic blue-light effect strongly facilitates the phytochrome action in the low-intensity response with an action spectrum similar to that in flavin absorption (Büttner *et al.*, 1982). It is reasonable to assume that all these effects are based on a common photoreceptor pigment, but the substantial contribution of membranes to these effects cannot yet be concluded in general.

2.3. Perception of Light Direction

In the preceding sections we have considered perception of light with its qualities intensity, wavelength, and polarization. For a light-oriented response,

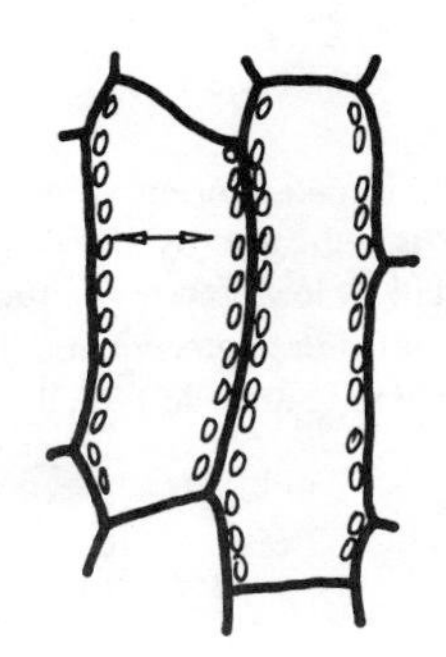

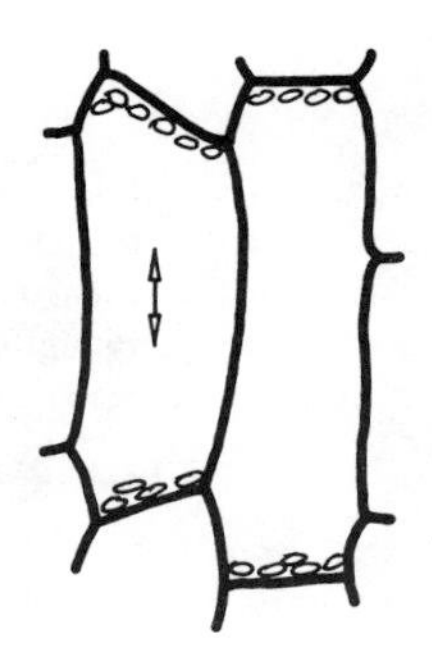

Figure 7. Action dichroism of light-oriented chloroplast rearrangement in *Vallisneria spiralis* (cells shown in surface view). High-intensity arrangement after irradiation with polarized blue light, the E vector of which is shown by the double-headed arrows. (After Seitz, 1967*a*.)

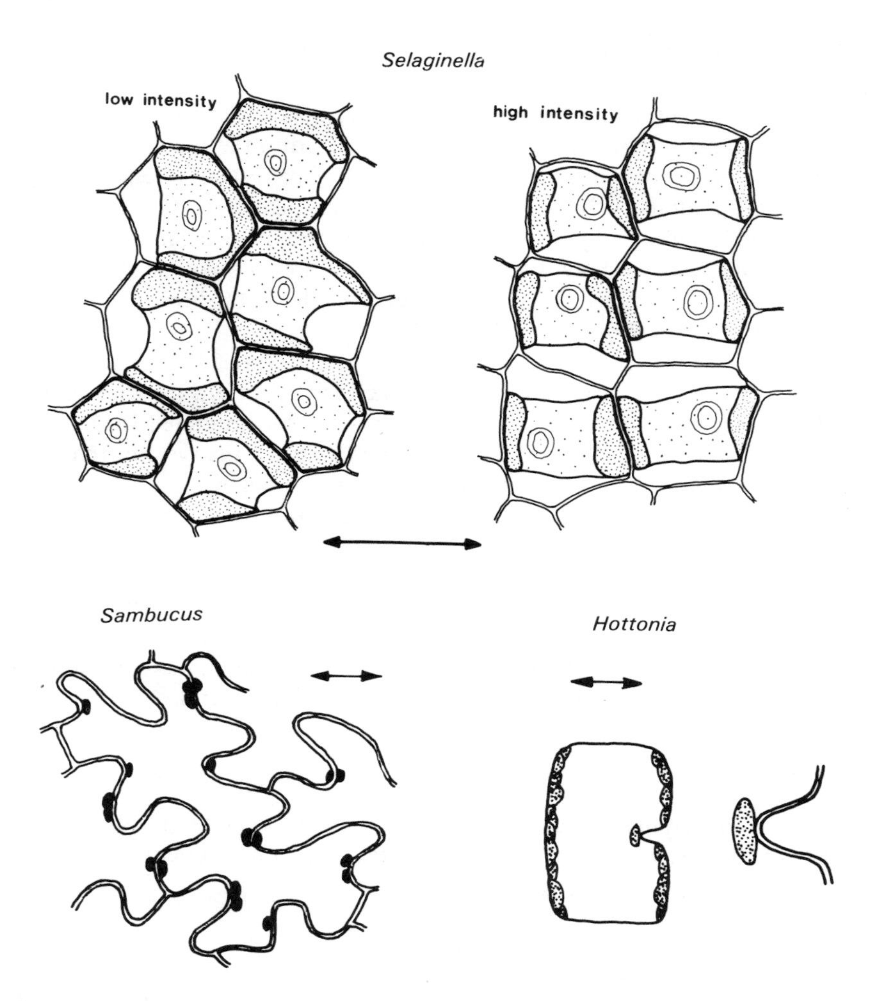

Figure 8. Action dichroism of chloroplast orientation; all cells shown in surface view. The **E** vector of linearly polarized light is indicated by the double-headed arrows. Epidermal cells of *Selaginella martensii* are irradiated by low- and high-intensity light, respectively, and *Sambucus* and *Hottonia* by high-intensity light. For *Hottonia,* a detail is given to the right in higher magnification. (Redrawn from Haupt, 1973, after Mayer, 1964 and Zurzycki and Lełatko, 1969.)

such as rearrangement of chloroplasts, perception of light direction is necessary as well. Chloroplasts orient to the regions with the highest or lowest light absorption, respectively. The obvious question is how unidirectional light is transformed into an absorption gradient in the cell (Haupt, 1965).

One possibility seems trivial and can be found in a system that does not belong strictly to the typical chloroplast rearrangements: Germination of spores of *Equisetum* or of zygotes of *Fucus* starts with an asymmetrical cell division, which is oriented by light (Weisenseel, 1979). The earliest visible effect of light is a redistribution of chloroplasts, which accumulate at the side toward the light. It can be shown by partial illumination that the chloroplasts approach the region with highest light absorption. In unidirectional light, this region is the half-sphere toward the light, and the gradient is simply attributable to the strong absorption of light during its path across the cell. This mechanism can operate reliably because these cells are tightly packed with organelles (Haupt, 1983).

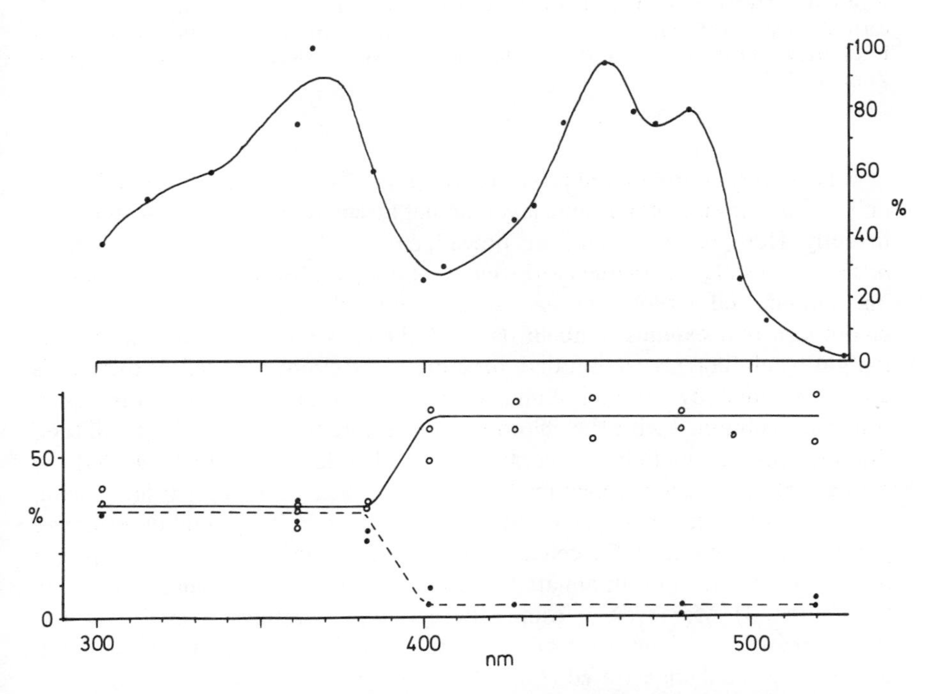

Figure 9. Action spectrum and wavelength-dependent action dichroism for chloroplast rearrangement in *Funaria hygrometrica.* (Above) Relative quantum effectiveness for 50% response in low-intensity movement (action spectrum). (Below) Percentage of chloroplasts in high-intensity movement that occupy a unit length at anticlinal walls perpendicular (○) and parallel (●) to the **E** vector of polarized light (action dichroism). (Redrawn from Zurzycki, 1967*a*, 1972.)

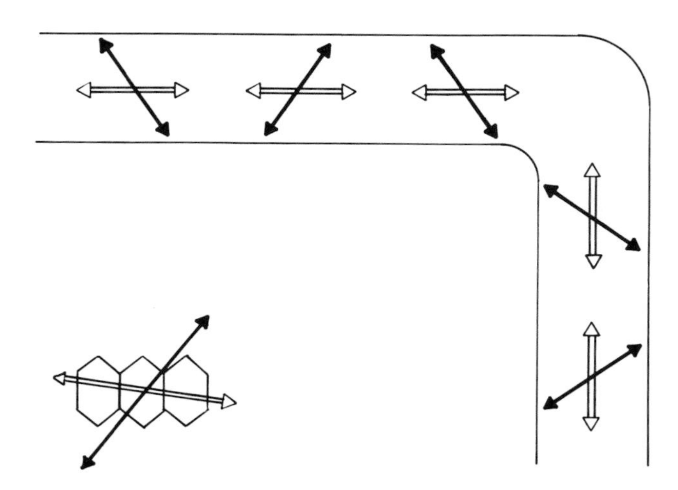

Figure 10. Hypothetical orientation of the isoalloxazine structure of riboflavin molecules in the cortical cytoplasm. Transition moments are shown by the white arrows for the blue range, by black arrows for the UV range. The isoalloxazine structure is shown in the lower left. (After Zurzycki, 1972.)

However, differentiated cells consisting mainly of the central vacuole with only a thin parietal cytoplasmic layer cannot attenuate the impinging light sufficiently. Here, other mechanisms prevail, one of which can be shown in *Hormidium.* This alga is unique (with the exception of a few other organisms having similar cell structures) insofar as each cell contains only one large chloroplast that extends to about one-half the circumference and that orients to light by gliding in an azimuthal direction of the cylindrical cell. If these cells are surrounded by air and illuminated by moderate light, a low-intensity response is obtained with the chloroplasts orienting to the rear of the cell (i.e., the side opposite the light source; Fig. 1C). It had long been assumed that the cylindrical cell acts as a collecting lens, focusing the light to its rear half (Senn, 1908). This can be confirmed by illuminating the cells in the highly refractive paraffin oil; in this case, the cell acts as a diverging lens, the region toward the light source is the most illuminated, and the chloroplast now orients toward the front (Fig. 13). Moreover, by using different oil preparations, a linear relationship of rear versus front orientation on the refractive index of the surrounding medium can be demonstrated (Fig. 14; Scholz, 1976*a*).

Besides this effect of light refraction, which closely resembles the well-known lens effect in phototropism of *Phycomyces,* light refraction can result in an absorption gradient in a different way. At the beginning of this century, Senn (1908) computed the pathway of the light rays through a *Vaucheria* cell surrounded by air. Again, a lens effect has been found, but the difference in illuminance between front and rear seems to have little importance for the chloroplast rearrangement. It has been shown, however, that light refraction also results in the light bypassing a peripheral region just behind the flanks, which therefore receives much less light than either front or rear. Accordingly, in a low-intensity arrangement, the chloroplasts are found at front and rear, with an empty space mainly behind the flanks; in a high-intensity arrangement, the tightest accumulation is located at and behind the flanks (Fig. 1B).

This pattern of light absorption, similarly computed by Gabryś-Mizera (1976) for a model cell, can be found only in the peripheral layer of the "lens." The observed chloroplast pattern, matching the predicted absorption pattern, is most compatible with the photoreceptor molecules being localized close to the cell membrane. Nevertheless, an absolute necessity of their membrane association for perception of light direction cannot be derived from these results. Only the fourth mechanism, discussed next, has this association as an absolute requirement.

The mechanisms for establishing an absorption gradient, which are based on light refraction, obviously require a sufficient difference between refractive indices of cells and the surrounding medium. Accordingly, as we have seen, these mechanisms work perfectly in air as the medium. In water, however, the difference is very small, and only weak gradients can be expected. Nevertheless, many submersed algae exhibit conspicuous and very precise chloroplast reorientations. Here, the dichroic orientation of photoreceptor molecules becomes important; this has been thoroughly investigated in the green alga *Mougeotia* (Haupt and Schönbohm, 1970).

As stated earlier, the large single chloroplast can rotate to expose its face

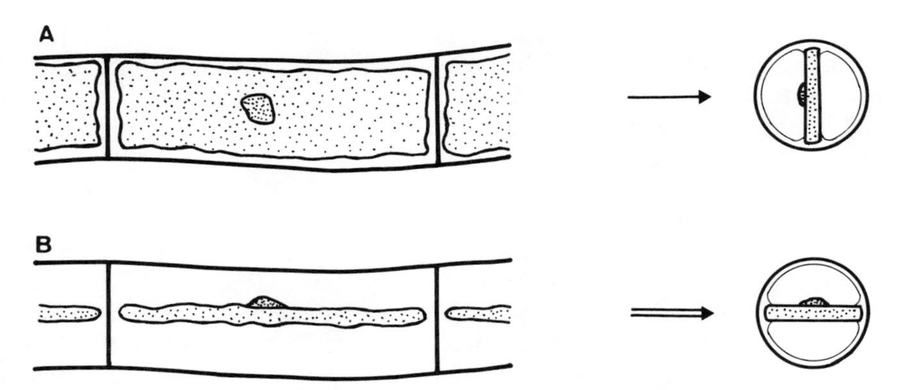

Figure 11. *Mougeotia* cell in surface view and in cross section, showing chloroplast in (A) low-intensity and (B) high-intensity arrangement. (Redrawn from Haupt, 1973.)

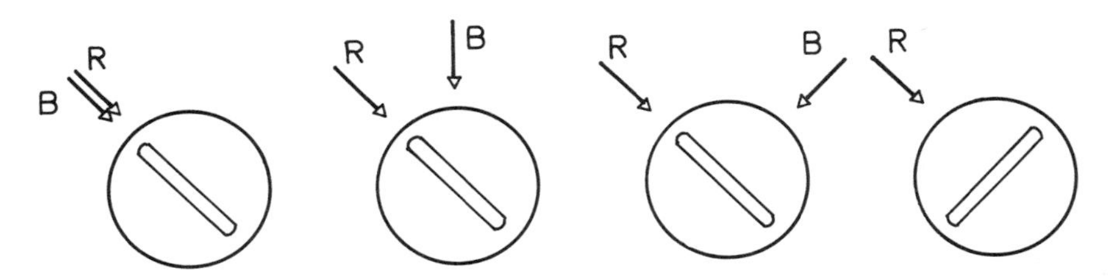

Figure 12. Schematic cross section through a *Mougeotia* cell showing the tonic effect of blue light. The chloroplast is oriented by combinations of unidirectional red and blue light from various directions. In the cross section at the far right, only red light is applied. (After Haupt, 1972, according to data of Schönbohm, 1967.)

or its edge to the light. The former movement, i.e., the low-intensity response, will now be analyzed. In contrast to other chloroplast reorientations, this response is a phytochrome effect, with its spectral maximum in the red region and with far-red reversing an induction by red light (Haupt, 1959*c*). By partial irradiation with microbeams, the photoreceptor pigment has been located in the parietal cytoplasmic layer rather than in the chloroplast, and the latter orients in a cytoplasmic gradient of the "active phytochrome" P_{fr} (Bock and Haupt, 1961).

Most important for the perception of light direction is the discovery of an action dichroism; in linearly polarized light, only those cells show response for which the electrical vector of the light (the vibration plane) is perpendicular to the long axis, or for which this light at least has a component in that direction (Fig. 15). This finding led to the conclusion that the phytochrome molecules serving as photoreceptors for the response are oriented in a dichroic pattern (Haupt, 1960). This pattern has been analyzed by proper combinations of irradiations with polarized light, especially to small areas. Accordingly, the transition moments of the red-absorbing phytochrome (P_r) molecules are oriented parallel to the cell surface and aligned along a helical pattern. Such a strong and obviously stable orientation is suggested to demonstrate close association to the cell membrane (Haupt and Bock, 1962).

The importance of this dichroic orientation for the perception of light direction can be seen in terms of a theoretical model (Fig. 16, Haupt and Schönbohm, 1970). Assume photoreceptors oriented parallel to the surface in two populations: one parallel and one perpendicular to the axis of the cylindrical cell. If such a cell is irradiated by polarized light, vibrating parallel to the axis, all receptors of the parallel population have an equal chance for light absorption, hence light is absorbed uniformly around the periphery, no absorption gradient is established, and light direction cannot be detected. If, on the other hand, the vibration plane is oriented normal to the cell axis, the perpendicular population of receptors becomes involved in absorption. However, in this case the photoreceptors are not equally oriented in relationship to the vibration plane of the light. Rather, optimal chance for absorption is restricted to front and rear, whereas at the flanks absorption can hardly take place. As a result, a strong absorption gradient ensues, light direction is perceived, and the chloroplast orients accordingly (cf. Fig. 15).

Finally, unpolarized light can be considered equivalent to a mixture of parallel and perpendicular vibrating polarized light. In this case, absorption is obviously stronger at front and rear than at the flanks. As a final generalization, in our theoretical model the two populations of receptors can be replaced by a pattern which, upon vectorial decomposition, yields vectors with parallel and perpendicular orientation, as is true for the helical pattern in *Mougeotia.* In this organism, detection of light direction is based on the dichroic orientation of the photoreceptor molecules, in turn requiring their stable association with the cell membrane (Haupt, 1982*b*).

In fact, the perception system is still more complicated in *Mougeotia;* this complication is a necessity as will be shown. Unlike other photoreceptor pigments, which upon irradiation are transformed into a short-lived excited state and which after a fast dark relaxation are immediately available for absorbing the next photon, P_r is transformed to the metastable far-red absorbing phyto-

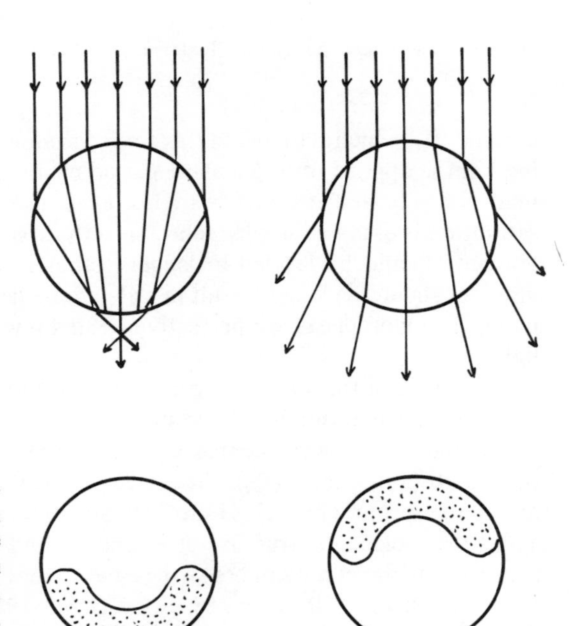

Figure 13. Schematic cross section through a *Hormidium* cell with the light refraction (above) and the chloroplast in low-intensity position (below). Cell in air (left) or paraffin oil (right). (Redrawn from Haupt, 1983.)

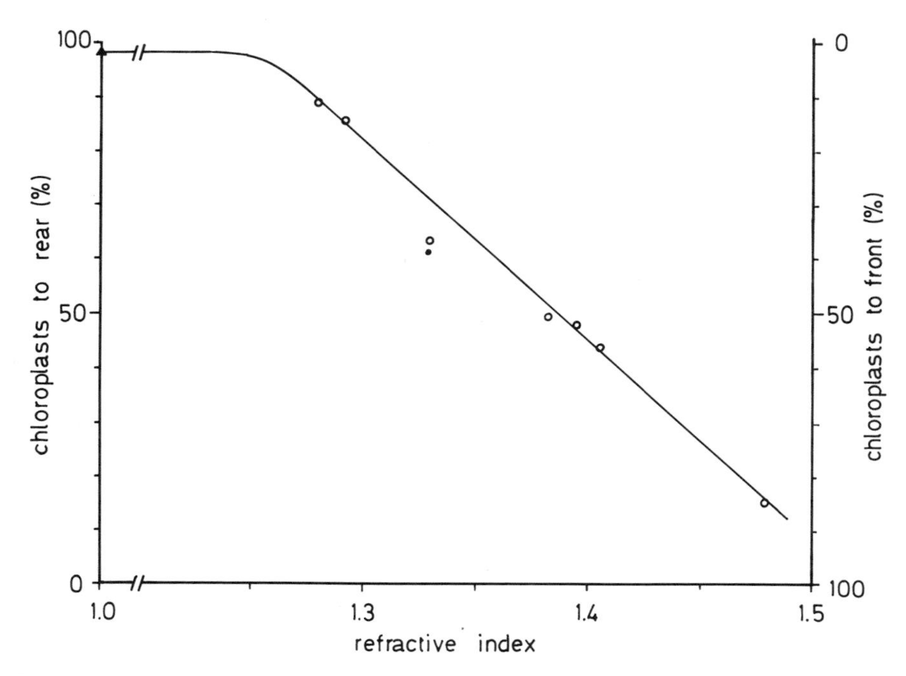

Figure 14. Low-intensity chloroplast position in *Hormidium* as a function of the refractive index of the surrounding medium. (Redrawn from Scholz, 1976*a*.) (○) Oil, (●) water, (▲) air.

chrome, P_{fr}, which remains in this form for minutes or hours. Thus, if saturating light is applied during a short period of time, P_{fr} formation becomes maximal not only at front and rear, but eventually at the flanks as well. Since saturation is obtained in phytochrome with rather low fluences already, any P_{fr} gradient should be leveled in longer-lasting irradiations, i.e., the absorption gradient should no longer result in a P_{fr} gradient. Yet, in their natural environment, the chloroplasts are perfectly oriented toward the continuous saturating light.

To explain this apparent contradiction, the absorption dichroism of phytochrome has to be analyzed one step further. The surface-parallel, helical pattern of the transition moments is true for the red-absorbing form only (P_r). The far-red absorbing form (P_{fr}), the active phytochrome, however, is oriented normal to the surface (Fig. 17, Haupt, 1968*a*). Interestingly, recent considerations about the molecular structure of P_r and P_{fr} and about the mode of P_{fr} action suggest that the chromophore swings away from the protein moiety upon photoconversion $P_r \rightarrow P_{fr}$ (Fig. 18; Song *et al.,* 1979). Moreover, the changing absorption dichroism between P_r and P_{fr}, originally concluded from physiological experiments, has recently been demonstrated *in vitro* as well (Sundqvist and Björn, 1983*a,b*). It is true, in these investigations that the orientation changed by only about 32° rather than by 90°. But Björn (1984) has shown that the main observations in Mougeotia can be explained as well, if P_r forms an angle of about 20° with the surface, and P_{fr} forms an angle of about 50° (i.e., the latter deviates from the normal direction by 40°).

On this basis, we can explain the P_{fr} gradient in saturating light. Since the phytochrome molecule is supposed to be fixed with respect to the cell surface, the changing absorption dichroism has to cause fundamental differences in the absorption probability, depending on the azimuth. Accordingly, the photo-equilibrium $P_r \rightleftharpoons P_{fr}$, which in randomly distributed phytochrome is only wavelength dependent, also becomes dependent on the azimuth, ensuring a gradient of P_{fr} even in saturation (Fig. 19; Haupt, 1972).

In summary, as a first step dichroic orientation results in the establishment of an absorption gradient in unidirectional light. As a second step, the change in dichroism (the flip-flop dichroism) is necessary to guarantee a gradient of the metastable product P_{fr} of light absorption even in saturation. This latter step seems to be unique to phytochrome, yet unnecessary in other cases. The first step, however, is widely distributed in plant cells.

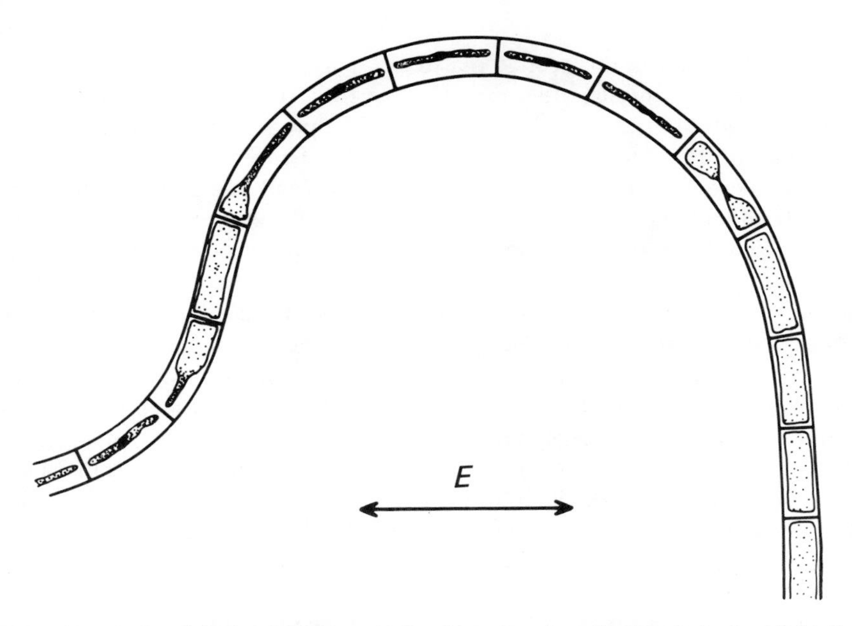

Figure 15. Action dichroism in *Mougeotia*. Low-intensity orientation in polarized red light, the E vector of which is indicated by the double-headed arrow. The experiment started with all the chloroplasts in the profile orientation. (From Haupt, 1970.)

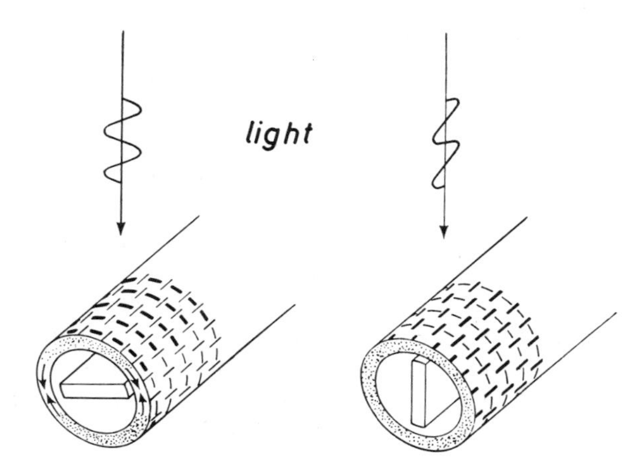

Figure 16. Absorption of linearly polarized light in a model cell of *Mougeotia*. The transition moments of the dichroic photoreceptor molecules are shown by dashes. Heavy and light dashes denote those photoreceptors that can or cannot absorb the light, respectively, owing to their dichroic orientation. The resulting absorption is indicated by dots near the periphery and its gradient by arrows. The waves denote the electrical vector of the light. (Redrawn from Haupt and Schönbohm, 1970.)

Recall earlier in this section that the response in *Hormidium,* i.e., movement of the chloroplast to the rear or to the front of the cell, depends on the refractive index of the surrounding medium. In a properly balanced mixture of different oils, the refractive index of the cell can be matched by the medium so that chloroplasts go with equal probability to front and rear. But interestingly, front and rear are clearly preferred over the flanks; therefore, an absorption gradient front-and-rear versus flanks has to be postulated (Scholz, 1976*a*). Since light refraction is out of operation in this case, the gradient has to be attributable to oriented photoreceptor molecules. Indeed, the blue-light photoreceptor in *Hormidium* has been shown to be oriented dichroically in a surface-parallel pattern (Scholz, 1976*b*). Thus, in *Hormidium* dichroic orientation of photoreceptor molecules can be used for perception of light direction in addition to the common lens mechanism.

Another example is *Selaginella martensii,* the epidermal cells of which have only one chloroplast each. In low- or moderate-intensity light, this chloroplast moves to the rear; it is reasonably assumed that the same focusing effect that occurs in *Hormidium* at the air–cell interface is responsible for establishing the orienting absorption gradient in *Selaginella* (Senn, 1908). Moreover, the photoreceptor pigment is localized outside the chloroplast (Mayer, 1964). In high-intensity light, the chloroplast moves to one of the less-absorbing anticlinal walls, exposing its edge to the light. There is no preferred anticlinal wall for this arrangement in unpolarized light. If, however, the cell is irradiated with polarized high-intensity light, an obvious action dichroism is observed: The chloroplast selectively occupies those anticlinal walls that are oriented normal toward the vibration plane. Interestingiy, the opposite pattern has been observed for the low-intensity arrangement (Fig. 8A,B). This clearly points to a surface-parallel orientation of the photoreceptor molecules, hence to a close association with the cell membrane (Mayer, 1964). This absorption dichroism, however, may not have a significance for perception of light direction in *Selaginella* under natural conditions.

In cells of "normal" architecture, i.e., containing many small chloroplasts, the action dichroism in low- and in high-intensity light, respectively, basically resembles that of *Selaginella* (Section 2.2.). Again, there is not much evidence for a substantial contribution of this action dichroism to the perception of light direction in the natural environment, although it is not without effect (Gabryś-Mizera, 1976).

A final consideration may again demonstrate the importance of association of photoreceptor pigments with cell structures, especially membranes. Whenever an intracellular rearrangement of organelles is oriented by an absorption gradient of light, it is necessary that the effect of the absorption remain localized as long as the response proceeds, i.e., it must not spread through the cell. It is therefore important that the very first step of the transduction chain, i.e., perception of light, remain localized; this is ensured by the inability of the photoreceptor pigment molecules to move through the cytoplasm, being firmly associated instead with the cell membrane. This is especially important in the case of phytochrome, for which a molecule may respond to light absorption by triggering secondary effects over a long period of ≥ 1 hr (Haupt, 1959*c*; Fetzer, 1963; Wagner, 1974). Thus, perception of light direction requires association with the cell membrane, if it is to result in a localized reaction as the basis of orientation movement. But association with

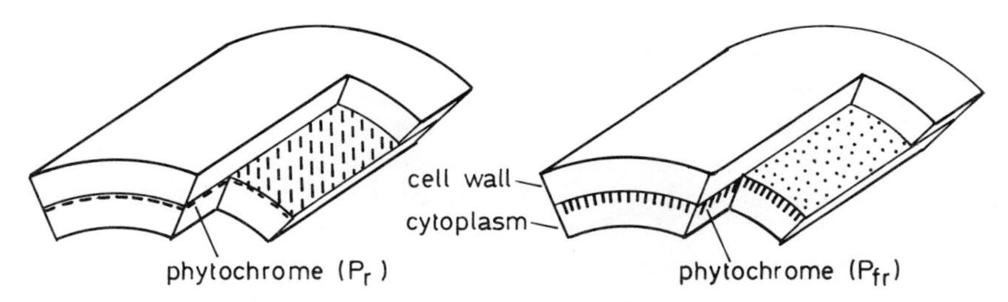

Figure 17. Schematic drawing of the outer part of a *Mougeotia* cell. The transition moments of the phytochrome molecules in the inactive (P_r) and active form (P_{fr}), respectively, are indicated by dashes and dots. (Redrawn from Haupt, 1970.)

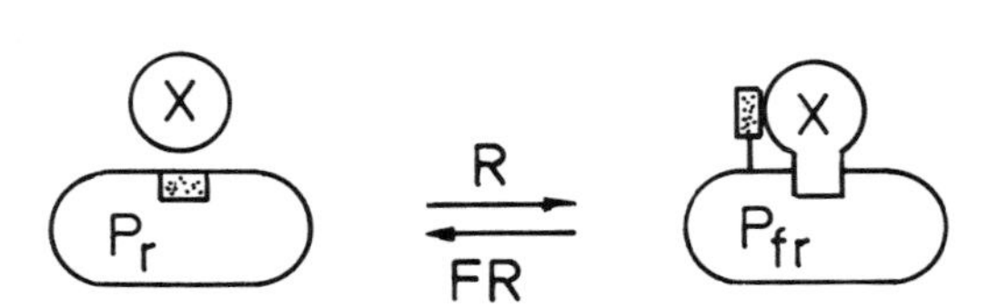

Figure 18. Model of the phytochrome molecule with the changing conformation of its chromophore (bar) as resulting from phototransformation. R and FR denote irradiation with red and far-red light, respectively. X represents the P_{fr} receptor, which may be a membrane effector. (After Song *et al.*, 1979.)

the membrane does not necessarily imply interaction, and indeed pigment–membrane interaction has not yet been proved as part of the perception process in light-controlled organelle movement. It is the transduction which, according to current knowledge, is connected with cell membranes not only topographically, but causally as well. Before the transduction processes can be analyzed, however, it is necessary to describe the mechanism of movement proper, with its motive force and its energetics.

3. MECHANICS OF MOVEMENT

Cytoplasmic transport of cell organelles including chloroplasts is based on a complex array of structural proteins, notably the microfilaments and microtubules, collectively referred to as the plant cytoskeleton. A third class of structural proteins known from animal cells, i.e., intermediate filaments of thicker diameter than that of actin but thinner than that of myosin, has not yet been reported in plants.

Microtubules, on the one hand, are the main structures underlying movement of chromosomes during mitosis and movement of cytoplasmic vesicles during cellulose cell wall formation. However, for chloroplast reorientation, their function seems to be limited. Microfilaments of actin and myosin, on the other hand, appear to be the common structural basis for cytoplasmic streaming and chloroplast movement in most cases investigated so far.

Microtubules are a class of structural proteins of strikingly similar morphology in the kingdom of eukaryotes, including cilia and flagella. A microtubule is built from 13 individual protofilaments in parallel rows to form a hollow cylinder of $\sim$ 25 nm in diameter. Each protofilament consists of globular dimers of tubulin (α and β, 55,000 daltons each) connected end on.

Microtubules in the cytoplasm are far less stable than those in the orderly array of parallel microtubules, e.g., in cilia. Cytoplasmic microtubules disintegrate upon cleavage of the two anhydride bonds in GTP; disintegration may be triggered *in vivo* by external factors, such as exposure to low temperatures, high pressure, or antimitotic drugs. The drug colchicine binds rather specifically to unpolymerized microtubule protein and finally leads to microtubule disintegration. Hence, blockage or alteration of a movement by colchicine is taken as strong evidence that the movement is dependent on microtubules (Filner and Yadav, 1979).

Indeed, rhythmic chloroplast movement in *Ulva* is sensitive to colchicine (Fig. 20). While submillimolar concentrations of colchicine shift *Ulva* chloroplasts more into profile position without inhibiting the extent of movement, millimolar concentrations of the drug actually stop profile-to-face movement (Britz, 1979). This inhibition is not reversed with washing, but (as expected for an effect on microtubules) the inhibition is reversed by irradiation with UV light, which forms the non-tubulin-binding isomer of colchicine, lumicolchicine.

In contrast to this circadian movement, involvement of microtubules in light-induced and light-oriented chloroplast movement has not been found thus far. Although in *Mougeotia* microtubules have been identified in the cortical cytoplasm, i.e., close to the plasmalemma (Foos, 1970), where they form a helical array (Marchant, 1978), photomovement of the chloroplast is completely insensitive to colchicine (Foos, 1971; Schönbohm, 1975; Tendel and Haupt, 1981). Consistently, the antimitotic drug, amiprophosmethyl, in concentrations of ≤ 0.1 mmole/liter turned out to be without effect (Klein, 1981).

The first action discovered for microfilaments has been reported for the rotational streaming (cyclosis) in characean cells. This is now generally under-

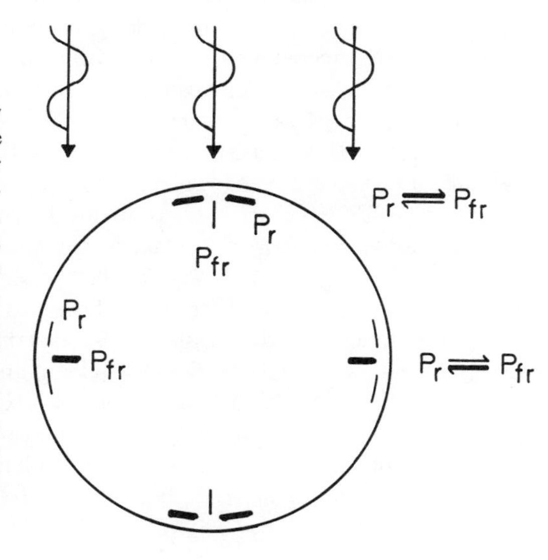

Figure 19. Absorption of linearly polarized light by phytochrome molecules, which change their dichroic orientation upon photoconversion $P_r \rightleftharpoons P_{fr}$. Cross section of a *Mougeotia* cell with the dashes indicating the transition moments of the photoreceptor molecules (cf. Fig. 17). Favorable and unfavorable orientation for absorption are shown by heavy and light dashes, respectively. The resulting shift of the photostationary state $P_r \rightleftharpoons P_{fr}$ is shown at the righthand side for the cell region facing the light and for the flank. (Redrawn from Haupt, 1972.)

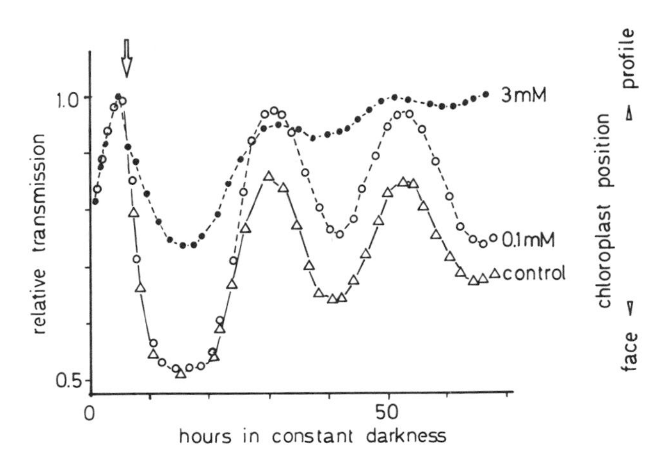

Figure 20. Circadian chloroplast movement in *Ulva lactuca* as measured by integrated light transmittance of the thallus. Colchicine added to the solution at the time indicated by the arrow, with final concentrations given at the curves. (After Britz, 1979.)

stood to be caused by unidirectional sliding of endoplasmic organelles along bundles of microfilaments anchored to the cortical layer of stationary chloroplasts. The streaming under natural conditions is unidirectional and continuous, but decreases upon irradiation by red light or stops during action potential. In cell-free preparations of *Nitella,* the presence of actin has been demonstrated biochemically (cf. Kamiya, 1981). Moreover, cytochalasin B (CB) inhibits the streaming in this organism and in *Chara* (Williamson, 1976). Finally, microfilaments have been demonstrated in the cortical cytoplasm by the electron microscope; their identity with actin has been proved by arrowhead decoration with heavy meromyosin (HMM) or subfragment S_1 (Palevitz *et al.,* 1974; Williamson, 1974; Palevitz and Hepler, 1975). Still more convincingly, the array of these structures coincides with the slightly helical pattern of streaming, and the arrowheads always point against the direction of streaming (Kersey *et al.,* 1976). Thus, there is no doubt about the function of actomyosin in cytoplasmic streaming of characean algae, as is also true for *Elodea* and *Vallisneria* (Yamaguchi and Nagai, 1981).

Generally, there are different methods to test the involvement of actin–myosin interaction in a given response. Several drugs can be used to inhibit this interation, e.g., cytochalasin B (CB), *N*-ethylmaleimide (NEM), *p*-chloromercuribenzoic acid (PCMB), or *Amanita* toxins (e.g., phallotoxin, phalloidin). CB, an isolate of the fungus *Helminthosporium dematoideum,* turned out to interfere rather specifically with the polymerization process of globular actin (42,000 daltons) into the filamentous F form. Actually, the polymerization process was shown *in vitro* to be interrupted by binding of CB onto the growing end of actin (Brown and Spudich, 1981).

Phalloidin, the well-known poison of the fungus *Amanita phalloides,* was introduced by Wieland (1977) also as an actin-specific drug. Similar to CB, phalloidin acts stoichiometrically with muscle F actin, but antagonizes CB by strongly promoting actin polymerization into stabilized F-actin filaments.

The sulfhydril (SH) reagents, *p*-chloromercuribenzoic acid (PCMB) and *N*-ethylmaleimide (NEM), bind to myosin rather than to actin. Presumably, this is because of the SH groups of the myosin–ATPase. Reactive SH groups, however, are also fairly abundant elsewhere in the cell and are well known in chloroplast and mitochondrial ATPases. Therefore, an inhibitory effect from PCMB or NEM could be attributable to either inactivation of myosin or/and inhibition of the energy supply.

Moreover, a highly specific test is the application of actin antibodies. This elegant technique was successfully applied in nonmuscle animal cells and also permits the introduction of fluorescent labels. In plant cells, fluorescent phallotoxin has so far been preferred in detecting actin at the light microscopic level, because of its low molecular weight of 1250 (Wulf *et al.,* 1979) and its high penetration rate of cell membranes. Digestion of the plant cell wall and permeabilization of the cytoplasmic membranes often lead to improved results (cf. Fig. 25).

There are several examples of light-induced and light-oriented chloroplast movement, which according to these tests make use of actin–myosin interaction as the mechanism of chemomechanical energy transduction. These examples, however, are more complicated than the rotational streaming in characean algal cells or in *Elodea* and in mesophyll cells of *Vallisneria.*

In the siphonaceous alga *Vaucheria,* the blue-light-induced chloroplast aggregation is not entirely hooked onto the bulk stream of cytoplasm: In darkness or when the alga is fully exposed to light, the chloroplasts swim with the stream of cytoplasm. Upon local irradiation of the alga, however, cortical fibers can be seen to reticulate, and concomitantly chloroplasts aggregate in this region. Local light therefore acts as a trap to the passing chloroplasts while part of the cytoplasm continues to move.

Several lines of evidence indicate that the cortical fibers enclose actin filaments and filament bundles of actin (Blatt *et al.,* 1980). CB was found both to inhibit streaming and to alter the organization of the cortical fibers (Fig. 21), while treatment of the cells with phalloidin, which stabilizes the F actin, protects the organellar movement from the action of CB. When decorated by a preparation of purified subfragment (S_1) from rabbit muscle myosin, the filament bundles appear fuzzy; in single filaments, the distinct arrowhead pattern characteristic of S_1-bound F actin is clearly visible.

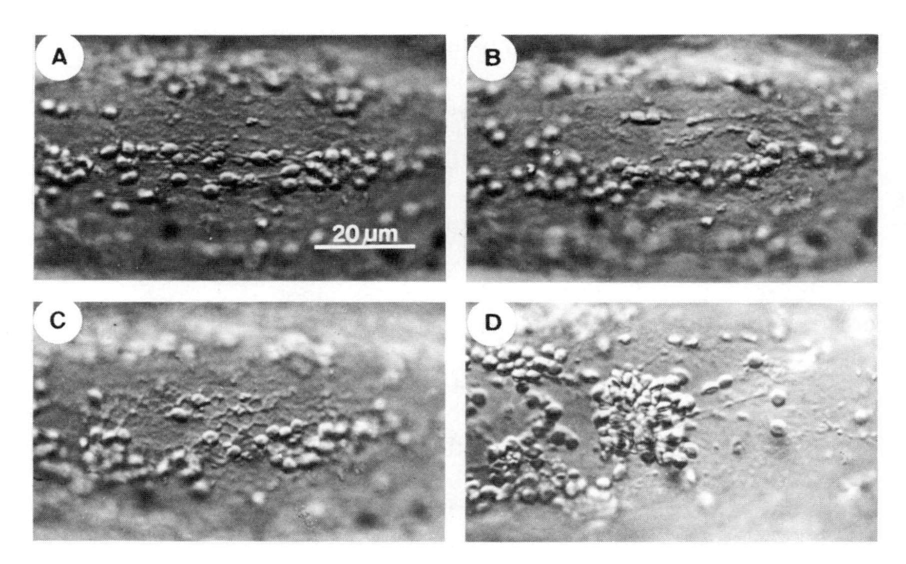

Figure 21. Influence of cytochalasin B (CB) on the cortical fibers of *Vaucheria sessilis*. (A) Before CB treatment streaming of organelles follows longitudinally oriented cortical fibers just visible as horizontal striations between the chloroplasts. (B–D) Following the addition of 100 μM CB, the chloroplasts begin to cluster (B), and within 10–15 min the fiber reticulum visibly expands (C). Then, 30 min after CB addition, large, slowly rotating masses of organelles are formed (D), which extend well into the vacuolar region of the cell. All movement ceased in this algal filament after 45-min exposure to CB (A–D ×650). (From Blatt *et al.*, 1980.)

The organelles may move along the cortical actin fibers (Fig. 22) in much the same way as proposed for characean algae (Williamson, 1975), but light apparently disrupts the cable assembly. Thus, the organelles are trapped in the irradiated region, where actin–myosin interaction is interrupted (Blatt *et al.*, 1980).

Finally, the most advanced mechanism of chloroplast reorientation, permanently independent of cytoplasmic streaming, is realized in *Mougeotia*. The cell of this organism has been shown to orient its chloroplast in a tetrapolar gradient of phytochrome. After phytochrome phototransformation in the cell has reached a certain threshold value, e.g., in response to a light pulse even as short as microseconds (Haupt, 1982), light is no longer needed, and the chloroplast reorients in darkness. Thus, the gradient of the metastable active form of phytochrome (P_{fr}) seems to enable the cell to memorize the direction of light. Indeed, persistence of this memory may be judged from inhibitor experiments with CB. When movement is blocked by CB over a certain period of time, the information proper of chloroplast reorientation is lost very slowly by the cell.

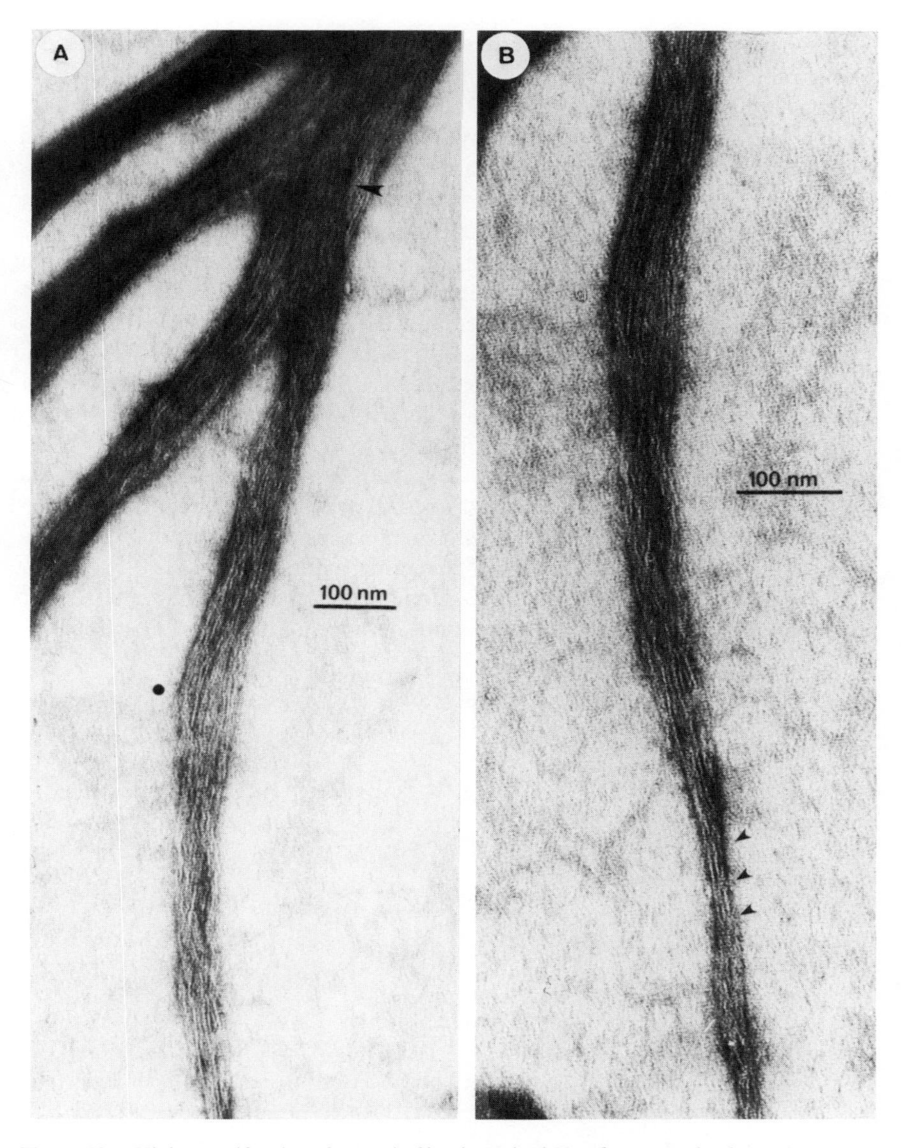

Figure 22. High magnification of an actin fiber branch of *Vaucheria sessilis* shows the manner in which bundles of actin filaments join to form a larger cable. (A) The arrow points to an unusual interweaving pattern of filaments, which may be caused by a background of underlying filaments (×110,000). (B) Distant to the black dot in (A), a region of the right filament bundle is shown, in which the beaded pattern is lost at intervals of 35–40 nm (arrows) (×140,000). (From Blatt *et al.*, 1980.)

The time constant of this first-order reaction of $\tau/2 \approx 90$ min (Fig. 23) is comparable to the half-life of dark reversion or of dark destruction of P_{fr} in other organisms.

These experiments with CB (Wagner *et al.*, 1972; Wagner and Klein, 1981; cf. also Schönbohm, 1975) are also consistent with the inhibitory action of PCMB (Schönbohm, 1972*a*) and of NEM (Fig. 24; Klein, 1981); altogether they indicate the involvement of actomyosin in chloroplast movement in *Mougeotia.* Recently, filamentous structures of 5–10 nm in diameter, running between the chloroplast edge and the plasmalemma, have been shown by transmission electron microscopy (Wagner and Klein, 1978). Moreover, actin was identified by HMM decoration in a cell homogenate (Marchant, 1976) and in spread protoplasts of *Mougeotia* (Klein *et al.*, 1980). Finally, fluorescent phallotoxin heavily stains the chloroplast edge merging into the parietal cytoplasm (Fig. 25; cf. also Fig. 11A).

There apparently is little doubt as to the role of actin and possibly myosin in the movement of *Mougeotia* chloroplasts. However, the mechanism of actin–myosin regulation and interaction with membranes remains an open question, as unfortunately seems to be true in general in plants.

Polymerization and depolymerization of microfilament actin and anchorage to membrane structures are functions of associated proteins. An adequate

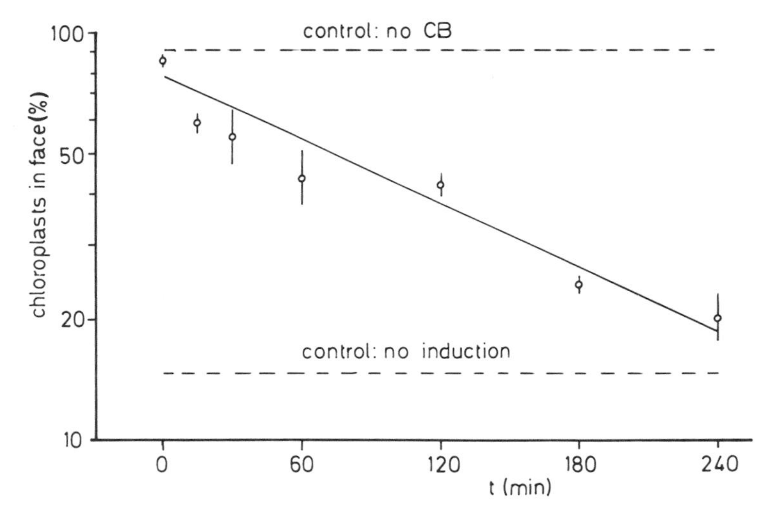

Figure 23. "Memory" of *Mougeotia,* treated with cytochalasin B (CB). Cells have been incubated in CB (0.015 m*M*) at $t = -15$ min, and induced to chloroplast orientation to face position at $t = 0$. After various times, as indicated by the abscissa, CB has been washed out in darkness, and the response evaluated 30 min thereafter. The bars show the standard error of the mean. (Redrawn from Wagner and Klein, 1981.)

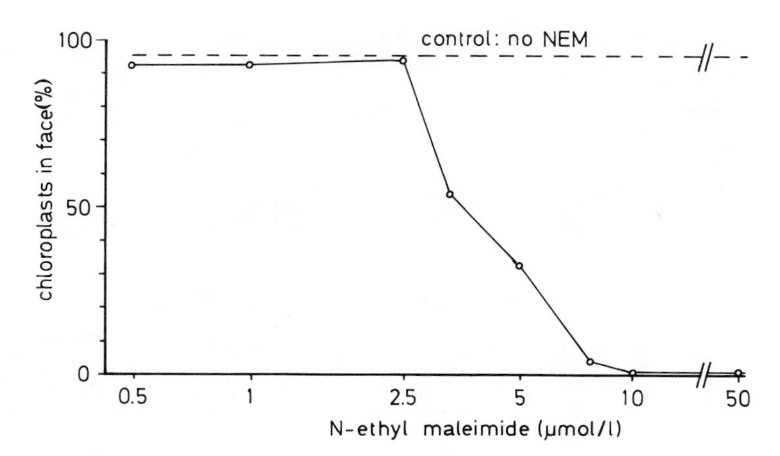

Figure 24. Chloroplast orientation under the influence of *N*-ethyl maleimide. Cells have been incubated in the concentrations given at the abscissa, and induced by white light to orient from profile to face position. The percent response after 60 min is shown at the ordinate. (After Klein, 1981.)

amount of detailed biochemical information on the various components in the prospective relay system is available only in the red blood cell and is referred to here as a possible guideline to plant systems. The cytoskeleton of an erythrocyte consists mainly of the proteins spectrin, bands 2.1 and 4.1, and actin. Direct contact to the membrane is made via the protein ankyrin (Koch, 1981). From plant material, *Physarum* actinin is known to affect the filament structure of actin; it may also serve as a membrane anchorage protein.

In conclusion, the cortical cytoplasm together with the adjacent cellular or organelle membranes seems to be the structure at which cytoplasmic movement often initiates. This is true in particular for well-studied organisms, e.g., characean algae, *Vallisneria* and *Elodea, Vaucheria,* and *Mougeotia.* In these particular cases, actin–myosin interaction provides the mechanical force for chloroplast movement in the cell. Since this movement is under the control of light, we need to know the connecting link between light absorption and activation or inactivation of actin–myosin interaction.

4. TRANSDUCTION PROCESSES

In considering current hypotheses of transduction processes, one has to be aware of the dual role of microfilaments (Seitz, 1982), as has been discussed for *Vallisneria* and *Mougeotia.* On the one hand, anchoring of chloroplasts to

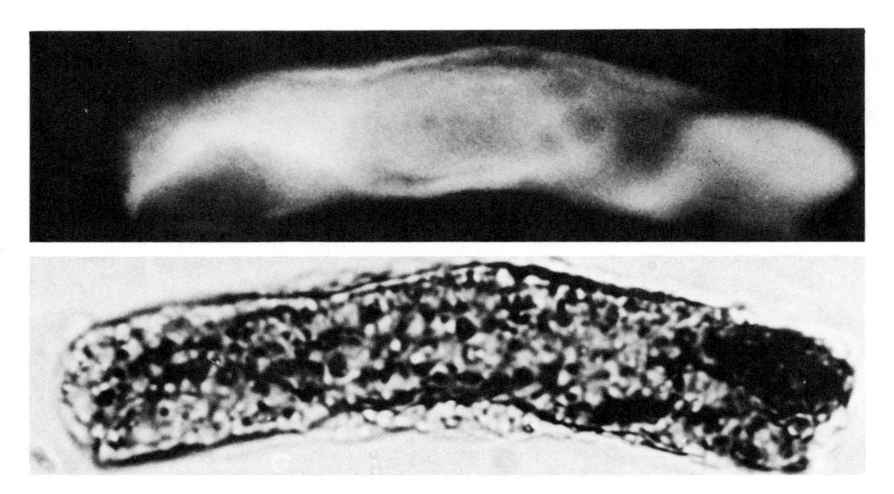

Figure 25. *Mougeotia* protoplast as stained by fluorescent phallotoxin. After digestion of the cell wall by incubation at 16°C for 16 hr with the enzyme cellulysin (1% w/v Calbiochem) in Waris' (1953) medium ("MXS") containing 0.4 mole/liter mannitol, released protoplasts were washed in MXS–mannitol medium without enzyme (Klein *et al.*, 1981), rendered permeable by a 30-min exposure to 1% v/v glutaraldehyde in MXS–mannitol medium, and washed again. Fluorescent phallotoxin (Wulf *et al.*, 1979) was applied to the protoplasts at a concentration of 0.2 mg/ml in MXS–mannitol medium and allowed to react for at least 30 min at room temperature in a humid atmosphere. The protoplasts were washed with several changes of MXS–mannitol medium and examined in a Leitz microscope Dialux with epifluorescence (HBO 50 W, interference band pass filter 390–490 nm in filter set PLOEMOPAK H 2). (Top) Epifluorescent microscopic view. (Bottom) Transmission light microscopic view. [Photographs were taken with Kodak Ektachrome 400 film, × 1500. Color prints are available upon request. (G. Wagner, unpublished).]

the stationary phase of cytoplasm (ectoplasm, cortical cytoplasm) is a function of permanent actin–myosin interaction, yet the regular and periodic actin–myosin cross-bridge cycle is necessary to generate the motive force (gliding force). Thus, any factor controlling actin–myosin interaction simultaneously influences cytoplasmic viscosity as well as the motive force. This makes it very difficult to establish a hypothesis for transduction that is consistent throughout.

4.1. The ATP Hypothesis

The ATP hypothesis has been proposed as the most direct transduction chain. Evidence in favor of it has been presented mainly for *Vallisneria* (Seitz, 1972, 1982). In this plant, chloroplast mobility was determined by centrifugation and can be shown to depend on the ATP level, i.e., anchoring is weakened by ATP; this effect can be compared with the ATP-dependent release of the rigor state of muscle actomyosin (Seitz, 1971).

Thus, the triphasic fluence rate–response curve of the light effect on chloroplast mobility (see Fig. 27) can be explained (Seitz, 1979*a*):

1. *Low-intensity white light:* In white light of very low intensity, chloroplast mobility is enhanced. This points to the availability of additional ATP under those conditions. Indeed, stimulation of oxygen consumption by low-intensity blue light has been observed in several plants, an effect that seems to be mediated by a flavin photoreceptor. Accordingly, this increase in respiration is assumed to produce additional ATP.
2. *Moderate-intensity white light:* In white light of moderate intensity (which corresponds to the range of low intensity in experiments on chloroplast orientation), chloroplast mobility is decreased in comparison with darkness, indicating that the cells are depleted of ATP. This latter can be explained by the photosynthetic activity of the chloroplasts, which uses more ATP for the Calvin cycle than it produces by the noncyclic electron flow. Accordingly, this light effect is abolished by inhibitors of photosynthesis, but is not influenced by inhibitors of oxidative phosphorylation; its spectral sensitivity points to chlorophyll as the photoreceptor pigment (Fig. 26, lower curve).
3. *High-intensity white light:* In high-intensity light, chloroplast mobility is increased again, pointing to a surplus of available ATP. Since in

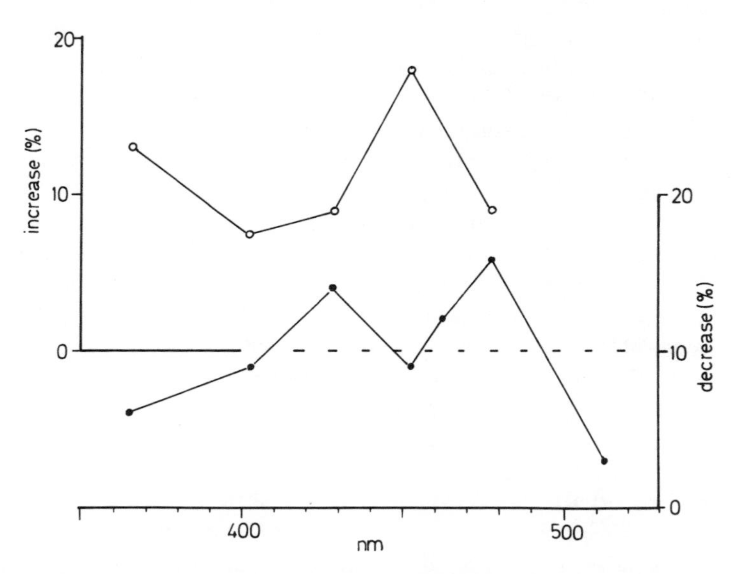

Figure 26. Spectral dependence of light-induced changes in passive motility in *Vallisneria* leaves. Increase in motility by high-intensity (750 pmoles cm^{-2} sec^{-1}, ○—○) and decrease by low-intensity light (0.8 pmole cm^{-2} sec^{-1}, ●—●), respectively. (Redrawn from Seitz, 1980.)

saturation of photosynthesis, cyclic photophosphorylation compensates for the ATP-draining effect of the Calvin cycle, photosynthetically active light no longer decreases the overall ATP level in the cell, and thus the blue light-enhanced respiratory ATP can again become effective. Indeed, this strong light effect is insensitive to inhibitors of photosynthesis, but sensitive to the uncoupler of oxidative phosphorylation, DNP; moreover, the effect is restricted to the short-wavelength region, and the shape of the action spectrum is compatible with a flavin as the photoreceptor pigment (Fig. 26, upper curve).

Photodinesis in *Vallisneria,* i.e., acceleration of cytoplasmic streaming and induction of cyclosis of chloroplasts, has an action spectrum identical to the high-intensity effect of light on chloroplast mobility (cf. Fig. 5); moreover, the intensity range of both effects is identical, and finally they share a common action dichroism. Thus, increased mobility obviously facilitates chloroplast displacement as a result of cytoplasmic streaming (Seitz, 1967*a*, 1971).

If this hypothesis were true, a corresponding negative photodinesis should be expected in moderate light intensity, which has been shown to reduce the chloroplast mobiltiy in *Vallisneria.* However, since the chloroplasts are resting in darkness, no inhibitory effect compared with dark movement can be observed. Only in etiolated *Avena* coleoptiles has a negative photodinesis been found in the appropriate light intensity (Bottelier, 1933), but this cannot be explained according to the above model, because there is no photosynthetic ATP consumption in etiolated tissue. Finally, no photodinetic effect has yet been reported in *Vallisneria* for very low-intensity light, which enhances chloroplast mobility (Seitz, 1979*b*).

So far the transduction chain, involving ATP as the main link, can reasonably explain the photodinetic effect of light. In order to interpret chloroplast orientation accordingly, additional information is needed (Seitz, 1971). If chloroplast mobility is measured at anticlinal and periclinal walls separately, the triphasic fluence rate–response curves are found to be shifted with respect to each other; much less light is required for an effect at the periclinal walls as compared with the anticlinal walls, the factor being more than 10:1 (Fig. 27). This nicely reflects the absorption gradient within the cell, as explained in Section 2. On this basis, the chloroplast orientation can easily be explained (cf. also Seitz, 1979*b*). Whenever a gradient of chloroplast mobility is found in the cell, chloroplasts are easily taken with the cytoplasmic streaming from those sites at which they are only loosely anchored, but they come to rest as soon as they are transported to the regions characterized by low mobility, i.e., showing strong anchoring of the chloroplasts. As a result, precisely those chloroplast distributions have to be expected that are observed in the corresponding light intensities. Finally, this interpretation agrees with the earlier observations that

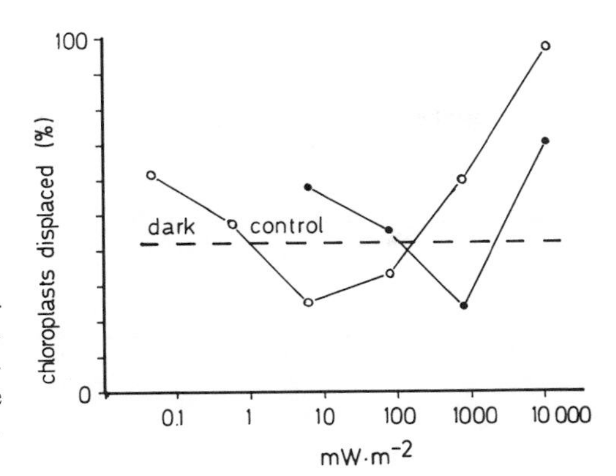

Figure 27. Chloroplast mobility in *Vallisneria* leaves at periclinal (○) and anticlinal (●) walls, as tested by centrifugation. Dependence of the effect on the fluence rate of blue light (454 nm). (Redrawn from Seitz, 1971.)

the action spectrum of the high-intensity chloroplast orientation is identical to the action spectra of photodinesis and of light-induced chloroplast mobility, and that an action dichroism is found as well (Seitz, 1976*a*).

Stimulating as this hypothesis is, it might be too simple, and it can hardly be generalized. Thus, it is not easy to imagine how an ATP gradient should orient the single large chloroplast in *Selaginella, Hormidium,* or *Mougeotia,* according to the hypothesis. Even if the hypothesis is extended to assume a gradient of ATP-dependent motive force in addition to that of ATP-dependent mobility, the very direct paths of chloroplasts in *Lemna* cannot be explained, as observed after changing high-intensity to low-intensity conditions (Zurzycka and Zurzycki, 1957). Finally, the fact that the action spectra for high- and low-intensity responses are identical in several species (Section 2.2) is hard to explain on the basis of this hypothesis.

4.2. The Glyoxylate Hypothesis

In *Lemna* and *Funaria,* glyoxylate is suggested as linking the light-perception process with the final response. According to this hypothesis, glyoxylate acts as an attractant for the chloroplasts, which would therefore orient in a gradient of glyoxylate (Zurzycki, 1972). The generation of this gradient is explained by a dual role of light (Fig. 28).

1. Glyoxylate is supposed to be formed from glycine under the control of a membrane-bound oxidative glycine deaminase, which is activated by blue light via flavin mononucleotide (FMN). This light effect recalls the blue-light stimulation of respiration, which has the same action spectrum as chloroplast orientation in *Lemna,* and which can exhibit an action dichroism as well (Zurzycki, 1970, 1971). Probably the oxi-

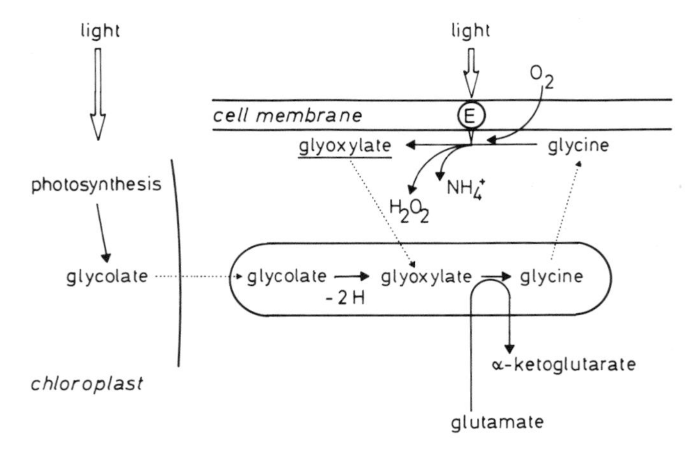

Figure 28. Glycolate pathway and glycine–glyoxylate cycle in different compartments of a plant cell. Light-dependent glycolate formation and light-stimulated glycine oxidation in combination result in glyoxylate accumulation close to the cell membrane; E, enzyme (oxidative glycin deaminase). (After Zurzycki, 1972.)

dative glycine deaminase and a rate-limiting step in respiration both make use of the same flavin photoreceptor in being activated by light.

2. Glycine, the precursor of glyoxylate according to the present model, is formed in the peroxisomes; its precursor, glycolate, is supplied by the Calvin cycle of photosynthesis. Thus, light is necessary to initiate formation of the substrate upon which light acts a second time via FMN and glycine oxidase, finally resulting in the production of the attractant. It is obvious that in this way an absorption gradient is transduced into a gradient of attractant. But nothing is known or speculated as to how the attraction can be understood at a molecular level (Zurzycki, 1980).

One experiment in *Funaria* represents an attempt to confirm the hypothesis by a rather direct approach (Godziemba-Czyz, 1973). The leaf, which contains only one layer of cells, is placed between two solutions, touching the upper and lower sides, respectively. One of the solutions serves as the control, while the other is the test solution. If the latter contains glyoxylate, chloroplasts preferably approach the cell walls facing this solution, in light as well as in darkness. If, instead, glycine is contained in the test solution, preference of chloroplasts to this side can be observed only in light, i.e., if the assumed light-induced activation of the glyoxylate-forming enzyme is realized. These experiments appear to be qualitatively convincing, but unfortunately the differences between numbers of chloroplasts approaching the test solution and the control solution, respectively, are very small. And again, it seems improbable to explain orientation of a single large chloroplast by this hypothesis. Thus, it is hardly conceivable that the chloroplast of *Hormidium,* occupying one-half the circumference of the cell (180°), should respond to a chemical gradient in the cytoplasm, e.g., between flanks and rear (90°). Hence, generalization of the glyoxylate hypothesis is certainly premature.

4.3. The Proton Hypothesis

In *Vaucheria* illumination of cytoplasm with a microbeam results in local stopping of cytoplasmic streaming (Section 2.1). It has been reported, moreover, that this effect is very probably caused by an inactivation of actin microfilaments, this latter being recognized as a disintegration and reticulation of fiber structures, slightly preceding cessation of movement. Hence we need to know the link between light absorption and actin destabilization.

With a new method, the vibrating electrode system, it has been found that locally illuminated areas of *Vaucheria* reveal an electrical outward current, which results from an efflux of protons (Fig. 29; Blatt *et al.,* 1981). As a result, a local hyperpolarization is generated. This proton efflux shows the same fluence–response dependency and the spectral sensitivity as the chloroplast aggregation, and it slightly precedes the latter. Moreover, when the light is switched off, the previous resting potential is reestablished.

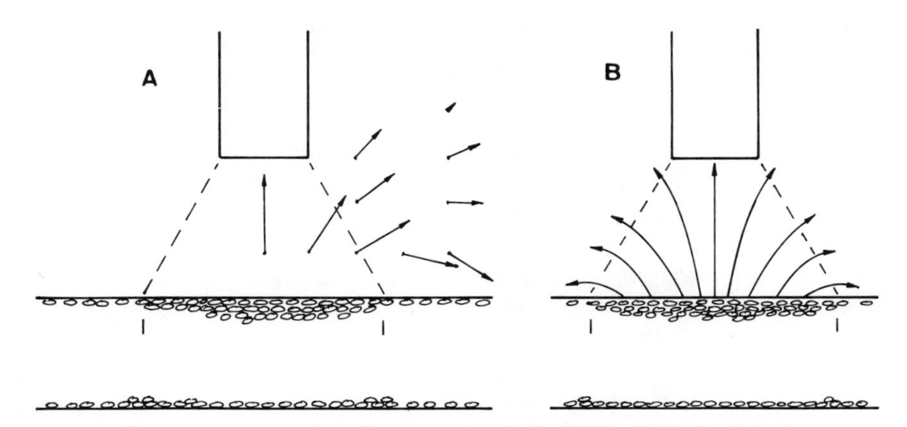

Figure 29. (A) Vector diagram of the blue-light-induced current in *Vaucheria.* (B) Current profile as derived from the vector diagram. Optical longitudinal sections of the filament are given, with the chloroplast accumulation resulting from the point irradiation. The optical fiber and its aperture are shown above. The irradiated region is limited by the dashes. (Redrawn from Blatt *et al.,* 1981.)

It seems reasonable to attribute the function of a transduction step to the light-induced proton efflux. Nevertheless, several problems remain unanswered. First, we have no knowledge of how light absorption in the blue-light photoreceptor affects proton transport through the cell membrane, although we may be aware that such effects are well known for other light-mediated responses (e.g., vectorial electron/proton transport in photosynthesis according to the Mitchell theory, or vectorial proton transport in light-harvesting *Halobacterium* cells). Second, the question of how actin is inactivated as a consequence of proton efflux has not yet been answered. It is suggested that this proton efflux starts a redistribution of other ions; in particular, a change in local calcium concentration has been discussed as the missing link (Blatt *et al.,* 1981). Since calcium is known to control actin–myosin interaction, it seems reasonable to assume that this ion is an integral part of the transduction mechanism. Evidence for this assumption is presented in Section 4.4.

A more fundamental problem in *Vaucheria* becomes evident from a comparison of local versus total illumination. As has been reported previously, movement of cytoplasm and chloroplasts is stopped locally by light, if small areas are illuminated. Yet, no significant reduction of overall streaming is found in totally illuminated cells (Fischer-Arnold, 1963). Thus, if proton extrusion can be considered one of the transduction steps, the question arises as to why this effect stops movement only upon local irradiation. One might speculate that the outward current is more important than the hyperpolarization resulting from it; indeed, a significant current requires that it may return to the cell in an unilluminated region. However, no measurements of either membrane potential or proton extrusion are available for totally illuminated cells.

4.4. The Calcium Hypothesis

A predominant factor in the regulation of actomyosin is the cytoplasmic concentration of the physiologically active ion calcium. This is well known for striated muscle, wherein calcium induces the sliding interaction of actin with myosin.

Action of calcium has been supposed early in cyclosis of characean alga cells and assigned to a sudden influx of calcium across the plasmalemma upon action potential. Cytoplasmic perfusion experiments and techniques employing the calcium-sensitive fluorescent protein aequorin have convincingly demonstrated that calcium stops cyclosis in characean cells (Williamson and Ashley, 1982). This finding is peculiar because in this case calcium acts in just the opposite way described for its action in muscle.

In light-oriented chloroplast movements, however, the calcium effect appears to be consistent with muscle physiology. In fact, chloroplast aggregation in *Vaucheria* as a result of actin inactivation is supposed to be coupled to a light-induced depletion of cytoplasmic calcium together with cytoplasmic H^+ concentration (Blatt *et al.,* 1981). Moreover, chloroplast movement in *Mougeotia* decreases much in parallel to the concentration of an H_2O-insoluble fraction of intracellular calcium, which can be manipulated by growing the cells in calcium-free or in calcium-containing medium (Wagner and Klein, 1978).

The electron microprobe and vital fluorescence staining have demonstrated that significant amounts of calcium are restricted to particular membrane vesicles, generally called physodes or tannin vesicles (Wagner and Rossbacher, 1980). These organelles are most abundant in the cytoplasm close to the chloroplast's edge, where the plastid merges into the parietal cytoplasm. In addition, this is precisely the region in which actin filaments have been demonstrated to be present (cf. Fig. 25).

Thus, a hypothetical photosensory reaction chain has been proposed, consisting of membrane-bound phytochrome, release of calcium from membrane-coated calcium stores, and activation of actin microfilaments (Wagner and Klein, 1981). In this chain, two steps have to be considered in more detail: phytochrome control of calcium transport through membranes, and the mode of calcium effect on cytoplasmic actin microfilaments.

Phytochrome control of calcium transport has been tested in *Mougeotia* by Dreyer and Weisenseel (1979), who measured the uptake of ^{45}Ca from the medium after different irradiations (Fig. 30). Red-light-pretreated cells accumulate much more radioactivity than do unirradiated cells, but this increase is not observed when red light is followed by far-red light. Calcium uptake is clearly controlled by active phytochrome.

Recently, it has been possible to measure *in vitro* calcium fluxes through the calcium vesicular membrane from *Mougeotia* (F. Grolig and G. Wagner, unpublished data). It remains an open question, however, as to whether and how calcium accumulation in, or release from, the vesicles is affected by P_{fr}. As a consequence, only the general statement that phytochrome is able to con-

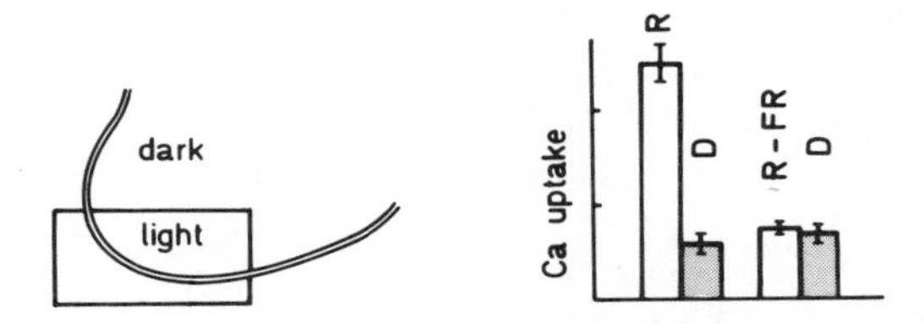

Figure 30. Uptake of $^{45}Ca^{2+}$ into *Mougeotia* cells. (Left) Draft of partial irradiation of a filament. (Right) Radioactivity taken up during 1 min after the irradiations as indicated, compared in the irradiated and dark area. R = 659 nm; FR > 700 nm; 30 seconds each. (After Dreyer and Weisenseel, 1979.)

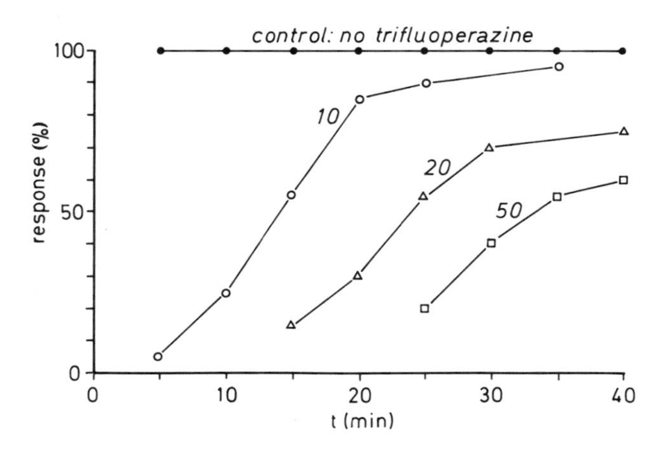

Figure 31. Chloroplast orientation from edge to face position in *Mougeotia* under the influence of trifluoperazine at different concentrations (μmole/liter). Cells were incubated in trifluoperazine, dissolved in MXS medium (Waris, 1953) at $t = -15$ min, and induced to chloroplast reorientation at $t = 0$. On the ordinate, the percentage of cells is given in which chloroplast movement has started at the times shown. (G. Wagner, unpublished).

trol calcium fluxes through a membrane can be made. But it seems possible to link such an effect directly to those calcium-containing compartments that are supposed to play a key role in controlling and orienting the movement of the chloroplast.

The action of calcium on the actin–myosin system can either concern assembly of G actin to F actin or activation of the actin–myosin interaction. Several calcium-dependent regulatory mechanisms have been reported (Wagner, 1979). A multifunctional regulatory protein well known in animals but also found in eukaryotic plants is the troponin C-like polypeptide calmodulin (17,000 daltons). Plant calmodulin has so far been shown to regulate Ca^{2+}-NAD-kinase and Ca^{2+}-ATPase in higher plants (Marmé and Dieter, 1983). It would not be surprising, however, to find a number of additional Ca^{2+}-calmodulin-dependent functions in plants, as known from animals, including myosin kinase. Calmodulin was recently purified from *Mougeotia*, and chloroplast movement in this alga is affected by the calcium-competitive inhibitor of calmodulin, trifluoperazine (Fig. 31) (Grolig *et al.*, 1983).

5. GENERAL CONCLUSIONS

There are several pieces of evidence that support the idea of membranes being essential to light-induced or light-oriented chloroplast movement. The light signal is perceived by photoreceptor pigments that are closely associated with membranes, as has been concluded from the nearly ubiquitous action dichroism of the responses in question. It is not clear whether those pigments are actually part of the membrane, are firmly bound to the membrane, or are bound to cell structures that are in a fixed position in relationship to the membrane. The neutral term "association" has therefore been used.

None of the ion-transport hypotheses of transduction has been generally accepted. It is not yet clear whether membranes play an integral role in light perception proper, as is known, e.g., from studies on vision in man and animals or from photosynthesis in plants. However, perception of light direction has been shown, in several cases, to require a fixed and well-determined molecular orientation of the photoreceptor molecules. Thus, at least for the perception of direction of light, membrane association of the pigment appears to be essential.

All the possible mechanisms of signal transduction referred to in Section 4 require the participation of membranes for their functioning. This is immediately evident for the two hypotheses involving ions in signal transduction (Sections 4.3 and 4.4): the proton hypothesis and the calcium hypothesis. Membranes are an integral part of the ATP hypothesis as well: oxidative and photosynthetic ATP production proceed according to the Mitchell theory at the mitochondrial and thylakoid membranes, respectively. Finally, the glyoxylate hypothesis invokes a membrane-bound, photoregulated deaminase for an important step in the transduction chain.

The underlying principle of the response proper, i.e., interaction of actin with myosin, superficially seems not to be based on the functions of membrane. Nevertheless, such a correlation may still be assumed: Whether organelles such as chloroplasts are pulled, pushed, or shifted by actin–myosin interaction can be understood only if one of the two substances has close mechanical contact with the organelle surface, i.e., with its membrane, and/or the other one is fixed at a resting structure of the cytoplasm, again a membrane. Indeed, examples have been reported that demonstrate such a structural association.

In addition to signal perception, signal transduction, and mechanics of response, an energy-providing system is necessary in order for a response to occur. Whenever it has been investigated, oxidative and/or photosynthetic ATP has been found as the energy source (Fetzer, 1963; Zurzycki, 1965; Schönbohm, 1969, 1972*a*,*b*; Schönbohm and Schönbohm, 1983). As mentioned for the assumed transducing function of ATP, its formation always requires the action of mitochondrial or photosynthetic membranes.

Perception, transduction, energetics, and mechanics of light-controlled chloroplast redistribution are all in some way connected with biomembranes, although their functional involvement has not yet been elucidated in all details. However, one additional argument points strongly to an overall function of membranes in our responses.

Chloroplasts orient in a gradient of light absorption. This is also true in darkness after a short light pulse. In *Mougeotia,* the chloroplast orients in darkness to its final and predetermined position in 15 to 45 min (Haupt, 1959*c*; Weisenseel, 1968); proper experiments with inhibitors or low temperature can uncouple the response from the perception. It turns out that under these conditions the orientation stimulus can be stored in darkness for at least 1 h (Mugele, 1962; Fetzer, 1963; Wagner, 1974). Similarly, in *Vallisneria* and *Lemna,* as well as in some terrestrial plants, an aftereffect of a short orienting light pulse can be observed in 10–20 min (Seitz, 1967*a*; Gabryś *et al.,* 1981; Zurzycki *et al.,* 1983). Thus, even after such long, dark intervals the whole transduction chain is still perfectly localized in the cell, i.e., no significant spreading of the information throughout the cell has occured during measurable periods of time. This is only possible under either of two conditions: The main steps of the transduction chain are states of a strongly localized structure or transducing substances are firmly associated with such a localized structure. Again, there is nearly no escape from attributing this function to a biomembrane.

In summary, the study of light-oriented chloroplast movement provides an excellent means of exploring the involvement of membranes in perception and transduction of, and in response to specific signals. By virtue of the slowness of these responses, they might turn out to have some experimental advantages over transduction chains in animals as processed by the neural system.

Acknowledgments. Professor Wieland of the Max-Planck-Institut für Medizinische Forschung, Heidelberg, kindly provided us with a fluorescent phallotoxin. Work in the authors' laboratories was supported by the Deutsche Forschungsgemeinschaft.

REFERENCES

Björn, L. O., 1984, Light-induced linear dichroism in photoreversibly photochromic sensor pigments. IV Reinterpretation of the experiments on *in vivo* action dichroism of phytochrome, *Physiol. Plantarum* (in press).

Blatt, M. R., Wessells, N. K., and Briggs, W. R., 1980, Actin and cortical fiber reticulation in the siphonaceous alga *Vaucheria sessilis, Planta* **147**:363–375.

Blatt, M. R., Weisenseel, M. H., and Haupt, W., 1981, A light-dependent current associated with chloroplast aggregation in the alga *Vaucheria sessilis, Planta* **152**:513–526.

Bock, G., and Haupt, W., 1961, Die Chloroplastendrehung bei *Mougeotia.* III. Die Frage der Lokalisierung des Hellrot-Dunkelrot-Pigmentsystems in der Zelle, *Planta* **57**:518–530.

Bottelier, H. P., 1933, Über den Einflu des Lichtes auf die Protoplasmaströmung von *Avena, Proc. K. Ned. Akad. Wet.* **36**:790–794.

Britz, S. J., 1979, Chloroplast and nuclear migration, in: *Encyclopedia of Plant Physiology, new ser.,* Vol. 7 (W. Haupt and M. E. Feinleib, eds.), pp. 170–205, Springer-Verlag, Berlin.

Brown, S. S., and Spudich, J. A., 1981, Mechanism of action of cytochalasin: evidence that it binds to actin filament ends, *J. Cell Biol.* **88**:487–491.
Büttner, G., Kraml, M., and Haupt, W., 1982, Blaulicht realisiert die P_{fr}-Wirkung bei der Chloroplastenbewegung von *Mesotaenium,* Tagung der Deutschen Botanischen Gesellschaft Freiburg, Sept. 1982 (abst.).
Dreyer, E. M., and Weisenseel, M. H., 1979, Phytochrome-mediated uptake of calcium in *Mougeotia* cells, *Planta* **146**:31–39.
Fetzer, J., 1963, Über die Beteiligung energieliefernder Stoffwechselprozesse an den lichtinduzierten Chloroplastenbewegungen von *Mougeotia, Z. Bot.* **51**:468–506.
Filner, Ph., and Yadav, N. S., 1979, Role of microtubules in intracellular movements, in: *Encyclopedia of Plant Physiology,* new ser., vol. 7 (W. Haupt and M. E. Feinleib, eds.), pp. 95–113, Springer-Verlag, Berlin.
Fischer-Arnold, G., 1963, Untersuchungen über die Chloroplastenbewegung bei *Vaucheria sessilis, Protoplasma* **56**:495–520.
Foos, K., 1970, Mikrotubuli bei *Mougeotia* spec., *Z. Pflanzenphysiol.* **62**:201–203.
Foos, K., 1971, Untersuchungen zur Feinstruktur von *Mougeotia* spec. und zum Bewegungsmechanismus des Chloroplasten, *Z. Pflanzenphysiol.* **64**:369–386.
Gabryś-Mizera, H., 1976, Model considerations of the light conditions in noncylindrical plant cells, *Photochem. Photobiol.* **24**:453–461.
Gabryś, H., Walczak, T., and Zurzycki, J., 1981, Chloroplast translocations induced by light pulses. Effects of single light pulses, *Planta* **152**:553–556.
Gärtner, R., 1970, Die Bewegung des *Mesotaenium*-Chloroplasten im Starklichtbereich. II. Aktionsdichroismus und Wechselwirkungen des Photoreceptors mit Phytochrom, *Z. Pflanzenphysiol.* **63**:428–443.
Godziemba-Czyz, J., 1973, Certain aspects of the chemotaxis reaction of chloroplasts in *Funaria hygrometrica, Acta Soc. Bot. Pol.* **42**:453–459.
Grolig, F., Valentin, P. and Wagner, G., 1983, Calcium vesicles and calmodulin in the green alga *Mougeotia* sp., *Europ. J. Cell Biol.* **2** (Suppl.):12.
Haupt, W., 1959*a*, Chloroplastenbewegung, in: *Encyclopedia of Plant Physiology,* Vol. 17/1 (E. Bünning, ed.), pp. 278–317, Springer-Verlag, Berlin.
Haupt, W., 1959*b*, Photodinese, in: *Encylopedia of Plant Physiology,* Vol. 17/1 (E. Bünning, ed.), pp. 388–398, Springer-Verlag, Berlin.
Haupt, W., 1959*c*, Die Chloroplastendrehung bei *Mougeotia.* I. Über den quantitativen und qualitativen Lichtbedarf der Schwachlichtbewegung, *Planta* **53**:484–501.
Haupt, W., 1960, Die Chloroplastendrehung bei *Mougeotia.* II. Die Induktion der Schwachlichtbewegung durch linear polarisiertes Licht, *Planta* **55**:465–479.
Haupt, W., 1965, Perception of environmental stimuli orienting growth and movement in lower plants, *Annu. Rev. Plant Physiol.* **16**:267–290.
Haupt, W., 1968*a*, Die Orientierung der Phytochrom-Moleküle in der *Mougeotia*zelle: Ein neues Modell zur Deutung der experimentellen Befunde, *Z. Pflanzenphysiol.* **58**:331–346.
Haupt, W., 1968*b*, Die Orientierungsbewegungen der Chloroplasten, *Biol. Rund.* **6**:121–136.
Haupt, W., 1970, Localization of phytochrome in the cell, *Physiol. Veg.* **8**:551–563.
Haupt, W., 1972, Localization of phytochrome within the cell, in: *Phytochrome* (K. Mitrakos, and W. Shropshire, Jr., eds.), pp. 553–569, Academic Press, London.
Haupt, W., 1973, Role of light in chloroplast movement, *BioScience* **23**:289–296.
Haupt, W., 1982, Light-mediated movement of chloroplasts, *Annu. Rev. Plant Physiol.* **33**:205–233.
Haupt, W., 1983, The perception of light direction and orientation responses in chloroplasts, in: *The Biology of Photoreception, Symposium of the Society of Experimental Biology* (D. Cosens and D. Vince-Prue, eds.), pp. 423–442, Cambridge University Press, Cambridge.

Haupt, W., and Bock, G., 1962, Die Chloroplastendrehung bei *Mougeotia*. IV. Die Orientierung der Phytochrom-Moleküle im Cytoplasma, *Planta* **59**:38–48.
Haupt, W., and Schönbohm, E., 1970, Light-oriented chloroplast movements, in: *Photobiology of Microorganisms* (P. Halldal, ed.), pp. 283–307, Wiley, London.
Haupt, W., and Schönfeld, I., 1962, Über das Wirkungsspektrum der "negativen Phototaxis" der *Vaucheria*-Chloroplasten, *Ber. Dtsch. Bot. Ges.* **75**:14–23.
Inoue, Y., and Shibata, K., 1973, Light-induced chloroplast rearrangements and their action spectra as measured by absorption spectrophotometry, *Planta* **114**:341–358.
Kamiya, N., 1962, Protoplasmic streaming, in: *Encyclopedia of Plant Physiology,* Vol. 17/2 (E. Bünning, ed.), pp. 979–1035, Springer-Verlag, Berlin.
Kamiya, N., 1981, Physical and chemical basis of cytoplasmic streaming, *Annu. Rev. Plant Physiol.* **32**:205–236.
Kersey, Y. M., Hepler, P. K., Palevitz, B. A., and Wessels, N. K., 1976, Polarity of actin filaments in Characean algae, *Proc. Natl. Acad. Sci. USA* **73**:165–167.
Klein, K., 1981, Feinstrukturelle Untersuchungen zur Bewegung des *Mougeotia*-Chloroplasten, Ph.D. thesis, pp. 1–167, University of Erlangen-Nürnberg, Erlangen, West Germany.
Klein, K., Wagner, G., and Blatt, M. R., 1980, Heavy-meromyosin-decoration of microfilaments from *Mougeotia* protoplasts, *Planta* **150**:354–356.
Koch, G. L. E., 1981, The anchorage of cell surface receptors to the cytoskeleton, in: *International Cell Biology 1980–1981* (H. G. Schweiger, ed.), pp. 321–345, Springer-Verlag, Berlin.
Lechowski, Z., 1972, Action spectrum of chloroplast displacements in the leaves of land plants, *Acta Protozool.* **11**:201–209.
Lenci, F., and Colombetti, G. (eds.), 1980, *Photoreception and Sensory Transduction in Aneural Organisms,* Plenum Press, New York.
Marchant, H. J., 1976, Actin in the green algae *Coleochaete* and *Mougeotia, Planta* **131**:119–120.
Marchant, H. J., 1978, Microtubules associated with the plasma membrane isolated from protoplasts of the green alga *Mougeotia, Exp. Cell Res.* **115**:25–30.
Marmé, D., and Dieter, P., 1983, Role of Ca^{2+} and calmodulin in plants, in: *Calcium and Cell Function,* Vol. 4 (W. Y. Cheung, ed.), pp. 263–311, Academic Press, New York.
Mayer, F., 1964, Lichtorientierte Chloroplasten-Verlagerungen bei *Selaginella martensii, Z. Bot.* **52**:346–381.
Mayer, F., 1971, Light-induced chloroplast contraction and movement, in: *Structure and Function of Chloroplasts* (M. Gibbs, ed.), pp. 35–49, Springer-Verlag, New York.
Mugele, F., 1962, Der Einfluß der Temperatur auf die lichtinduzierte Chloroplastenbewegung, *Z. Bot.* **50**:368–388.
Nultsch, W., and Pfau, J., 1979, Occurence and biological role of light-induced chromatophore displacements in seaweeds, *Marine Biol.* **51**:77–82.
Nultsch, W., Pfau, J., and Rüffer, U., 1981, Do correlations exist between chromatophore arrangement and photosynthetic activity in seaweeds? *Marine Biol.* **62**:111–117.
Palevitz, B. A., and Hepler, P. K., 1975, Identification of actin in situ at the ectoplasm–endoplasm interface of *Nitella.* Microfilament–chloroplast association, *J. Cell Biol.* **65**:29–38.
Palevitz, B. A., Ash, J. F., and Hepler, P. K., 1974, Actin in the green alga, *Nitella, Proc. Natl. Acad. Sci. USA* **71**:363–366.
Pfau, J., Rüffer, U., and Nultsch, W., 1979, Der Einfluß polarisierten Lichtes auf die Chromatophorenanordnung von *Dictyota dichotoma, Ber. Dtsch. Bot. Ges.* **92**:695–715.
Schmidt, W., 1980, Physiological Bluelight Reception, *Struct. Bond.* **41**:1–44.
Scholz, A., 1976*a*, Lichtorientierte Chloroplastenbewegung bei *Hormidium flaccidum:* Perception der Lichtrichtung mittels Sammellinseneffekt, *Z. Pflanzenphysiol.* **77**:406–421.
Scholz, A., 1976*b*, Lichtorientierte Chloroplastenbewegung bei *Hormidium flaccidum:* Ver-

schiedene Methoden der Lichtrichtungsperception und die wirksamen Pigmente, *Z. Pflanzenphysiol.* **77**:422–436.

Schönbohm, E., 1966, Der Einfluß von Rotlicht auf die negative Phototaxis des *Mougeotia*-Chloroplasten: Die Bedeutung eines Gradienten von P_{730} für die Orientierung, *Z. Pflanzenphysiol.* **55**:278–286.

Schönbohm, E., 1969, Untersuchungen über den Einfluß von Photosynthesehemmstoffen und Halogeniden auf die Starklicht- und Schwachlichtbewegung des Chloroplasten von *Mougeotia spec., Z. Pflanzenphysiol.* **60**:255–269.

Schönbohm, E., 1971*a*, Über die Lokalisierung des Photoreceptors für den tonischen Blaulicht-Effekt bei der Verlagerung des *Mougeotia*-Chloroplasten im Starklicht, *Z. Pflanzenphysiol.* **65**:453–457.

Schönbohm, E., 1971*b*, Untersuchungen zum Photoreceptorproblem beim tonischen Blaulicht-Effekt der Starklichtbewegung des *Mougeotia*-Chloroplasten, *Z. Pflanzenphysiol.* **66**:20–33.

Schönbohm, E., 1972*a*, Die Wirkung von SH-Blockern sowie von Licht und Dunkel auf die Verankerung der *Mougeotia*-Chloroplasten im cytoplasmatischen Wandbelag, *Z. Pflanzenphysiol.* **66**:113–132.

Schönbohm, E., 1972*b*, Experiments on the mechanism of chloroplast movement in light-oriented chloroplast arrangement. *Acta Protozool.* **11**:211–224.

Schönbohm, E., 1975, Der Einfluss von Colchicin sowie von Cytochalasin B auf fädige Plasmastrukturen, auf die Verankerung der Chloroplasten sowie auf die orientierte Chloroplastenbewegung. 4. Mitteilung: Zur Mechanik der Chloroplastenbewegung, *Ber. Dtsch. Bot. Ges.* **88**:211–224.

Schönbohm, E., 1980, Phytochrome and non-phytochrome dependent blue light effects on intracellular movements in freshwater algae, in: *The Blue Light Syndrome* (H. Senger, ed.), pp. 69–96, Springer-Verlag, Berlin.

Schönbohm, E., and Schönbohm, E. 1983, Die Bedeutung der Photosysteme I und II sowie der oxidativen Phosphorylierung für die lichtorientierte Bewegung des *Mougeotia*-Chloroplasten, *Biochem. Physiol. Pflanzen* **178**:157–176.

Seitz, K., 1964, Das Wirkungsspektrum der Photodinese bei *Elodea canadensis, Protoplasma* **58**:621–640.

Seitz, K., 1967*a*, Wirkungsspektren für die Starklichtbewegung der Chloroplasten, die Photodinese und die lichtabhängige Viskositätsänderung bei *Vallisneria spiralis ssp. torta, Z. Pflanzenphysiol.* **56**:246–261.

Seitz, K., 1967*b*, Eine Analyse der für die lichtabhängigen Bewegungen der Chloroplasten verantwortlichen Photorezeptorsysteme bei *Vallisneria spiralis ssp. torta, Z. Pflanzenphysiol.* **57**:96–104.

Seitz, K., 1971, Die Ursache der Phototaxis der Chloroplasten: ein ATP-Gradient? *Z. Pflanzenphysiol.* **64**:241–256.

Seitz, K., 1972, Primary processes controlling the light induced movement of chloroplasts, *Acta Protozool.* **11**:226–235.

Seitz, K., 1979*a*, Light induced changes in the centrifugability of chloroplasts: different action spectra and different influence of inhibitors in the low and high intensity range, *Z. Pflanzenphysiol.* **95**:1–12.

Seitz, K., 1979*b*, Cytoplasmic streaming and cyclosis of chloroplasts, in: *Encyclopedia of Plant Physiology,* new ser., Vol 7 (W. Haupt and M. E. Feinleib, eds.), pp. 150–169, Springer-Verlag, Berlin.

Seitz, K., 1980, Light induced changes in the centrifugability of chloroplasts mediated by an irradiance dependent interaction of respiratory and photosynthetic processes, in: *The Blue Light Syndrome* (H. Senger, ed), pp. 637–642, Springer-Verlag, Berlin.

Seitz, K., 1982, Chloroplast motion in response to light in aquatic vascular plants, in: *Studies on*

Aquatic Vascular Plants (J. J. Symoens, S. S. Hooper, and P. Compère, eds.), pp. 89–101, Royal Botanical Society of Belgium, Brussels.

Senn, G., 1908, *Die Gestalts- und Lageveränderung der Pflanzen-Chromatophoren,* Engelmann, Stuttgart.

Song, P.-S., Chae, Q., and Gardner, J. D., 1979, Spectroscopic properties and chromophore conformations of the photomorphogenic receptor: phytochrome, *Biochim. Biophys. Acta* **576**:479–495.

Sundqvist, C., and Björn, L. O., 1983*a*, Light-induced linear dichroism in photoreversibly photochromic sensor pigments. II. Chromophore rotation in immobilized phytochrome, *Photochem. Photobiol.* **37**:69–75.

Sundqvist, C., and Björn, L. O., 1983*b*, Light-induced linear dichroism in photoreversibly photochromic sensor pigments. III. Chromophore rotation estimated by polarized light reversal of dichroism, *Physiol. Plantarum* **59**:263–269.

Tendel, J., and Haupt, W., 1981, Mechanische und energetische Grundlagen der lichtabhängigen Gestaltänderung des *Mougeotia*-Chloroplasten, *Z. Pflanzenphysiol.* **104**:169–185.

Wagner, G., 1974, Some physiological properties of phytochrome in the alga *Mougeotia* as studied by cytochalasin B and aminophylline, in: *Proceedings of the Annual European Symposium on Plant Photomorphogenesis* (J. A. de Greef, ed.), pp. 22–25, University Press, Antwerp.

Wagner, G., 1979, Actomyosin as a basic mechanism of movement in animals and plants, in: *Encyclopedia of Plant Physiology,* new ser., vol. 7 (W. Haupt and M. E. Feinleib, eds.), pp. 114–126, Springer-Verlag, Berlin.

Wagner, G., and Klein, K., 1978, Differential effect of calcium on chloroplast movement in *Mougeotia, Photochem. Photobiol.* **27**:137–140.

Wagner, G., and Klein, K., 1981, Mechanism of chloroplast movement in *Mougeotia, Protoplasma* **109**:169–185.

Wagner, G., and Rossbacher, R., 1980, X-ray microanalysis and chlorotetracycline staining of calcium vesicles in the green alga *Mougeotia, Planta* **149**:298–305.

Wagner, G., Haupt, W., and Laux, A., 1972, Reversible inhibition of chloroplast movement by cytochalasin B in the green alga *Mougeotia, Science* **176**:808–809.

Waris, H., 1953, The significance for algae of chelating substances in the nutrient solutions, *Physiol. Plant.* **6**:538–543.

Weisenseel, M., 1968, Vergleichende Untersuchungen zum Einfluss der Temperatur auf lichtinduzierte Chloroplastenverlagerungen. II. Die statistische Bewegungsgeschwindigkeit der Chloroplasten und ihre Abhängigkeit von der Temperatur, *Z. Pflanzenphysiol.* **59**:153–171.

Weisenseel, M. H., 1979, Induction of polarity, in: *Encyclopedia of Plant Physiology,* new ser., Vol. 7 (W. Haupt and M. E. Feinleib, eds.), pp. 485–505, Springer-Verlag, Berlin.

Wieland, T., 1977, Modification of actins by phallotoxins, *Naturwissenschaften* **64**:303–309.

Williamson, R. E., 1974, Actin in the alga, *Chara corallina, Nature (Lond.)* **248**:801–802.

Williamson, R., 1975, Cytoplasmic streaming in *Chara:* a cell model activated by ATP and inhibited by cytochalasin B, *J. Cell Sci.* **17**:655–668.

Williamson, R. E., 1976, Cytoplasmic streaming in characean algae, in: *Transport and Transfer Processes in Plants* (I. F. Wardlaw and J. B. Passioura, eds.), pp. 51–58, Academic Press, New York.

Williamson, R. E. and Ashley, C. C., 1982, Free Ca^{2+} and cytoplasmic streaming in the alga *Chara, Nature* **296**:647–651.

Wulf, E., Deboben, A., Bautz, F. A., Faulstich, H., and Wieland, T., 1979, Fluorescent phallotoxin, a tool for the visualization of cellular actin, *Proc. Natl. Acad. Sci. USA* **76**:4498–4502.

Yamaguchi, Y., and Nagai, R., 1981, Motile apparatus in *Vallisneria* leaf cells. I. Organization of microfilaments, *J. Cell Sci.* **48**:193–205.

Zurzycka, A., and Zurzycki, J., 1957, Cinematographic studies on phototactic movements of chloroplasts, *Acta Soc. Bot. Pol.* **26**:177–206.
Zurzycki, J., 1962*a*, The action spectrum for the light dependent movements of chloroplasts in *Lemna trisulca L., Acta Soc. Bot. Pol.* **31**:489–538.
Zurzycki, J., 1962*b*, The mechanism of the movements of plastids, in: *Encyclopedia of Plant Physiology,* Vol. 17/2 (E. Bünning, ed.), pp. 940–978, Springer-Verlag, Berlin.
Zurzycki, J., 1965, The energy of chloroplast movement in *Lemna trisulca* L., *Acta Soc. Bot. Pol.* **34**:637–666.
Zurzycki, J., 1967*a*, Properties and localization of the photoreceptor active in displacement of chloroplasts in *Funaria hygrometrica.* I. Action spectrum, *Acta Soc. Bot. Pol.* **36**:133–142.
Zurzycki, J., 1967*b*, Properties and localization of the photoreceptor active in displacement of chloroplasts in *Funaria hygrometrica.* II. Studies with polarized light, *Acta Soc. Bot. Pol.* **36**:143–152.
Zurzycki, J., 1968, Properties and localization of the photoreceptor active in displacement of chloroplasts in *Funaria hygrometrica.* V. Studies on plasmolyzed cells, *Acta Soc. Bot. Pol.* **37**:11–17.
Zurzycki, J., 1969, Experimental modification of the reaction pattern of *Lemna* leaf cells to polarized light, *Protoplasma* **68**:193–207.
Zurzycki, J., 1970, Light respiration in *Lemna trisulca, Acta Soc. Bot. Pol.* **39**:485–495.
Zurzycki, J., 1971, Effect of linear polarized light on the O_2 uptake in leaves, *Biochem. Physiol. Pflanzen* **162**:310–317.
Zurzycki, J., 1972, Primary reactions in the chloroplast rearrangements, *Acta Protozool.* **11**:189–200.
Zurzycki, J., 1980, Blue light-induced intracellular movement, in: *The Blue Light Syndrome* (H. Senger, Ed.), pp. 50–68, Springer-Verlag, Berlin.
Zurzycki, J., and Lełatko, Z., 1969, Action dichroism in the chloroplast rearrangements in various plant species, *Acta Soc. Bot. Pol.* **38**:493–506.
Zurzycki, J., Walczak, T., Gabryś, H., and Kajfosz, J., 1983, Chloroplast translocations in *Lemna trisulca* L. induced by continuous irradiation and by light pulses. Kinetic analysis, *Planta* **157**:502–510.

Index